AN INTRODUCTION TO ANALOG AND DIGITAL COMMUNICATIONS

AN INTRODUCTION TO ANALOG AND DIGITAL COMMUNICATIONS

SIMON HAYKIN
MCMASTER UNIVERSITY

WILEY-INDIA

AN INTRODUCTION TO ANALOG AND DIGITAL COMMUNICATIONS

Authorized reprint by Wiley India (P.) Ltd., 4435/7, Ansari Road, Daryaganj, New Delhi 110 002.

Bicentennial logo designer: Richard J. Pacifico

Reprint : 2009

Printed at : Gurukripa Enterprises, Delhi

ISBN : 978-81-265-0932-4

TO

NANCY

ABOUT THE AUTHOR

Simon Haykin received his B.Sc. (First Class Honors), Ph.D., and D.Sc., all in electrical engineering, from the University of Birmingham, England. He is a Fellow of the Institute of Electrical and Electronics Engineers (IEEE), and a Fellow of the Royal Society of Canada. He is a recipient of the McNaughton Gold Medal, awarded by the IEEE, Region 7.

He is Professor of Electrical and Computer Engineering, and founding Director of the Communications Research Laboratory, McMaster University, Hamilton, Ontario, Canada.

Professor Haykin's research interests include detection and estimation, spectrum analysis, adaptive systems, and radar systems. He has taught undergraduate courses on Communication Systems and Digital Communications, and graduate courses on Detection and Estimation, Adaptive Systems, and Digital Signal Processing. He is the author of *Communication Systems,* Second Edition, (Wiley, 1983), *Introduction to Communication Systems* (Wiley, 1988), *Adaptive Filter Theory* (Prentice-Hall, 1986), and the editor of *Array Signal Processing* (Prentice-Hall, 1985), and *Nonlinear Methods of Spectral Analysis,* Second Edition (Springer-Verlag, 1983).

PREFACE

In this book I present an introductory treatment of communication theory as applied to the transmission of information-bearing signals, with attention given to both analog and digital communications. Numerous examples, worked out in detail, have been included to help students develop an intuitive grasp of the theory under discussion. Each chapter begins with introductory remarks and (except for Chapter 1) ends with a set of problems. Also, except for Chapter 1, each chapter includes exercises that are intended to help students improve their understanding of the material covered in the chapter. Answers to the exercises are given at the end of the book.

Chapter 1 is an introduction to communication theory and a review of some basic concepts.

In Chapter 2, I review two classic methods for frequency analysis, namely, the Fourier series and the Fourier transform. In Chapter 3, I discuss the time-domain and frequency-domain descriptions of signal transmission through linear filters and channels. In Chapter 4, I present a unified treatment of the spectral density and correlation functions of energy signals and power signals. Thus, Chapters 2 through 4 constitute the first major part of the book, pertaining to the *characterization of signals and systems.*

In Chapter 5, I describe pulse-code modulation (PCM) and related techniques for the conversion of information-bearing analog signals into digital form. In Chapter 6, I cover issues that arise in baseband data transmission. In Chapter 7, I describe various modulation techniques, with attention given to both analog and digital signals as the modulating wave. Thus, Chapters 5 through 7 constitute the second major part of the book, concerned with the *transmission of message (information-bearing) signals over communication channels.*

The third major part of the book, made up of Chapters 8 through 10, deals with *noise in analog and digital communications.* In Chapter 8, I present a review of probability theory and random processes. This is followed by a discussion of the effects of channel (receiver) noise on the performance of analog modulation schemes in Chapter 9. Finally, in Chapter 10, I discuss the design of optimum receivers for data transmission over a noisy communication channel.

The book also includes four appendixes. In Appendix A, I review power ratios and the decibel. In Appendix B, I review Bessel functions of the first kind and their properties. In Appendix C, I describe various sources of noise and present a treatment of noise calculations in communication systems. Appendix D includes various mathematical tables. To assist the student, I have also included a glossary of notations and abbreviations.

Footnotes are included in the text to add historical notes, to cite reference material, and to highlight points of interest. A list of references and bibliography is appended at the end of the book.

The book is essentially self-contained and suited for a one-semester course in communication systems taken by electrical engineering juniors or seniors. For juniors, the course may be organized as follows:

1. Chapter 1 for an overview of the issues involved in communications.
2. Chapter 2 on Fourier analysis, emphasizing properties of the Fourier transform and the interplay between time-domain and frequency-domain descriptions of signals.
3. Chapter 3 on filtering and signal distortion, with particular attention given to the transmission of signals through band-pass filters and channels.
4. Chapter 4 on spectral densities and correlation functions, emphasizing the application of these concepts to both energy and power signals.
5. Material taken from Chapter 7 on standard amplitude modulation, double-sideband suppressed-carrier modulation, and frequency modulation.

For seniors, the course may be organized as follows:

1. An overview of communications, and a review of Fourier analysis of signals and systems, emphasizing the important roles of Fourier transformation and linear filtering in communications; material from Chapters 1 through 4 may be used for this review.
2. Chapter 5 on pulse-code modulation and related techniques for analog-to-digital conversion.
3. Chapter 6, to provide an appreciation for intersymbol interference in data transmission and practical cures for it.
4. Chapter 7 on modulation techniques for both analog and digital signals.
5. Sections 8.1 through 8.4 for a review of probability theory, the extent of which depends on the background of students taking the course. This review is followed by a study of random processes (Section 8.5 through 8.14), with emphasis on the characterization of wide-sense stationary processes and narrow-band Gaussian processes.
6. Chapter 9 on noise in (continuous-wave) amplitude and frequency modulation.
7. Chapter 10 on optimum receivers for data transmission, to develop an understanding of the matched filter and the role it plays in digital communications.

The book offers enough flexibility for organizing the course material to suit the interests of course teachers and students in other ways.

Attention is given to both analog and digital communications at an *introductory* level. Readers who wish to avail themselves of a more advanced treatment of digital communications, may refer to a companion book written by the author on this subject: S. Haykin, *Digital Communications,* Wiley 1988.

As aids to teachers of the course, the following material is available from the publishers of the book:

A complete manual, containing detailed solutions of all the problems in the book, master transparencies of all the figures and tables in the book and six computer-oriented experiments, with detailed results for all the experiments.

I am grateful to the following reviewers for their helpful suggestions: F. Carden, New Mexico State University; S. Kesler, Drexel University; M. Siegel, Michigan State University; and Dr. M. C. Wernicki, an independent consultant. I thank S. Rappaport, State University of New York at Stony Brook, for some helpful inputs.

I am indebted to my colleagues, Peter Weber and Norman Secord, Communications Research Laboratory, McMaster University, for reading the entire book and for many constructive suggestions.

I also wish to thank my editor, Christina Mediate, for her guidance and support, Richard J. Koreto, Dawn Reitz, David Levy, and other staff members of Wiley for their help in the production of the book. The fine work done by Ernest Kohlmetz in editing the manuscript of the book is deeply appreciated.

Finally, I thank my secretary, Lola Brooks, for typing many versions of the manuscript.

SIMON HAYKIN

CONTENTS

CHAPTER 1 INTRODUCTION 1

1.1 Communications 1
1.2 Signals and Their Classifications 2
1.3 Fourier Analysis of Signals and Systems 5
1.4 Elements of a Communication System 6
1.5 Transmission of Message Signals 7
1.6 Limitations and Resources of Communication Systems 9

CHAPTER 2 FOURIER ANALYSIS 13

2.1 Fourier Series 14
2.2 Fourier Transform 20
2.3 Properties of the Fourier Transform 29
2.4 Interplay Between Time-Domain and Frequency-Domain Descriptions 50
2.5 Dirac Delta Function 53
2.6 Fourier Transforms of Periodic Signals 63
2.7 Sampling Theorem 66
2.8 Numerical Computation of the Fourier Transform 71
2.9 Relationship Between the Fourier and Laplace Transforms 73
Problems 74

CHAPTER 3 FILTERING AND SIGNAL DISTORTION 83

3.1 Time Response 84
3.2 Frequency Response 91
3.3 Linear Distortion and Equalization 96
3.4 Ideal Low-Pass Filters 102
3.5 Band-Pass Transmission 106
3.6 Phase Delay and Group Delay 114
3.7 Nonlinear Distortion 116
Problems 120

CHAPTER 4 SPECTRAL DENSITY AND CORRELATION 127

4.1 Energy Spectral Density 128
4.2 Correlation of Energy Signals 137

4.3 Power Spectral Density 141
4.4 Correlation of Power Signals 146
4.5 Flowchart Summaries 150
4.6 Spectral Characteristics of Periodic Signals 151
4.7 Spectral Characteristics of Random Signals and Noise 158
4.8 Noise-Equivalent Bandwidth 166
Problems 168

CHAPTER 5 DIGITAL CODING OF ANALOG WAVEFORMS 177

5.1 Digital Pulse Modulation 178
5.2 Pulse-Code Modulation 179
5.3 Sampling 180
5.4 Quantizing 187
5.5 Coding 194
5.6 Regeneration 201
5.7 Differential Pulse-Code Modulation 202
5.8 Delta Modulation 206
5.9 Discussion 210
5.10 Time-Division Multiplexing 213
5.11 Application: Digital Multiplexers for Telephony 215
Problems 221

CHAPTER 6 INTERSYMBOL INTERFERENCE AND ITS CURES 227

6.1 Baseband Transmission of Binary Data 228
6.2 The Intersymbol Interference Problem 230
6.3 Ideal Solution 231
6.4 Raised Cosine Spectrum 233
6.5 Correlative Coding 237
6.6 Baseband Transmission of *M*-ary Data 247
6.7 Eye Pattern 248
6.8 Adaptive Equalization 250
Problems 254

CHAPTER 7 MODULATION TECHNIQUES 259

7.1 Amplitude Modulation 260
7.2 Double-Sideband Suppressed-Carrier Modulation 274
7.3 Quadrature-Carrier Multiplexing 283
7.4 Single-Sideband Modulation 284
7.5 Vestigial Sideband Modulation 295

7.6 Comparison of Amplitude Modulation Techniques 303
7.7 Frequency Translation 306
7.8 Frequency-Division Multiplexing 309
7.9 Application I: Radio Broadcasting 310
7.10 Angle Modulation: Basic Concepts 322
7.11 Frequency Modulation 327
7.12 Phase-Locked Loop 353
7.13 Limiting of FM Waves 363
7.14 Application II: FM Radio 364
7.15 Digital Modulation Techniques 368
7.16 Application III: Digital Communications by Satellite 382
Problems 384

CHAPTER 8 PROBABILITY THEORY AND RANDOM PROCESSES ***403***

8.1 Probability Theory 404
8.2 Random Variables 414
8.3 Gaussian Distribution 425
8.4 Transformation of Random Variables 430
8.5 Random Processes 434
8.6 Stationarity 437
8.7 Mean, Correlation, and Covariance Functions 439
8.8 Random Process Transmission Through Linear Filters 450
8.9 Power Spectral Density 452
8.10 Cross-Correlation Functions 463
8.11 Cross-Spectral Densities 465
8.12 Gaussian Process 468
8.13 Narrow-band Random Process 471
8.14 Envelope and Phase of Narrow-band Random Process 478
Problems 481

CHAPTER 9 NOISE IN ANALOG MODULATION ***493***

9.1 Signal-to-Noise Ratios 494
9.2 AM Receiver Model 495
9.3 Signal-to-Noise Ratios for Coherent Reception 497
9.4 Noise in AM Receivers Using Envelope Detection 503
9.5 FM Receiver Model 508
9.6 Noise in FM Reception 509
9.7 FM Threshold Effect 517
9.8 Pre-emphasis and De-emphasis in FM 524
9.9 Discussion 529
Problems 532

CHAPTER 10 OPTIMUM RECEIVERS FOR DATA COMMUNICATION 539

10.1 Formulation of the Optimum Receiver Problem 540
10.2 Maximization of Output Signal-to-Noise Ratio 541
10.3 Properties of Matched Filters 545
10.4 Approximations in Matched Filter Design 553
10.5 Probabilistic Approach 558
10.6 Probability of Error for Binary PCM 567
10.7 Noise in Digital Modulation Schemes 574
10.8 Coherent Detection of Binary Modulated Waves 575
10.9 Noncoherent Detection of Binary Modulated Waves 581
10.10 Coherent Detection of Quaternary Modulated Waves 583
10.11 Discussion 585
10.12 Tradeoffs in *M*-ary Data Transmission 588
Problems 591

APPENDIX A POWER RATIOS AND DECIBELS 597
APPENDIX B BESSEL FUNCTION 598
APPENDIX C SYSTEM NOISE AND CALCULATIONS 603
APPENDIX D TABLES 623
GLOSSARY 631
REFERENCES AND BIBLIOGRAPHY 635
ANSWERS TO EXERCISES 640
INDEX 645

CHAPTER ONE

INTRODUCTION

1.1 COMMUNICATIONS

Communication enters our daily lives in so many different ways that it is easy to overlook the multitude of its facets. The telephones in our homes and offices make it possible for us to communicate with others, no matter how far away. The radio and television sets in our living rooms bring us entertainment from near as well as far-away places. Communication by radio or satellite provides the means for ships on the high seas, aircraft in flight, and rockets and exploratory probes in space to maintain contact with their home bases. Communication keeps the weather forecaster informed of atmospheric conditions that are measured by a multitude of sensors. Communication makes it possible for computers to interact. The list of applications involving the use of communications in one way or another goes on.[1]

In the most fundamental sense, *communication* involves implicitly the transmission of information from one point to another through a succession of processes, as described here:

1. The generation of a thought pattern or image in the mind of an originator.
2. The description of that thought pattern or image, with a certain measure of precision, by a set of aural or visual symbols.
3. The encoding of these symbols in a form that is suitable for transmission over a physical medium (channel) of interest.
4. The transmission of the encoded symbols to the desired destination.
5. The decoding and reproduction of the initial symbols.
6. The re-creation of the original thought pattern or image—with a definable degradation in quality—in the mind of a recipient, with the degradation being caused by imperfections in the system.

The form of communication just described involves a thought pattern or image originating in a human mind. Of course, there are many other forms of communication that do not directly involve the human mind in real time. In space exploration, for example, human decisions may enter only the commands sent to the space probe or to the computer responsible for processing images of far-away planets (e.g., Mars, Jupiter, Saturn) that are sent back by the probe. In computer communications, human decisions enter only in setting up the computer programs or in monitoring the results of computer processing.

Whatever form of communication is used, some basic signal-processing operations are involved in the transmission of information. The next section describes the different types of signals encountered in the study of communication systems. The signal-processing operations of interest are highlighted later in the chapter.

1.2 SIGNALS AND THEIR CLASSIFICATIONS

For our purposes, a *signal* is defined as a single-valued function of time that conveys information. Consequently, for every instant of time there is a unique value of the function. This value may be a real number, in which case we have a *real-valued signal*, or it may be a complex number, in which case we have a *complex-valued signal*. In either case, the independent variable (namely, time) is real-valued.

[1]For an essay on communications, see Berkner (1962).

For a given situation, the most useful method of signal representation hinges on the particular type of signal being considered. Depending on the feature of interest, we may identify four different methods of dividing signals into two classes:

1. PERIODIC SIGNALS, NONPERIODIC SIGNALS

A *periodic signal* $g(t)$ is a function that satisfies the condition

$$g(t) = g(t + T_0) \tag{1.1}$$

for all t, where t denotes time and T_0 is a constant. The smallest value of T_0 that satisfies this condition is called the *period* of $g(t)$. Accordingly, the period T_0 defines the duration of one complete cycle of $g(t)$.

Any signal for which there is no value of T_0 to satisfy the condition of Eq. 1.1 is called a *nonperiodic* or *aperiodic signal.*

2. DETERMINISTIC SIGNALS, RANDOM SIGNALS

A *deterministic signal* is a signal about which there is no uncertainty with respect to its value at any time. Accordingly, we find that deterministic signals may be modeled as completely specified functions of time.

On the other hand, a *random signal* is a signal about which there is uncertainty before its actual occurrence. Such a signal may be viewed as belonging to an ensemble of signals, with each signal in the collection having a different waveform. Moreover, each signal within the ensemble has a certain *probability* of occurrence.

3. ENERGY SIGNALS, POWER SIGNALS

In communication systems, a signal may represent a voltage or a current. Consider a voltage $v(t)$ developed across a resistor R, producing a current $i(t)$. The *instantaneous power* dissipated in this resistor is defined by

$$p = \frac{v^2(t)}{R} \tag{1.2}$$

or, equivalently,

$$p = Ri^2(t) \tag{1.3}$$

In both cases, the instantaneous power p is proportional to the squared amplitude of the signal. Furthermore, for a resistance R of 1 ohm, we see that Eqs. 1.2 and 1.3 take on the same mathematical form. Accordingly, in signal analysis it is customary to work with a 1-ohm resistor, so that, regardless of whether a given signal $g(t)$ represents a voltage or a current, we may express the instantaneous power associated with the signal as

$$p = g^2(t) \tag{1.4}$$

Based on this convention, we define the *total energy* of a signal $g(t)$ as

$$\begin{aligned} E &= \lim_{T\to\infty} \int_{-T}^{T} g^2(t)dt \\ &= \int_{-\infty}^{\infty} g^2(t)dt \end{aligned} \tag{1.5}$$

and its *average power* as

$$P = \lim_{T\to\infty} \frac{1}{2T} \int_{-T}^{T} g^2(t)dt \tag{1.6}$$

We say that the signal $g(t)$ is an *energy signal* if and only if the total energy of the signal satisfies the condition

$$0 < E < \infty$$

We say that the signal $g(t)$ is a *power signal* if and only if the average power of the signal satisfies the condition

$$0 < P < \infty$$

The energy and power classifications of signals are mutually exclusive. In particular, an energy signal has zero average power, whereas a power signal has infinite energy. Also, it is of interest to note that, usually, periodic signals and random signals are power signals, whereas signals that are both deterministic and nonperiodic are energy signals.

4. ANALOG SIGNALS, DIGITAL SIGNALS

An *analog signal* is a signal with an *amplitude* (i.e., value of the signal at some fixed time) that varies continuously for all time; that is, *both amplitude and time are continuous over their respective intervals.* Analog signals arise when a physical waveform such as an acoustic wave or a light wave is converted into an electrical signal. The conversion is effected by means of a *transducer;* examples include the microphone, which converts sound pressure variations into corresponding voltage or current variations, and the photodetector cell, which does the same for light-intensity variations.

On the other hand, a *discrete-time signal* is defined only at discrete instants of time. Thus, in this case, the independent variable takes on only discrete values, which are usually uniformly spaced. Consequently, discrete-time signals are described as sequences of samples that may take on a continuum of values. When each sample of a discrete-time signal is *quantized* (i.e., it is only allowed to take on a finite set of discrete values) and then *coded,* the resulting signal is referred to as a *digital signal.* The output of a digital computer is an example of a digital signal. Naturally, an analog signal may be converted into digital form by *sampling in time, then quantizing and coding.*

1.3 FOURIER ANALYSIS OF SIGNALS AND SYSTEMS

In theory, there are many possible methods for the representation of signals. In practice, however, we find that *Fourier analysis,* involving the resolution of signals into *sinusoidal components,* overshadows all other methods in usefulness. Basically, this is a consequence of the well-known fact that the output of a system to a sine-wave input is another sine wave of the same frequency[2] (but with a different phase and amplitude) under two conditions:

1. The system is *linear* in that it obeys the *principle of superposition.* That is, if $y_1(t)$ and $y_2(t)$ denote the responses of a system to the inputs $x_1(t)$ and $x_2(t)$, respectively, the system is linear if the response to the composite input $a_1x_2(t) + a_2x_2(t)$ is equal to $a_1y_1(t) + a_2y_2(t)$, where a_1 and a_2 are arbitrary constants.
2. The system is *time-invariant.* That is, if $y(t)$ is the response of a system to the input $x(t)$, the system is time-invariant if the response to the time-shifted input $x(t - t_0)$ is equal to $y(t - t_0)$, where t_0 is constant.

Given a linear time-invariant system, the *response* of the system to a single-frequency *excitation* represented by the complex exponential time function $A\exp(j2\pi ft)$ is equal to $AH(f)\exp(j2\pi ft)$, where $H(f)$ is the *transfer function* of the system; the complex exponential $\exp(j2\pi ft)$ contains the cosine function $\cos(2\pi ft)$ as its real part and the sine function $\sin(2\pi ft)$ as its imaginary part..Thus, the response of the system exhibits exactly the same variation with time as the excitation applied to the system. This remarkable property of linear time-invariant systems is realized only by using the complex exponential time function.

In the study of communication systems, we are usually interested in a *band of frequencies.* For example, although the average voice spectrum extends well beyond 10 kHz, most of the energy is concentrated in the range of 100 to 600 Hz, and a voice signal lying inside the band from 300 to 3400 Hz gives good articulation. Accordingly, we find that telephone circuits that respond well to the band of frequencies from 300 to 3400 Hz give satisfactory commercial telephone service.

To talk meaningfully about the *frequency-domain description* or *spectrum* of a signal, we need to know the *amplitude* and *phase* of each frequency component contained in the signal. We get this information by performing a Fourier analysis on the signal. However, there are several methods of Fourier analysis available for the representation of signals. The particular version that is used in practice depends on the type of signal being considered. For example, if the signal is periodic, then the logical choice is to use the *Fourier series* to represent the signal as a set of harmonically related sine waves. On the other hand, if the signal is an energy signal, then it is customary to use the *Fourier transform* to represent the signal. Irrespective of the type of signal being considered, Fourier methods are invertible. Specifically, if we are given the complete spectrum of a signal, then the original signal (as a function of time) can be reconstructed exactly. The Fourier analysis of signals and systems is considered in Chapters 2 through 4.

1.4 ELEMENTS OF A COMMUNICATION SYSTEM

The purpose of a *communication system* is to transmit information-bearing signals from a *source*, located at one point, to a *user destination*, located at another point some distance away. When the message produced by the source is not electrical in nature, which is often the case, an input transducer is used to convert it into a time-varying electrical signal called the *message signal*. By using another transducer connected to the output end of the system, a "distorted" version of the message is re-created in its original form, so that it is suitable for delivery to the user destination. The distortion mentioned here is due to inherent limitations in the communication system.

Figure 1.1 is a block diagram of a communication system consisting of three basic components: transmitter, channel, and receiver. The *transmitter* has the function of processing the message signal into a form suitable for

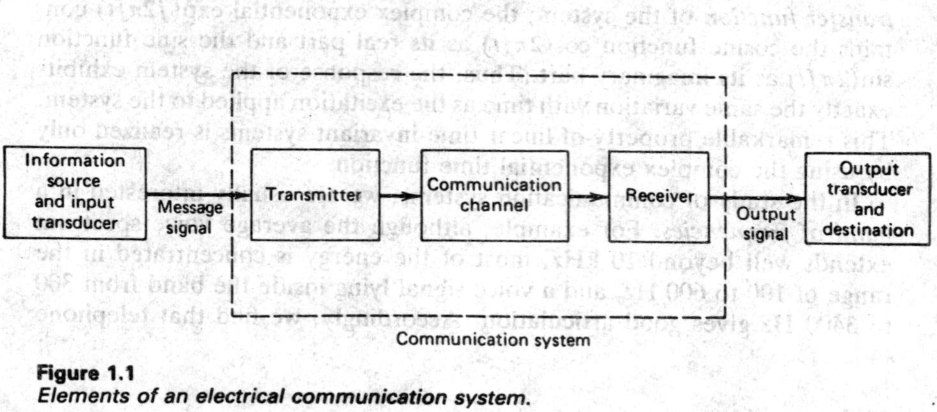

Figure 1.1
Elements of an electrical communication system.

transmission over the channel; such an operation is called *modulation*. The function of the *channel* is to provide a physical connection between the transmitter output and the receiver input. The function of the *receiver* is to process the *received signal* so as to produce an "estimate" of the original message signal; this second operation is called *detection* or *demodulation*.

There are two types of channels, namely, point-to-point channels and broadcast channels. Examples of *point-to-point channels* include wire lines, microwave links, and optical fibers. *Wire lines* operate by guided electromagnetic waves; they are used for local telephone transmission. In *microwave links*, the transmitted signal is radiated as an electromagnetic wave in free space; microwave links are used in long-distance telephone transmission. An *optical fiber* is a low-loss, well-controlled, guided optical medium; optical fibers are used in optical communications.[3] Although these three channels operate differently, they all provide a physical medium for the transmission of signals from one point to another point; hence, the term "point-to-point channels."

Broadcast channels, on the other hand, provide a capability where many receiving stations may be reached simultaneously from a single transmitter. An example of a broadcast channel is a *satellite in geostationary orbit*, which covers about one third of the earth's surface. Thus, three such satellites provide a complete coverage of the earth's surface, except for the polar regions.

1.5 TRANSMISSION OF MESSAGE SIGNALS

To transmit a message (information-bearing) signal over a communication channel, we may use *analog* or *digital* methods. The use of digital methods offers several important operational advantages over analog methods, which include the following:

1. Increased *immunity* to channel noise and external interference.
2. *Flexible operation* of the system.
3. A *common format* for the transmission of different kinds of message signals (e.g., voice signals, video signals, computer data).
4. Improved *security* of communication through the use of encryption.

These advantages are attained, however, at the cost of *increased transmission (channel) bandwidth and increased system complexity*. The first requirement is catered to by the availability of *wideband communication channels* (e.g., optical fibers, satellite channels). The second requirement is taken care of by the use of *very large-scale integration* (VLSI) technology, which offers a cost-effective way of building hardware. Accordingly, there

[3]For a discussion of electronic and photonic (optical) communication systems, see Williams (1987).

is an ever-increasing trend toward the use of digital communications and away from analog communications. This trend is being accelerated by the pervasive influence of digital computers in so many facets of our daily lives. Nevertheless, analog communications remain a force to be reckoned with. Most of the broadcasting systems and a large part of the telephone networks in use today are analog in nature and, moreover, they will remain in service for some time yet. It is therefore important that we understand the operations and requirements of both analog and digital communications.

Notable among the digital methods that may be used for the transmission of message signals over a communication channel is *pulse-code modulation* (PCM). In PCM, the message signal is *sampled, quantized,* and then *encoded.* The sampling operation permits representation of the message signal by a sequence of samples taken at uniformly spaced instants of time. Quantization trims the amplitude of each sample to the nearest value selected from a finite set of representation levels. The combination of sampling and quantization permits the use of a *code* (e.g., binary code) for the transmission of a message signal. Pulse-code modulation and related methods of *analog-to-digital conversion* are covered in Chapter 5.

When digital data are transmitted over a band-limited channel, a form of interference known as *intersymbol interference* may result. The effect of intersymbol interference, if left uncontrolled, is to severely limit the rate at which digital data may be transmitted over the channel. The cure for controlling the effects of intersymbol interference lies in *shaping* the transmitted pulse representing a binary symbol 1 or 0. Intersymbol interference is considered in Chapter 6.

To transmit a message signal (be it in analog or digital form) over a *band-pass* communication channel (e.g., telephone channel, microwave radio link, satellite channel) we need to modify the message signal into a form suitable for *efficient transmission* over the channel. Modification of the message signal is achieved by means of a process known as *modulation.* This process involves varying some parameter of a carrier wave in accordance with the message signal in such a way that the spectrum of the modulated wave matches the assigned channel bandwidth. Correspondingly, the receiver is required to re-create the original message signal from a degraded version of the transmitted signal after propagation through the channel. The re-creation is accomplished by using a process known as *demodulation,* which is the inverse of the modulation process used in the transmitter.

There are other reasons for performing modulation. In particular, the use of modulation permits *multiplexing,* that is, the simultaneous transmission of signals from several message sources over a common channel. Also, modulation may be used to convert the message signal into a form less susceptible to noise and interference.

A carrier wave commonly used to perform modulation is the sinusoidal wave. Such a carrier wave has three independent parameters that can be varied in accordance with the message signal; they are the carrier amplitude, phase, and frequency. The corresponding forms of modulation are known as *amplitude modulation, phase modulation,* and *frequency modulation,* respectively. Amplitude modulation offers simplicity of implementation and a transmission bandwidth requirement equal to twice the message bandwidth; the *message bandwidth* is defined as the extent of significant frequencies contained in the message signal. With special processing, the transmission bandwidth requirement may be reduced to a value equal to the message bandwidth, which is the minimum possible. Phase and frequency modulation, on the other hand, are more complex, requiring transmission bandwidths greater than that of amplitude modulation. In exchange, they offer a superior noise immunity, compared to amplitude modulation. Modulation techniques for analog and digital forms of message signals are studied in Chapter 7.

1.6 LIMITATIONS AND RESOURCES OF COMMUNICATION SYSTEMS

Typically, in propagating through a channel, the transmitted signal is distorted because of *nonlinearities* and *imperfections in the frequency response of the channel.* Other sources of degradation are *noise* and *interference* picked up by the signal during the course of transmission through the channel. Noise and distortion constitute two basic *limitations* in the design of communication systems.

There are various sources of noise, internal as well as external to the system. Although noise is random in nature, it may be described in terms of its *statistical properties* such as the *average power* or the spectral distribution of the average power. The mathematical discipline that deals with the statistical characteristics of noise and other random signals is *probability theory.* A discussion of probability theory and the related subject of *random processes* is presented in Chapter 8. Sources of noise and related system calculations are covered in Appendix C.

In any communication system, there are two primary communication resources to be employed, namely, *average transmitted power* and *channel bandwidth.* The average transmitted power is the average power of the transmitted signal. The channel bandwidth defines the range of frequencies that the channel can handle for the transmission of signals with satisfactory fidelity. A general system design objective is to use these two resources as efficiently as possible. In most channels, one resource may be considered more important than the other. Hence, we may also classify communication channels as *power-limited* or *band-limited.* For example, the telephone circuit is a typical band-limited channel, whereas a deep-space communication link or a satellite channel is typically power-limited.

The transmitted power is important because, for a receiver of prescribed *noise figure,* it determines the allowable separation between the transmitter and receiver. Stated in another way, for a receiver of prescribed noise figure and a prescribed distance between it and the transmitter, the available transmitted power determines the *signal-to-noise ratio* at the receiver input. This, in turn, determines the *noise performance* of the receiver. Unless this performance exceeds a certain design level, the transmission of message signals over the channel is not considered to be satisfactory.

The effects of noise in analog communications are evaluated in Chapter 9. This evaluation is traditionally done in terms of signal-to-noise ratios. In the case of digital communications, however, the preferred method of assessing the noise performance of a receiver is in terms of the *average probability of symbol error.* Such an approach leads to considerations of optimum receiver design. In this context, the *matched filter* offers optimum performance for the detection of pulses in an idealized form of receiver (channel) noise known as *additive white Gaussian noise.* As such, the matched-filter receiver or its equivalent, the *correlation receiver,* plays a key role in the design of digital communication systems. The matched filter and related issues are studied in Chapter 10.

Turning next to the other primary communication resource, channel bandwidth, it is important because, for a prescribed band of frequencies characterizing a message signal, the channel bandwidth determines the number of such message signals that can be *multiplexed* over the channel. Stated in another way, for a prescribed number of independent message signals that have to share a common channel, the channel bandwidth determines the band of frequencies that may be allotted to the transmission of each message signal without discernible distortion.

There is another important role for channel bandwidth, which is not that obvious. Specifically, channel bandwidth and transmitted (signal) power are *exchangeable* in that we may trade off one for the other for a prescribed system performance. The choice of one modulation scheme over another or the transmission of a message signal is often dictated by the nature of this trade-off. Indeed, the interplay between channel bandwidth and signal-to-noise ratio, and the limitation that they impose on communication, is highlighted most vividly by Shannon's famous *channel capacity theorem.*[4] Let B denote the channel bandwidth, and SNR denote the received signal-to-noise ratio. The channel capacity theorem states that ideally these two parameters are related by

$$C = B \log_2(1 + \mathrm{SNR}), \text{ bits/s} \tag{1.7}$$

where C is the *channel capacity,* and a *bit* refers to a *bi*nary digi*t*. The channel capacity is defined as the maximum rate at which information may be transmitted without error through the channel; it is measured in *bits per second.* Equation 1.7 clearly shows that for a prescribed channel capacity, we may reduce the required SNR by increasing the channel band-

[4] In 1948, Shannon published a paper that laid the foundations of communication theory (Shannon, 1948). The channel capacity theorem is one of three theorems presented in that classic paper.

width B. Moreover, it provides an idealized framework for comparing the noise performance of one modulation system against another.

Finally, mention should be made of the issue of *system complexity*. We usually find that the efficient exploitation of channel bandwidth or transmitted power or both is achieved at the expense of increased system complexity. We therefore have to keep the issue of system complexity in mind, alongside that of channel bandwidth and transmitted power when considering the various trade-offs involved in the design of communication systems.

CHAPTER TWO

FOURIER ANALYSIS

In this chapter, we begin our study of Fourier analysis. We first review the *Fourier series,* by means of which we are able to represent a periodic signal as an infinite sum of sine-wave components. Next, we develop the *Fourier transform,* which performs a similar role in the analysis of nonperiodic signals. The Fourier transform is more general in application than the Fourier series.[1] The primary motivation for using the Fourier series or the Fourier transform is to obtain the *spectrum* of a

[1]The origin of the theory of Fourier series and Fourier transform is found in J. B. J. Fourier, *The Analytical Theory of Heat* (trans. A. Freeman), Cambridge University Press, London, 1878.

given signal, which describes the frequency content of the signal. In effect, this transformation provides an alternative method of viewing the signal that is often more revealing than the original description of the signal as a function of time.

2.1 FOURIER SERIES

Let $g_p(t)$ denote a periodic signal with period T_0. By using a *Fourier series expansion* of this signal, we are able to resolve the signal into an infinite sum of sine and cosine terms. This expansion may be expressed in the form

$$g_p(t) = a_0 + 2\sum_{n=1}^{\infty}\left[a_n \cos\left(\frac{2\pi nt}{T_0}\right) + b_n \sin\left(\frac{2\pi nt}{T_0}\right)\right] \tag{2.1}$$

where the coefficients a_n and b_n represent the unknown amplitudes of the cosine and sine terms, respectively. The quantity n/T_0 represents the nth harmonic of the *fundamental frequency* $f_0 = 1/T_0$. Each of the cosine and sine functions in Eq. 2.1 is called a *basis function*. These basis functions form an *orthogonal set* over the interval T_0 in that they satisfy the following set of relations:

$$\int_{-T_0/2}^{T_0/2} \cos\left(\frac{2\pi mt}{T_0}\right)\cos\left(\frac{2\pi nt}{T_0}\right) dt = \begin{cases} T_0/2, & m = n \\ 0, & m \neq n \end{cases} \tag{2.2}$$

$$\int_{-T_0/2}^{T_0/2} \cos\left(\frac{2\pi mt}{T_0}\right)\sin\left(\frac{2\pi nt}{T_0}\right) dt = 0 \qquad \text{for all } m \text{ and } n \tag{2.3}$$

$$\int_{-T_0/2}^{T_0/2} \sin\left(\frac{2\pi mt}{T_0}\right)\sin\left(\frac{2\pi nt}{T_0}\right) dt = \begin{cases} T_0/2, & m = n \\ 0, & m \neq n \end{cases} \tag{2.4}$$

To determine the coefficient a_0, we integrate both sides of Eq. 2.1 over a complete period. We thus find that a_0 is the *mean value* of the periodic signal $g_p(t)$ over one period, as shown by the *time average*

$$a_0 = \frac{1}{T_0}\int_{-T_0/2}^{T_0/2} g_p(t)\, dt \tag{2.5}$$

To determine the coefficient a_n, we multiply both sides of Eq. 2.1 by the cosine function $\cos(2\pi nt/T_0)$ and integrate over the interval $-T_0/2$ to $T_0/2$. Then, using Eqs. 2.2 and 2.3, we find that

$$a_n = \frac{1}{T_0}\int_{-T_0/2}^{T_0/2} g_p(t)\cos\left(\frac{2\pi nt}{T_0}\right) dt, \qquad n = 1, 2, \ldots \tag{2.6}$$

Similarly, we find that

$$b_n = \frac{1}{T_0}\int_{-T_0/2}^{T_0/2} g_p(t)\sin\left(\frac{2\pi nt}{T_0}\right)dt, \qquad n = 1, 2, \ldots \tag{2.7}$$

To apply the Fourier series representation of Eq. 2.1, it is sufficient that inside the interval $-(T_0/2) \leqslant t \leqslant (T_0/2)$ the function $g_p(t)$ satisfies the following conditions:

1. The function $g_p(t)$ is single-valued.
2. The function $g_p(t)$ has a finite number of discontinuities.
3. The function $g_p(t)$ has a finite number of maxima and minima.
4. The function $g_p(t)$ is absolutely integrable, that is,

$$\int_{-T_0/2}^{T_0/2} |g_p(t)|\, dt < \infty$$

where $g_p(t)$ is assumed to be complex valued.

These conditions are known as *Dirichlet's conditions*. They are satisfied by the periodic signals usually encountered in communication systems.

COMPLEX EXPONENTIAL FOURIER SERIES

The Fourier series of Eq. 2.1 can be put into a much simpler and more elegant form with the use of complex exponentials. We do this by substituting in Eq. 2.1 the exponential form for the cosine and sine, namely:

$$\cos\left(\frac{2\pi nt}{T_0}\right) = \frac{1}{2}\left[\exp\left(\frac{j2\pi nt}{T_0}\right) + \exp\left(-\frac{j2\pi nt}{T_0}\right)\right]$$

$$\sin\left(\frac{2\pi nt}{T_0}\right) = \frac{1}{2j}\left[\exp\left(\frac{j2\pi nt}{T_0}\right) - \exp\left(-\frac{j2\pi nt}{T_0}\right)\right]$$

We thus obtain

$$g_p(t) = a_0 + \sum_{n=1}^{\infty}\left[(a_n - jb_n)\exp\left(\frac{j2\pi nt}{T_0}\right) + (a_n + jb_n)\exp\left(-\frac{j2\pi nt}{T_0}\right)\right] \tag{2.8}$$

The two product terms inside the square brackets in Eq. 2.8 are the complex conjugate of each other. We may also note the following relation:

$$\sum_{n=1}^{\infty} (a_n + jb_n) \exp\left(-\frac{j2\pi nt}{T_0}\right) = \sum_{n=-\infty}^{-1} (a_n - jb_n) \exp\left(\frac{j2\pi nt}{T_0}\right)$$

Let c_n denote a complex coefficient related to a_n and b_n by

$$c_n = \begin{cases} a_n - jb_n, & n > 0 \\ a_0, & n = 0 \\ a_n + jb_n, & n < 0 \end{cases} \tag{2.9}$$

Accordingly, we may simplify Eq. 2.8 as follows:

$$g_p(t) = \sum_{n=-\infty}^{\infty} c_n \exp\left(\frac{j2\pi nt}{T_0}\right) \tag{2.10}$$

where

$$c_n = \frac{1}{T_0} \int_{-T_0/2}^{T_0/2} g_p(t) \exp\left(-\frac{j2\pi nt}{T_0}\right) dt, \qquad n = 0, \pm 1, \pm 2, \ldots \tag{2.11}$$

The series expansion of Eq. 2.10 is referred to as the *complex exponential Fourier series*. The c_n are called the *complex Fourier coefficients*. Equation 2.11 states that, given a periodic signal $g_p(t)$, we may determine the complete set of complex Fourier coefficients. On the other hand, Eq. 2.10 states that, given this set of values, we may reconstruct the original periodic signal exactly.

According to this representation, a periodic signal contains all frequencies (both positive and negative) that are harmonically related to the fundamental. The presence of negative frequencies is simply a result of the fact that the mathematical model of the signal as described by Eq. 2.10 requires the use of negative frequencies. Indeed, this representation also requires the use of a complex-valued basis function $\exp(j2\pi nt/T_0)$, which has no physical meaning either. The reason for using complex-valued basis functions and negative frequency components is merely to provide a compact mathematical description of a periodic signal, which is well-suited for both theoretical and practical work.

DISCRETE SPECTRUM

The representation of a periodic signal by a Fourier series is equivalent to the resolution of the signal into its various harmonic components. Thus, using the complex exponential Fourier series, we find that a periodic signal $g_p(t)$ with period T_0 has components of frequencies 0, $\pm f_0$, $\pm 2f_0$, $\pm 3f_0$, . . . , and so forth, where $f_0 = 1/T_0$ is the fundamental frequency. That is, while the signal $g_p(t)$ exists in the time domain, we may say that its frequency-domain description consists of components of frequencies, 0, $\pm f_0$, $\pm 2f_0$, . . . , called the *spectrum*.[2] If we specify the periodic signal $g_p(t)$, we can determine its spectrum; conversely, if we specify the spectrum, we can determine the corresponding signal. This means that a periodic signal $g_p(t)$ can be specified in two equivalent ways: (1) the time-domain representation where $g_p(t)$ is defined as a function of time, and (2) the frequency-domain representation where the signal is defined in terms of its spectrum. Although the two descriptions are separate aspects of a given phenomenon, they are not independent of each other, but are related, as Fourier theory shows.

In general, the Fourier coefficient c_n is a complex number; so we may express it in the form

$$c_n = |c_n| \exp[j \arg(c_n)] \tag{2.12}$$

The term $|c_n|$ defines the amplitude of the nth harmonic component of the periodic signal $g_p(t)$, so that a plot of $|c_n|$ versus frequency yields the *discrete amplitude spectrum* of the signal. A plot of $\arg(c_n)$ versus frequency yields the *discrete phase spectrum* of the signal. We refer to the spectrum as a *discrete spectrum* because both the amplitude and phase of c_n have nonzero values only for discrete frequencies that are integer (both positive and negative) multiples of the fundamental frequency.

For a real-valued periodic function $g_p(t)$, we find from the definition of the Fourier coefficient c_n given by Eq. 2.11 that

$$c_{-n} = c_n^* \tag{2.13}$$

where c_n^* is the complex conjugate of c_n. We therefore have

$$|c_{-n}| = |c_n| \tag{2.14}$$

and

$$\arg(c_{-n}) = -\arg(c_n) \tag{2.15}$$

That is, the amplitude spectrum of a real-valued periodic signal is *symmetric* (an even function of n) and the phase spectrum is *asymmetric* (an odd function of n) about the vertical axis passing through the origin.

[2]The term "spectrum" comes from the Latin word for "image." It was originally introduced by Sir Isaac Newton. For a historical account of spectrum analysis, see Gardner (1987).

EXAMPLE 1 PERIODIC PULSE TRAIN

Consider a periodic train of rectangular pulses of duration T and period T_0, as shown in Fig. 2.1. For convenience, the origin has been chosen to coincide with the center of the pulse. This signal may be described analytically over one period, $-(T_0/2) \leqslant t \leqslant (T_0/2)$, as follows

$$g_p(t) = \begin{cases} A, & -\dfrac{T}{2} \leqslant t \leqslant \dfrac{T}{2} \\ 0, & \text{for the remainder of the period} \end{cases} \tag{2.16}$$

Using Eq. 2.11 to evaluate the complex Fourier coefficient c_n, we get

$$\begin{aligned} c_n &= \frac{1}{T_0} \int_{-T/2}^{T/2} A \exp\left(-\frac{j2\pi nt}{T_0}\right) dt \\ &= \frac{A}{n\pi} \sin\left(\frac{n\pi T}{T_0}\right), \quad n = 0, \pm 1, \pm 2, \ldots \end{aligned} \tag{2.17}$$

To simplify notation in the foregoing and subsequent results, we will use the *sinc function* defined by

$$\text{sinc}(\lambda) = \frac{\sin(\pi\lambda)}{\pi\lambda} \tag{2.18}$$

where λ is the independent variable. The sinc function plays an important role in communication theory. As shown in Fig. 2.2, it has its maximum value of unity at $\lambda = 0$, and approaches zero as λ approaches infinity, oscillating through positive and negative values. It goes through zero at $\lambda = \pm 1, \pm 2, \ldots$, and so on. Thus, in terms of the sinc function we may rewrite Eq. 2.17 as follows

$$c_n = \frac{TA}{T_0} \text{sinc}\left(\frac{nT}{T_0}\right) \tag{2.19}$$

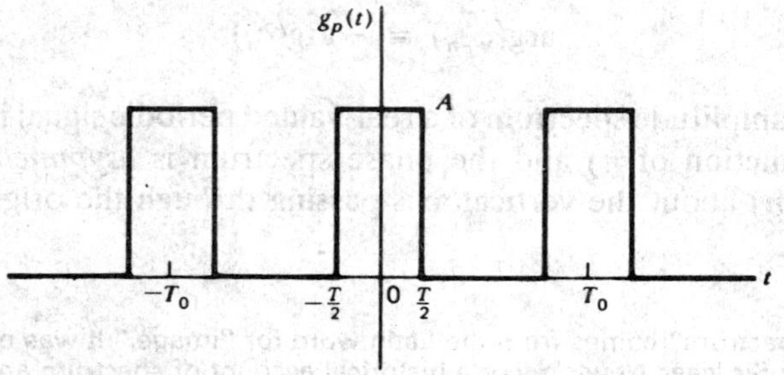

Figure 2.1
Periodic train of rectangular pulses of amplitude A, duration T, and period T_0.

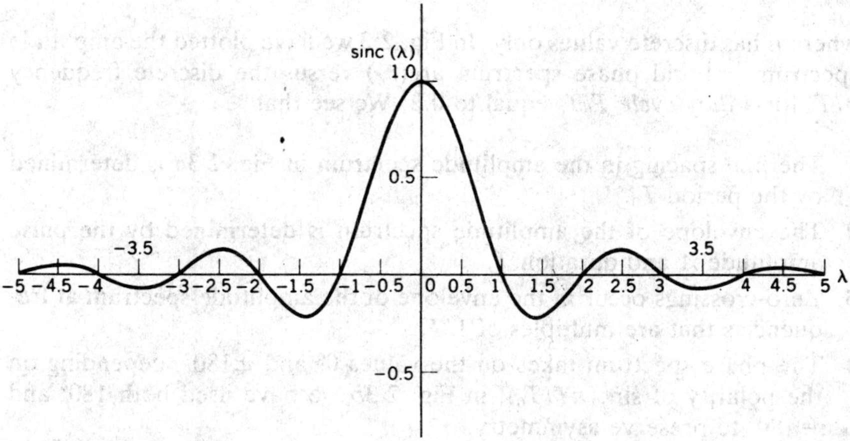

Figure 2.2
The sinc function.

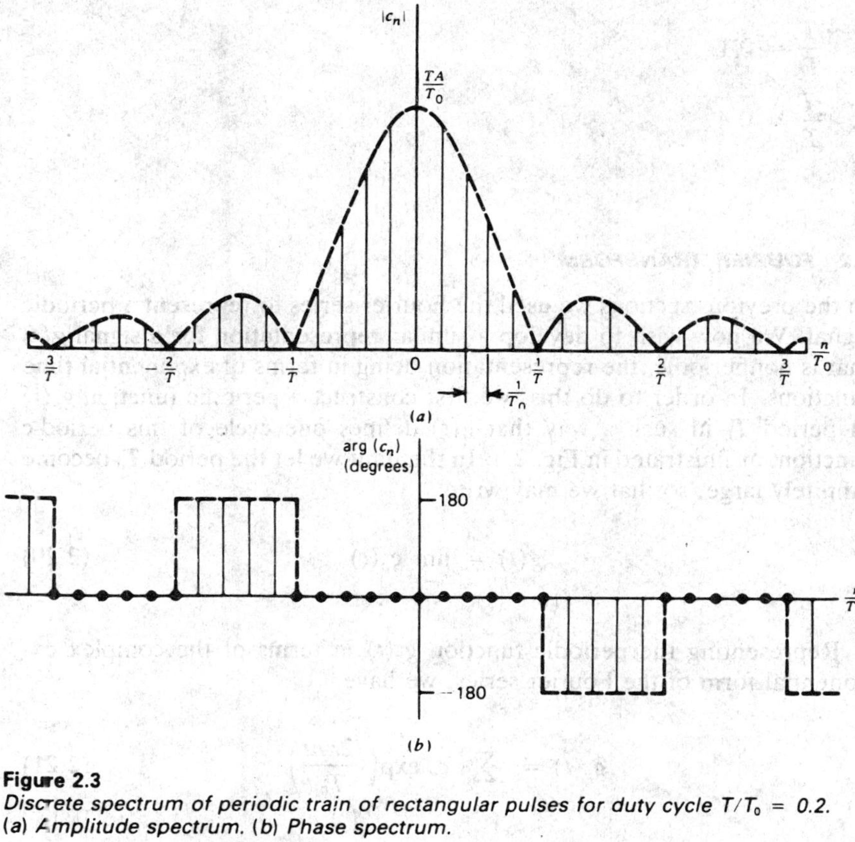

Figure 2.3
Discrete spectrum of periodic train of rectangular pulses for duty cycle $T/T_0 = 0.2$. (a) Amplitude spectrum. (b) Phase spectrum.

where n has discrete values only. In Fig. 2.3 we have plotted the amplitude spectrum $|c_n|$ and phase spectrum $\arg(c_n)$ versus the discrete frequency n/T_0 for a *duty cycle* T/T_0 equal to 0.2. We see that

1. The line spacing in the amplitude spectrum in Fig. 2.3*a* is determined by the period T_0.
2. The envelope of the amplitude spectrum is determined by the pulse amplitude A and duration T.
3. Zero-crossings occur in the envelope of the amplitude spectrum at frequencies that are multiples of $1/T$.
4. The phase spectrum takes on the values 0° and ±180°, depending on the polarity of $\text{sinc}(nT/T_0)$; in Fig. 2.3*b* we have used both 180° and −180° to preserve asymmetry.

EXERCISE 1 Plot the amplitude spectra of rectangular pulses of unit amplitude and the following two values of duty cycle:

a. $\dfrac{T}{T_0} = 0.1$

b. $\dfrac{T}{T_0} = 0.4$

2.2 FOURIER TRANSFORM

In the previous sections we used the Fourier series to represent a periodic signal. We now wish to develop a similar representation for a signal $g(t)$ that is nonperiodic, the representation being in terms of exponential time functions. In order to do this, we first construct a periodic function $g_p(t)$ of period T_0 in such a way that $g(t)$ defines one cycle of this periodic function, as illustrated in Fig. 2.4. In the limit we let the period T_0 become infinitely large, so that we may write

$$g(t) = \lim_{T_0 \to \infty} g_p(t) \tag{2.20}$$

Representing the periodic function $g_p(t)$ in terms of the complex exponential form of the Fourier series, we have

$$g_p(t) = \sum_{n=-\infty}^{\infty} c_n \exp\left(\frac{j2\pi nt}{T_0}\right) \tag{2.21}$$

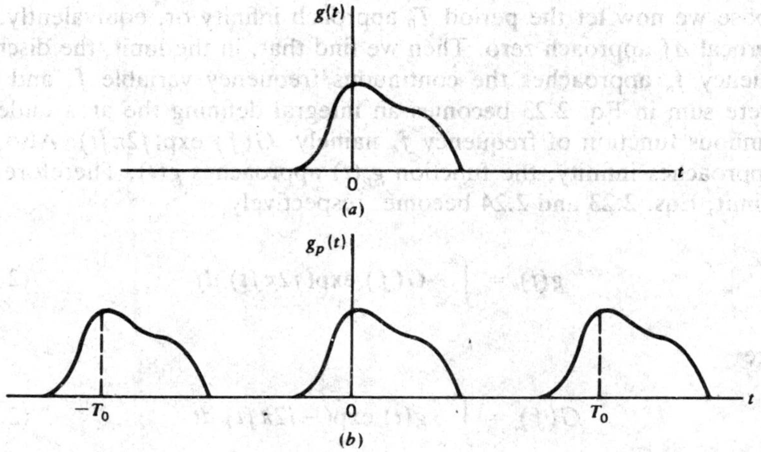

Figure 2.4
The construction of a periodic function from an arbitrarily defined function of time.

where

$$c_n = \frac{1}{T_0}\int_{-T_0/2}^{T_0/2} g_p(t)\exp\left(-\frac{j2\pi nt}{T_0}\right) dt \tag{2.22}$$

Define

$$\Delta f = \frac{1}{T_0}$$

$$f_n = \frac{n}{T_0}$$

and

$$G(f_n) = c_n T_0$$

Thus, making this change of notation in the Fourier series representation of $g_p(t)$, given by Eqs. 2.21 and 2.22, we get the following relations for the interval $-(T_0/2) \leqslant t \leqslant (T_0/2)$,

$$g_p(t) = \sum_{n=-\infty}^{\infty} G(f_n)\exp(j2\pi f_n t)\,\Delta f \tag{2.23}$$

where

$$G(f_n) = \int_{-T_0/2}^{T_0/2} g_p(t)\exp(-j2\pi f_n t)\,dt \tag{2.24}$$

Suppose we now let the period T_0 approach infinity or, equivalently, its reciprocal Δf approach zero. Then we find that, in the limit, the discrete frequency f_n approaches the continuous frequency variable f, and the discrete sum in Eq. 2.23 becomes an integral defining the area under a continuous function of frequency f, namely, $G(f) \exp(j2\pi ft)$. Also, as T_0 approaches infinity, the function $g_p(t)$ approaches $g(t)$. Therefore, in the limit, Eqs. 2.23 and 2.24 become, respectively,

$$g(t) = \int_{-\infty}^{\infty} G(f) \exp(j2\pi ft)\, df \tag{2.25}$$

where

$$G(f) = \int_{-\infty}^{\infty} g(t) \exp(-j2\pi ft)\, dt \tag{2.26}$$

We have thus achieved our aim of representing an arbitrarily defined signal $g(t)$ in terms of exponential time functions over the entire time interval from $-\infty$ to ∞. Note that in Eqs. 2.25 and 2.26 we have used a lowercase letter to denote the time function and an uppercase letter to denote the corresponding frequency function.

Equation 2.26 states that, given a time function $g(t)$, we can determine a new function $G(f)$ of the frequency variable f. Equation 2.25 states that, given this new or transformed function $G(f)$, we can recover the original time function $g(t)$. Thus, since from $g(t)$ we can define the function $G(f)$ and from $G(f)$ we can reconstruct $g(t)$, the time function is also specified by $G(f)$. The function $G(f)$ can be thought of as a transformed version of $g(t)$ and is referred to as the *Fourier transform* of $g(t)$. The time function $g(t)$ is similarly referred to as the *inverse Fourier transform* of $G(f)$. The functions $g(t)$ and $G(f)$ are said to constitute a *Fourier transform pair.*

DIRICHLET'S CONDITIONS

For a signal $g(t)$ to be Fourier transformable, it is sufficient that $g(t)$ satisfies *Dirichlet's conditions:*

1. The function $g(t)$ is single-valued, with a finite number of maxima and minima and a finite number of discontinuities in any finite time interval.
2. The function $g(t)$ is absolutely integrable, that is,

$$\int_{-\infty}^{\infty} |g(t)|\, dt < \infty$$

The Dirichlet conditions are not strictly necessary but sufficient for the Fourier transformability of a signal. These conditions include all energy signals, for which we have[3]

$$\int_{-\infty}^{\infty} |g(t)|^2\, dt < \infty$$

In the two conditions described herein, the signal $g(t)$ is assumed to be complex.

NOTATIONS

The formulas for the Fourier transform and the inverse Fourier transform presented in Eqs. 2.25 and 2.26 are written in terms of *time* t and *frequency* f, with t measured in *seconds* (s) and f measured in *hertz* (Hz). The frequency f is related to the *angular frequency* ω as $\omega = 2\pi f$, which is measured in *radians per second* (rad/s). We may simplify the expressions for the exponents in the integrands of Eqs. 2.25 and 2.26 by using ω instead of f. However, the use of f is preferred over ω for two reasons. First, we have the mathematical *symmetry* of Eqs. 2.25 and 2.26 with respect to each other. Second, the frequency contents of communication signals (i.e., speech and video signals) are usually expressed in hertz.

A convenient *shorthand* notation for the transform relations of Eqs. 2.26 and 2.25 is

$$G(f) = \mathrm{F}[g(t)] \tag{2.27a}$$

$$g(t) = \mathrm{F}^{-1}[G(f)] \tag{2.27b}$$

Another convenient shorthand notation for the Fourier transform pair, represented by $g(t)$ and $G(f)$, is

$$g(t) \rightleftharpoons G(f) \tag{2.28}$$

The shorthand notations described herein are used in the text where appropriate.

SPECTRUM

By using Fourier transformation, an energy signal $g(t)$ is represented by the Fourier transform $G(f)$, which is a function of the frequency variable

[3]If the function $g(t)$ is such that the value of $\int_{-\infty}^{\infty} |g(t)|^2\, dt$ is defined and finite, then the Fourier transform $G(f)$ of the function $g(t)$ exists and

$$\lim_{A\to\infty} \left[\int_{-\infty}^{\infty} \left| g(t) - \int_{-A}^{A} G(f) \exp(j2\pi ft)\, df \right|^2 dt \right] = 0$$

This result is known as *Plancherel's theorem.*

f. A plot of the Fourier transform $G(f)$ versus the frequency f is called the *spectrum* of the signal $g(t)$. The spectrum is continuous in the sense that it is defined for all frequencies. In general, the Fourier transform $G(f)$ is a complex function of the frequency f. We may therefore express it in the form

$$G(f) = |G(f)| \exp[j\theta(f)] \tag{2.29}$$

where $|G(f)|$ is called the *amplitude spectrum* of $g(t)$, and $\theta(f)$ is called the *phase spectrum* of $g(t)$.

For the special case of a real-valued function $g(t)$, we have

$$G(f) = G^*(-f)$$

Therefore, it follows that if $g(t)$ is a real-valued function of time t, then

$$|G(-f)| = |G(f)| \tag{2.30}$$

and

$$\theta(-f) = -\theta(f) \tag{2.31}$$

Accordingly, we may make the following statements on the spectrum of a *real-valued signal:*

1. The amplitude spectrum of the signal is an even function of the frequency; that is, the amplitude spectrum is *symmetric* about the vertical axis.
2. The phase spectrum of the signal is an odd function of the frequency; that is, the phase spectrum is *antisymmetric* about the vertical axis.

These two statements are often summed up by saying that the spectrum of a real-valued signal exhibits *conjugate symmetry.*

EXAMPLE 2 RECTANGULAR PULSE

Consider a *rectangular pulse* of duration T and amplitude A, as shown in Fig. 2.5. To define this pulse mathematically in a convenient form, we use the following notation

$$\text{rect}(t) = \begin{cases} 1, & -\frac{1}{2} < t < \frac{1}{2} \\ 0, & |t| > \frac{1}{2} \end{cases} \tag{2.32}$$

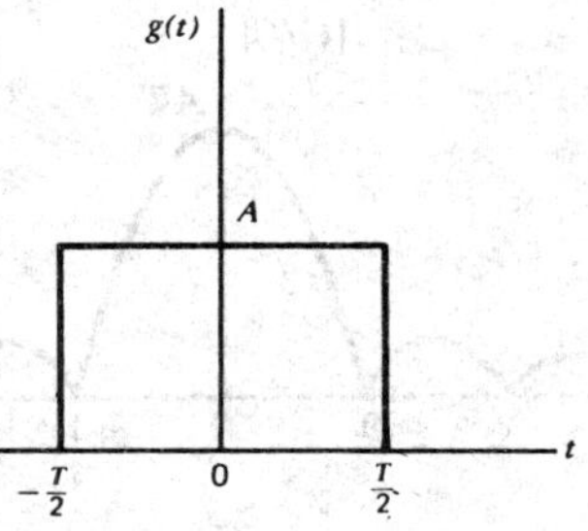

Figure 2.5
Rectangular pulse.

which stands for a *rectangular function* of unit amplitude and unit duration centered at $t = 0$. Then, in terms of this function, we may express the rectangular pulse of Fig. 2.5 simply as follows:

$$g(t) = A \operatorname{rect}\left(\frac{t}{T}\right)$$

The Fourier transform of this rectangular pulse is given by

$$\begin{aligned} G(f) &= \int_{-T/2}^{T/2} A \exp(-j2\pi ft)\, dt \\ &= AT\left[\frac{\sin(\pi fT)}{\pi fT}\right] \\ &= AT \operatorname{sinc}(fT) \end{aligned}$$

We thus have the Fourier transform pair

$$A \operatorname{rect}\left(\frac{t}{T}\right) \rightleftharpoons AT \operatorname{sinc}(fT) \tag{2.33}$$

The amplitude spectrum $|G(f)|$ of the rectangular pulse $g(t)$ is shown plotted in Fig. 2.6*a*. From this spectrum, we may make the following observations:

1. The amplitude spectrum has a *main lobe* of total width $2/T$, centered on the origin.

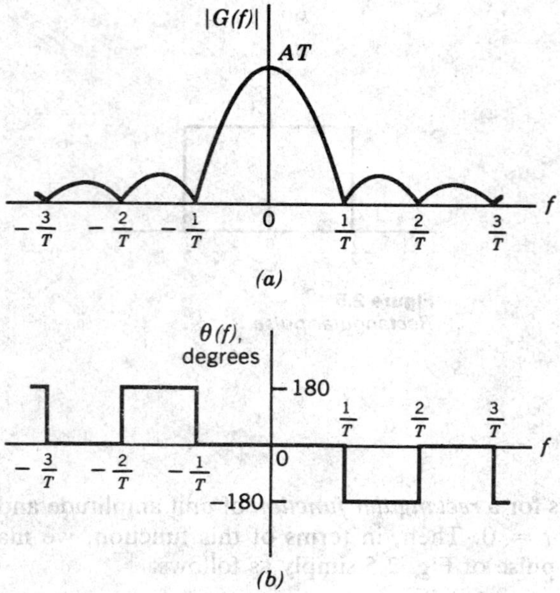

Figure 2.6
Spectrum of rectangular pulse. (a) Amplitude spectrum. (b) Phase spectrum.

2. The *side lobes,* on either side of the main lobe, decrease in amplitude with increasing $|f|$. *Indeed, the amplitudes of the side lobes are bounded by the curve* $1/|f|$.
3. The *zero crossings* of the spectrum occur at $f = \pm 1/T, \pm 2/T, \ldots$

The phase spectrum $\theta(f)$ of the rectangular pulse $g(t)$ is shown plotted in Fig. 2.6*b*. Depending on the sign of the sinc function $\text{sinc}(fT)$, the phase spectrum takes on the values 0° and ±180° in an asymmetric fashion.

EXAMPLE 3 EXPONENTIAL PULSE

A truncated form of decaying *exponential pulse* is shown in Fig. 2.7*a*. We may define this pulse mathematically in a convenient form by using the *unit step function:*

$$u(t) = \begin{cases} 1, & t > 0 \\ \frac{1}{2}, & t = 0 \\ 0, & t < 0 \end{cases} \tag{2.34}$$

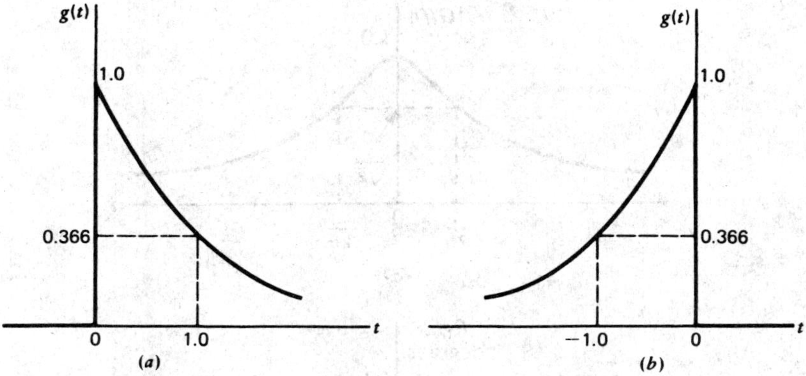

Figure 2.7
(a) Decaying exponential pulse. (b) Rising exponential pulse.

We may then express the exponential pulse of Fig. 2.7*a* as

$$g(t) = \exp(-t)u(t) \tag{2.35}$$

The Fourier transform of this pulse is

$$\begin{aligned} G(f) &= \int_0^\infty \exp(-t)\exp(-j2\pi ft)\,dt \\ &= \int_0^\infty \exp[-t(1 + j2\pi f)]\,dt \\ &= \frac{1}{1 + j2\pi f} \end{aligned} \tag{2.36}$$

Thus, combining Eqs. 2.35 and 2.36, we obtain the Fourier transform pair:

$$\exp(-t)u(t) \rightleftharpoons \frac{1}{1 + j2\pi f} \tag{2.37}$$

Figure 2.8 shows the spectrum of the decaying exponential pulse.

A truncated rising exponential pulse is shown in Fig. 2.7*b*, which is defined by

$$g(t) = \exp(t)u(-t) \tag{2.38}$$

Note that $u(-t)$ is equal to unity for $t < 0$, one-half at $t = 0$, and zero

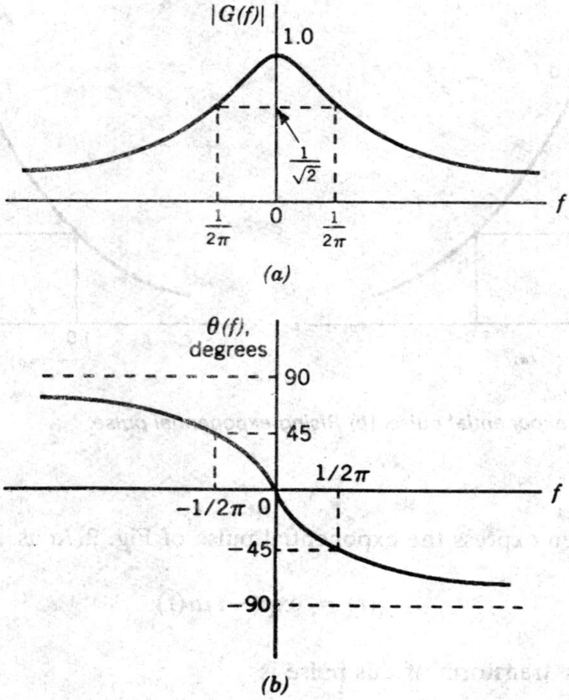

Figure 2.8
Spectrum of decaying exponential pulse. (a) Amplitude spectrum. (b) Phase spectrum.

for $t > 0$. The Fourier transform of this pulse is given by

$$G(f) = \int_{-\infty}^{0} \exp(t) \exp(-j2\pi ft)\, dt$$
$$= \int_{-\infty}^{0} \exp[t(1 - j2\pi f)]\, dt$$
$$= \frac{1}{1 - j2\pi f} \tag{2.39}$$

We thus have the Fourier transform pair:

$$\exp(t)u(-t) \rightleftharpoons \frac{1}{1 - j2\pi f} \tag{2.40}$$

Figure 2.9 shows the spectrum of the rising exponential pulse.

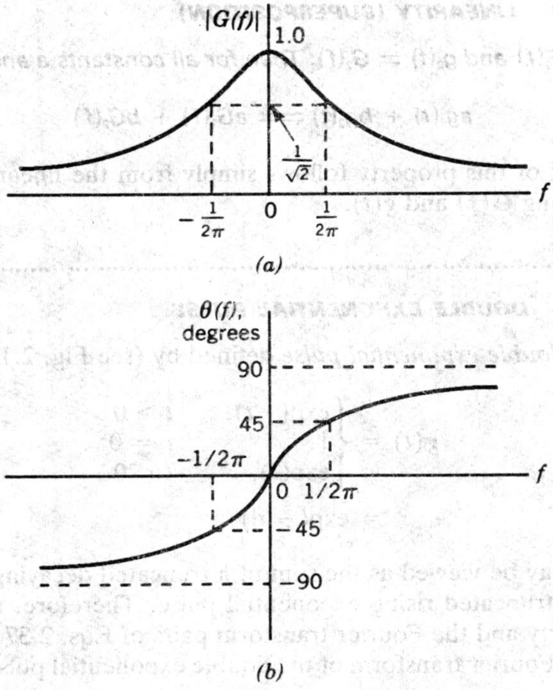

Figure 2.9
Spectrum of rising exponential pulse. (a) Amplitude spectrum. (b) Phase spectrum.

Comparing the spectra of Figs. 2.8 and 2.9, we may make the following two observations:

1. The decaying and rising exponentials of Fig. 2.7 have the same amplitude spectrum.
2. The phase spectrum of the rising exponential is the negative of that of the decaying exponential.

2.3 *PROPERTIES OF THE FOURIER TRANSFORM*

It is useful to have a feeling for the relationship between a function $g(t)$ and its Fourier transform $G(f)$, and for the effect that various operations on the function $g(t)$ have on the transform $G(f)$. This may be achieved by examining certain properties of the Fourier transform. This section describes 10 of these properties, which will be proved, one by one. These properties are summarized in Table 1 of Appendix D.

PROPERTY 1 LINEARITY (SUPERPOSITION)

Let $g_1(t) \rightleftharpoons G_1(f)$ and $g_2(t) \rightleftharpoons G_2(f)$. Then for all constants a and b, we have

$$ag_1(t) + bg_2(t) \rightleftharpoons aG_1(f) + bG_2(f) \tag{2.41}$$

The proof of this property follows simply from the linearity of the integrals defining $G(f)$ and $g(t)$.

EXAMPLE 4 DOUBLE EXPONENTIAL PULSE

Consider a *double exponential pulse* defined by (see Fig. 2.10)

$$g(t) = \begin{cases} \exp(-t), & t > 0 \\ 1, & t = 0 \\ \exp(t), & t < 0 \end{cases}$$
$$= \exp(-|t|) \tag{2.42}$$

This pulse may be viewed as the sum of a truncated decaying exponential pulse and a truncated rising exponential pulse. Therefore, using the linearity property and the Fourier-transform pairs of Eqs. 2.37 and 2.40, we find that the Fourier transform of the double exponential pulse of Fig. 2.10 is as follows

$$G(f) = \frac{1}{1 + j2\pi f} + \frac{1}{1 - j2\pi f}$$
$$= \frac{2}{1 + (2\pi f)^2}$$

We thus have the Fourier transform pair

$$\exp(-|t|) \rightleftharpoons \frac{2}{1 + (2\pi f)^2} \tag{2.43}$$

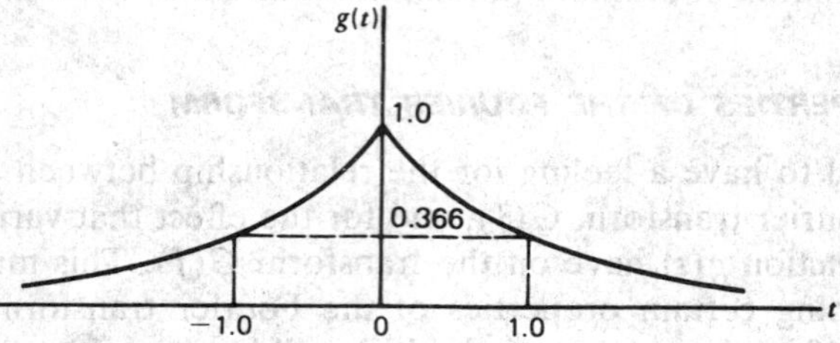

Figure 2.10
Double exponential pulse.

PROPERTY 2 TIME SCALING

Let $g(t) \rightleftharpoons G(f)$. Then,

$$g(at) \rightleftharpoons \frac{1}{|a|} G\left(\frac{f}{a}\right) \tag{2.44}$$

where a is a time-scaling factor that may be positive or negative.

To prove this property, we note that

$$F[g(at)] = \int_{-\infty}^{\infty} g(at) \exp(-j2\pi f t)\, dt$$

Set $\tau = at$. There are two cases that can arise, depending on whether the scaling factor a is positive or negative. If $a > 0$, we get

$$\begin{aligned} F[g(at)] &= \frac{1}{a} \int_{-\infty}^{\infty} g(\tau) \exp\left[-j2\pi \left(\frac{f}{a}\right) \tau\right] d\tau \\ &= \frac{1}{a} G\left(\frac{f}{a}\right) \end{aligned}$$

On the other hand, if $a < 0$, the limits of integration are interchanged so that we have the multiplying factor $-(1/a)$ or, equivalently, $1/|a|$. This completes the proof of Eq. 2.44.

Note that the function $g(at)$ represents $g(t)$ compressed in time by a factor a, whereas the function $G(f/a)$ represents $G(f)$ expanded in frequency by the same factor a. Thus the scaling property states that the compression of a function $g(t)$ in the time domain is equivalent to the expansion of its Fourier transform $G(f)$ in the frequency domain by the same factor, and vice versa.

EXAMPLE 5 RECTANGULAR PULSE (CONTINUED)

Example 2 evaluated the Fourier transform of a rectangular pulse; the result of the evaluation is given by the Fourier transform pair of Eq. 2.33. For convenience of presentation, let the rectangular pulse be normalized to have unit amplitude and unit duration. Then, putting $A = 1$ and $T = 1$ in Eq. 2.33, we have

$$\text{rect}(t) \rightleftharpoons \text{sinc}(f)$$

Hence, applying the time-scaling property to this Fourier transform pair, we get

$$\mathrm{rect}(at) \rightleftharpoons \frac{1}{|a|}\,\mathrm{sinc}\left(\frac{f}{a}\right)$$

Figure 2.11 shows the rectangular pulse and its amplitude spectrum for three different values of the time-scaling factor a, namely $a = 1/2, 1, 2$. With $a = 1$ regarded as the frame of reference, we may view the use of $a = 1/2$ as expansion in time, and $a = 2$ as compression in time. These

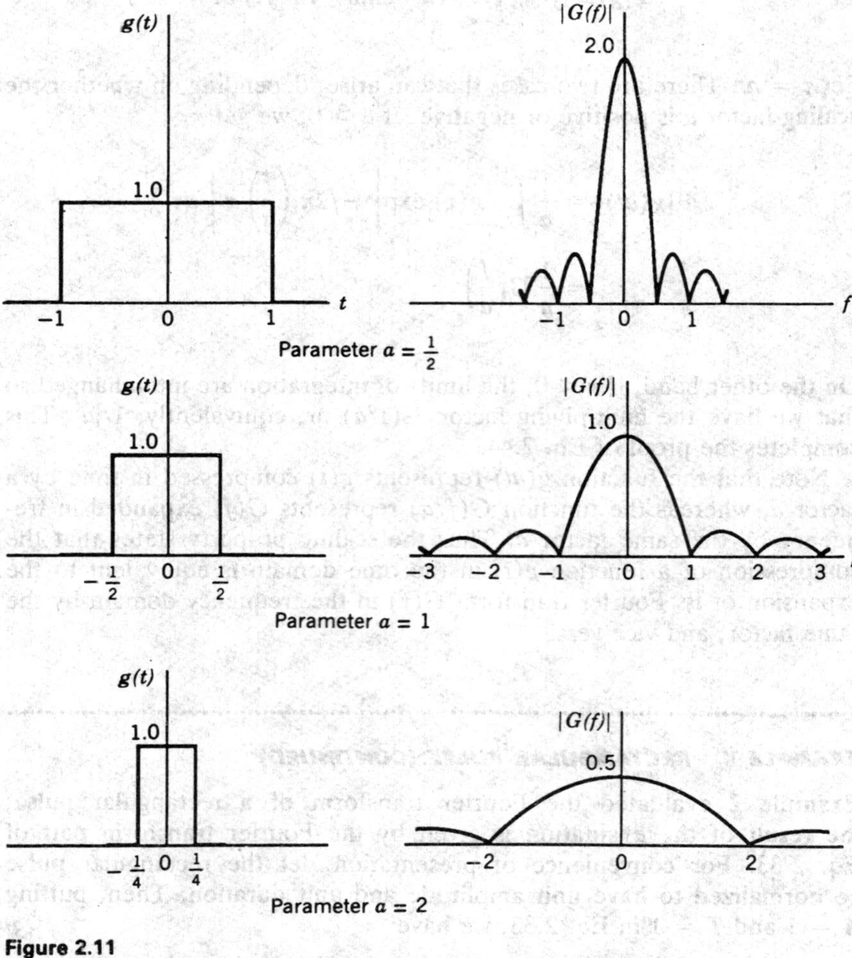

Figure 2.11
The inverse relation between time- and frequency-domain descriptions of rectangular pulse $g(t) = \mathrm{rect}(at)$.

observations are confirmed by the three time-domain descriptions depicted on the left side of Fig. 2.11. The corresponding effects of these time-scale changes on the amplitude spectrum of the rectangular pulse are shown on the right side of Fig. 2.11. The two sets of plots depicted in this figure clearly show that the relationship between the time-domain and frequency-domain descriptions of a signal is an *inverse* one. That is, a narrow pulse (in time) has a wide spectrum (in frequency), and vice versa.

EXERCISE 2 Example 3 showed that the decaying exponential pulse and rising exponential pulse of Fig. 2.7 have the same amplitude spectra but opposite phase spectra. Use the time-scaling property of the Fourier transform to explain this behavior.

PROPERTY 3 DUALITY

If $g(t) \rightleftharpoons G(f)$, then

$$G(t) \rightleftharpoons g(-f) \tag{2.45}$$

This property follows from the relation defining the inverse Fourier transform by writing it in the form

$$g(-t) = \int_{-\infty}^{\infty} G(f) \exp(-j2\pi ft)\, df$$

and then interchanging t and f. Note that $G(t)$ is obtained from $G(f)$ by using t in place of f, and $g(-f)$ is obtained from $g(t)$ by using $-f$ in place of t.

EXAMPLE 6 SINC PULSE

Consider a signal $g(t)$ in the form of a sinc function, as shown by

$$g(t) = A\,\text{sinc}(2Wt) \tag{2.46}$$

To evaluate the Fourier transform of this function, we apply the duality and time-scaling properties to the Fourier transform pair of Eq. 2.33. Then, recognizing that the rectangular function is an even function, we obtain the following result:

$$A\,\text{sinc}(2Wt) \rightleftharpoons \frac{A}{2W}\,\text{rect}\left(\frac{f}{2W}\right) \tag{2.47}$$

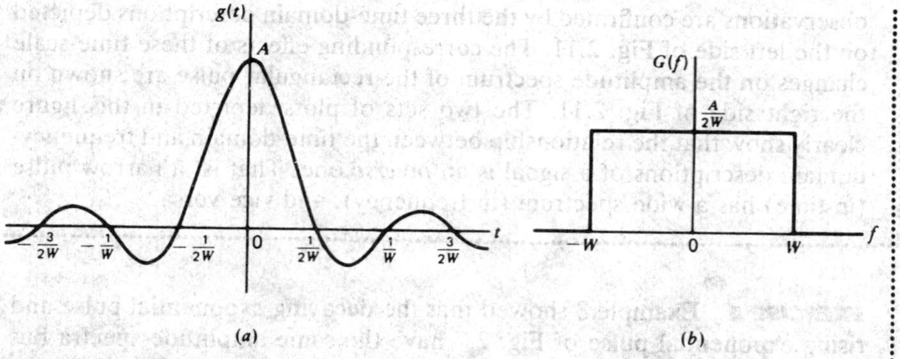

Figure 2.12
(a) Sinc pulse g(t). (b) Fourier transform G(f).

which is illustrated in Fig. 2.12. We thus see that the Fourier transform of a sinc pulse is zero for $|f| > W$. Note also that the sinc pulse itself is only asymptotically limited in time.

EXERCISE 3 Show that the total area under the curve of the sinc function equals one; that is,

$$\int_{-\infty}^{\infty} \operatorname{sinc}(t)\, dt = 1$$

EXERCISE 4 Consider a one-sided frequency function $G(f)$, defined by

$$G(f) = \begin{cases} \exp(-f), & f > 0 \\ \dfrac{1}{2}, & f = 0 \\ 0, & f < 0 \end{cases}$$

Applying the duality property to the Fourier transform pair of Eq. 2.40, write the inverse Fourier transform of $G(f)$.

PROPERTY 4 TIME SHIFTING

If $g(t) \rightleftharpoons G(f)$, then for a constant time shift t_0,

$$g(t - t_0) \rightleftharpoons G(f)\exp(-j2\pi f t_0). \qquad (2.48)$$

To prove this property, we take the Fourier transform of $g(t - t_0)$ and then set $\tau = t - t_0$ to obtain

$$\begin{aligned} F[g(t - t_0)] &= \exp(-j2\pi f t_0) \int_{-\infty}^{\infty} g(\tau) \exp(-j2\pi f \tau)\, d\tau \\ &= \exp(-j2\pi f t_0) G(f) \end{aligned}$$

The time-shifting property states that if a function $g(t)$ is shifted in the positive direction by an amount t_0, the effect is equivalent to multiplying its Fourier transform $G(f)$ by the factor $\exp(-j2\pi f t_0)$. This means that the amplitude of $G(f)$ is unaffected by the time shift but its phase is changed by the amount $-2\pi f t_0$.

EXAMPLE 7 RECTANGULAR PULSE (CONTINUED)

Consider the rectangular pulse $g_a(t)$ of Fig. 2.13*a*, which starts at time $t = 0$ and terminates at $t = T$. This pulse is defined by

$$g_a(t) = A \operatorname{rect}\left(\frac{t - T/2}{T}\right) \tag{2.49}$$

This pulse is obtained by shifting the rectangular pulse of Fig. 2.5 to the right by $T/2$ seconds. Therefore, applying the time-shifting property to the Fourier transform pair of Eq. 2.33, we find that the Fourier transform $G_a(f)$ of the rectangular pulse $g_a(t)$ defined in Eq. 2.49 is given by

$$G_a(f) = AT \operatorname{sinc}(fT) \exp(-j\pi f T) \tag{2.50}$$

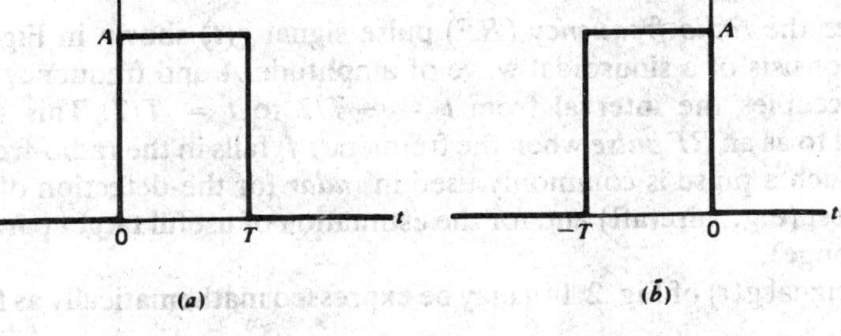

Figure 2.13
Time-shifted versions of a rectangular pulse.

Consider next the rectangular pulse $g_b(t)$ of Fig. 2.13*b*, which starts at time $t = -T$ and terminates at $t = 0$. This second pulse is defined by

$$g_b(t) = A \operatorname{rect}\left(\frac{t + T/2}{T}\right) \tag{2.51}$$

The pulse $g_b(t)$ is obtained by shifting the rectangular pulse of Fig. 2.5 to the left by $T/2$ seconds. Therefore, applying the time-shifting property to the Fourier transform pair of Eq. 2.33, we find that the Fourier transform $G_b(f)$ of the rectangular pulse $g_b(t)$ defined in Eq. 2.51 is given by

$$G_b(f) = AT \operatorname{sinc}(fT) \exp(j\pi fT) \tag{2.52}$$

PROPERTY 5 FREQUENCY SHIFTING

If $g(t) \rightleftharpoons G(f)$, then for a constant frequency shift f_c,

$$\exp(j2\pi f_c t)\, g(t) \rightleftharpoons G(f - f_c). \tag{2.53}$$

This property follows from the fact that

$$\begin{aligned} F[\exp(j2\pi f_c t) g(t)] &= \int_{-\infty}^{\infty} g(t) \exp[-j2\pi t(f - f_c)]\, dt \\ &= G(f - f_c) \end{aligned}$$

That is, multiplication of a function $g(t)$ by the factor $\exp(j2\pi f_c t)$ is equivalent to shifting its Fourier transform $G(f)$ in the positive direction by the amount f_c. Note the duality between the time-shifting and frequency-shifting operations.

EXAMPLE 8 RADIO FREQUENCY PULSE

Consider the *radio frequency* (*RF*) pulse signal $g(t)$ shown in Fig. 2.14*a*, which consists of a sinusoidal wave of amplitude A and frequency f_c. The pulse occupies the interval from $t = -T/2$ to $t = T/2$. This signal is referred to as an *RF pulse* when the frequency f_c falls in the radio-frequency band. Such a pulse is commonly used in *radar* for the detection of targets of interest (e.g., aircraft) and for the estimation of useful target parameters (e.g., range).

The signal $g(t)$ of Fig. 2.14*a* may be expressed mathematically as follows:

$$g(t) = A \operatorname{rect}\left(\frac{t}{T}\right) \cos(2\pi f_c t) \tag{2.54}$$

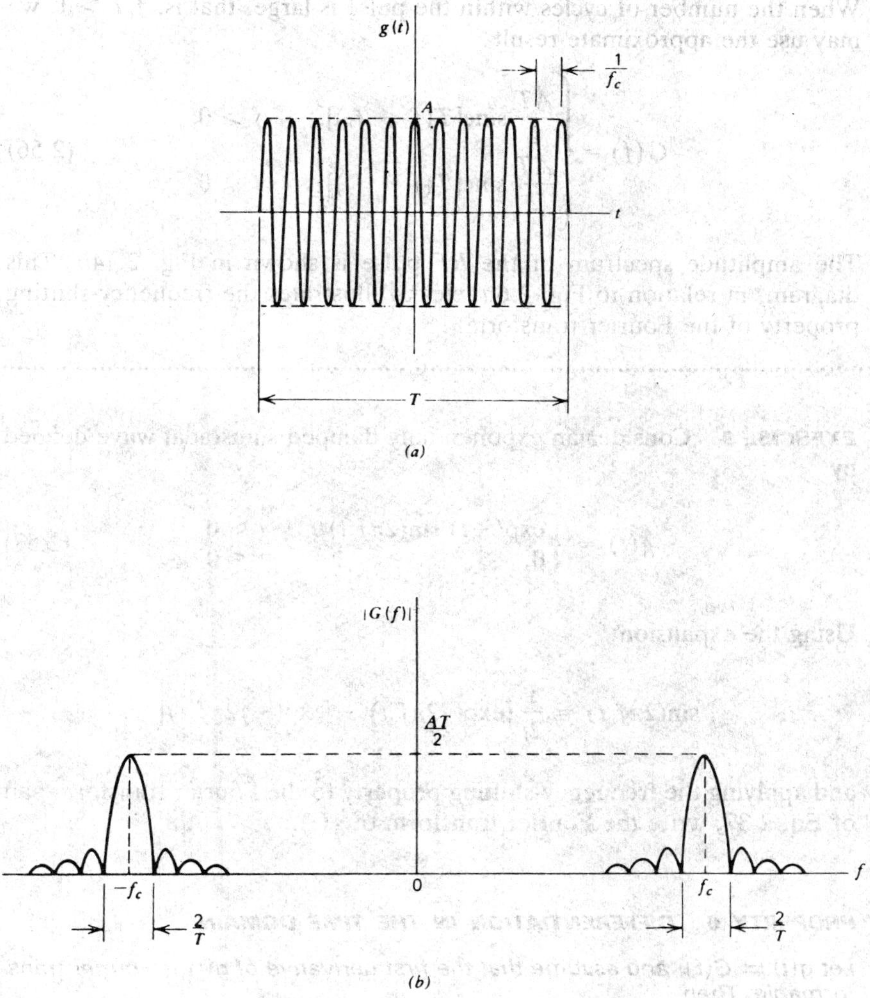

Figure 2.14
(a) RF pulse. (b) Amplitude spectrum.

To find the Fourier transform of this signal, we note that

$$\cos(2\pi f_c t) = \tfrac{1}{2}[\exp(j2\pi f_c t) + \exp(-j2\pi f_c t)]$$

Therefore, applying the frequency-shifting property to the Fourier transform pair of Eq. 2.33, we get the desired result

$$G(f) = \frac{AT}{2}\{\text{sinc}[T(f - f_c)] + \text{sinc}[T(f + f_c)]\} \qquad (2.55)$$

When the number of cycles within the pulse is large, that is, $f_c T \gg 1$, we may use the approximate result

$$G(f) \simeq \begin{cases} \dfrac{AT}{2}\,\mathrm{sinc}[T(f - f_c)], & f > 0 \\ \dfrac{AT}{2}\,\mathrm{sinc}[T(f + f_c)], & f < 0 \end{cases} \tag{2.56}$$

The amplitude spectrum of the *RF* pulse is shown in Fig. 2.14*b*. This diagram, in relation to Fig. 2.6*a*, clearly illustrates the frequency-shifting property of the Fourier transform.

EXERCISE 5 Consider an exponentially damped sinusoidal wave defined by

$$g(t) = \begin{cases} \exp(-t)\sin(2\pi f_c t), & t > 0 \\ 0, & t \le 0 \end{cases} \tag{2.57}$$

Using the expansion

$$\sin(2\pi f_c t) = \frac{1}{2j}[\exp(j2\pi f_c t) - \exp(-j2\pi f_c t)]$$

and applying the frequency-shifting property to the Fourier transform pair of Eq. 2.37, write the Fourier transform of $g(t)$.

PROPERTY 6 DIFFERENTIATION IN THE TIME DOMAIN

Let $g(t) \rightleftharpoons G(f)$, and assume that the first derivative of $g(t)$ is Fourier transformable. Then

$$\frac{d}{dt}g(t) \rightleftharpoons j2\pi f G(f) \tag{2.58}$$

That is, differentiation of a time function $g(t)$ has the effect of multiplying its Fourier transform $G(f)$ by the factor $j2\pi f$.

This result is obtained simply by taking the first derivative of both sides of the relation defining the inverse Fourier transform of $G(f)$, namely, Eq. 2.25, and then interchanging the operations of integration and differentiation; we are justified to make this interchange because integration and differentiation are both linear operations.

Multiplication of the Fourier transform $G(f)$ by the factor $j2\pi f$ on the right side of Eq. 2.58 implies that differentiation of $g(t)$ with respect to time enhances the high frequency components of the signal $g(t)$.

We may generalize Eq. 2.58 as follows:

$$\frac{d^n}{dt^n} g(t) \rightleftharpoons (j2\pi f)^n G(f) \tag{2.59}$$

EXAMPLE 9 GAUSSIAN PULSE

In this example we wish to use the differentiation property of the Fourier transform to derive the pulse signal $g(t)$ whose Fourier transform $G(f)$ has the same form.

Let $g(t)$ denote the pulse as a function of time, and $G(f)$ its Fourier transform. We note that by differentiating the formula for the Fourier transform $G(f)$ with respect to f, we have

$$-j2\pi t g(t) \rightleftharpoons \frac{d}{df} G(f) \tag{2.60}$$

which expresses the effect of differentiation in the frequency domain. Equation 2.60 is the dual of Eq. 2.58 that describes the time-differentiation property. Dividing both sides of Eq. 2.60 by j, we may also write

$$-2\pi t g(t) \rightleftharpoons \frac{1}{j} \frac{d}{df} G(f) \tag{2.61}$$

Suppose that the pulse-signal $g(t)$ satisfies the first-order differential equation

$$\frac{d}{dt} g(t) = -2\pi t g(t) \tag{2.62}$$

The imposition of this condition on the pulse signal $g(t)$ is equivalent to equating the left-hand members of Eqs. 2.58 and 2.61. Correspondingly, we may equate the right-hand members of Eqs. 2.58 and 2.61, and thus write

$$\frac{1}{j} \frac{d}{df} G(f) = j2\pi f G(f) \tag{2.63}$$

Since $j^2 = -1$, we may rewrite Eq. 2.63 as

$$\frac{d}{df} G(f) = -2\pi f G(f) \tag{2.64}$$

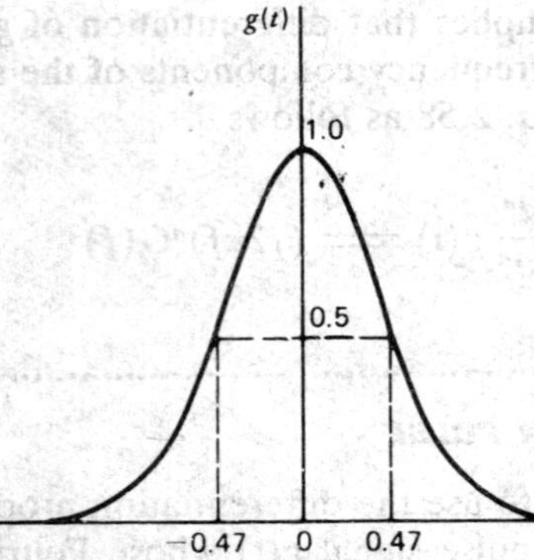

Figure 2.15
Gaussian pulse.

We may now state that if a pulse signal $g(t)$ satisfies the first-order differential equation (2.62), then its Fourier transform $G(f)$ must satisfy the first-order differential equation (2.64). However, these two differential equations have exactly the same mathematical form. Hence, the pulse signal and its transform are the same function. In other words, provided that the pulse signal $g(t)$ satisfies the differential equation (2.62), then $G(f) = g(f)$. Solving Eq. 2.62 for $g(t)$, we obtain

$$g(t) = \exp(-\pi t^2) \tag{2.65}$$

This result is shown plotted in Fig. 2.15.

The pulse defined by Eq. 2.65 is called a *Gaussian pulse,* the name being derived from the similarity of the function to the Gaussian probability density function. We conclude therefore that the Gaussian pulse is its own Fourier transform as shown by

$$\exp(-\pi t^2) \rightleftharpoons \exp(-\pi f^2) \tag{2.66}$$

EXERCISE 6 Show that

$$\int_{-\infty}^{\infty} \exp(-\pi t^2)\, dt = 1 \tag{2.67}$$

Hint: Consider the formula for the Fourier transform of $\exp(-\pi t^2)$ evaluated at $f = 0$.

PROPERTY 7 INTEGRATION IN THE TIME DOMAIN

Let $g(t) \rightleftharpoons G(f)$. Then, provided $G(0) = 0$, we have

$$\int_{-\infty}^{t} g(t)\, dt \rightleftharpoons \frac{1}{j2\pi f} G(f) \tag{2.68}$$

That is, integration of a time function $g(t)$ has the effect of dividing its Fourier transform $G(f)$ by the factor $j2\pi f$, assuming that $G(0)$ is zero.

To prove this property, we write the Fourier transform of the integrated signal as follows

$$\mathrm{F}\left[\int_{-\infty}^{t} g(\tau)\, d\tau\right] = \int_{-\infty}^{\infty} \exp(-j2\pi ft) \int_{-\infty}^{t} g(\tau)\, d\tau\, dt \tag{2.69}$$

On the right side of this relation, we have a definite integral with respect to the variable t. Clearly, we may view the corresponding integrand as the product of two time functions: the exponential $\exp(-j2\pi ft)$ and the integrated signal $\int_{-\infty}^{t} g(\tau)\, d\tau$. Hence, using the formula for integration by parts and assuming that

$$G(0) = \int_{-\infty}^{\infty} g(\tau)\, d\tau = 0$$

and then simplifying the result, we get the relation of Eq. 2.68. The condition $G(0) = 0$ ensures that $g(\tau)$ integrates out to zero as τ approaches infinity. The more general case, for which $G(0) \neq 0$, is treated later in Section 2.5.

Division of the Fourier transform $G(f)$ by the factor $j2\pi f$ on the right side of Eq. 2.68 implies that integration of $g(t)$ with respect to time suppresses the high-frequency components of $g(t)$. As expected, this effect is the opposite of that produced by differentiation of $g(t)$.

EXAMPLE 10 TRIANGULAR PULSE

Consider the *doublet pulse* $g_1(t)$ shown in Fig. 2.16*a*. By integrating this pulse with respect to time, we obtain the *triangular pulse* $g_2(t)$ shown in Fig. 2.16*b*. The duration of this triangular pulse at the half-amplitude points is the same as the duration of the rectangular pulse of Fig. 2.5. We note that the doublet pulse $g_1(t)$ consists of two rectangular pulses: one of amplitude A, defined for the interval $-T \leq t \leq 0$, and the other of amplitude $-A$, defined for the interval $0 \leq t \leq T$. Therefore, using the results

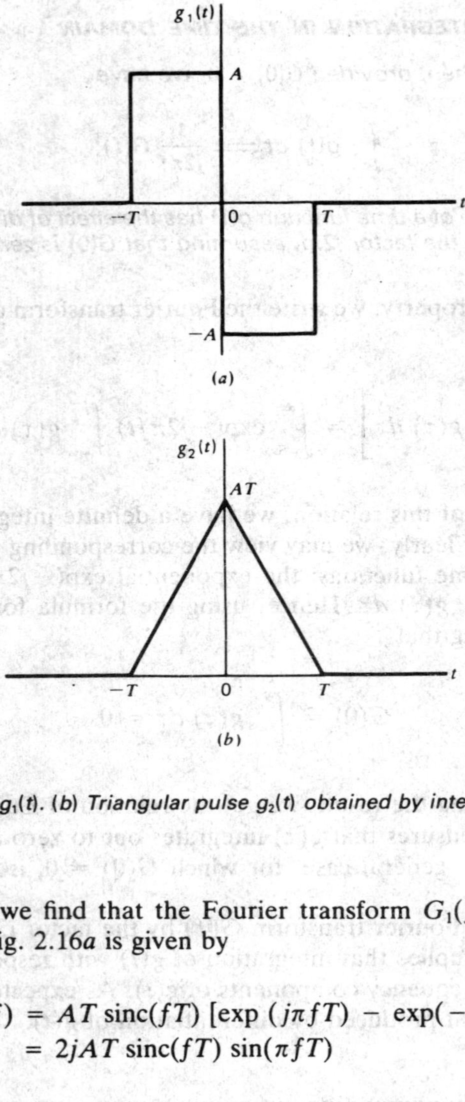

Figure 2.16
(a) Doublet pulse $g_1(t)$. (b) Triangular pulse $g_2(t)$ obtained by integrating $g_1(t)$.

of Example 7, we find that the Fourier transform $G_1(f)$ of the doublet pulse $g_1(t)$ of Fig. 2.16*a* is given by

$$\begin{aligned} G_1(f) &= AT\,\mathrm{sinc}(fT)\,[\exp(j\pi fT) - \exp(-j\pi fT)] \\ &= 2jAT\,\mathrm{sinc}(fT)\sin(\pi fT) \end{aligned} \tag{2.70}$$

We further note that $G_1(0)$ is zero. Hence, using Eqs. 2.68 and 2.70, we find that the Fourier transform $G_2(f)$ of the triangular pulse $g_2(t)$ of Fig. 2.16*b* is given by

$$\begin{aligned} G_2(f) &= \frac{1}{j2\pi f} G_1(f) \\ &= AT \frac{\sin(\pi fT)}{\pi f}\,\mathrm{sinc}(fT) \\ &= AT^2\,\mathrm{sinc}^2(fT) \end{aligned} \tag{2.71}$$

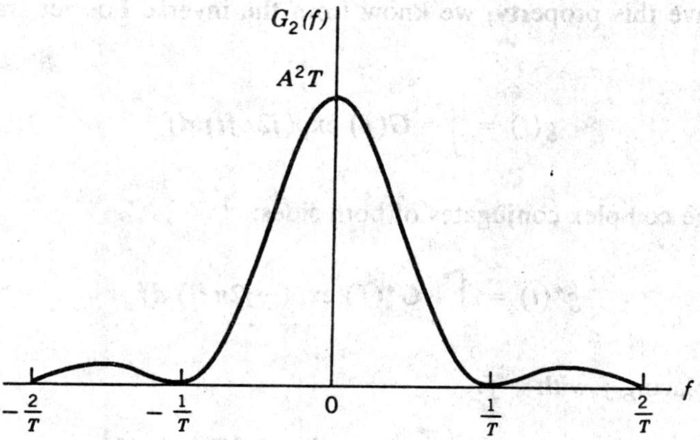

Figure 2.17
Spectrum of triangular pulse.

The Fourier transform $G_2(f)$ is a positive real function of f, which means that the amplitude spectrum of $g_2(t)$ is the same as $G_2(f)$, and its phase spectrum is zero for all f. The Fourier transform $G_2(f)$ is plotted in Fig. 2.17. Note that the spectrum of the triangular pulse is more tightly centered around the origin than the spectrum of the rectangular pulse. Also, the spectrum of the triangular pulse decreases as $1/f^2$, whereas the spectrum of the rectangular pulse is discontinuous and decreases as $1/|f|$.

EXERCISE 7

a. Show that the Fourier transform of a triangular pulse of unit amplitude and unit duration (measured at the half-amplitude points) is equal to $\text{sinc}^2(f)$.

b. Using the result in part a, show that

$$\int_{-\infty}^{\infty} \text{sinc}^2(f)\, df = 1$$

Hint: For part b, consider the formula for the inverse Fourier transform of $\text{sinc}^2(f)$ evaluated at $t = 0$.

PROPERTY 8 CONJUGATE FUNCTIONS

If $g(t) \rightleftharpoons G(f)$, then for a complex-valued time function $g(t)$ we have

$$g^*(t) \rightleftharpoons G^*(-f) \tag{2.72}$$

where the asterisk denotes the complex conjugate operation.

To prove this property, we know from the inverse Fourier transform that

$$g(t) = \int_{-\infty}^{\infty} G(f) \exp(j2\pi ft)\, df$$

Taking the complex conjugates of both sides:

$$g^*(t) = \int_{-\infty}^{\infty} G^*(f) \exp(-j2\pi ft)\, df$$

Next, replacing f with $-f$:

$$g^*(t) = -\int_{\infty}^{-\infty} G^*(-f) \exp(j2\pi ft)\, df$$

$$= \int_{-\infty}^{\infty} G^*(-f) \exp(j2\pi ft)\, df$$

That is, $g^*(t$ is the inverse Fourier transform of $G^*(-f)$, which is the desired resu

EXAMPLE 11 REAL AND IMAGINARY PARTS OF A TIME FUNCTION

Expressing a complex-valued function $g(t)$ in terms of its real and imaginary parts, we may write

$$g(t) = \text{Re}[g(t)] + j\,\text{Im}[g(t)] \tag{2.73}$$

where Re denotes the "real part of" and Im denotes the "imaginary part of." The complex conjugate of $g(t)$ is

$$g^*(t) = \text{Re}[g(t)] - j\,\text{Im}[g(t)] \tag{2.74}$$

Adding Eqs. 2.73 and 2.74:

$$\text{Re}[g(t)] = \frac{1}{2}[g(t) + g^*(t)] \tag{2.75}$$

and subtracting them:

$$\text{Im}[g(t)] = \frac{1}{2j}[g(t) - g^*(t)] \tag{2.76}$$

Therefore, applying Property 8, we obtain the following two Fourier transform pairs:

$$\mathrm{Re}[g(t)] \rightleftharpoons \frac{1}{2}[G(f) + G^*(-f)] \tag{2.77}$$

$$\mathrm{Im}[g(t)] \rightleftharpoons \frac{1}{2j}[G(f) - G^*(-f)] \tag{2.78}$$

From Eq. 2.78, it is apparent that in the case of a real-valued time function $g(t)$, we have $G(f) = G^*(-f)$; that is, $G(f)$ exhibits *conjugate symmetry*. This result is a restatement of Eqs. 2.30 and 2.31.

EXERCISE 8 Show that for a real-valued signal $g(t)$, Eq. 2.72 may be rewritten in the equivalent form:

$$g(-t) \rightleftharpoons G^*(f)$$

PROPERTY 9 MULTIPLICATION IN THE TIME DOMAIN

Let $g_1(t) \rightleftharpoons G_1(f)$ and $g_2(t) \rightleftharpoons G_2(f)$. Then

$$g_1(t)\, g_2(t) \rightleftharpoons \int_{-\infty}^{\infty} G_1(\lambda) G_2(f - \lambda)\, d\lambda \tag{2.79}$$

To prove this property, we first denote the Fourier transform of the product $g_1(t)g_2(t)$ by $G_{12}(f)$, so that we may write

$$g_1(t)\, g_2(t) \rightleftharpoons G_{12}(f)$$

where

$$G_{12}(f) = \int_{-\infty}^{\infty} g_1(t) g_2(t) \exp(-j2\pi f t)\, dt$$

For $g_2(t)$, we next substitute the inverse Fourier transform

$$g_2(t) = \int_{-\infty}^{\infty} G_2(f') \exp(j2\pi f' t)\, df'$$

in the integral defining $G_{12}(f)$ to obtain

$$G_{12}(f) = \int_{-\infty}^{\infty} \int_{-\infty}^{\infty} g_1(t) G_2(f') \exp[-j2\pi(f - f')t]\, df'\, dt$$

Define $\lambda = f - f'$. Then, interchanging the order of integration, we obtain

$$G_{12}(f) = \int_{-\infty}^{\infty} d\lambda\, G_2(f - \lambda) \int_{-\infty}^{\infty} g_1(t) \exp(-j2\pi\lambda t)\, dt$$

The inner integral is recognized simply as $G_1(\lambda)$, so we may write

$$G_{12}(f) = \int_{-\infty}^{\infty} G_1(\lambda) G_2(f - \lambda)\, d\lambda$$

which is the desired result. This integral is known as the *convolution integral* expressed in the frequency domain, and the function $G_{12}(f)$ is referred to as the *convolution* of $G_1(f)$ and $G_2(f)$. We conclude that *the multiplication of two signals in the time domain is transformed into the convolution of their individual Fourier transforms in the frequency domain*. This property is known as the *multiplication theorem*.

In a discussion of convolution, the following shorthand notation is frequently used:

$$G_{12}(f) = G_1(f) \star G_2(f)$$

where the star $\star$ denotes convolution. Note that convolution is commutative, that is,

$$G_{12}(f) = G_{21}(f)$$

or

$$G_1(f) \star G_2(f) = G_2(f) \star G_1(f)$$

EXAMPLE 12 TRUNCATED SINC PULSE

Consider the truncation of the sinc pulse $\text{sinc}(2Wt)$, so that the resulting signal $g(t)$ is zero outside the interval $-(T/2) \leq t \leq (T/2)$, as shown in Fig. 2.18*a*. This signal may be expressed as the product of a sinc pulse and a rectangular pulse, as shown by

$$g(t) = \text{sinc}(2Wt)\, \text{rect}\left(\frac{t}{T}\right) \tag{2.80}$$

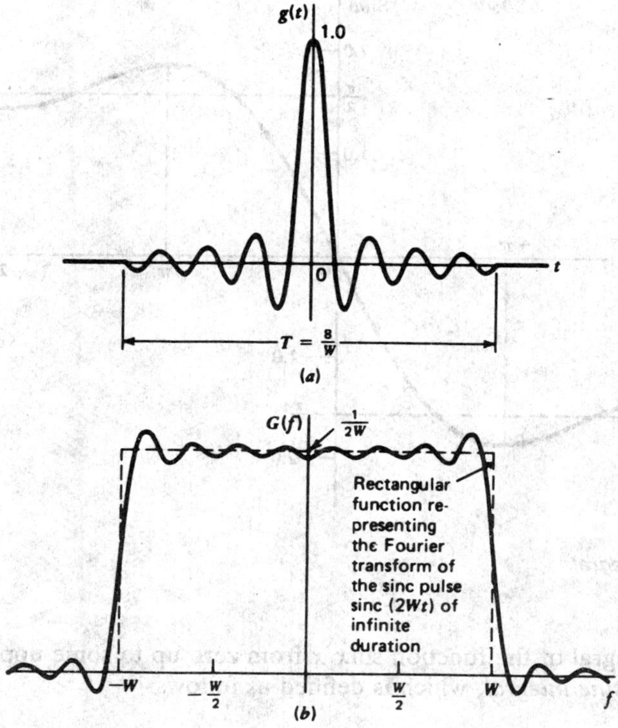

Figure 2.18
The Gibbs phenomenon. (a) A truncated sinc function g(t). (b) Fourier transform G(f).

From Eqs. 2.33 and 2.47, we have

$$F\left[\operatorname{rect}\left(\frac{t}{T}\right)\right] = T\operatorname{sinc}(fT)$$

$$F[\operatorname{sinc}(2Wt)] = \frac{1}{2W}\operatorname{rect}\left(\frac{f}{2W}\right)$$

Therefore, using Eq. 2.79, we find that the Fourier transform of the truncated sinc pulse $g(t)$ is given by

$$\begin{aligned} G(f) &= \frac{T}{2W}\int_{-\infty}^{\infty}\operatorname{rect}\left(\frac{\lambda}{2W}\right)\operatorname{sinc}[(f-\lambda)T]\,d\lambda \\ &= \frac{T}{2W}\int_{-W}^{W}\operatorname{sinc}[(f-\lambda)T]\,d\lambda \\ &= \frac{T}{2W}\int_{-W}^{W}\frac{\sin[\pi(f-\lambda)T]}{\pi(f-\lambda)T}\,d\lambda \end{aligned} \tag{2.81}$$

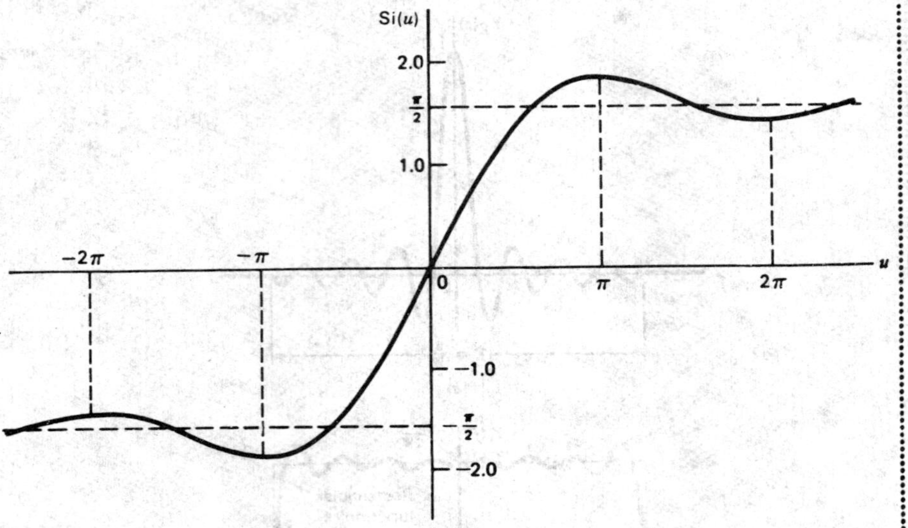

Figure 2.19
The sine integral.

The integral of the function $\sin x/x$ from zero up to some upper limit is called the *sine integral*, which is defined as follows

$$\mathrm{Si}(u) = \int_0^u \frac{\sin x}{x}\, dx \tag{2.82}$$

The sine integral $\mathrm{Si}(u)$ cannot be integrated in closed form in terms of elementary functions, but it can be integrated as a power series.[4] It is plotted in Fig. 2.19. We see that: (1) the sine integral $\mathrm{Si}(u)$ is odd symmetric about $u = 0$; (2) it has its maxima and minima at multiples of π; and (3) it approaches the limiting value $\pi/2$ for large values of u.

Substituting $x = \pi(f - \lambda)T$ in Eq. 2.81, we find that the Fourier transform $G(f)$ of the truncated sinc pulse may be expressed conveniently in terms of the sine integral as follows:

$$G(f) = \frac{1}{2\pi W}[\mathrm{Si}(\pi WT - \pi fT) + \mathrm{Si}(\pi WT + \pi fT)] \tag{2.83}$$

This relation is plotted in Fig. 2.18*b* for the case when $T = 8/W$. We see that $G(f)$ approximates the Fourier transform of a sinc pulse $\mathrm{sinc}(2Wt)$ of infinite duration in an oscillatory fashion, with a maximum deviation of about 9%. Furthermore, for a given value of W, as the pulse duration T

[4]See Goldman (1948, pp. 76–79).

is increased, the ripples in the vicinities of the discontinuities of the rectangular function show a proportionately increased rate of oscillation versus the frequency, f, whereas their amplitudes relative to the magnitude of the discontinuity remain the same. This effect is an example of *Gibbs phenomenon* in Fourier transforms.

EXERCISE 9 Using Eq. 2.79, show that

$$\int_{-\infty}^{\infty} g_1(t)g_2(t)\,dt = \int_{-\infty}^{\infty} G_1(f)G_2(-f)\,df$$

How is the left side of this relation affected by replacing $G_2(-f)$ with $G_2(f)$ in the integral on the right side of the relation?

PROPERTY 10 CONVOLUTION IN THE TIME DOMAIN

Let $g_1(t) \rightleftharpoons G_1(f)$ and $g_2(t) \rightleftharpoons G_2(f)$. Then

$$\int_{-\infty}^{\infty} g_1(\tau)g_2(t-\tau)\,d\tau \rightleftharpoons G_1(f)G_2(f) \tag{2.84}$$

This result follows directly by combining Property 3 (duality) and Property 9 (time-domain multiplication). We may thus state that *the convolution of two signals in the time domain is transformed into the multiplication of their individual Fourier transforms in the frequency domain.* This property is known as the *convolution theorem.* Its use permits us to exchange a convolution operation for a transform multiplication, an operation that is ordinarily easier to manipulate.

Using the shorthand notation for convolution, we may rewrite Eq. 2.84 in the form

$$g_1(t) \star g_2(t) \rightleftharpoons G_1(f)G_2(f) \tag{2.85}$$

where the star $\star$ denotes convolution.

EXAMPLE 13 DERIVATIVE OF A CONVOLUTION INTEGRAL

Let $g_{12}(t)$ denote the result of convolving two signals $g_1(t)$ and $g_2(t)$. Then the derivative of $g_{12}(t)$ is equal to the convolution of $g_1(t)$ with the derivative of $g_2(t)$, or vice versa. That is, if

$$g_{12}(t) = g_1(t) \star g_2(t)$$

then

$$\frac{d}{dt} g_{12}(t) = \left[\frac{d}{dt} g_1(t)\right] ☆ g_2(t)$$

To prove this result, we use the differentiation property (i.e., Eq. 2.58) in conjunction with the convolution property (i.e., Eq. 2.85), obtaining

$$\frac{d}{dt}[g_1(t) ☆ g_2(t)] \rightleftharpoons j2\pi f[G_1(f)G_2(f)]$$

Associating the factor $j2\pi f$ with $G_1(f)$, we may write

$$\left[\frac{d}{dt} g_1(t)\right] ☆ g_2(t) \rightleftharpoons [j2\pi f G_1(f)]G_2(f)$$

which yields the desired result:

$$\frac{d}{dt}[g_1(t) ☆ g_2(t)] = \left[\frac{d}{dt} g_1(t)\right] ☆ g_2(t) \tag{2.86}$$

Equation 2.86 shows that the derivative of the convolution of two time functions is equal to the convolution of one function with the derivative of the other.

EXERCISE 10 Using Eq. 2.84, show that

$$\int_{-\infty}^{\infty} g_1(t)g_2(-t)\, dt = \int_{-\infty}^{\infty} G_1(f)G_2(f)\, df$$

How is the right side of this relation affected by replacing $g_2(-t)$ with $g_2^*(t)$ in the integral on the left side of the relation? How does this result compare with that of Exercise 9?

2.4 INTERPLAY BETWEEN TIME-DOMAIN AND FREQUENCY-DOMAIN DESCRIPTIONS

The properties of the Fourier transform and the various examples used to illustrate them clearly show that the time-domain and frequency-domain descriptions of a signal are *inversely* related. In particular, we may make the following statements:

1. If the time-domain description of a signal is changed, the frequency-domain description of the signal is changed in an *inverse* manner, and

vice versa. This inverse relationship prevents arbitrary specifications of a signal in both domains. In other words, *we may specify an arbitrary function of time or an arbitrary spectrum, but we cannot specify both of them together*.

2. If a signal is strictly limited in frequency, the time-domain description of the signal will trail on indefinitely, even though its amplitude may assume a progressively smaller value. We say a signal is *strictly limited in frequency* or *strictly band-limited* if its Fourier transform is exactly zero outside a finite band of frequencies. The sinc pulse is an example of a strictly band-limited signal, as illustrated in Fig. 2.12. This figure also shows that the sinc pulse is only *asymptotically limited in time,* which confirms the opening statement we made for a strictly band-limited signal. In an inverse manner, if a signal is *strictly limited in time* (i.e., the signal is exactly zero outside a finite time interval), then the spectrum of the signal is infinite in extent, even though the amplitude spectrum may assume a progressively smaller value. This behavior is exemplified by both the rectangular pulse (described in Figs. 2.5 and 2.6) and the triangular pulse (described in Figs. 2.16*b* and 2.17). Accordingly, we may state that *a signal cannot be strictly limited in both time and frequency*.

BANDWIDTH

The *bandwidth* of a signal provides a measure of the *extent of significant spectral content of the signal for positive frequencies*. When the signal is strictly band-limited, the bandwidth is well defined. For example, the sinc pulse described in Fig. 2.12 has a bandwidth equal to W. When, however, the signal is not strictly band-limited, which is generally the case, we encounter difficulty in defining the bandwidth of the signal. The difficulty arises because the meaning of "significant" attached to the spectral content of the signal is mathematically imprecise. Consequently, there is no universally accepted definition of bandwidth.

Nevertheless, there are some commonly used definitions for bandwidth. In this section, we consider two such definitions;[5] the formulation of each definition depends on whether the signal is low-pass or band-pass. A signal is said to be *low-pass* if its significant spectral content is centered around the origin. A signal is said to be *band-pass* if its significant spectral content is centered around $\pm f_c$, where f_c is a nonzero frequency.

When the spectrum of a signal is symmetric with a *main lobe* bounded by well-defined *nulls* (i.e., frequencies at which the spectrum is zero), we may use the main lobe as the basis for defining the bandwidth of the signal. Specifically, if the signal is low-pass, the bandwidth is defined as one half

[5]Another definition for the bandwidth of a signal is presented in Section 4.8.

the total width of the main spectral lobe, since only one half of this lobe lies inside the positive frequency region. For example, a rectangular pulse of duration T seconds has a main spectral lobe of total width $2/T$ hertz centered at the origin, as depicted in Fig. 2.6*a*. Accordingly, we may define the bandwidth of this rectangular pulse as $1/T$ hertz. If, on the other hand, the signal is band pass with main spectral lobes centered around $\pm f_c$, where f_c is large, the bandwidth is defined as the width of the main lobe for positive frequencies. This definition of bandwidth is called the *null-to-null bandwidth*. For example, an RF pulse of duration T seconds and frequency f_c has main spectral lobes of width $2/T$ hertz centered around $\pm f_c$, as depicted in Fig. 2.14*b*. Hence, we may define the null-to-null bandwidth of this RF pulse as $2/T$ hertz.

Another popular definition of bandwidth is the *3-dB bandwidth*.[6] Specifically, if the signal is low-pass, the 3-dB bandwidth is defined as the separation between zero frequency, where the amplitude spectrum attains its peak value, and the positive frequency at which the amplitude spectrum drops to $1/\sqrt{2}$ of its peak value. For example, the decaying exponential and rising exponential pulses defined in Fig. 2.7 have a 3-dB bandwidth of $1/2\pi$ hertz. If, on the other hand, the signal is band pass, centered at $\pm f_c$, the 3-dB bandwidth is defined as the separation (along the positive frequency axis) between the two frequencies at which the amplitude spectrum of the signal drops to $1/\sqrt{2}$ of the peak value at f_c. The 3-dB bandwidth has the advantage in that it can be read directly from a plot of the amplitude spectrum. However, it has the disadvantage in that it may be misleading if the amplitude spectrum has slowly decreasing tails.

EXERCISE 11 Using the idea of a main spectral lobe, what is the bandwidth of a triangular pulse defined in Figs. 2.16*b* and 2.17?

EXERCISE 12 What is the 3-dB bandwidth of the decaying exponential pulse $\exp(-at)$ that is zero for negative time?

TIME–BANDWIDTH PRODUCT

For any family of pulse signals that differ by a time-scaling factor, the product of the signal's duration and its bandwidth is always a constant, as shown by

$$(\text{duration}) \cdot (\text{bandwidth}) = \text{constant}$$

[6] For a discussion of the decibel (dB), see Appendix A.

The product is called the *time–bandwidth product* or *bandwidth–duration product.* The constancy of the time–bandwidth product is another manifestation of the inverse relationship that exists between the time-domain and frequency-domain descriptions of a signal. In particular, if the duration of a pulse signal is decreased by reducing the time scale by a factor a, the frequency scale of the signal's spectrum, and therefore the bandwidth of the signal, is increased by the same factor a, by virtue of Property 2, and the time–bandwidth product of the signal is thereby maintained constant. For example, a rectangular pulse of duration T seconds has a bandwidth (defined on the basis of the positive-frequency part of the main lobe) equal to $1/T$ hertz, making the time–bandwidth product of the pulse equal unity. Whatever definition we use for the bandwidth of a signal, the time–bandwidth product remains constant over certain classes of pulse signals. The choice of a particular definition for bandwidth simply changes the value of the constant.

2.5 DIRAC DELTA FUNCTION

Strictly speaking, the theory of the Fourier transform, as described in Sections 2.2 and 2.3, is applicable only to time functions that satisfy the Dirichlet conditions. Such functions include energy signals. However, it would be highly desirable to extend this theory in two ways:

1. To combine the Fourier series and Fourier transform into a unified theory, so that the Fourier series may be treated as a special case of the Fourier transform.
2. To include power signals in the list of signals to which we may apply the Fourier transform.

It turns out that both these objectives can be met through the "proper use" of the *Dirac delta function* or *unit impulse.*

The Dirac delta function belongs to a special class of functions known as *generalized distributions* that are defined by the use of assignment rules given in Eqs. 2.87 and 2.88. In particular, the Dirac delta function,[7] denoted by $\delta(t)$, is defined as having zero amplitude everywhere except at $t = 0$, where it is infinitely large in such a way that it contains unit area under its curve, as shown by the pair of rules:

$$\delta(t) = 0, \qquad t \neq 0 \tag{2.87}$$

[7]For a detailed treatment of the delta function, see Bracewell (1978) or Lighthill (1959). The notation $\delta(t)$, which was first introduced into quantum mechanics by Dirac, is now in general use; see Dirac (1947).

and

$$\int_{-\infty}^{\infty} \delta(t)\, dt = 1 \tag{2.88}$$

It is important to realize that no function in the ordinary sense can satisfy the two rules of Eqs. 2.87 and 2.88. However, we can imagine a sequence of functions that have progressively taller and thinner peaks at $t = 0$, with the area under the curve remaining equal to unity, whereas the value of the function tends to zero at every point, except at $t = 0$ where it tends to infinity. That is, we may view the delta function as the limiting form of a *unit-area pulse as the pulse duration approaches zero*. It is immaterial what sort of pulse shape is used. For example, we may use a rectangular pulse of unit area, and thus write

$$\delta(t) = \lim_{\tau \to 0} \frac{1}{\tau} \operatorname{rect}\left(\frac{t}{\tau}\right) \tag{2.89}$$

The rectangular pulse is plotted in Fig. 2.20*a* for $\tau = 5, 1, 0.2$. For another example, we may use a Gaussian pulse of unit area and thus write

$$\delta(t) = \lim_{\tau \to 0} \frac{1}{\tau} \exp\left(-\frac{\pi t^2}{\tau^2}\right) \tag{2.90}$$

The Gaussian pulse is plotted in Fig. 2.20*b* for $\tau = 5, 1, 0.2$. From Fig. 2.20, we clearly see that both pulses take on an impulse-like appearance as the parameter τ becomes progressively smaller. Some other examples are considered in Problem 18.

EXERCISE 13 Plot the spectra for the rectangular and Gaussian pulses for the different values of parameter τ given in Fig. 2.20.

PROPERTIES OF THE DELTA FUNCTION

The delta function $\delta(t)$ has several useful properties that are consequences of the two rules defining it, namely, Eqs. 2.87 and 2.88. These properties are discussed here:

1. The delta function is an even function of time; that is,

$$\delta(t) = \delta(-t) \tag{2.91}$$

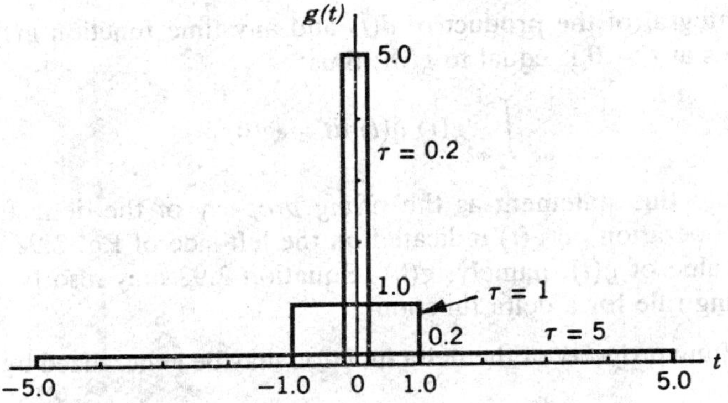

(a)

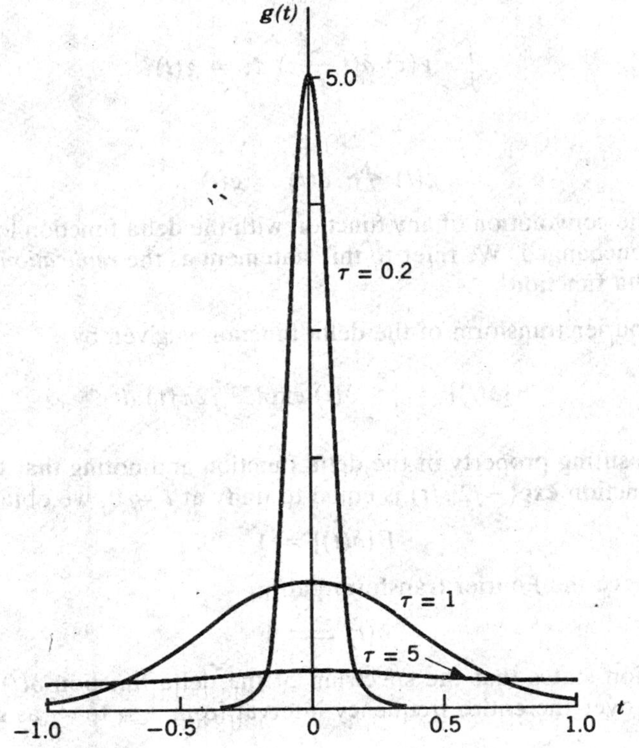

(b)

Figure 2.20
(a) Rectangular pulse $g(t) = 1/\tau \, \text{rect}(t/\tau)$ for varying τ. (b) Gaussian pulse $g(t) = 1/\tau \exp(-\pi t^2/\tau^2)$ for varying τ.

2. The integral of the product of $\delta(t)$ and any time function $g(t)$ that is continuous at $t = 0$ is equal to $g(0)$; thus

$$\int_{-\infty}^{\infty} g(t)\, \delta(t)\, dt = g(0) \tag{2.92}$$

We refer to this statement as the *sifting property* of the delta function, since the operation on $g(t)$ indicated on the left side of Eq. 2.92 sifts out a single value of $g(t)$, namely, $g(0)$. Equation 2.92 may also be used as the defining rule for a delta function.

3. The sifting property of the delta function may be generalized by writing

$$\int_{-\infty}^{\infty} g(t)\, \delta(t - t_0)\, dt = g(t_0) \tag{2.93}$$

Since the delta function $\delta(t)$ is an even function of t, we may rewrite Eq. 2.93 in a way emphasizing resemblance to the convolution integral, as follows:

$$\int_{-\infty}^{\infty} g(\tau)\, \delta(t - \tau)\, d\tau = g(t) \tag{2.94}$$

or

$$g(t) \star \delta(t) = g(t) \tag{2.95}$$

That is, the convolution of any function with the delta function leaves that function unchanged. We refer to this statement as the *replication property* of the delta function.

4. The Fourier transform of the delta function is given by

$$F[\delta(t)] = \int_{-\infty}^{\infty} \delta(t) \exp(-j2\pi f t)\, dt$$

Using the sifting property of the delta function and noting that the exponential function $\exp(-j2\pi ft)$ is equal to unity at $t = 0$, we obtain

$$F[\delta(t)] = 1$$

We thus have the Fourier transform pair:

$$\delta(t) \rightleftharpoons 1 \tag{2.96}$$

This relation states that the spectrum of the delta function $\delta(t)$ extends uniformly over the entire frequency interval from $-\infty$ to ∞, as shown in Fig. 2.21.

APPLICATIONS OF THE DELTA FUNCTION

dc Signal By applying the duality property to the Fourier transform pair of Eq. 2.96, and noting that the delta function is an even function, we

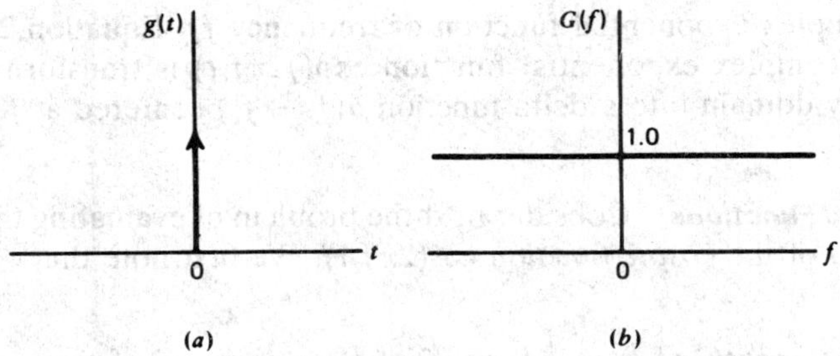

Figure 2.21
(a) Dirac delta function (b) Spectrum.

obtain

$$1 \rightleftharpoons \delta(f) \tag{2.97}$$

Equation 2.97 states that a *dc signal* is transformed in the frequency domain into a delta function $\delta(f)$ occurring at zero frequency, as shown in Fig. 2.22. Of course, this result is intuitively satisfying. From Eq. 2.97 we also deduce the useful relation

$$\int_{-\infty}^{\infty} \exp(-j2\pi ft)\, dt = \delta(f) \tag{2.98}$$

where the integral on the left side is simply the Fourier transform of a function equal to one for all time t.

Complex Exponential Function Next, by applying the frequency-shifting property to Eq. 2.97, we obtain the Fourier transform pair

$$\exp(j2\pi f_c t) \rightleftharpoons \delta(f - f_c) \tag{2.99}$$

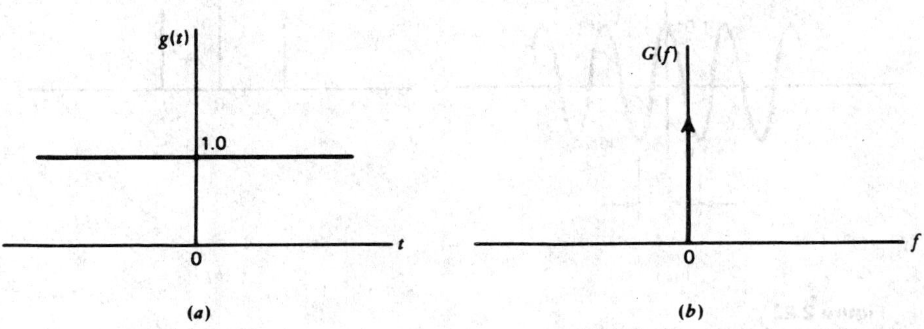

Figure 2.22
(a) dc signal. (b) Spectrum.

for a complex exponential function of frequency f_c. Equation 2.99 states that the complex exponential function $\exp(j2\pi f_c t)$ is transformed in the frequency domain into a delta function $\delta(f - f_c)$ centered at $f = f_c$.

Sinusoidal Functions Consider next the problem of evaluating the Fourier transform of the *cosine function* $\cos(2\pi f_c t)$. We first note that

$$\cos(2\pi f_c t) = \frac{1}{2}[\exp(j2\pi f_c t) + \exp(-j2\pi f_c t)]$$

Therefore, using Eq. 2.99, we find that the cosine function $\cos(2\pi f_c t)$ is represented by the Fourier transform pair

$$\cos(2\pi f_c t) \rightleftharpoons \frac{1}{2}[\delta(f - f_c) + \delta(f + f_c)] \tag{2.100}$$

In other words, the spectrum of the cosine function $\cos(2\pi f_c t)$ consists of a pair of delta functions centered at $f = \pm f_c$, each of which is weighted by the factor ½, as shown in Fig. 2.23.

Similarly, we may show that the *sine function* $\sin(2\pi f_c t)$ is represented by the Fourier transform pair

$$\sin(2\pi f_c t) \rightleftharpoons \frac{1}{2j}[\delta(f - f_c) - \delta(f + f_c)] \tag{2.101}$$

which is illustrated in Fig. 2.24.

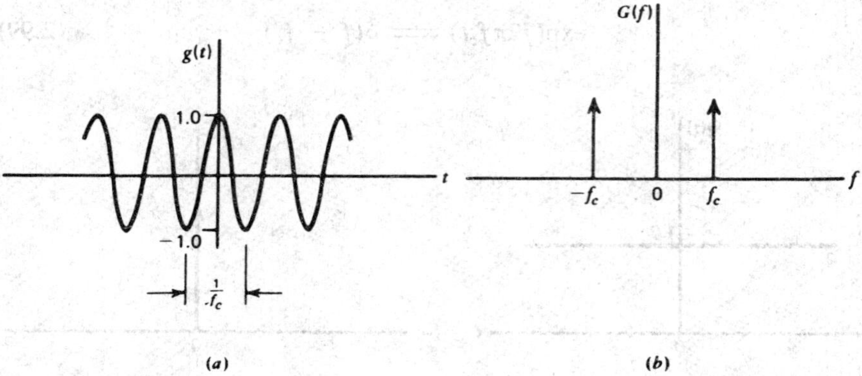

Figure 2.23
(a) Cosine function. (b) Spectrum.

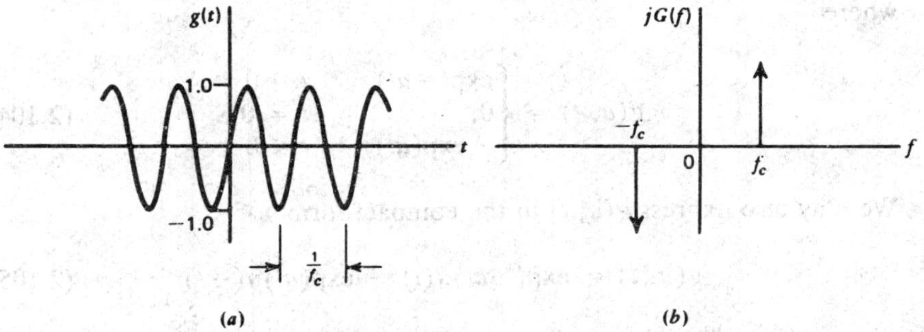

Figure 2.24
(a) Sine function. (b) Spectrum.

Signum Function The *signum function,* denoted by sgn(t), is an odd function of time defined as follows:

$$\text{sgn}(t) = \begin{cases} 1, & t > 0 \\ 0, & t = 0 \\ -1, & t < 0 \end{cases} \tag{2.102}$$

The waveform of the signum function is shown in Fig. 2.25*a*. We may view the signum function as the limiting form of a time function that consists of a positive decaying exponential for positive time and a negative rising exponential for negative time. That is, we write

$$\text{sgn}(t) = \lim_{a \to 0} g(a, t) \tag{2.103}$$

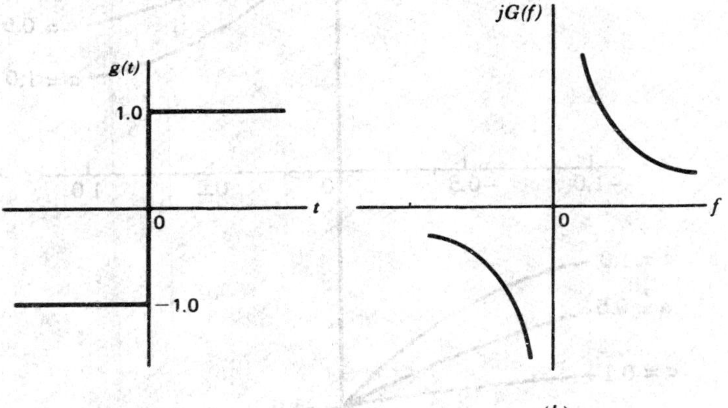

Figure 2.25
(a) Signum function. (b) Spectrum.

where

$$g(a, t) = \begin{cases} \exp(-at), & t > 0 \\ 0, & t = 0 \\ -\exp(at), & t < 0 \end{cases} \qquad (2.104)$$

We may also express $g(a, t)$ in the compact form

$$g(a, t) = \exp(-at)u(t) - \exp(at)u(-t) \qquad (2.105)$$

where $u(t)$ is the unit step function. The function $g(a, t)$ is plotted in Fig. 2.26 for the parameter $a = 1, 0.5, 0.1$. We clearly see that as the value of parameter a is progressively reduced, the function $g(a, t)$ becomes closer to the signum function in appearance. Applying the time-scaling property to the Fourier transform pairs of Eqs. 2.37 and 2.40, we get

$$\exp(-at)u(t) \rightleftharpoons \frac{1/a}{1 + (j2\pi f/a)}$$

and

$$\exp(at)u(-t) \rightleftharpoons \frac{1/a}{1 - (j2\pi f/a)}$$

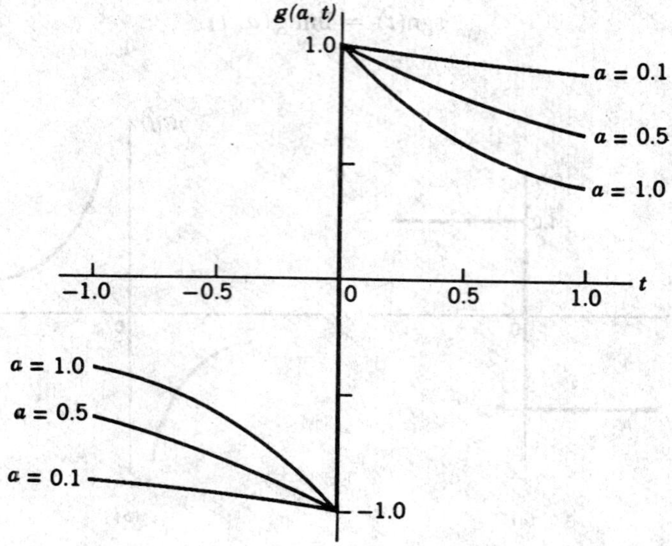

Figure 2.26
The function g(a, t) for varying a.

Subtracting the second Fourier transform pair from the first one, and then using the definition of Eq. 2.105, we get (after combining terms and simplifying):

$$g(a, t) \rightleftharpoons \frac{4\pi f}{j(a^2 + 4\pi^2 f^2)} \tag{2.106}$$

For the limiting condition when the parameter a approaches zero, the function $g(a, t)$ approaches the signum function, in accordance with Eq. 2.103. Therefore, putting $a = 0$ in Eq. 2.106, we obtain the desired Fourier transform pair for the signum function:

$$\text{sgn}(t) \rightleftharpoons \frac{1}{j\pi f} \tag{2.107}$$

The spectrum of the signum function is plotted in Fig. 2.25*b*.

Another useful Fourier transform pair, involving a signum function defined in the frequency domain, is obtained by applying Property 3 (duality) to Eq. 2.107. We thus obtain the following result:

$$\frac{1}{\pi t} \rightleftharpoons j\,\text{sgn}(f) \tag{2.108}$$

where the signum function sgn(f) is defined by

$$\text{sgn}(f) = \begin{cases} 1, & f > 0 \\ 0, & f = 0 \\ -1, & f < 0 \end{cases}$$

EXERCISE 14 Plot the spectrum of the function $g(a, t)$ for parameter $a = 1, 0.5, 0.1$, and compare your results with the spectrum of the signum function shown in Fig. 2.25*b*.

Unit Step Function The *unit step function*, $u(t)$, is defined in Eq. 2.34, reproduced here for convenience:

$$u(t) = \begin{cases} 1, & t > 0 \\ 1/2, & t = 0 \\ 0, & t < 0 \end{cases} \tag{2.109}$$

The waveform of the unit step function is shown in Fig. 2.27*a*. From Eqs. 2.102 and 2.109, or from the corresponding waveforms shown in Figs. 2.25*a*

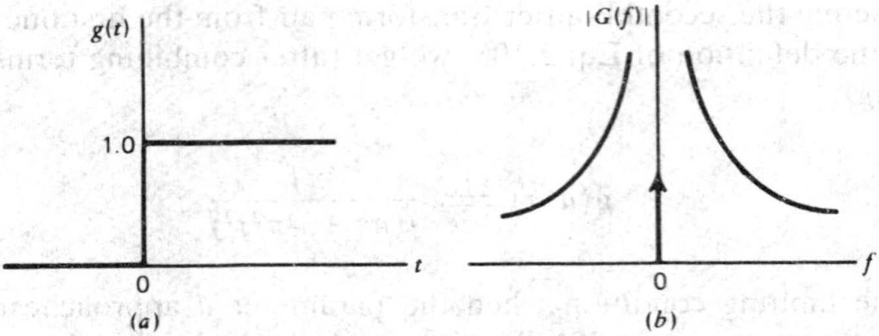

Figure 2.27
(a) Unit step function. (b) Amplitude spectrum.

and 2.27*a*, we see that the unit step function and signum function are related by

$$u(t) = \frac{1}{2}[\text{sgn}(t) + 1] \tag{2.110}$$

Hence, using the linearity property of the Fourier transform and the Fourier transform pairs of Eqs. 2.97, and 2.107, we find that the unit step function is represented by the Fourier transform pair

$$u(t) \rightleftharpoons \frac{1}{j2\pi f} + \frac{1}{2}\delta(f) \tag{2.111}$$

This means that the spectrum of the unit step function contains a delta function weighted by a factor of 1/2 and occurring at zero frequency, as shown in Fig. 2.27*b*.

EXERCISE 15 Using the frequency-shifting property, determine the Fourier transform of the signal

$$g(t) = u(t)\cos(2\pi f_c t)$$

where $u(t)$ is the unit step function.

Integration in the Time Domain (Revisited) The relation of Eq. 2.68 describes the effect of integration on the Fourier transform of a signal $g(t)$, assuming that $G(0)$ is zero. We now consider the more general case, with no such assumption made.

Let

$$y(t) = \int_{-\infty}^{t} g(\tau)\,d\tau \tag{2.112}$$

The integrated signal $y(t)$ can be viewed as the convolution of the original signal $g(t)$ and the unit step function $u(t)$, as shown by

$$y(t) = \int_{-\infty}^{\infty} g(\tau)u(t - \tau)\, d\tau \tag{2.113}$$

where the time-shifted unit step function $u(t - \tau)$ is defined by

$$u(t - \tau) = \begin{cases} 1, & \tau < t \\ \frac{1}{2}, & \tau = t \\ 0, & \tau > t \end{cases} \tag{2.114}$$

Recognizing that convolution in the time domain is transformed into multiplication in the frequency domain, and using the Fourier transform pair of Eq. 2.111 for the unit step function $u(t)$, we find that the Fourier transform of $y(t)$ is

$$Y(f) = G(f)\left[\frac{1}{j2\pi f} + \frac{1}{2}\delta(f)\right] \tag{2.115}$$

where $G(f)$ is the Fourier transform of $g(t)$. Since

$$G(f)\,\delta(f) = G(0)\,\delta(f)$$

we may rewrite Eq. 2.115 in the equivalent form:

$$Y(f) = \frac{1}{j2\pi f} G(f) + \frac{1}{2} G(0)\,\delta(f) \tag{2.116}$$

That is, the effect of integrating the signal $g(t)$ is described by the Fourier transform pair:

$$\int_{-\infty}^{t} g(\tau)\, d\tau \rightleftharpoons \frac{1}{j2\pi f} G(f) + \frac{1}{2} G(0)\,\delta(f) \tag{2.117}$$

which is the desired result.

This proof is indirect in that it relies on knowledge of the Fourier transform of the unit step function. For a direct proof from first principles, refer to Problem 20.

2.8 FOURIER TRANSFORMS OF PERIODIC SIGNALS

From Section 2.1 we recall that by using the Fourier series, a periodic signal $g_p(t)$ can be represented as a sum of complex exponentials. Also we know that, in a limiting sense, we can define Fourier transforms of complex exponentials. Therefore, it seems reasonable that a periodic signal can be represented in terms of a Fourier transform, provided that this transform is permitted to include delta functions.

Consider a periodic signal $g_p(t)$ of period T_0. We can represent $g_p(t)$ in terms of the complex exponential Fourier series as in Eq. 2.10, which is reproduced here for convenience,

$$g_p(t) = \sum_{n=-\infty}^{\infty} c_n \exp\left(\frac{j2\pi nt}{T_0}\right) \tag{2.118}$$

where c_n is the complex Fourier coefficient defined by

$$c_n = \frac{1}{T_0}\int_{-T_0/2}^{T_0/2} g_p(t) \exp\left(-\frac{j2\pi nt}{T_0}\right) dt \tag{2.119}$$

Let $g(t)$ be a pulse-like function, which equals $g_p(t)$ over one period and is zero elsewhere; that is,

$$g(t) = \begin{cases} g_p(t), & -\dfrac{T_0}{2} \leqslant t \leqslant \dfrac{T_0}{2} \\ 0, & \text{elsewhere} \end{cases} \tag{2.120}$$

The periodic signal $g_p(t)$ may be expressed in terms of the function $g(t)$ as an infinite summation, as shown by

$$g_p(t) = \sum_{m=-\infty}^{\infty} g(t - mT_0) \tag{2.121}$$

Based on this representation, we may view $g(t)$ as a *generating function*, which generates the periodic signal $g_p(t)$.

The function $g(t)$ is Fourier transformable. Accordingly, we may rewrite Eq. 2.119 as follows:

$$\begin{aligned} c_n &= \frac{1}{T_0}\int_{-\infty}^{\infty} g(t) \exp\left(-\frac{j2\pi nt}{T_0}\right) dt \\ &= \frac{1}{T_0} G\left(\frac{n}{T_0}\right) \end{aligned} \tag{2.122}$$

where $G(n/T_0)$ is the Fourier transform of $g(t)$, evaluated at the frequency n/T_0. We may thus rewrite Eq. 2.118 as

$$g_p(t) = \frac{1}{T_0}\sum_{n=-\infty}^{\infty} G\left(\frac{n}{T_0}\right) \exp\left(\frac{j2\pi nt}{T_0}\right) \tag{2.123}$$

or, equivalently,

$$\sum_{m=-\infty}^{\infty} g(t - mT_0) = \frac{1}{T_0} \sum_{n=-\infty}^{\infty} G\left(\frac{n}{T_0}\right) \exp\left(\frac{j2\pi nt}{T_0}\right) \tag{2.124}$$

Equation 2.124 is one form of *Poisson's sum formula.*

Finally, using Eq. 2.99, which defines the Fourier transform of a complex exponential function, and Eq. 2.124, we deduce the following Fourier transform pair for a periodic signal $g_p(t)$ with a generating function $g(t)$ and period T_0:

$$g_p(t) = \sum_{m=-\infty}^{\infty} g(t - mT_0) \rightleftharpoons \frac{1}{T_0} \sum_{n=-\infty}^{\infty} G\left(\frac{n}{T_0}\right) \delta\left(f - \frac{n}{T_0}\right) \tag{2.125}$$

This relation simply states that the Fourier transform of a periodic signal consists of delta functions occurring at integer multiples of the fundamental frequency $1/T_0$, including the origin, and that each delta function is weighted by a factor $G(n/T_0)$. Indeed, this relation merely provides an alternate way of displaying the frequency content of a periodic signal $g_p(t)$.

It is of interest to observe that the function $g(t)$, constituting one period of the periodic signal $g_p(t)$, has a continuous spectrum defined by $G(f)$. On the other hand, the periodic signal $g_p(t)$ itself has a discrete spectrum. We conclude, therefore, that *periodicity in the time domain has the effect of making the spectrum of the signal take on a discrete form, where the separation between adjacent spectral lines equals the reciprocal of the period.*

EXAMPLE 14 IDEAL SAMPLING FUNCTION

An *ideal sampling function,* or *Dirac comb,* consists of an infinite sequence of uniformly spaced delta functions, as shown in Fig. 2.28*a*. We will denote this waveform by

$$\delta_{T_0}(t) = \sum_{m=-\infty}^{\infty} \delta(t - mT_0) \tag{2.126}$$

We observe that the generating function $g(t)$ for the ideal sampling function $\delta_{T_0}(t)$ consists simply of the delta function $\delta(t)$. Therefore, $G(f) = 1$, so that

$$G\left(\frac{n}{T_0}\right) = 1, \quad \text{for all } n \tag{2.127}$$

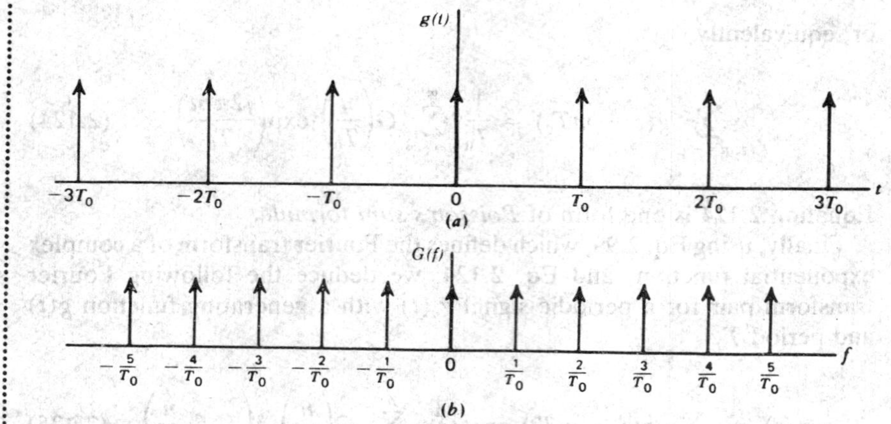

Figure 2.28
(a) Dirac comb. (b) Spectrum.

Thus the use of Eq. 2.125 yields the result

$$\sum_{m=-\infty}^{\infty} \delta(t - mT_0) \rightleftharpoons \frac{1}{T_0} \sum_{n=-\infty}^{\infty} \delta\left(f - \frac{n}{T_0}\right) \tag{2.128}$$

Equation 2.128 states that the Fourier transform of a periodic train of delta functions in the time domain consists of another periodic train of delta functions in the frequency domain as in Fig. 2.28*b*. In the special case of the period T_0 equal to 1 second, a periodic train of delta functions is, like a Gaussian pulse, its own Fourier transform.

We also deduce from Poisson's sum formula, Eq. 2.124, the following useful relation

$$\sum_{m=-\infty}^{\infty} \delta(t - mT_0) = \frac{1}{T_0} \sum_{n=-\infty}^{\infty} \exp\left(\frac{j2\pi nt}{T_0}\right)$$

The dual of this relation is

$$\sum_{m=-\infty}^{\infty} \exp(j2\pi mfT_0) = \frac{1}{T_0} \sum_{n=-\infty}^{\infty} \delta\left(f - \frac{n}{T_0}\right) \tag{2.129}$$

2.7 SAMPLING THEOREM

An operation that is basic to digital signal processing and digital communications is the *sampling process,* whereby an analog signal is converted into a corresponding sequence of samples that are usually spaced uniformly in time. For such a procedure to have practical utility, it is necessary that we choose the sampling rate properly, so that the sequence of samples uniquely defines the original analog signal. This is the essence of the sampling theorem, which is derived in the sequel.

Consider the arbitrary signal $g(t)$ of finite energy, which is specified for all time. A segment of the signal $g(t)$ is shown in Fig. 2.29*a*. Suppose that we sample the signal $g(t)$ instantaneously and at a uniform rate, once every T_s seconds. Consequently, we obtain an infinite sequence of samples spaced T_s seconds apart and denoted by $\{g(nT_s)\}$ where n takes on all possible integer values. We refer to T_s as the *sampling period,* and to its reciprocal $f_s = 1/T_s$ as the *sampling rate.* This ideal form of sampling is called *instantaneous sampling.*

Let $g_\delta(t)$ denote the signal obtained by individually weighting the elements of a periodic sequence of delta functions spaced T_s seconds apart

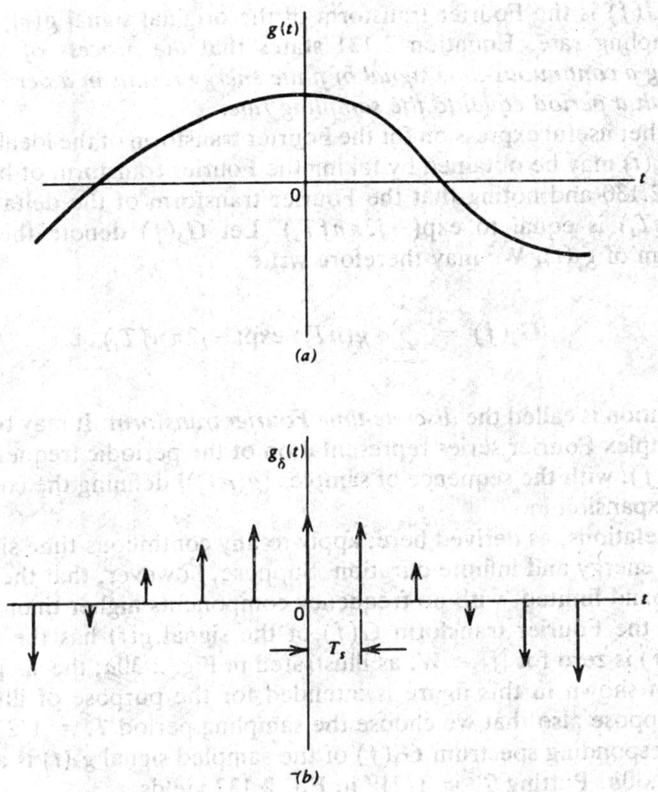

Figure 2.29
The sampling process. (a) Analog signal. (b) Instantaneously sampled version of the signal.

by the sequence of numbers $\{g(nT_s)\}$, as shown by (see Fig. 2.29b)

$$g_\delta(t) = \sum_{n=-\infty}^{\infty} g(nT_s)\,\delta(t - nT_s) \tag{2.130}$$

We refer to $g_\delta(t)$ as the *ideal sampled signal*. The ideal sampled signal $g_\delta(t)$ has a mathematical form similar to that of the Fourier transform of a periodic signal. This is readily established by comparing Eq. 2.130 for $g_\delta(t)$ with the Fourier transform of a periodic signal given in Eq. 2.125. This correspondence suggests that we may determine the Fourier transform of the ideal sampled signal $g_\delta(t)$ by applying the duality property to the Fourier transform of Eq. 2.125. By so doing, and using the fact that a delta function is an even function, we get the desired result:

$$g_\delta(t) \rightleftharpoons f_s \sum_{m=-\infty}^{\infty} G(f - mf_s) \tag{2.131}$$

where $G(f)$ is the Fourier transform of the original signal $g(t)$, and f_s is the sampling rate. Equation 2.131 states that *the process of uniformly sampling a continuous-time signal of finite energy results in a periodic spectrum with a period equal to the sampling rate.*

Another useful expression for the Fourier transform of the ideal sampled signal $g_\delta(t)$ may be obtained by taking the Fourier transform of both sides of Eq. 2.130 and noting that the Fourier transform of the delta function $\delta(t - nT_s)$ is equal to $\exp(-j2\pi nfT_s)$. Let $G_\delta(f)$ denote the Fourier transform of $g_\delta(t)$. We may therefore write

$$G_\delta(f) = \sum_{n=-\infty}^{\infty} g(nT_s)\exp(-j2\pi nfT_s) \tag{2.132}$$

This relation is called the *discrete-time Fourier transform*. It may be viewed as a complex Fourier series representation of the periodic frequency function $G_\delta(f)$, with the sequence of samples $\{g(nT_s)\}$ defining the coefficients of the expansion.

The relations, as derived here, apply to any continuous-time signal $g(t)$ of finite energy and infinite duration. Suppose, however, that the signal is strictly band-limited, with no frequency components higher than W hertz. That is, the Fourier transform $G(f)$ of the signal $g(t)$ has the property that $G(f)$ is zero for $|f| \geq W$, as illustrated in Fig. 2.30a; the shape of the spectrum shown in this figure is intended for the purpose of illustration only. Suppose also that we choose the sampling period $T_s = 1/2W$. Then the corresponding spectrum $G_\delta(f)$ of the sampled signal $g_\delta(t)$ is as shown in Fig. 2.30b. Putting $T_s = 1/2W$ in Eq. 2.132 yields

$$G_\delta(f) = \sum_{n=-\infty}^{\infty} g\left(\frac{n}{2W}\right)\exp\left(-\frac{j\pi nf}{W}\right) \tag{2.133}$$

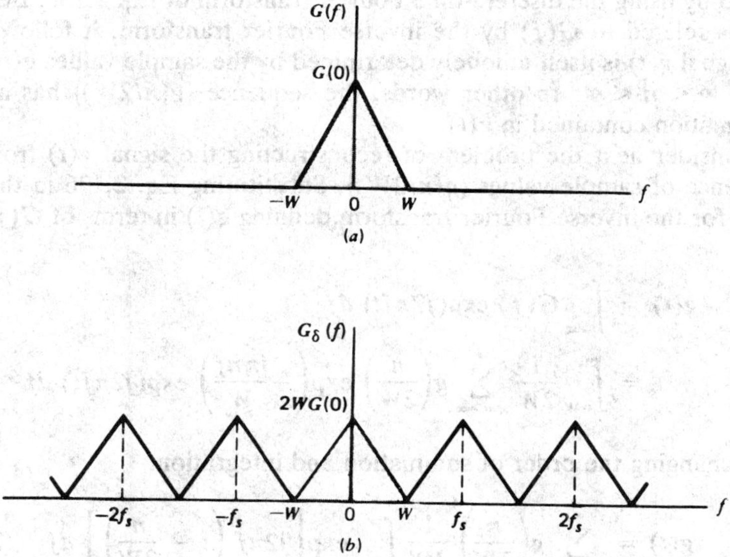

Figure 2.30
(a) Spectrum of a strictly band-limited signal g(t). (b) Spectrum of sampled version of g(t) for a sampling period $T_s = 1/2W$.

From Eq. 2.131, we have

$$G_\delta(f) = f_s G(f) + f_s \sum_{\substack{m=-\infty \\ m\neq 0}}^{\infty} G(f - mf_s) \tag{2.134}$$

Hence, under the following two conditions:

1. $G(f) = 0$ for $|f| \geq W$
2. $f_s = 2W$

we find from Eq. 2.134 that

$$G(f) = \frac{1}{2W} G_\delta(f), \qquad -W < f < W \tag{2.135}$$

Substituting Eq. 2.133 in Eq. 2.135, we may also write

$$G(f) = \frac{1}{2W} \sum_{n=-\infty}^{\infty} g\left(\frac{n}{2W}\right) \exp\left(-\frac{j\pi nf}{W}\right), \qquad -W < f < W \tag{2.136}$$

Therefore, if the sample values $g(n/2W)$ of a signal $g(t)$ are specified for all time, then the Fourier transform $G(f)$ of the signal is uniquely determined by using the discrete-time Fourier transform of Eq. 2.136. Because $g(t)$ is related to $G(f)$ by the inverse Fourier transform, it follows that the signal $g(t)$ is itself uniquely determined by the sample values $g(n/2W)$ for $-\infty < n < \infty$. In other words, the sequence $\{g(n/2W)\}$ has all the information contained in $g(t)$.

Consider next the problem of reconstructing the signal $g(t)$ from the sequence of sample values $\{g(n/2W)\}$. Substituting Eq. 2.136 in the formula for the inverse Fourier transform defining $g(t)$ in terms of $G(f)$, we get

$$\begin{aligned} g(t) &= \int_{-\infty}^{\infty} G(f) \exp(j2\pi ft)\, df \\ &= \int_{-W}^{W} \frac{1}{2W} \sum_{n=-\infty}^{\infty} g\left(\frac{n}{2W}\right) \exp\left(-\frac{j\pi nf}{W}\right) \exp(j2\pi ft)\, df \end{aligned}$$

Interchanging the order of summation and integration:

$$g(t) = \sum_{n=-\infty}^{\infty} g\left(\frac{n}{2W}\right) \frac{1}{2W} \int_{-W}^{W} \exp\left[j2\pi f\left(t - \frac{n}{2W}\right)\right] df \quad (2.137)$$

The integral term in Eq. 2.137 may be readily evaluated yielding

$$\begin{aligned} g(t) &= \sum_{n=-\infty}^{\infty} g\left(\frac{n}{2W}\right) \frac{\sin(2\pi Wt - n\pi)}{(2\pi Wt - n\pi)} \\ &= \sum_{n=-\infty}^{\infty} g\left(\frac{n}{2W}\right) \operatorname{sinc}(2Wt - n), \qquad -\infty < t < \infty \end{aligned} \quad (2.138)$$

Equation 2.138 provides an *interpolation formula* for reconstructing the original signal $g(t)$ from the sequence of sample values $\{g(n/2W)\}$, with the sinc function $\operatorname{sinc}(2Wt)$ playing the role of an *interpolation function*. Each sample is multiplied by a delayed version of the interpolation function, and all the resulting waveforms are added to obtain $g(t)$.

We may now state the *sampling theorem*[8] for band-limited signals of finite energy in two equivalent parts:

1. *A band-limited signal of finite energy, which has no frequency components higher than W hertz, is completely described by specifying the values of the signal at instants of time separated by 1/2W seconds.*

[8]The sampling theorem was introduced to communication theory by Shannon (1949). It is for this reason that the theorem is sometimes referred to in the literature as the "Shannon sampling theorem." However, the interest of communication engineers in the sampling theorem may be traced back to Nyquist (1928). Indeed, the sampling theorem was known to mathematicians at least since 1915. For historical notes on the sampling theorem, see the review paper by Jerri (1977).

2. *A band-limited signal of finite energy, which has no frequency components higher than W hertz, may be completely recovered from a knowledge of its samples taken at the rate of 2W samples per second.*

The sampling rate of $2W$ samples per second, for a signal bandwidth of W hertz, is called the *Nyquist rate;* its reciprocal $1/2W$ (measured in seconds) is called the *Nyquist interval.* The sampling theorem serves as the basis for the interchangeability of analog signals and digital sequences, which is so valuable in digital signal processing and digital communications.

The derivation of the sampling theorem, as described herein, is based on the assumption that the signal $g(t)$ is strictly band-limited. In practice, however, an information-bearing signal is not strictly band-limited. Hence, distortion may result from the application of the sampling theorem to such a signal. (More will be said on this issue in Chapter 5.)

EXERCISE 16 Apply the duality property to the Fourier transform pair of Eq 2.125 and thereby derive Eq. 2.131 for the ideal sampled signal $g_\delta(t)$.

2.8 NUMERICAL COMPUTATION OF THE FOURIER TRANSFORM

This section briefly describes a procedure for the computation of the Fourier transform, which is particularly well suited for use on a digital computer. We assume that the given signal $g(t)$ is of finite duration. The procedure involves first, the *uniform sampling* of $g(t)$ to obtain a finite sequence of samples denoted by $g(0), g(T_s), g(2T_s), \ldots, g(NT_s - T_s)$, where T_s is the *sampling period* and N is the number of samples. For a correct representation of the signal, the *sampling rate* $1/T_s$ must be equal to or greater than twice the highest frequency component of the signal. For the purpose of our present discussion, it is adequate to assume that this requirement has been satisfied. It is possible, of course, that the signal initially may be in the form of a sequence of samples. In any event, for this sequence of samples, we may define a *discrete Fourier transform* denoted by $\{G(kF_s)\}$, which consists of another sequence of N samples separated in frequency by F_s hertz, as shown by

$$G(kF_s) = T_s \sum_{n=0}^{N-1} g(nT_s) \exp\left(-j\frac{2\pi}{N}kn\right), \qquad k = 0, 1, 2, \ldots, N-1 \tag{2.139}$$

Equation 2.139 is precisely the formula that would be obtained by using the trapezoidal rule for approximating the integral that defines the Fourier transform of the given signal $g(t)$. The difference between the actual Fourier transform and the sequence $\{G(kF_s)\}$ obtained from Eq. 2.139 gives the integration error evaluated at $f = kF_s$. The parameters T_s and F_s are related by

$$T_s F_s = \frac{1}{N} \tag{2.140}$$

To derive the inverse relationship expressing the sequence $\{g(nT_s)\}$ in terms of the discrete spectrum $\{G(kF_s)\}$, we multiply both sides of Eq. 2.139 by $\exp(j2\pi km/N)$ and sum over k, obtaining

$$\sum_{k=0}^{N-1} G(kF_s) \exp\left(j\frac{2\pi}{N}km\right) = T_s \sum_{k=0}^{N-1}\sum_{n=0}^{N-1} g(nT_s) \exp\left[j\frac{2\pi}{N}k(m-n)\right] \tag{2.141}$$

Interchanging the order of summation on the right side of Eq. 2.141, and using the fact that

$$\sum_{k=0}^{N-1} \exp\left[j\frac{2\pi}{N}k(m-n)\right] = \begin{cases} N, & m = n \\ 0, & \text{otherwise} \end{cases} \tag{2.142}$$

we get

$$\sum_{k=0}^{N-1} G(kF_s) \exp\left(j\frac{2\pi}{N}km\right) = NT_s g(mT_s) \tag{2.143}$$

Next, substituting the index n for m and rearranging the terms in Eq. 2.143, we get the desired relation

$$g(nT_s) = F_s \sum_{k=0}^{N-1} G(kF_s) \exp\left(j\frac{2\pi}{N}kn\right), \qquad n = 0, 1, \ldots, N-1 \tag{2.144}$$

which defines the *inverse discrete Fourier transform*. Here again, it is of interest to note that Eq. 2.144 is precisely the formula that would be obtained by using the trapezoidal rule for approximating the integral that defines the inverse Fourier transform.

The discrete Fourier transform, as defined in Eq. 2.139, has properties that are analogous to those of the continuous Fourier transform.

An important feature of the discrete Fourier transform is that the signal $\{g(nT_s)\}$ and its spectrum $\{G(kF_s)\}$ are both in discrete form. Furthermore, they are both periodic, with the period of either one consisting of a finite number of samples N. That is,

$$g(nT_s) = g(nT_s + NT_s) \tag{2.145}$$

and

$$G(kF_s) = G(kF_s + NF_s) \tag{2.146}$$

We thus find that the numerical computation of the discrete Fourier transform is well suited for a digital computer or special-purpose digital processor. Indeed, it is this feature that makes the discrete Fourier transform so eminently useful in practice for spectral analysis and for the simulation of filters on digital computers. This is all the more so by virtue of the availability of an algorithm known as the *fast Fourier transform algorithm* (FFT), which provides a highly efficient procedure for computing the discrete Fourier transform of a finite-duration sequence. This algorithm takes advantage of the fact that the calculation of the coefficients of the discrete Fourier transform may be carried out in an iterative manner, thereby resulting in a considerable saving of computation time.[9] To compute the discrete Fourier transform of a sequence of N samples using the FFT algorithm, we require, in general, $N \log_2 N$ complex additions and $(N/2) \log_2 N$ complex multiplications. On the other hand, by using Eq. 2.139 to compute the discrete Fourier transform directly, we find that for each of the N output samples, we require $(N - 1)$ complex additions and N complex multiplications, so that the direct computation of the discrete Fourier transform requires a total of $N(N - 1)$ complex additions and N^2 complex multiplications. Accordingly, by using the FFT algorithm, the number of arithmetic operations is reduced by a factor of $N/\log_2 N$, which represents a considerable saving in computation effort for large N. For example, with $N = 1024$, we reduce the computation effort by about two orders of magnitude. Indeed, it is this kind of improvement that also makes it possible to use special-purpose digital processors for the hardware implementation of the FFT algorithm.

2.9 RELATIONSHIP BETWEEN THE FOURIER AND LAPLACE TRANSFORMS

The Fourier transform (as we have described it) is fully adequate for handling the frequency-domain description of signals encountered in the study of communication theory. Nevertheless, it can be helpful to briefly examine the relation between it and the Laplace transform, which is commonly used in transient analysis.

Consider the special case of a *causal signal* $g(t)$, defined as a signal that is zero for negative time. In other words, the signed $g(t)$ starts at or after $t = 0$. In such a case, the formula for the Fourier transform of $g(t)$ takes

[9] For a description of the FFT algorithm and its applications, see Roberts and Mullis (1987) or Oppenheim and Schafer (1975).

the form

$$G(f) = \int_0^\infty g(t) \exp(-j2\pi ft)\, dt \tag{2.147}$$

This integral bears a close resemblance to the *one-sided Laplace transform* of $g(t)$, as shown by

$$\bar{G}(s) = \int_0^\infty g(t) \exp(-st)\, dt \tag{2.148}$$

which implies that $g(t) = 0$ for $t < 0$. The quantity

$$s = \sigma + j\omega \tag{2.149}$$

is a *complex variable* whose real and imaginary parts are σ and ω, respectively. Comparing Eqs. 2.147 and 2.148, we see that the Fourier transform $G(f)$ may be obtained from the Laplace transform $\bar{G}(s)$ by putting

$$s = j\omega = j2\pi f$$

This is the link that connects the Fourier and Laplace transforms.

As mentioned previously, the Fourier transform is adequate for most purposes in communication theory. As such, we will use it exclusively in the rest of the book.

PROBLEMS

The problems are divided into sections that correspond to the major sections in the Chapter. For example, the problems in Section P2.1 pertain to Section 2.1. *This practice is followed in subsequent chapters.*

P2.1 Fourier Series

Problem 1 A signal that is sometimes used in communication systems is a *raised cosine pulse*. Figure P2.1 shows a signal $g_p(t)$ that is a periodic

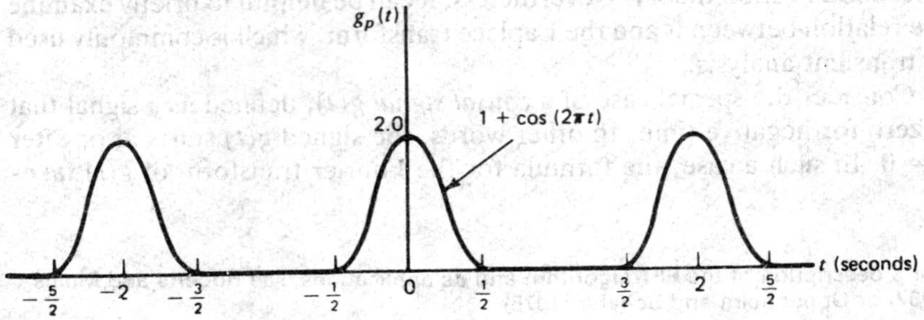

Figure P2.1

sequence of these pulses with equal spacing between them. Show that the first three terms in the Fourier series expansion of $g_p(t)$ are as follows:

$$g_p(t) = \tfrac{1}{2} + \frac{8}{3\pi}\cos(\pi t) + \tfrac{1}{2}\cos(2\pi t) + \cdots$$

Problem 2 Evaluate the amplitude spectrum of the periodic pulsed RF waveform shown in Fig. P2.2, assuming that $f_c T_0 \gg 1$.

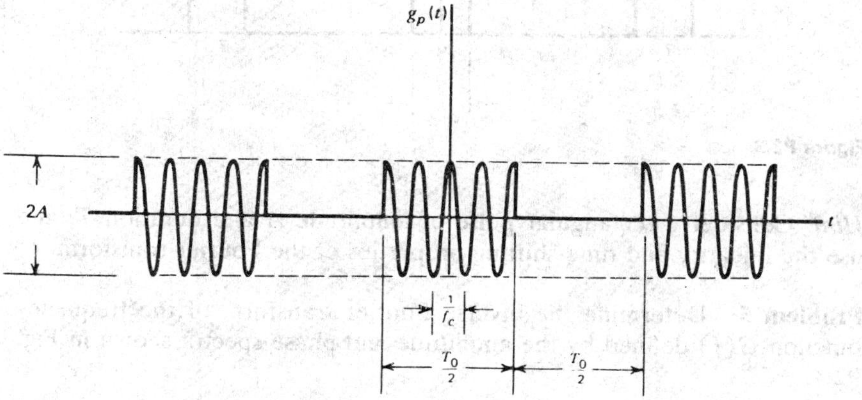

Figure P2.2

Problem 3 Prove the following properties of the Fourier series:

(a) If the periodic function $g_p(t)$ is even, that is,

$$g_p(-t) = g_p(t)$$

then the Fourier coefficients, the c_n, are purely real and even, that is, $c_{-n} = c_n$.

(b) If $g_p(t)$ is odd, that is,

$$g_p(-t) = -g_p(t)$$

then the c_n are purely imaginary and an odd function of n.

(c) If $g_p(t)$ has half-wave symmetry, that is,

$$g_p\left(t \pm \frac{1}{2}T_0\right) = -g_p(t)$$

where T_0 is the period of $g_p(t)$, then the Fourier series of such a signal consists of only odd-order terms.

P2.2 Fourier Transform

Problem 4 Determine the Fourier transform of the signal $g(t)$ consisting of three rectangular pulses, as shown in Fig. P2.3. Sketch the amplitude spectrum of this signal for the case when $T \ll T_0$.

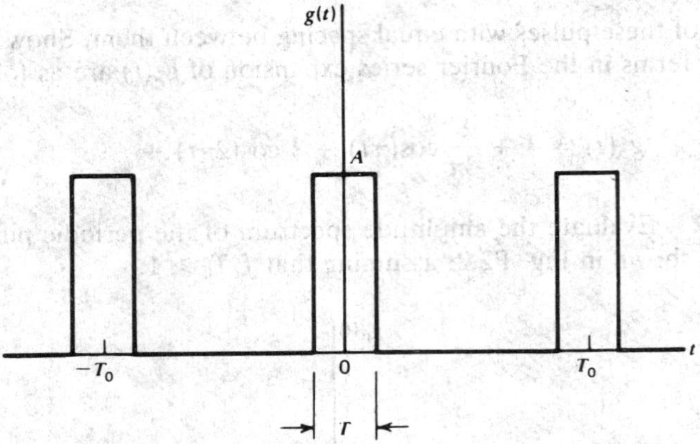

Figure P2.3

Hint: Consider a rectangular pulse of amplitude A and duration T, and use the linearity and time-shifting properties of the Fourier transform.

Problem 5 Determine the inverse Fourier transform of the frequency function $G(f)$ defined by the amplitude and phase spectra shown in Fig. P2.4.

Problem 6 Show that the spectrum of a real symmetric signal is either (a) purely real and even, or (b) purely imaginary and odd.

P2.3 Properties of the Fourier Transform

Problem 7 Let

$$g_1(t) = x\left(\frac{t}{5}\right)$$

$$g_2(t) = x(5t)$$

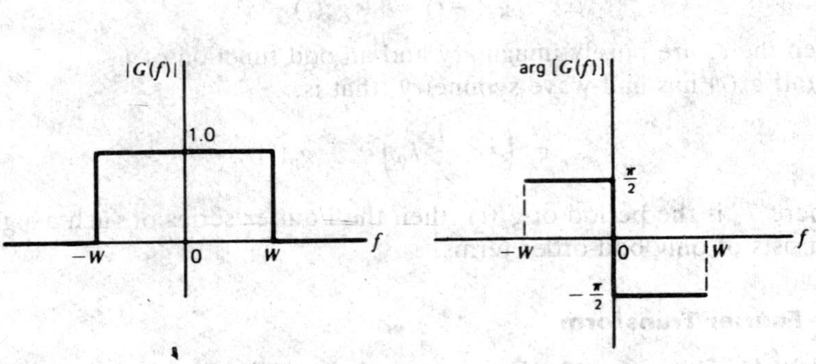

Figure P2.4

(a) Determine the Fourier transforms $G_1(f)$ and $G_2(f)$ in terms of the Fourier transform $X(f)$.
(b) Which of the two time functions, $g_1(t)$ and $g_2(t)$, corresponds to time compression, and which one to time expansion?
(c) Let

$$y(t) = a\, g_1(t)$$

Find the value of scaling factor a required to make $Y(0) = X(0)$, where $Y(f)$ is the Fourier transform of $y(t)$. Repeat your calculation for $g_2(t)$ in place of $g_1(t)$.

Problem 8

(a) Find the Fourier transform of the half-cosine pulse shown in Fig. P2.5*a*.
(b) Apply the time-shifting property to the result obtained in part (a) to evaluate the spectrum of the half-sine pulse shown in Fig. P2.5*b*.
(c) What is the spectrum of a half-sine pulse having a duration equal to aT?
(d) What is the spectrum of the negative half-sine pulse shown in Fig. P2.5*c*?
(e) Find the spectrum of the single sine pulse shown in Fig. P2.5*d*.

Problem 9 Any function $g(t)$ can be split unambiguously into an *even part* and an *odd part,* as shown by

$$g(t) = g_e(t) + g_o(t)$$

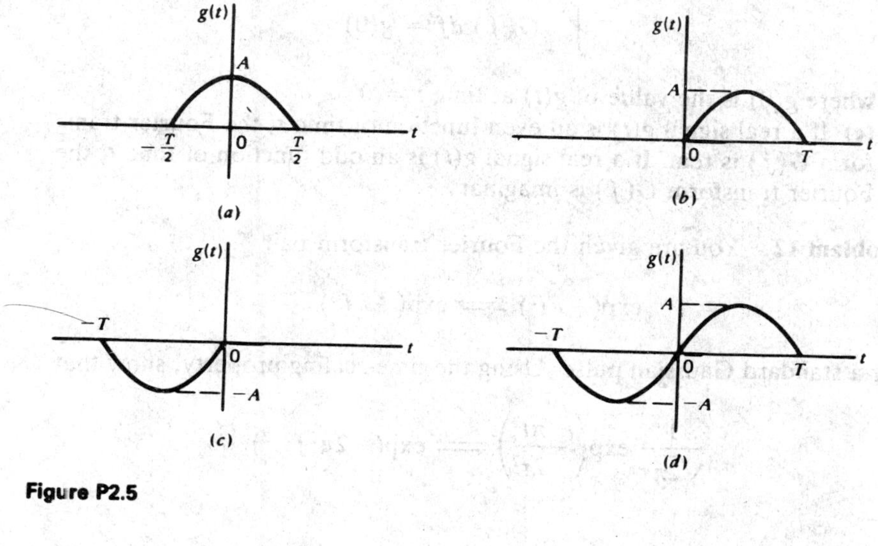

Figure P2.5

The even part is defined by

$$g_e(t) = \tfrac{1}{2}[g(t) + g(-t)]$$

and the odd part is defined by

$$g_o(t) = \tfrac{1}{2}[g(t) - g(-t)]$$

(a) Evaluate the even and odd parts of a rectangular pulse defined by

$$g(t) = A \operatorname{rect}\left(\frac{t}{T} - \frac{1}{2}\right)$$

(b) What are the Fourier transforms of these two parts of the pulse?

Problem 10 Assume the availability of a device that is capable of computing the Fourier transform of an energy signal $g(t)$ used as input. Explain the modifications that will have to be made to the input and output signals of such a device so that it may also be used to compute the inverse Fourier transform of the quantity $G(f)$, where $g(t) \rightleftharpoons G(f)$.

Problem 11 The Fourier transform of a signal $g(t)$ is denoted by $G(f)$. Prove the following properties of the Fourier transform:

(a) The total area under the curve of $g(t)$ is given by

$$\int_{-\infty}^{\infty} g(t)\, dt = G(0)$$

where $G(0)$ is the zero-frequency value of $G(f)$.

(b) The total area under the curve of $G(f)$ is given by

$$\int_{-\infty}^{\infty} G(f)\, df = g(0)$$

where $g(0)$ is the value of $g(t)$ at time $t = 0$

(c) If a real signal $g(t)$ is an even function of time t, the Fourier transform $G(f)$ is real. If a real signal $g(t)$ is an odd function of time t, the Fourier transform $G(f)$ is imaginary.

Problem 12 You are given the Fourier transform pair

$$\exp(-\pi t^2) \rightleftharpoons \exp(-\pi f^2)$$

for a standard Gaussian pulse. Using the time-scaling property, show that

$$\frac{1}{\sqrt{2\pi}\tau} \exp\left(-\frac{\pi t^2}{2\tau^2}\right) \rightleftharpoons \exp(-2\pi^2 f^2 \tau^2)$$

Problem 13 Prove the following properties of the convolution process:

(a) The *commutative* property:

$$g_1(t) \star g_2(t) = g_2(t) \star g_1(t)$$

(b) The *associative* property:

$$g_1(t) \star [g_2(t) \star g_3(t)] = [g_1(t) \star g_2(t)] \star g_3(t)$$

(c) The *distributive* property:

$$g_1(t) \star [g_2(t) + g_3(t)] = g_1(t) \star g_2(t) + g_1(t) \star g_3(t)$$

P2.4 Interplay Between Time-Domain and Frequency-Domain Descriptions

Problem 14 Consider a triangular pulse of height A and base $2T$. The duration of the pulse is measured at half-amplitude points. The bandwidth of the pulse is defined as one-half the main lobe of the pulse's spectrum. Show that the time–bandwidth product of the pulse equals unity.

Problem 15 Consider the sinc pulse

$$g(t) = A \operatorname{sinc}(2Wt)$$

The duration of the pulse is defined as the duration of the main lobe of the pulse. Hence, show that the time–bandwidth product of the sinc pulse equals unity.

Problem 16 Consider the Gaussian pulse

$$g(t) = \frac{1}{\sqrt{2\pi}\tau} \exp\left(-\frac{\pi t^2}{2\tau^2}\right)$$

The parameter τ provides one possible measure for the duration of the pulse. Defining the bandwidth of the pulse in a similar manner, show that the time–bandwidth product is 1/4.
Hint: Evaluate the Fourier transform of $g(t)$.

P2.5 Dirac Delta Function

Problem 17 Show that the effect of scaling the argument of the delta function by a constant a is described by

$$\delta(at) = \frac{1}{|a|}\delta(t)$$

Problem 18 The delta function may be considered as the limiting form of an ordinary function. Some useful representations are

$$\delta(\tau) = \lim_{\tau\to 0} \frac{1}{2\tau} \exp\left(-\frac{|t|}{\tau}\right)$$

$$= \lim_{\tau\to 0} \frac{\tau}{(t^2 + \tau^2)}$$

$$= \lim_{\tau\to 0} \frac{\sin(t/\tau)}{\pi t}$$

For each representation, plot the time function and its Fourier transform for different values of parameter τ. Hence, demonstrate that each time function approaches the delta function in the limit.

Problem 19 Determine the Fourier transform of the signal

$$g(t) = \cos^2(2\pi f_c t)$$

Problem 20 Let

$$g(t) \rightleftharpoons G(f)$$

and assume that $G(0)$ is nonzero. Starting with the Fourier transform of a signal, evaluate the Fourier transform of the integrated signal

$$\int_{-\infty}^{t} g(\tau)\, d\tau.$$

Hints:

(a) Use the formula for integration by parts.
(b) Use the limiting forms

$$\lim_{t\to 0} \frac{\sin(2\pi f t)}{\pi f} = \delta(f)$$

$$\lim_{t\to 0} \frac{\cos(2\pi f t)}{\pi f} = 0$$

P2.6 Fourier Transforms of Periodic Signals

Problem 21 Consider again the periodic signal $g_p(t)$ defined in Problem 1, which has a period of 2 seconds. The generating function of the signal is defined by

$$g(t) = \begin{cases} 1 + \cos(2\pi t), & -\frac{1}{2} \leq t \leq \frac{1}{2} \\ 0, & \text{for remainder of the period} \end{cases}$$

(a) Determine the Fourier transform of the generating function $g(t)$.
(b) Hence, using the formula of Eq. 2.125, determine the Fourier transform of the periodic signal $g_p(t)$. Compare your result with that of Problem 1.

P2.7 Sampling Theorem

Problem 22 Specify the Nyquist rate and the Nyquist interval for each of the following energy signals:

(a) $g(t) = \text{sinc}(200t)$
(b) $g(t) = \text{sinc}^2(200t)$
(c) $g(t) = \text{sinc}(200t) + \text{sinc}^2(200t)$

CHAPTER THREE

FILTERING AND SIGNAL DISTORTION

In Chapter 2 we used Fourier methods to study *spectral* properties of various kinds of *signals* and relationships between spectra and time-domain characteristics of the signals. We also studied the effects that various time-domain operations on a signal have on the spectrum of the signal. In this chapter we study *filtering* characteristics of *systems*. The system may be a linear time-invariant filter or communication channel. We also consider the linear and nonlinear forms of *signal distortion,* which result from transmission through linear and nonlinear systems, respectively. We begin the study by considering the time response of a linear time-invariant system.

3.1 TIME RESPONSE

A *system* refers to any physical device that produces an output signal in response to an input signal. It is customary to refer to the input signal as the *excitation* and to the output signal as the *response*. A system is said to be *linear* if the *principle of superposition* holds; that is, *the response of a linear system to a number of excitations applied simultaneously is equal to the sum of the responses of the system when the excitations are applied individually*. The system is said to be *time-invariant* if a time shift in the excitation applied to the system produces the same time shift in the response of the system. In this section, we study the *time response* of *linear time-invariant systems,* with particular reference to filters and channels. A *filter* refers to a frequency-selective device that is used to limit the spectrum of a signal to some band of frequencies. A *channel* refers to a physical medium that connects the transmitter of a communication system to the receiver. The operation of limiting the spectrum of a signal to some band of frequencies (by passing the signal through a filter or channel) is called *filtering*. In the time domain, a linear system is described in terms of its *impulse response, which is defined as the response of the system (with zero initial conditions) to a unit impulse or delta function* $\delta(t)$ *applied to the input of the system*. If the system is *time-invariant,* then the shape of the impulse response is the same no matter when the unit impulse is applied to the system. Thus, assuming that the unit impulse or delta function is applied at time $t = 0$, we may denote the impulse response of a linear time-invariant system by $h(t)$. Let this system be subjected to an arbitrary excitation $x(t)$, as in Fig. 3.1*a*. To determine the response $y(t)$ of the system, we begin by first approximating $x(t)$ by a staircase function composed of narrow rectangular pulses, each of duration $\Delta\tau$, as shown in Fig. 3.1*b*. Clearly the approximation becomes better for smaller $\Delta\tau$. As $\Delta\tau$ approaches zero, each pulse approaches, in the limit, a delta function weighted by a factor equal to the height of the pulse times $\Delta\tau$. Consider a typical pulse, shown shaded in Fig. 3.1*b*, which occurs at $t = \tau$. This pulse has an area equal to $x(\tau)\,\Delta\tau$. By definition, the response of the system to a unit impulse or delta function $\delta(t)$, occurring at $t = 0$, is $h(t)$. It follows, therefore, that the response of the system to a delta function, weighted by the factor $x(\tau)\,\Delta\tau$ and occurring at $t = \tau$, must be $x(\tau)h(t - \tau)\,\Delta\tau$. To find the total response $y(t)$ at some time t, we apply the principle of superposition. Thus, summing the various infinitesimal responses due to the various input pulses, we obtain in the limit, as $\Delta\tau$ approaches zero,

$$y(t) = \int_{-\infty}^{\infty} x(\tau)h(t - \tau)\,d\tau \tag{3.1}$$

This relation is called the *convolution integral*. Note that for the response $y(t)$ to have the same dimension as the excitation $x(t)$, *the impulse response* $h(t)$ must have a dimension that is the inverse of time.

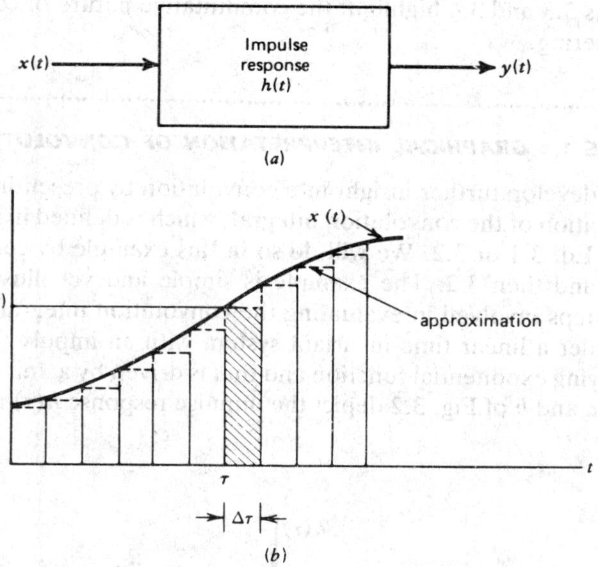

Figure 3.1
(a) Linear system. (b) Approximation of input x(t).

In Eq. 3.1, three different time scales are involved: *excitation time* τ, *response time* t, and *system-memory time* $t - \tau$. This relation is the basis of time-domain analysis of linear time-invariant systems. It states that the present value of the response of a linear time-invariant system is a weighted integral over the past history of the input signal, weighted according to the impulse response of the system. Thus the impulse response acts as a *memory function* for the system.

In Eq. 3.1, the excitation $x(t)$ is convolved with the impulse response $h(t)$ to produce the response $y(t)$. Since convolution is commutative, it follows that we may also write

$$y(t) = \int_{-\infty}^{\infty} h(\tau)x(t - \tau)\, d\tau \tag{3.2}$$

where $h(t)$ is convolved with $x(t)$.

Using the shorthand notation for convolution, we may rewrite Eq. 3.1 simply as

$$y(t) = x(t) \star h(t) \tag{3.3}$$

where $\star$ denotes convolution. Similarly, we may rewrite Eq. 3.2 as

$$y(t) = h(t) \star x(t) \tag{3.4}$$

Equations 3.3 and 3.4 highlight the commutative nature of convolution or linear filtering.

EXAMPLE 1 GRAPHICAL INTERPRETATION OF CONVOLUTION

We may develop further insight into convolution by presenting a graphical interpretation of the convolution integral, which is defined in mathematical terms in Eq. 3.1 or 3.2. We will do so in this example by considering Eq. 3.1 first and then 3.2. The example is simple and yet illustrative of the various steps involved in evaluating the convolution integral. Specifically, we consider a linear time-invariant system with an impulse response that is a decaying exponential function and that is driven by a unit step function.

Parts *a* and *b* of Fig. 3.2 depict the impulse response $h(\tau)$ and excitation

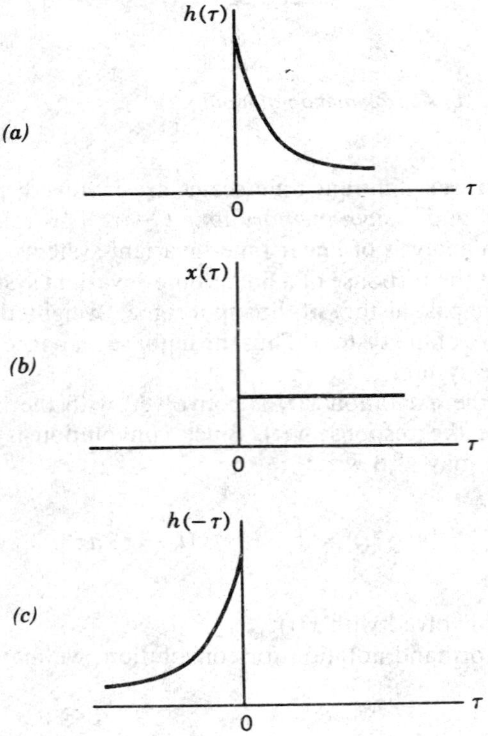

Figure 3.2
The steps involved in computing one form of the convolution integral. (a) Impulse response. (b) Excitation. (c) Image of the impulse response. (d) Time-shifted image of the impulse response. (e) Evaluation of the response.

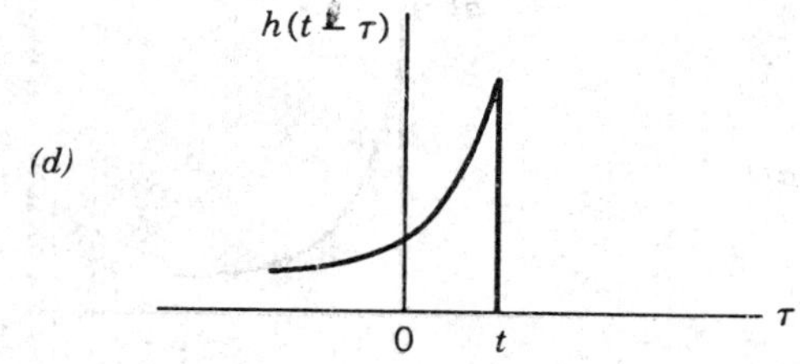

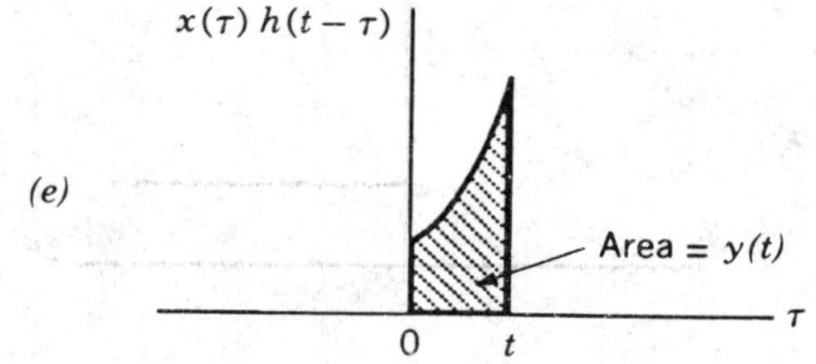

Figure 3.2 (*continued*)

$x(\tau)$, respectively. For reasons that will become apparent presently, the time variable in both cases is shown as τ. In accordance with Eq. 3.1, the integral consists of the product $x(\tau)h(t - \tau)$. We already have $x(\tau)$. To obtain $h(t - \tau)$, we proceed in two steps. First, we formulate $h(-\tau)$, which is the mirror image of $h(\tau)$ with respect to the vertical axis, as shown in Fig. 3.2*c*. Then, we shift $h(-\tau)$ to the right by an amount equal to the specified time t to obtain $h(t - \tau)$; this second step is shown in Fig. 3.2*d*. Next, we multiply $x(\tau)$ by $h(t - \tau)$, as in Fig. 3.2*e*, and thereby obtain the desired integrand $x(\tau)h(t - \tau)$ for the specified value of time t. Finally, we calculate the total area under $x(\tau)h(t - \tau)$, which is shown shaded in Fig. 3.2*e*. This area equals the value of the system response $y(t)$ at time t.

For the graphical interpretation of Eq. 3.2 we may proceed in a similar way, as illustrated in Fig. 3.3. In this second case, the integrand equals $h(\tau)x(t - \tau)$. The first multiplying factor $h(\tau)$ is already available, as in Fig. 3.3*a*. The second multiplying factor $x(t - \tau)$ is obtained by forming the image $x(-\tau)$ of the specified excitation $x(\tau)$, and then shifting the image $x(-\tau)$ to the right by an amount equal to the specified time t. The functions $x(\tau)$, $x(-\tau)$, and $x(t - \tau)$ are depicted in Figs. 3.3*b*, *c*, and *d*, respectively. The resulting product $h(\tau)x(t - \tau)$ is shown in Fig. 3.3*e*. Comparing Figs. 3.2*e* and 3.3*e*, we see that the products $x(\tau)h(t - \tau)$ and $h(\tau)x(t - \tau)$ are reversed with respect to each other. Naturally, they both have the same total area under their individual curves, which confirms the commutative property of convolution.

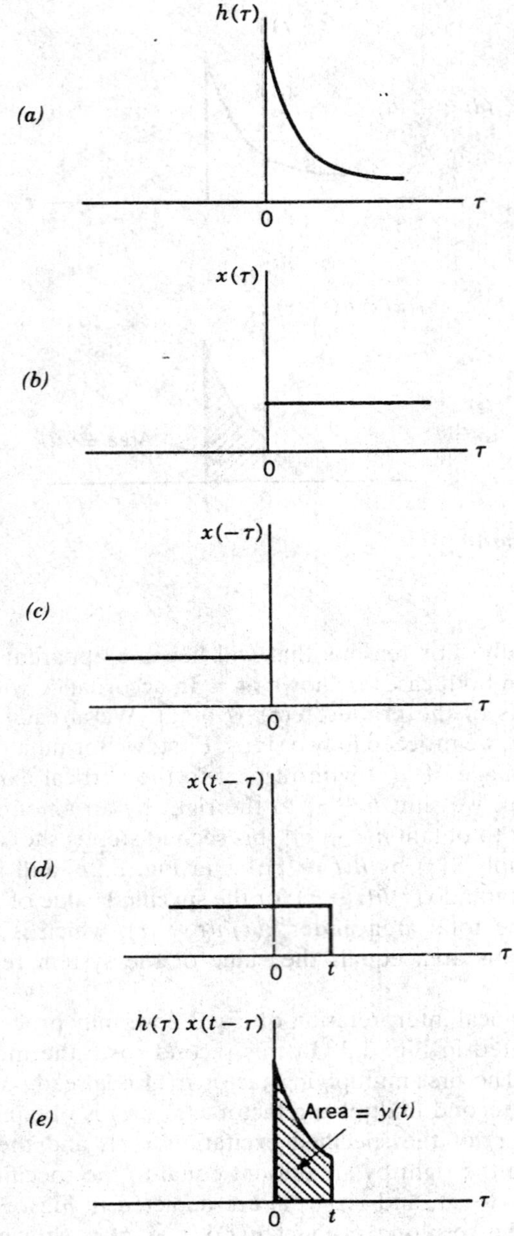

Figure 3.3
The steps involved in computing the second form of the convolution integral. (a) Impulse response. (b) Excitation. (c) Image of the excitation. (d) Time-shifted image of the excitation. (e) Evaluation of the response y(t).

EXAMPLE 2 TAPPED-DELAY-LINE FILTER

Consider a linear time-invariant filter with impulse response $h(t)$. We assume that

1. The impulse response $h(t) = 0$ for $t < 0$.
2. The impulse response of the filter is of finite duration, so that we may write $h(t) = 0$ for $t \geq T_f$.

Then we may express the filter output $y(t)$ produced in response to the input $x(t)$ as follows:

$$y(t) = \int_0^{T_f} h(\tau)x(t - \tau)\, d\tau \tag{3.5}$$

Let the input $x(t)$, impulse response $h(t)$, and output $y(t)$ be *uniformly sampled* at the rate $1/\Delta\tau$ samples per second, so that we may put

$$t = n\,\Delta\tau \tag{3.6}$$

and

$$\tau = k\,\Delta\tau \tag{3.7}$$

where k and n are integers, and $\Delta\tau$ is the *sampling period.* We assume that $\Delta\tau$ is small enough for the product $h(\tau)x(t - \tau)$ to remain essentially constant for $k\,\Delta\tau \leq \tau \leq (k + 1)\,\Delta\tau$ for all values of k and t of interest. Then we can approximate Eq. 3.5 by the *convolution sum:*

$$y(n\,\Delta\tau) = \sum_{k=0}^{N-1} h(k\,\Delta\tau)x(n\,\Delta\tau - k\,\Delta\tau)\,\Delta\tau \tag{3.8}$$

where $N\,\Delta\tau = T_f$. Defining

$$w_k = h(k\,\Delta\tau)\,\Delta\tau$$

we may rewrite Eq. 3.8 as

$$y(n\,\Delta\tau) = \sum_{k=0}^{N-1} w_k x(n\,\Delta\tau - k\,\Delta\tau) \tag{3.9}$$

Equation 3.9 is realized using the circuit shown in Fig. 3.4, which consists of a set of *delay elements* (each producing a delay of $\Delta\tau$ seconds), a set of

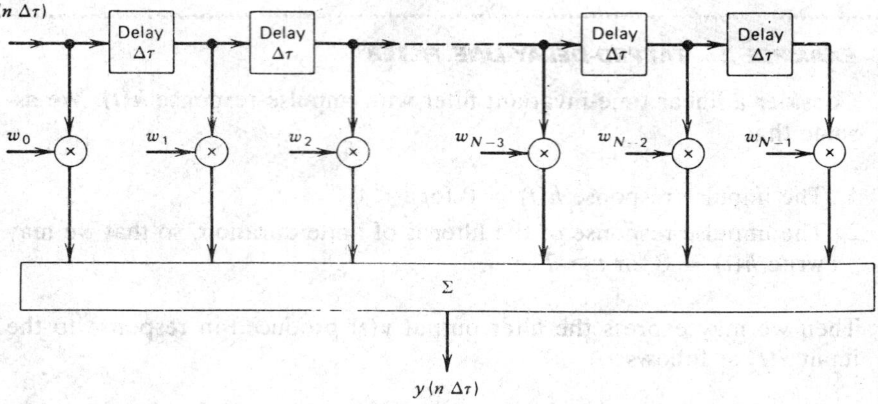

Figure 3.4
Tapped-delay-line filter.

multipliers connected to the *delay-line taps,* a corresponding set of *weights* applied to the multipliers, and a *summer* for adding the multiplier outputs. This circuit is known as a *tapped-delay-line filter* or *transversal filter*. Note that in Fig. 3.4 the tap-spacing or basic increment of delay is equal to the sampling period of the input sequence $\{x(n\,\Delta\tau)\}$.

When a tapped-delay-line filter is implemented using digital hardware, it is commonly referred to as a *finite-duration impulse response* (FIR) *digital filter*. The required delay is provided by means of a *shift register,* with the basic increment of delay, $\Delta\tau$, equal to the clock period. An important feature of a digital filter is that it is programmable, thereby offering a high degree of flexibility in design.[1]

CAUSALITY AND STABILITY

A system is said to be *causal* if it does not respond before the excitation is applied. For a linear time-invariant system to be causal, it is clear that the impulse response $h(t)$ must vanish for negative time. That is, the necessary and sufficient condition for causality is

$$h(t) = 0, \qquad t < 0 \tag{3.10}$$

Clearly, for a system operating in *real time* to be physically realizable, it must be causal. However, there are many applications in which the signal

[1]For a detailed treatment of the theory and design of digital filters, see Roberts and Mullis (1987) or Oppenheim and Schaffer (1975).

to be processed is available in stored form; in these situations the system can be noncausal and yet physically realizable.

The system is said to be *stable* if the output signal is bounded for all bounded input signals. Let the input signal $x(t)$ be bounded, as shown by

$$|x(t)| \leq M, \qquad -\infty < t < \infty \tag{3.11}$$

where M is a positive real finite number. Using Eqs. 3.2 and 3.11, we may write

$$|y(t)| \leq \int_{-\infty}^{\infty} |h(\tau)|\,|x(t-\tau)|\,d\tau = M \int_{-\infty}^{\infty} |h(\tau)|\,d\tau$$

It follows therefore that for a linear time-invariant system to be stable, the impulse response $h(t)$ must be absolutely integrable. That is, the necessary and sufficient condition for stability is

$$\int_{-\infty}^{\infty} |h(t)|\,dt < \infty \tag{3.12}$$

EXERCISE 1 The impulse response of a linear time-invariant system is defined by

$$h(t) = \exp(at)\,u(-t)$$

where $u(-t)$ is the time-reversed version of the unit step function $u(t)$. Is this system casual? Is it stable? Give reasons for your answers.

3.2 FREQUENCY RESPONSE

Consider a linear time-invariant system of impulse response $h(t)$ driven by a complex exponential input of unit amplitude and frequency f, that is,

$$x(t) = \exp(j2\pi ft) \tag{3.13}$$

Using Eq. 3.2, the response of the system is obtained as

$$\begin{aligned} y(t) &= \int_{-\infty}^{\infty} h(\tau) \exp[j2\pi f(t-\tau)]\,d\tau \\ &= \exp(j2\pi ft) \int_{-\infty}^{\infty} h(\tau) \exp(-j2\pi f\tau)\,d\tau \end{aligned} \tag{3.14}$$

Define

$$H(f) = \int_{-\infty}^{\infty} h(\tau) \exp(-j2\pi f\tau)\, d\tau \tag{3.15}$$

Then we may rewrite Eq. 3.14 in the form

$$y(t) = H(f) \exp(j2\pi ft) \tag{3.16}$$

The response of a linear time-invariant system to a complex exponential function of frequency f is, therefore, the same complex exponential function multiplied by a constant coefficient $H(f)$. The quantity $H(f)$ is called the *transfer function* of the system. The transfer function $H(f)$ and impulse response $h(t)$ form a Fourier transform pair, as shown by the pair of relations:

$$H(f) = \int_{-\infty}^{\infty} h(t) \exp(-j2\pi ft)\, dt \tag{3.17}$$

and

$$h(t) = \int_{-\infty}^{\infty} H(f) \exp(j2\pi ft)\, df \tag{3.18}$$

An alternative definition of the transfer function may be deduced by dividing Eq. 3.16 by 3.13 to obtain

$$H(f) = \left.\frac{y(t)}{x(t)}\right|_{x(t)=\exp(j2\pi ft)} \tag{3.19}$$

Consider next an arbitrary signal $x(t)$ applied to the system. The signal $x(t)$ may be expressed in terms of its Fourier transform as

$$x(t) = \int_{-\infty}^{\infty} X(f) \exp(j2\pi ft)\, df \tag{3.20}$$

or, equivalently, in the limiting form

$$x(t) = \lim_{\substack{\Delta f \to 0 \\ f = k\Delta f}} \sum_{k=-\infty}^{\infty} X(f) \exp(j2\pi ft)\, \Delta f \tag{3.21}$$

That is, the input signal $x(t)$ may be viewed as a superposition of complex exponentials of incremental amplitude. Because the system is linear, the response to this superposition of complex exponential inputs is

$$y(t) = \lim_{\substack{\Delta f \to 0 \\ f = k\Delta f}} \sum_{k=-\infty}^{\infty} H(f)X(f)\exp(j2\pi ft)\,\Delta f$$

$$= \int_{-\infty}^{\infty} H(f)X(f)\exp(j2\pi ft)\,df \tag{3.22}$$

The Fourier transform of the output is therefore

$$Y(f) = H(f)X(f) \tag{3.23}$$

A linear time-invariant system may thus be described simply in the frequency domain by noting that the Fourier transform of the output is equal to the product of the transfer function of the system and the Fourier transform of the input.

The result of Eq. 3.23 may, of course, be deduced directly by recognizing that the response $y(t)$ of a linear time-invariant system of impulse response $h(t)$ to an arbitrary input $x(t)$ is obtained by convolving $x(t)$ with $h(t)$, or vice versa, and by the fact that the convolution of a pair of time functions is transformed into the multiplication of their Fourier transforms. The foregoing derivation is presented primarily to develop an understanding of why the Fourier representation of a time function as a superposition of complex exponentials is so useful in analyzing the behavior of linear time-invariant systems.

AMPLITUDE RESPONSE AND PHASE RESPONSE

The transfer function $H(f)$ is a characteristic property of a linear time-invariant system. It is, in general, a complex quantity, so that we may express it in the form

$$H(f) = |H(f)|\exp[j\beta(f)] \tag{3.24}$$

where $|H(f)|$ is called the *amplitude response*, and $\beta(f)$ is called the *phase response*. The phase response is related to the transfer function $H(f)$ by

$$\beta(f) = \arg[H(f)] \tag{3.25}$$

In the case of a linear system with a real-valued impulse response $h(t)$, the transfer function $H(f)$ exhibits *conjugate symmetry*, which means that

$$|H(f)| = |H(-f)| \tag{3.26}$$

and

$$\beta(f) = -\beta(-f) \tag{3.27}$$

That is, the amplitude response $|H(f)|$ is an even function of frequency, whereas the phase response $\beta(f)$ is an odd function of frequency. Plots of the amplitude response $|H(f)|$ and the phase response $\beta(f)$ versus frequency f represent the frequency-domain description of the system. Hence, we may also refer to $H(f)$ as the *frequency response* of the system.

In some applications it is preferable to work with the logarithm of $H(f)$ rather than with $H(f)$ itself. Define

$$\ln H(f) = \alpha(f) + j\beta(f) \tag{3.28}$$

where

$$\alpha(f) = \ln|H(f)| \tag{3.29}$$

The function $\alpha(f)$ is called the *gain* of the system. It is measured in *nepers,* whereas $\beta(f)$ is measured in *radians*. Equation 3.28 indicates that the gain $\alpha(f)$ and phase response $\beta(f)$ are the real and imaginary parts of the logarithm of the transfer function $H(f)$, respectively. The squared amplitude response $|H(f)|^2$ is identified with power. Accordingly, we may also apply the *decibel* (dB) measure to the gain by writing

$$\alpha'(f) = 20 \log_{10}|H(f)| \tag{3.30}$$

The two gain functions $\alpha(f)$ and $\alpha'(f)$ are related by

$$\alpha'(f) = 8.69\alpha(f) \tag{3.31}$$

That is, 1 neper is equal to 8.69 dB.

EXAMPLE 3

Consider a linear time-invariant device with a transfer function defined by

$$\begin{aligned} H(f) &= \begin{cases} -j, & f > 0 \\ 0, & f = 0 \\ j, & f < 0 \end{cases} \\ &= -j\,\text{sgn}(f) \end{aligned} \tag{3.32}$$

where sgn(f) is the signum function.

The amplitude response and phase response of the device are shown in

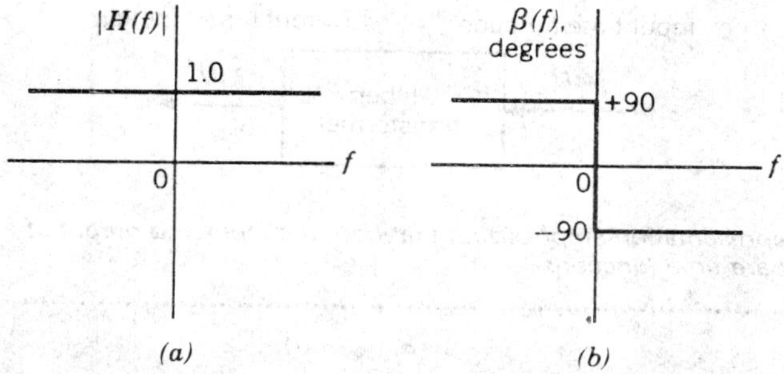

Figure 3.5
Characteristics of a Hilbert transformer. (a) Amplitude response. (b) Phase response.

Fig. 3.5*a* and *b*, respectively. That is, the device produces a phase shift of $-90°$ for all positive frequencies and a phase shift of $+90°$ for all negative frequencies. The amplitudes of all frequency components of the input signal are unaffected by transmission through the device. Such an ideal device is called a *Hilbert transformer*.

Figure 3.6 shows a black-box representation of the Hilbert transformer with a Fourier transformable signal $x(t)$ acting as input, and the resulting output[2] denoted by $\hat{x}(t)$. We wish to determine the output $\hat{x}(t)$, given the input $x(t)$. To do so, we first determine the impulse response of the device. Specifically, we use the Fourier transform pair of Eq. 2.107 to express the impulse response of the Hilbert transformer as

$$h(t) = \frac{1}{\pi t} \tag{3.33}$$

Hence, the convolution of this impulse response with a signal $x(t)$ applied to the input of the Hilbert transformer yields the resulting output $\hat{x}(t)$ as

$$\begin{aligned} \hat{x}(t) &= x(t) \star \left(\frac{1}{\pi t}\right) \\ &= \frac{1}{\pi}\int_{-\infty}^{\infty} \frac{x(\tau)}{t-\tau}\, d\tau \end{aligned} \tag{3.34}$$

According to this formula, $\hat{x}(t)$ is the *Hilbert transform* of $x(t)$.

[2]When dealing with Hilbert transformation, it is customary to denote the output by placing a circumflex (or "hat") over the symbol for the input; this explains the reason for using $\hat{x}(t)$ rather than $y(t)$ to denote the output.

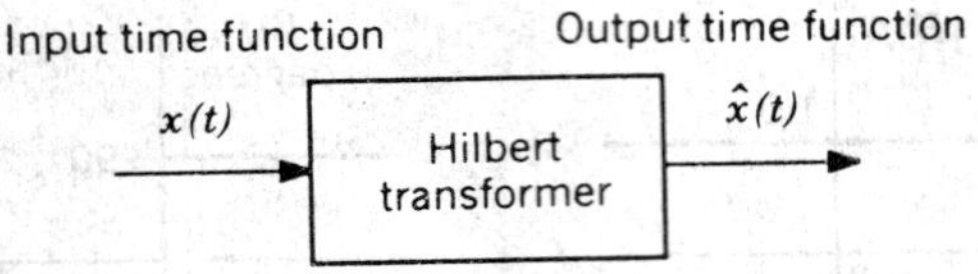

Figure 3.6
Black-box representation emphasizing that both the input and output of a Hilbert transformer are time functions.

EXERCISE 2 The *inverse Hilbert transform,* defining $x(t)$ in terms of $\hat{x}(t)$, is described by

$$x(t) = -\frac{1}{\pi}\int_{-\infty}^{\infty}\frac{\hat{x}(\tau)}{t-\tau}\,d\tau \qquad (3.35)$$

Starting with the transfer function of Eq. 3.32, derive the formula of Eq. 3.35.

SYSTEM BANDWIDTH

To specify the degree of dispersion of the amplitude response or gain of a system, we use a parameter called the *system bandwidth*. A common definition of system bandwidth is the *3-dB bandwidth,* the exact formulation of which depends on the type of system being considered. In the case of a *low-pass system,* the 3-dB bandwidth is defined as the difference between zero frequency, at which the amplitude response attains its peak value $|H(0)|$, and the frequency at which the amplitude response drops to a value equal to $|H(0)|/\sqrt{2}$, as illustrated in Fig. 3.7*a*. In the case of a *band-pass system,* the 3-dB bandwidth is defined as the difference between the frequencies at which the amplitude response drops to a value equal to $1/\sqrt{2}$ times the peak value $|H(f_c)|$ at the midband frequency f_c, as illustrated in Fig. 3.7*b*. Note that in both cases, the system bandwidth is defined for *positive* frequencies. Note also that an amplitude response value equal to $1/\sqrt{2}$ times the peak value of the amplitude response is equivalent to a drop in the gain of 3-dB below its peak value; hence, the name "3-dB bandwidth."

3.3 LINEAR DISTORTION AND EQUALIZATION

Two basic forms of *signal distortion* result from the transmission of a signal through a physical system: *linear distortion* and *nonlinear distortion*. In the context of telecommunications, the system of interest is comprised of all the components that constitute the path from the source of information to the desired destination. When the system is viewed as being linear and

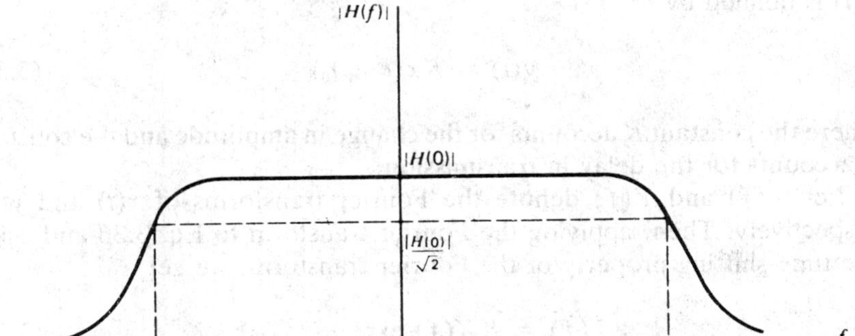

Figure 3.7
The definition of system bandwidth. (a) Low-pass system. (b) Band-pass system.

time invariant, linear distortion arises owing to *imperfections in the frequency response* of the system. On the other hand, nonlinear distortion arises owing to the presence of *nonlinearities* in the makeup of the system. In this section, we discuss the linear distortion problem; nonlinear distortion is considered in Section 3.7. We begin the discussion by formulating the conditions for distortionless transmission of a signal through a linear time-invariant system.

CONDITIONS FOR DISTORTIONLESS TRANSMISSION

By *distortionless transmission* we mean that the output signal of a system is an exact replica of the input signal, except for a possible change of amplitude and a constant time delay. We may therefore say that a signal $x(t)$ is transmitted through the system without distortion if the output signal

$y(t)$ is defined by

$$y(t) = Kx(t - t_0) \tag{3.36}$$

where the constant K accounts for the change in amplitude and the constant t_0 accounts for the delay in transmission.

Let $X(f)$ and $Y(f)$ denote the Fourier transforms of $x(t)$ and $y(t)$, respectively. Then, applying the Fourier transform to Eq. 3.36 and using the time-shifting property of the Fourier transform, we get

$$Y(f) = KX(f) \exp(-j2\pi f t_0) \tag{3.37}$$

The transfer function of a distortionless system is therefore

$$\begin{aligned} H(f) &= \frac{Y(f)}{X(f)} \\ &= K \exp(-j2\pi f t_0) \end{aligned} \tag{3.38}$$

Correspondingly, the impulse response of the system is given by

$$h(t) = K\delta(t - t_0) \tag{3.39}$$

where $\delta(t - t_0)$ is a Dirac delta function shifted by t_0 seconds.

Equation 3.38 indicates that in order to achieve distortionless transmission through a system, the transfer function of the system must satisfy two conditions:

1. The amplitude response $|H(f)|$ is constant for all frequencies, as shown by

$$|H(f)| = K \tag{3.40}$$

2. The phase $\beta(f)$ is linear with frequency, passing through zero as shown by

$$\beta(f) = -2\pi f t_0 \tag{3.41}$$

These two conditions are illustrated in parts a and b of Fig. 3.8, respectively.

EXERCISE 3 Using the impulse response of Eq. 3.39 in the convolution integral, show that the input–output relation of a distortionless system is as defined in Eq. 3.36.

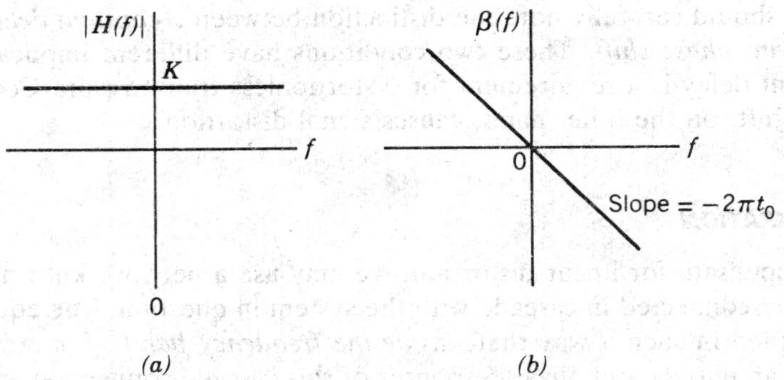

Figure 3.8
Frequency response for distortionless transmission. (a) Amplitude response. (b) Phase response.

EXERCISE 4 Show that the condition of Eq. 3.41 on the phase response $\beta(f)$ for distortionless transmission may be modified by adding a constant equal to a positive or negative integer multiple of 180°. How can such a modification arise in practice?

AMPLITUDE DISTORTION AND PHASE DISTORTION

In practice, the conditions for distortionless transmission, as just described, can only be satisfied approximately. That is to say, there is always a certain amount of linear distortion present in the output signal. In particular, we may distinguish two components of signal distortion produced by transmission through a linear time-invariant system:

1. When the amplitude response $|H(f)|$ of the system is not constant with frequency inside the frequency band of interest, the frequency components of the input signal are transmitted with different amounts of gain or attenuation. This effect is called *amplitude distortion*. The most common form of amplitude distortion is excess gain or attenuation at one or both ends of the frequency band of interest.
2. The second form of distortion arises when the phase response $\beta(f)$ of the system is not linear with frequency. Then if the input signal is divided into a set of components, each one of which occupies a narrow band of frequencies, we find that each of them is subject to a different delay in passing through the system, with the result that the output signal has a different waveform from the input. This form of distortion is called *phase* or *delay distortion*. We will have more to say on this issue in Section 3.6.

You should carefully note the distinction between a *constant delay* and a *constant phase shift.* These two conditions have different implications. Constant delay is a requirement for distortionless transmission. Constant phase shift, on the other hand, causes signal distortion.

EQUALIZATION

To compensate for linear distortion, we may use a network known as an *equalizer* connected in cascade with the system in question. The equalizer is designed in such a way that, *inside the frequency band of interest,* the overall amplitude and phase responses of this cascade connection approximate the conditions for distortionless transmission to within prescribed limits.

Consider, for example, a communication channel with transfer function $H_c(f)$. Let an equalizer of transfer function $H_{eq}(f)$ be connected in cascade with the channel, as in Fig. 3.9. The overall transfer function of this combination is equal to $H_c(f)H_{eq}(f)$. For overall transmission through the cascade connection of Fig. 3.9 to be distortionless, we require that (see Eq. 3.38)

$$H_c(f)H_{eq}(f) = K \exp(-j2\pi f t_0) \tag{3.42}$$

where K is a scaling factor and t_0 is a constant time delay. Ideally, therefore, the transfer function of the equalizer is *inversely related* to that of the channel, as shown by

$$H_{eq}(f) = \frac{K \exp(-j2\pi f t_0)}{H_c(f)} \tag{3.43}$$

In practice, the equalizer is designed such that its transfer function approximates the ideal value of Eq. 3.43 closely enough for the linear distortion to be reduced to a satisfactory level.

A network structure that is well-suited for the design of equalizers is the *tapped-delay-line filter,* depicted in Fig. 3.4. From the time-shifting property of the Fourier transform, we know that when a signal is shifted

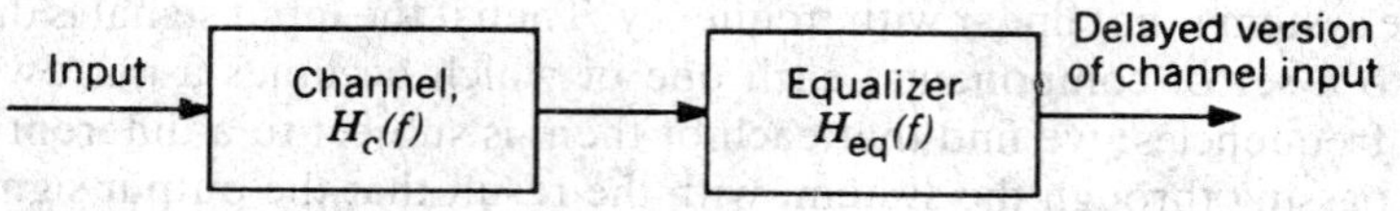

Figure 3.9
Block diagram of equalization.

in time by $\Delta\tau$ seconds, its Fourier transform is multiplied by the complex exponential $\exp(-j2\pi f\,\Delta\tau)$. Accordingly, the transfer function of this tapped-delay-line filter, used as an equalizer, is given by

$$H_{eq}(f) = \sum_{k=0}^{N-1} w_k \exp(-j2\pi k f\,\Delta\tau) \tag{3.44}$$

For convenience of analysis, let the number of taps be *odd*, as shown by

$$N = 2M + 1 \tag{3.45}$$

where M is an integer. Also, setting

$$k = m + M \qquad \begin{matrix} k = 0, \ldots, N-1 \\ m = -M, \ldots, -1, 0, 1, \ldots M \end{matrix} \tag{3.46}$$

and

$$w_k = c_m \tag{3.47}$$

we may rewrite Eq. 3.44 as

$$H_{eq}(f) = \left[\sum_{m=-M}^{M} c_m \exp(-j2\pi m f\,\Delta\tau)\right] \exp(-j2\pi M f\,\Delta\tau) \tag{3.48}$$

The expression inside the square brackets on the right side of Eq. 3.48 represents the discrete-time Fourier transform of the sequence of tap coefficients $c_{-M}, \ldots, c_{-1}, c_0, c_1, \ldots, c_M$, with a tap spacing (sampling interval) of $\Delta\tau$ seconds. This discrete-time Fourier transform may be viewed as a truncated version of the complex Fourier series with a frequency periodicity of $1/\Delta\tau$ hertz; note that in this interpretation, the usual roles of time and frequency in the complex Fourier series are interchanged.

We may now describe a procedure for designing the equalizer. Specifically, given a channel of transfer function $H_c(f)$ to be equalized over the interval $-B \leq f \leq B$, we first approximate the reciprocal transfer function $1/H_c(f)$ by a complex Fourier series with periodicity $(1/\Delta\tau) = B$. Typically, $H_c(f)$ is specified numerically in terms of its amplitude and phase components, in which case numerical integration is used to compute the complex Fourier coefficients. The total number of significant terms, $2M + 1$, is chosen to be just big enough to produce a satisfactory approximation to the prescribed $H_c(f)$. The tap coefficients of the equalizer, namely, $c_{-M}, \ldots, c_{-1}, c_0, c_1, \ldots, c_M$ are then matched to the complex Fourier coefficients.

EXERCISE 5 Write the formula for evaluating the coefficients of the complex Fourier series used to approximate $1/H_c(f)$ with periodicity $(1/\Delta\tau) = B$.

3.4 IDEAL LOW-PASS FILTERS

As previously mentioned, a *filter* is a frequency-selective device that is used to limit the spectrum of a signal to some specified band of frequencies. Its frequency response is characterized by a *passband* and a *stopband*, which are separated by a *guardband*. The frequencies inside the passband are transmitted with little or no distortion, whereas those in the stopband are rejected. The filter may be of the *low-pass*, *high-pass*, *band-pass*, or *band-stop* type, depending on whether it transmits low, high, intermediate, or all but intermediate frequencies, respectively.

In this section we study the time response of the *ideal low-pass filter*, which transmits, without any distortion all frequencies inside the passband and completely rejects all frequencies inside the stopband, as illustrated in Fig. 3.10. Note that the conditions for distortionless transmission need only be satisfied inside the pass band of the filter. The transfer function of the ideal low-pass filter so illustrated is defined by

$$H(f) = \begin{cases} \exp(-j2\pi f t_0), & -B \leq f \leq B \\ 0, & |f| > B \end{cases} \tag{3.49}$$

where, for convenience, we have set $K = 1$. The parameter B defines the bandwidth of the filter. For a finite t_0, the ideal low-pass filter is noncasual,

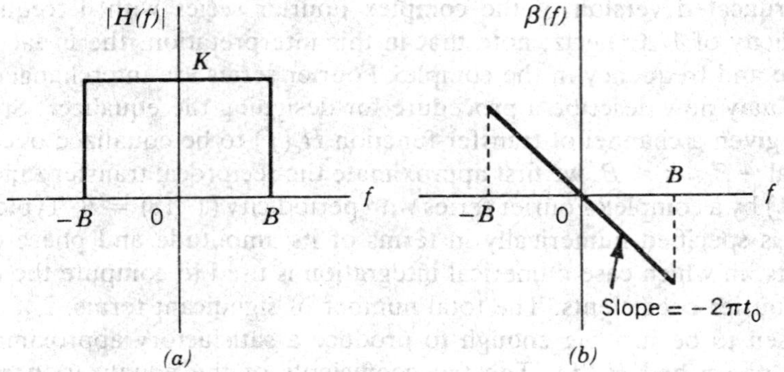

Figure 3.10
Frequency response of ideal low-pass filter. (a) Amplitude response. (b) Phase response.

which may be confirmed by examining the impulse response $h(t)$. Specifically, by evaluating the inverse Fourier transform of the transfer function of Eq. 3.49, we get

$$h(t) = \int_{-B}^{B} \exp[j2\pi f(t - t_0)]\, df \tag{3.50}$$

where the limits of integration have been reduced to the frequency band inside which $H(f)$ does not vanish. Equation 3.50 is readily integrated, yielding

$$\begin{aligned} h(t) &= \frac{\sin[2\pi B(t - t_0)]}{\pi(t - t_0)} \\ &= 2B \operatorname{sinc}[2B(t - t_0)] \end{aligned} \tag{3.51}$$

This impulse response has a peak amplitude of $2B$ centered on time t_0, as shown in Fig. 3.11. The duration of the main lobe of the impulse response is $1/B$, and the build-up time from the zero at the beginning of the main lobe to the peak value is $1/2B$. We see from Fig. 3.11 that, for any finite value of t_0, there is some response from the filter before the time $t = 0$ at which the unit impulse is applied to the input, confirming that the ideal low-pass filter is noncausal. However, despite its noncausality, the ideal low-pass filter serves as a useful standard against which the response of causal filters may be measured.

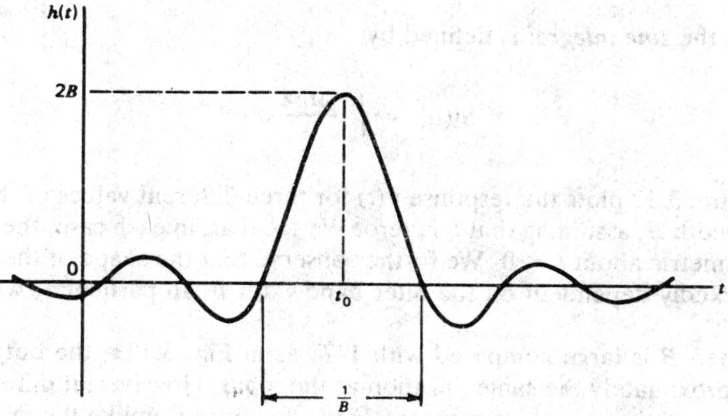

Figure 3.11
Impulse response of ideal low-pass filter.

EXAMPLE 4 PULSE RESPONSE OF IDEAL LOW-PASS FILTER

Consider a rectangular pulse $x(t)$ of unit amplitude and duration T, which is applied to an ideal low-pass filter of bandwidth B. The problem is to determine the response $y(t)$ of the filter.

The impulse response $h(t)$ of the filter is defined by Eq. 3.51. Its response is therefore given by the convolution integral

$$y(t) = \int_{-\infty}^{\infty} x(\tau)h(t-\tau)\,d\tau$$
$$= 2B\int_{-T/2}^{T/2} \frac{\sin[2\pi B(t-t_0-\tau)]}{2\pi B(t-t_0-\tau)}\,d\tau \qquad (3.52)$$

Define

$$\lambda = 2\pi B(t-t_0-\tau)$$

Then, changing the integration variable from τ to λ, we may rewrite Eq. 3.52 as

$$y(t) = \frac{1}{\pi}\int_{2\pi B(t-t_0-T/2)}^{2\pi B(t-t_0+T/2)} \frac{\sin\lambda}{\lambda}\,d\lambda$$
$$= \frac{1}{\pi}\left[\int_0^{2\pi B(t-t_0+T/2)} \frac{\sin\lambda}{\lambda}\,d\lambda - \int_0^{2\pi B(t-t_0-T/2)} \frac{\sin\lambda}{\lambda}\,d\lambda\right]$$
$$= \frac{1}{\pi}\{\mathrm{Si}[2\pi B(t-t_0+T/2)] - \mathrm{Si}[2\pi B(t-t_0-T/2)]\} \qquad (3.53)$$

where the *sine integral* is defined by

$$\mathrm{Si}(u) = \int_0^u \frac{\sin\lambda}{\lambda}\,d\lambda \qquad (3.54)$$

Figure 3.12 plots the response $y(t)$ for three different values of the filter bandwidth B, assuming that t_0 is zero. We see that, in each case, the output is symmetric about $t = 0$. We further observe that the shape of the output is markedly dependent on the filter bandwidth B. In particular, we note:

1. When B is large compared with $1/T$, as in Fig. 3.12*a*, the output has approximately the same duration as the input. However, it differs from the input in two major respects. First, the output, unlike the input, has nonzero rise and fall times that are inversely proportional to the filter bandwidth. Second, the output exhibits *ringing* at both the leading and trailing edges.

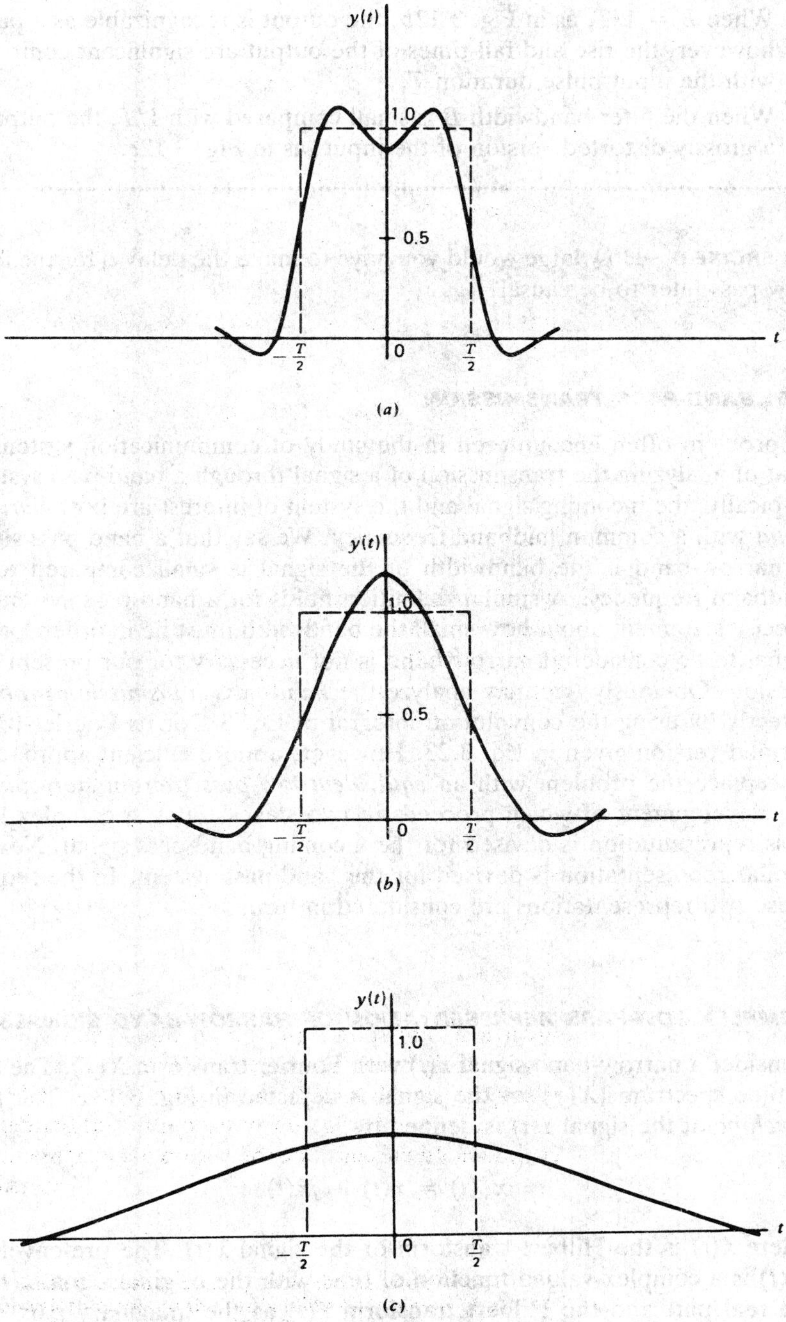

Figure 3.12
Pulse response of ideal low-pass filter for varying filter bandwidth. (a) $B = 2/T$. (b) $B = 1/T$. (c) $B = 1/4T$. The dashed rectangles represent the input signal.

2. When $B = 1/T$, as in Fig. 3.12*b*, the output is recognizable as a pulse; however, the rise and fall times of the output are significant compared with the input pulse duration T.
3. When the filter bandwidth B is small compared with $1/T$, the output is a grossly distorted version of the input, as in Fig. 3.12*c*.

EXERCISE 6 How large would you have to make the delay t_0 for the ideal low-pass filter to be causal?

3.5 BAND-PASS TRANSMISSION

A problem often encountered in the study of communication systems is that of analyzing the transmission of a signal through a band-pass system. Typically, the incoming signal and the system of interest are both *narrow-band* with a common midband frequency. We say that a band-pass signal is narrow-band if the bandwidth of the signal is small compared to its midband frequency. A similar definition holds for a band-pass system. A precise statement about how small the bandwidth must be in order for the signal to be considered narrow-band is not necessary for our present discussion. Obviously, we may analyze the *band-pass transmission problem* directly by using the convolution integral of Eq. 3.1 or its Fourier-transformed version given in Eq. 3.23. However, a more efficient approach is to replace the problem with an *equivalent low-pass transmission model,* the development of which proceeds in two stages. First, a complex low-pass representation is devised for the incoming band-pass signal. Next, a similar representation is devised for the band-pass system. In the sequel, these two representations are considered in turn.

COMPLEX LOW-PASS REPRESENTATION OF NARROW-BAND SIGNALS

Consider a narrow-band signal $x(t)$ with Fourier transform $X(f)$. The amplitude spectrum $|X(f)|$ of the signal is depicted in Fig. 3.13*a*. The *pre-envelope* of the signal $x(t)$ is defined by

$$x_+(t) = x(t) + j\hat{x}(t) \tag{3.55}$$

where $\hat{x}(t)$ is the Hilbert transform of the signal $x(t)$. The pre-envelope $x_+(t)$ is a complex-valued function of time with the original signal $x(t)$ as the real part and the Hilbert transform $\hat{x}(t)$ as the imaginary part. Let $X_+(f)$ denote the Fourier transform of the pre-envelope $x_+(t)$. We may thus write, in the frequency domain,

$$X_+(f) = X(f) + j\hat{X}(f) \tag{3.56}$$

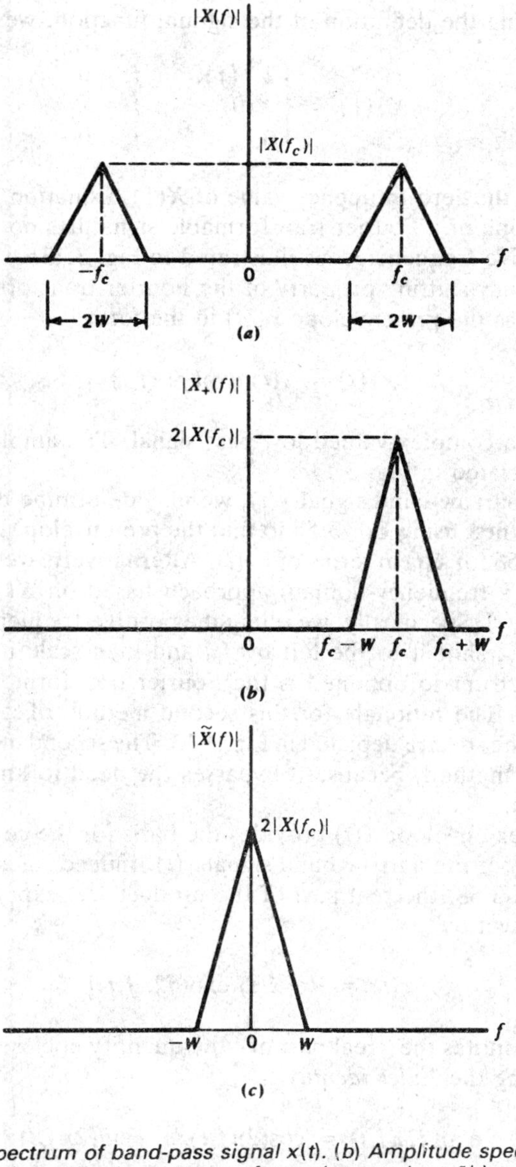

Figure 3.13
(a) Amplitude spectrum of band-pass signal $x(t)$. *(b) Amplitude spectrum of pre-envelope* $x_+(t)$. *(c) Amplitude spectrum of complex envelope* $\tilde{x}(t)$.

where $\hat{X}(f)$ is the Fourier transform of $\hat{x}(t)$. From Example 3, we deduce that $\hat{X}(f)$ equals the product $-j \operatorname{sgn}(f)X(f)$, where $\operatorname{sgn}(f)$ is the signum function. Accordingly, we may rewrite Eq. 3.56 as

$$\begin{aligned} X_+(f) &= X(f) + j[-j \operatorname{sgn}(f)]X(f) \\ &= X(f) + \operatorname{sgn}(f)X(f) \end{aligned}$$

Moreover, using the definition of the signum function, we get

$$X_+(f) = \begin{cases} 2X(f), & f > 0 \\ X(0), & f = 0 \\ 0, & f < 0 \end{cases} \tag{3.57}$$

where $X(0)$ is the zero-frequency value of $X(f)$. Equation 3.57 states that the pre-envelope of a Fourier transformable signal has no frequency content for negative frequencies, as illustrated in Fig. 3.13*b*.

The frequency-shifting property of the Fourier transform suggests that we may express the pre-envelope $x_+(t)$ in the form

$$x_+(t) = \tilde{x}(t) \exp(j2\pi f_c t) \tag{3.58}$$

where $\tilde{x}(t)$ is a complex-valued low-pass signal. The amplitude spectrum of $\tilde{x}(t)$ is illustrated in Fig. 3.13*c*.

Given the narrow-band signal $x(t)$, we may determine the *complex envelope* $\tilde{x}(t)$ by first using Eq. 3.55 to find the pre-envelope $x_+(t)$, and then solving Eq. 3.58 for $\tilde{x}(t)$ in terms of $x_+(t)$. Alternatively, we may determine $\tilde{x}(t)$ by using a frequency-domain approach based on $X(f)$, the Fourier transform of $x(t)$. Specifically, we retain the positive-frequency half of $X(f)$ centered on f_c, shift it to the left by f_c, and then scale it by a factor of two. The spectrum so obtained is the Fourier transform of the complex envelope $\tilde{x}(t)$. The rationale for this second method of determining $\tilde{x}(t)$ follows from the spectra depicted in Fig. 3.13. The second method is usually the preferred method, because it bypasses the need to know the Hilbert transform $\hat{x}(t)$.

The complex envelope $\tilde{x}(t)$ provides the basis for the complex low-pass representation of the narrow-band signal $x(t)$. Indeed, in accordance with Eqs. 3.55 and 3.58, the real part of the product $\tilde{x}(t) \exp(j2\pi f_c t)$ is equal to $x(t)$, as shown by

$$x(t) = \text{Re}[\tilde{x}(t) \exp(j2\pi f_c t)] \tag{3.59}$$

where Re[•] denotes the "real part of" the quantity enclosed in the square brackets. Using the *Euler identity*

$$\exp(j2\pi f_c t) = \cos(2\pi f_c t) + j \sin(2\pi f_c t)$$

and the definition for the complex envelope $\tilde{x}(t)$, we readily find from Eq. 3.59 that $x(t)$ may be expressed as[3]

$$x(t) = x_I(t) \cos(2\pi f_c t) - x_Q(t) \sin(2\pi f_c t) \tag{3.60}$$

[3]Equation 3.60 follows directly from the following rule. Let *a*, *b*, and *c* denote three complex numbers related to one another as

$c = ab$

This is the *canonical representation* for a narrow-band signal in terms of the *in-phase component* $x_I(t)$ and *quadrature component* $x_Q(t)$ of the complex envelope associated with the signal. Indeed, it is a representation basic to all linear modulation schemes; more will be said on this issue in Chapter 7.

The complex envelope $\tilde{x}(t)$ is defined in terms of the in-phase component $x_I(t)$ and the quadrature component $x_Q(t)$ as follows:

$$\tilde{x}(t) = x_I(t) + j\, x_Q(t) \tag{3.61}$$

In other words, $x_I(t)$ is the real part of $\tilde{x}(t)$, and $x_Q(t)$ is its imaginary part.

EXERCISE 7 Consider a narrow-band signal $x(t)$ with Fourier transform $X(f)$. Show that the value of $X_+(f)$, the Fourier transform of the pre-envelope of $x(t)$, at frequency $f = 0$ is $X(0)$.

EXERCISE 8 Let $x(t) = m(t)\cos(2\pi f_c t)$, where $m(t)$ is an information-bearing signal. What are the in-phase and quadrature components of $x(t)$? What is the complex envelope of $x(t)$?

COMPLEX LOW-PASS REPRESENTATION OF NARROW-BAND SYSTEM

Consider next a narrow-band system defined by the impulse response $h(t)$ or, equivalently, the transfer function $H(f)$. To develop a complex low-pass representation for this system, we may perform time-domain operations on $h(t)$ or frequency-domain operations on $H(f)$. From the previous discussion of narrow-band signals, we expect the second approach to be the preferred one, as it is computationally less intensive. Accordingly, from analogy with the complex low-pass representation of a narrow-band signal, we may develop the desired complex low-pass representation of the narrow-band system by retaining the positive-frequency half of the transfer function $H(f)$ centered on f_c, and shifting it to the left by f_c. Let $\tilde{H}(f)$ denote the transfer function of the complex low-pass system so defined. Figure 3.14 illustrates the relationship between $H(f)$ and $\tilde{H}(f)$, shown in parts *a* and *b* of the figure, respectively. Note, however, that in going from $H(f)$ to $\tilde{H}(f)$, we have purposely avoided amplitude scaling (see Exercise 9). Note also that for the frequency-domain transformation depicted in Fig. 3.14 to hold, the midband frequency f_c must be larger than half the bandwidth of the narrow-band system.

Then, the real part of c is given by

$$\text{Re}[c] = \text{Re}[a]\,\text{Re}[b] - \text{Im}[a]\,\text{Im}[b]$$

where Re[•] denotes the "real part of" and Im[•] denotes the "imaginary part of" the respective quantities enclosed in the square brackets.

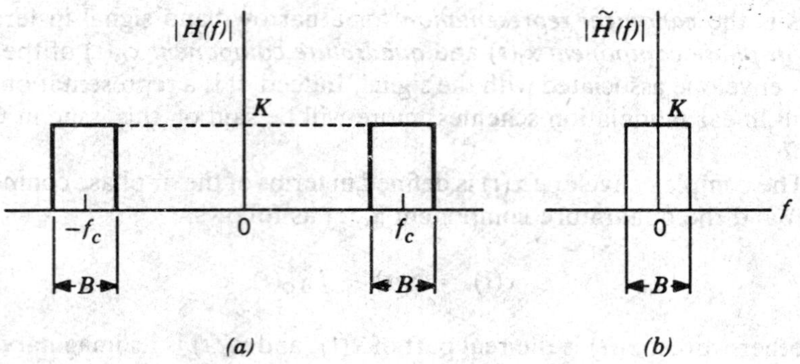

Figure 3.14
(a) Amplitude response of narrow-band system. (b) Amplitude response of complex low-pass system.

EQUIVALENT LOW-PASS TRANSMISSION MODEL

We are now equipped with the tools we need to formulate the equivalent low-pass transmission model for solving the band-pass transmission problem. Specifically, the analysis of a narrow-band system with transfer function $H(f)$ driven by a narrow-band signal with Fourier transform $X(f)$, as depicted in Fig. 3.15*a*, is replaced by an equivalent but simpler analysis of a complex low-pass system with transfer function $\tilde{H}(f)$ driven by a complex low-pass input with Fourier transform $\tilde{X}(f)$, as depicted in Fig. 3.15*b*. This *band-pass to low-pass transformation* completely retains the essence of the filtering process.

According to Fig. 3.15*a*, the Fourier transform of the output of the narrow-band system is given by

$$Y(f) = H(f)X(f)$$

The narrow-band output $y(t)$ itself is given by the inverse Fourier transform of $Y(f)$.

According to Fig. 3.15*b*, the Fourier transform of the output of the complex low-pass system is given by

$$\tilde{Y}(f) = \tilde{H}(f)\tilde{X}(f) \qquad (3.62)$$

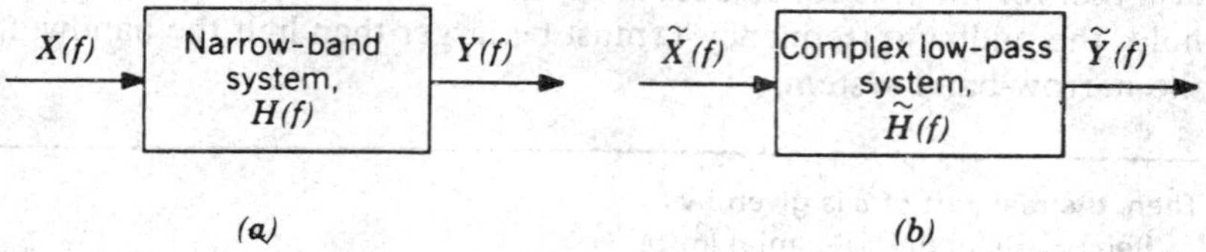

Figure 3.15
Transformation of narrow-band to complex low-pass system.

The complex low-pass output $\tilde{y}(t)$ itself is given by the inverse Fourier transform of $\tilde{Y}(f)$. Having determined $\tilde{y}(t)$, we may find the desired narrow-band output $y(t)$ simply by using the relation:

$$y(t) = \text{Re}[\tilde{y}(t)\exp(j2\pi f_c t)] \tag{3.63}$$

The low-pass transmission model of Fig. 3.15*b* is said to be the *baseband equivalent* of the narrow-band system in Fig. 3.15*a*. The equivalence is in the sense that the model of Fig. 3.15*b* completely *preserves the information content of the incoming narrow-band signal* $x(t)$ *and also that of the outgoing narrow-band signal* $y(t)$. In general, the term "baseband" is used to designate the band of frequencies representing a signal of interest as delivered by a source of information. In the context of our present situation, the term baseband refers to both input and output.

EXERCISE 9 Evaluate $y(0)$ using the two models of Fig. 3.15. Hence, justify the need for scaling the spectrum of the complex low-pass input $\tilde{x}(t)$ by a factor of two, as depicted in Fig. 3.13*c*.

EXAMPLE 5 RESPONSE OF AN IDEAL BAND-PASS FILTER TO A PULSED RF WAVE

Consider an ideal band-pass filter of midband frequency f_c, and bandwidth B as in Fig. 3.16*a*, with $f_c > B/2$. Note that the conditions for distortionless transmission need only be satisfied for the pass band of the filter. Note also that the phase response of the filter is zero at the mid-band frequence f_c. We wish to determine the response of this filter to an RF pulse of duration T and frequency f_c defined by (see Fig. 3.17*a*)

$$x(t) = A\,\text{rect}\left(\frac{t}{T}\right)\cos(2\pi f_c t)$$

where $f_c T \gg 1$.

Retaining the positive-frequency half of the transfer function $H(f)$, defined in Fig. 3.16*a*, and then shifting it to the origin, we find that the transfer function $\tilde{H}(f)$ of the low-pass equivalent filter is given by [see Fig. 3.16*b*]

$$\tilde{H}(f) = \begin{cases} \exp(-j2\pi f t_0), & -B/2 < f < B/2 \\ 0, & |f| > B/2 \end{cases} \tag{3.64}$$

The complex impulse response in this example has only a real component, as shown by

$$\tilde{h}(t) = B\,\text{sinc}[B(t - t_0)] \tag{3.65}$$

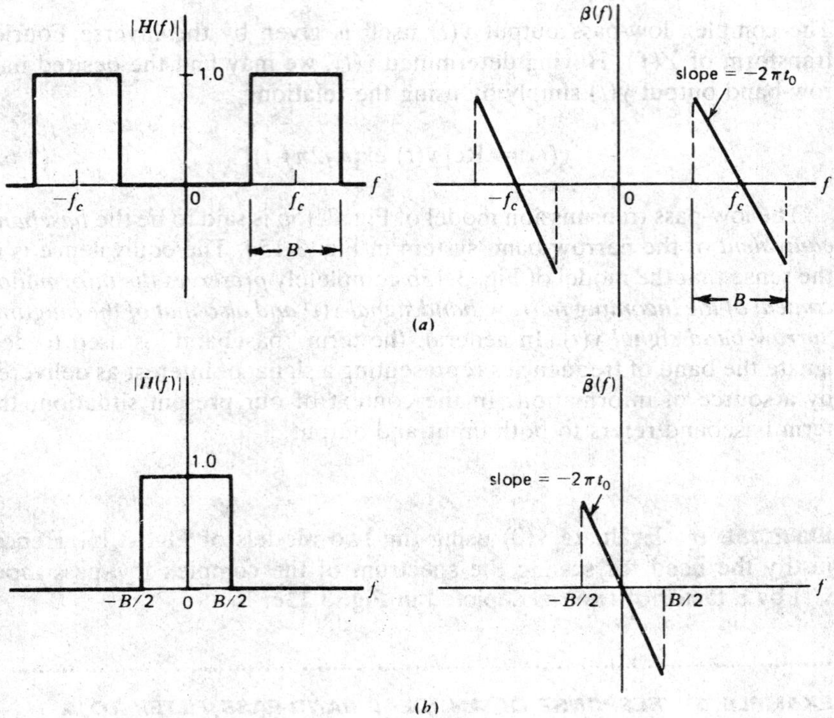

Figure 3.16
(a) Amplitude response $|H(f)|$ and phase response $\beta(f)$ of an ideal band-pass filter.
(b) Corresponding components of complex transfer function $\tilde{H}(f)$.

From Example 3 we recall that the complex envelope $\tilde{x}(t)$ of the input RF pulse also has only a real component, as shown by (see Fig. 3.17*b*):

$$\tilde{x}(t) = A \operatorname{rect}\left(\frac{t}{T}\right) \tag{3.66}$$

The complex envelope $\tilde{y}(t)$ of the filter output is obtained by convolving the $\tilde{h}(t)$ of Eq. 3.65 with the $\tilde{x}(t)$ of Eq. 3.66. This convolution is exactly the same as the low-pass filtering operation that we studied in Example 3. Thus, using Eq. 3.53 we may write

$$\tilde{y}(t) = \frac{A}{\pi}\left\{\operatorname{Si}\left[\pi B\left(t + \frac{T}{2} - t_0\right)\right] - \operatorname{Si}\left[\pi B\left(t - \frac{T}{2} - t_0\right)\right]\right\} \tag{3.67}$$

As expected, the complex envelope $\tilde{y}(t)$ of the output has only a real component. Accordingly, from Eqs. 3.63 and 3.67, the output is obtained

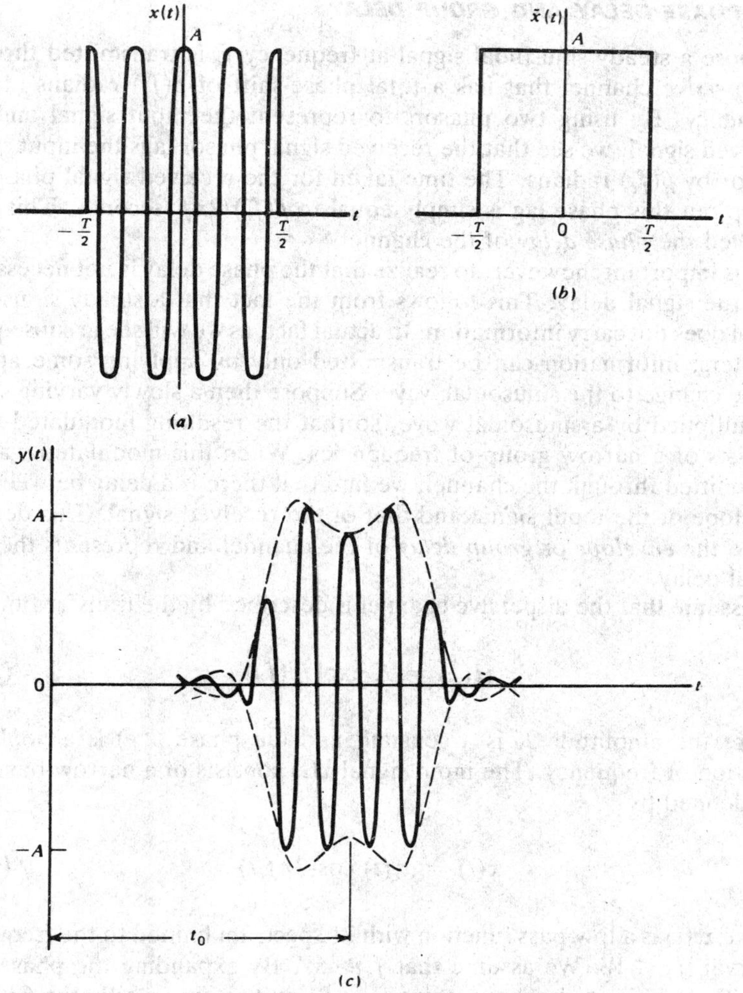

Figure 3.17
The response of an ideal band-pass filter to RF pulse input. (a) RF pulse input x(t). (b) Complex envelope $\tilde{x}(t)$ of RF pulse. (c) Response y(t).

as

$$y(t) = \frac{A}{\pi}\left\{\mathrm{Si}\left[\pi B\left(t + \frac{T}{2} - t_0\right)\right] - \mathrm{Si}\left[\pi B\left(t - \frac{T}{2} - t_0\right)\right]\right\}\cos(2\pi f_c t) \tag{3.68}$$

which is the desired result. Equation 3.68 is shown sketched in Fig. 3.17*c* for the case when the band-pass filter bandwidth $B = 1/T$.

3.6 PHASE DELAY AND GROUP DELAY

Suppose a steady sinusoidal signal at frequency f_c is transmitted through a *dispersive* channel that has a total phase-shift of $\beta(f_c)$ radians at that frequency. By using two phasors to represent the input signal and the received signal, we see that the received signal phasor lags the input signal phasor by $\beta(f_c)$ radians. The time taken for the received signal phasor to sweep out this phase lag is simply equal to $\beta(f_c)/2\pi f_c$ seconds. This time is called the *phase delay* of the channel.

It is important, however, to realize that the phase delay is not necessarily the true signal delay. This follows from the fact that a steady sinusoidal signal does not carry information. In actual fact, as we will see in subsequent chapters, information can be transmitted only by applying some appropriate change to the sinusoidal wave. Suppose then a slowly varying signal is multiplied by a sinusoidal wave, so that the resulting modulated wave consists of a narrow group of frequencies. When this modulated wave is transmitted through the channel, we find that there is a delay between the envelope of the input signal and that of the received signal. This delay is called the *envelope* or *group delay* of the channel and represents the true signal delay.

Assume that the dispersive channel is described by the transfer function

$$H(f) = K \exp[j\beta(f)] \tag{3.69}$$

where the amplitude K is a constant and the phase $\beta(f)$ is a nonlinear function of frequency. The input signal $x(t)$ consists of a narrow-band signal defined by

$$x(t) = x_c(t) \cos(2\pi f_c t) \tag{3.70}$$

where $x_c(t)$ is a low-pass function with its spectrum limited to the frequency interval $|f| \leq W$. We assume that $f_c \gg W$. By expanding the phase $\beta(f)$ in a *Taylor series*[4] about the point $f = f_c$, and retaining only the first two terms, we may approximate $\beta(f)$ as

$$\beta(f) \simeq \beta(f_c) + (f - f_c) \left.\frac{\partial \beta(f)}{\partial f}\right|_{f=f_c} \tag{3.71}$$

Define

$$\tau_p = -\frac{\beta(f_c)}{2\pi f_c} \tag{3.72}$$

[4]For a general definition of the Taylor series, see Appendix D, Table 4.

and

$$\tau_g = -\frac{1}{2\pi}\frac{\partial \beta(f)}{\partial f}\bigg|_{f=f_c} \tag{3.73}$$

Then we may rewrite Eq. 3.71 in the form

$$\beta(f) \simeq -2\pi f_c \tau_p - 2\pi(f - f_c)\tau_g \tag{3.74}$$

Correspondingly, the transfer function of the channel takes the form

$$H(f) \simeq K \exp[-j2\pi f_c \tau_p - j2\pi(f - f_c)\tau_g] \tag{3.75}$$

Following the procedure described in Section 3.5, we may replace the channel described by $H(f)$ by an equivalent low-pass filter with complex transfer function

$$\tilde{H}(f) \simeq K \exp(-j2\pi f_c \tau_p - j2\pi f \tau_g) \tag{3.76}$$

Similarly, we may replace the input narrow-band signal $x(t)$ by its low-pass complex envelope $\tilde{x}(t)$, which is

$$\tilde{x}(t) = x_c(t) \tag{3.77}$$

The Fourier transform of $\tilde{x}(t)$ is simply

$$\tilde{X}(f) = X_c(f) \tag{3.78}$$

where $X_c(f)$ is the Fourier transform of $x_c(t)$. Therefore, the Fourier transform of the complex envelope of the received signal is given by

$$\begin{aligned} \tilde{Y}(f) &= \tilde{H}(f)\tilde{X}(f) \\ &\simeq K \exp(-j2\pi f_c \tau_p) \exp(-j2\pi f \tau_g) X_c(f) \end{aligned} \tag{3.79}$$

We note that the multiplying factor $K \exp(-j2\pi f_c \tau_p)$ is a constant. We also note, from the time-shifting property of the Fourier transform, that the term $\exp(-j2\pi f \tau_g) X_c(f)$ represents the Fourier transform of the delayed signal $x_c(t - \tau_g)$. Accordingly, the complex envelope of the received signal equals

$$\tilde{y}(t) \simeq K \exp(-j2\pi f_c \tau_p) x_c(t - \tau_g) \tag{3.80}$$

Finally, we find that the received signal is itself given by

$$\begin{aligned} y(t) &= \text{Re}[\tilde{y}(t) \exp(j2\pi f_c t)] \\ &= K x_c(t - \tau_g) \cos[2\pi f_c(t - \tau_p)] \end{aligned} \tag{3.81}$$

Equation 3.81 shows that, as a result of transmission through the channel, two delay effects occur:

1. The sinusoidal carrier wave $\cos(2\pi f_c t)$ is delayed by τ_p seconds; hence τ_p represents the phase delay. Sometimes, τ_p is also referred to as the *carrier delay*.
2. The envelope $x_c(t)$ is delayed by τ_g seconds; hence, τ_g represents the envelope or group delay. Note that τ_g is related to the slope of the phase $\beta(f)$, measured at $f = f_c$, as in Eq. 3.73.

Note also that when the phase response $\beta(f)$ is linear with frequency, and $\beta(0) = 0$, the phase delay and group delay assume a common value.

EXERCISE 10 Explain why a linear time-invariant system with a phase response equal to a constant suffers from phase distortion.

3.7 NONLINEAR DISTORTION

Up to this point in our study of signal transmission through a system, we have assumed linearity. In practice, however, we find that the system connecting a source of information to its destination inevitably exhibits some form of *nonlinear* behavior. This occurs whenever the output is increased beyond a limit prescribed by the power that the system is capable of supplying. In such a situation, we say that the system is *overloaded*. When the system is overloaded, a change in the input signal does not produce a corresponding change in the output signal.

Figure 3.18 shows a typical input–output relation, called the *transfer characteristic*, that may give rise to nonlinear distortion. For the purpose of our discussion here, we assume that the system is *memoryless* in the sense that the output $y(t)$ depends only on the input $x(t)$ at time t. We may consider the transfer characteristic of Fig. 3.18 to be composed of the following parts:

1. A reasonably *linear region* centered at the origin, where a change in the input produces a proportional change in the output.
2. Two *saturation regions,* where the output is not affected by the input.
3. Two "knees" that join the linear region to the saturation regions. The useful amplitude range of operation of the system is defined by points P and Q that lie somewhere on the knees of the curve. Their precise locations are determined by the extent of nonlinear distortion that is considered to be tolerable. We may thus view P and Q as *overload points*.

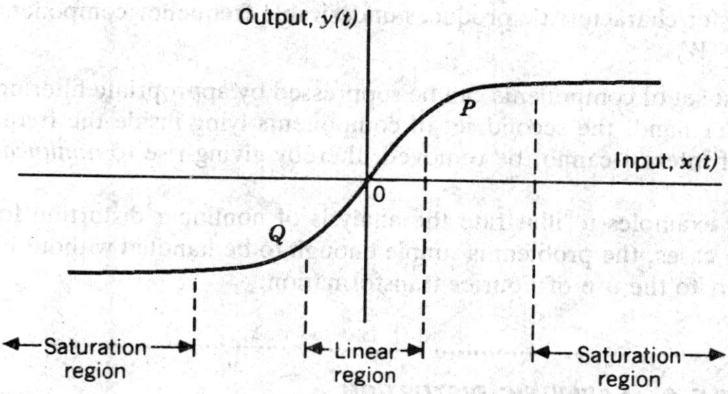

Figure 3.18
Transfer characteristic.

To evaluate the nonlinear distortion, the common procedure is to express the transfer characteristic mathematically by writing the output $y(t)$ as a power series of the input $x(t)$:

$$y(t) = a_1 x(t) + a_2 x^2(t) + a_3 x^3(t) + \cdots \tag{3.82}$$

The first term, $a_1 x(t)$, represents the linear response of the system. The second term, $a_2 x^2(t)$, accounts for a *lack of symmetry* that may exist between the positive and negative parts of the transfer characteristic. (This term would be zero for the symmetric curve shown in Fig. 3.18.) The third term, $a_3 x^3(t)$, provides a first approximation to the flattening of the transfer characteristic due to overloading. Higher order terms on the right side of Eq. 3.82 are usually neglected when operation of the system is bounded by the overload points (P and Q in Fig. 3.18).

Let $X(f)$ denote the Fourier transform of the input $x(t)$. Then, the Fourier transform of the output $y(t)$ is

$$\begin{aligned} Y(f) &= a_1 X(f) + a_2 X(f) \star X(f) \\ &\quad + a_3 X(f) \star X(f) \star X(f) + \cdots \end{aligned} \tag{3.83}$$

where $\star$ denotes convolution. Thus, $X(f) \star X(f)$ denotes the convolution of $X(f)$ with itself, and so on. Let $x(t)$ be band-limited in W, such that $X(f) = 0$ for $|f| \geq W$. Then, $x^2(t)$ is band-limited in $2W$, such that $X(f) \star X(f) = 0$ for $|f| \geq 2W$. Similarly, $x^3(t)$ is band-limited in $3W$, such that $X(f) \star X(f) \star X(f) = 0$ for $|f| \geq 3W$, and so on. We may therefore make two observations:

1. The output of a nonlinear system contains new frequency components for $f > W$, which are not present in the input.
2. The presence of nonlinearities (second order, third order, etc.) in the

transfer characteristic produces undesirable frequency components for $|f| \leq W$.

The first set of components can be suppressed by appropriate filtering. On the other hand, the second set of components lying inside the frequency band of interest cannot be removed, thereby giving rise to *nonlinear distortion.*

Two examples to illustrate the analysis of nonlinear distortion follow. In both cases, the problem is simple enough to be handled without having to resort to the use of Fourier transformation.

EXAMPLE 6 HARMONIC DISTORTION

Let the input consist of a single sinusoidal wave:

$$x(t) = A\cos(2\pi f t) \tag{3.84}$$

We assume that only second- and third-order nonlinearities in the transfer characteristic of Fig. 3.18 are of concern, so that fourth- and higher-order terms in Eq. 3.82 may be ignored. Then, substitution of Eq. 3.84 in 3.82 yields the output

$$y(t) = \frac{1}{2}a_2A^2 + \left(a_1A + \frac{3}{4}a_3A^3\right)\cos(2\pi f t) + \frac{1}{2}a_2A^2\cos(4\pi f t) + \frac{1}{4}a_3A^3\cos(6\pi f t) \tag{3.85}$$

Since we are concerned primarily with distortion (i.e., changes in the shape of the waveform), we may ignore the dc component, $\frac{1}{2}a_2A^2$. The components of interest in the output waveform are therefore as follows, with their respective amplitudes shown:

Fundamental:	$a_1A + \frac{3}{4}a_3A^3$
Second harmonic:	$\frac{1}{2}a_2A^2$
Third harmonic:	$\frac{1}{4}a_3A^3$

Accordingly, we define the *second-harmonic distortion,* D_2, as the ratio of the amplitude of the second-harmonic component in the output to that of the fundamental:

$$D_2 = \frac{\frac{1}{2}a_2A}{a_1 + \frac{3}{4}a_3A^2} \tag{3.86}$$

Similarly, we define the *third-harmonic distortion*, D_3, as the ratio of the amplitude of the third-harmonic component in the output to that of the fundamental:

$$D_3 = \frac{\frac{1}{4} a_3 A^2}{a_1 + \frac{3}{4} a_3 A^2} \tag{3.87}$$

The harmonic distortion factors D_2 and D_3 are usually expressed as percentages.

EXAMPLE 7 INTERMODULATION DISTORTION

Let the input $x(t)$ consist of the sum of two sinusoidal waves:

$$x(t) = A_1 \cos(2\pi f_1 t) + A_2 \cos(2\pi f_2 t) \tag{3.88}$$

Here again we assume that fourth- and higher-order terms in Eq. 3.82 may be ignored. Then, substituting this expression for $x(t)$ in Eq. 3.82, we find that the effects produced by the second- and third-order nonlinearities in the transfer characteristic are:

1. The second-order term, $a_2 x^2(t)$, produces a dc and a second-harmonic component corresponding to the single-frequency input, as expected. In addition, however, it produces new components at $f_1 + f_2$ and $f_1 - f_2$ that are the *sum* and *difference frequencies,* respectively. Such components are referred to as *second-order intermodulation products*.

TABLE 3.1

Type of Intermodulation Product	Frequency	Amplitude
Second-order	$f_1 + f_2$	$a_2 A_1 A_2$
	$f_1 - f_2$	$a_2 A_1 A_2$
Third-order	$2f_1 + f_2$	$\frac{3}{4} a_3 A_1^2 A_2$
	$2f_1 - f_2$	$\frac{3}{4} a_3 A_1^2 A_2$
	$2f_2 + f_1$	$\frac{3}{4} a_3 A_1 A_2^2$
	$2f_2 - f_1$	$\frac{3}{4} a_3 A_1 A_2^2$

2. The third-order term, $a_3x^3(t)$, produces the expected fundamental and third-harmonic components. In addition, it gives rise to intermodulation products of its own at the frequencies $2f_1 \pm f_2$ and $2f_2 \pm f_1$, which are referred to as *third-order* intermodulation products.

Table 3.1 presents a summary of the frequencies and amplitudes of the various second- and third-order intermodulation products.

PROBLEMS

P3.1 Time Response

Problem 1 The excitation applied to a linear time-invariant system with impulse response $h(t)$ consists of two delta functions, as shown by

$$x(t) = \delta(t + t_0) + \delta(t - t_0)$$

where t_0 is a constant time shift. Find the response of the system.

Problem 2 The impulse response of a linear time-invariant system is defined by

$$h(t) = \exp(-at)u(t)$$

where $u(t)$ is the unit step function. Determine the response of the system produced by an excitation consisting of the unit step function $u(t)$.

Problem 3 A periodic signal $x_p(t)$ of period T_0 is applied to a linear time-invariant system of impulse response $h(t)$. Use the complex Fourier series representation of $x_p(t)$ and the convolution integral to evaluate the response of the system.

Problem 4 The impulse response of a linear time-invariant system is defined by the Gaussian function:

$$h(t) = \exp[-\pi(t - t_0)^2]$$

where t_0 is a constant.

(a) Is this system causal?
(b) Is it stable?

Give reasons for your answers.

P3.2 Frequency Response

Problem 5 Continuing with the linear time-invariant system described in Problem 2, do the following:

(a) Determine the transfer function of the system.

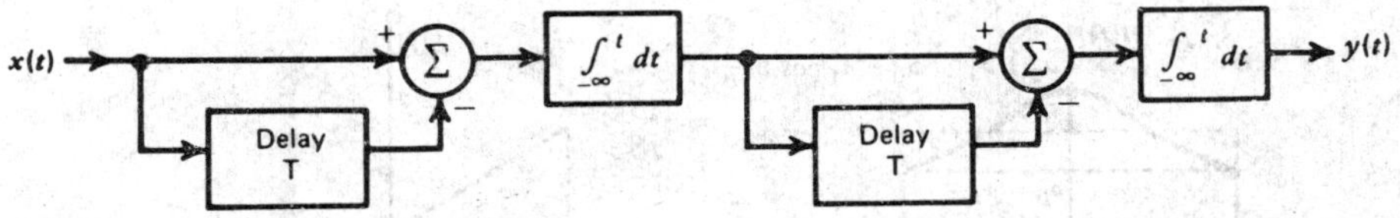

Figure P3.1

(b) Plot the amplitude response and phase response of the system.
(c) Find the 3-dB bandwidth of the system.

Problem 6 Find the transfer function of the linear time-invariant system with its impulse response defined in Problem 4. Hence, plot the amplitude response and phase response of the system. Indicate the 3-dB bandwidth of the system on the plot of the amplitude response.

Problem 7 Evaluate the transfer function of a linear system represented by the block diagram shown in Fig. P3.1.

Problem 8

(a) Determine the overall amplitude response of the cascade connection shown in Fig. P3.2 consisting of N identical stages, each with a time constant RC equal to τ_0.
(b) Show that as N approaches infinity, the amplitude response of the cascade connection approaches the Gaussian function $\exp(-\frac{1}{2}f^2T^2)$, where for each value of N, the time constant τ_0 is selected so that

$$\tau_0^2 = \frac{T^2}{4\pi^2 N}$$

P3.3 Linear Distortion and Equalization

Problem 9

(a) Consider a signal $x(t)$ with Fourier transform $X(f)$ limited to the band $-B \leqslant f \leqslant B$. This signal is applied to a linear time-invariant system with an amplitude response $|H(f)|$ and linear phase, as in Fig. P3.3a. Determine the resulting output of the system.

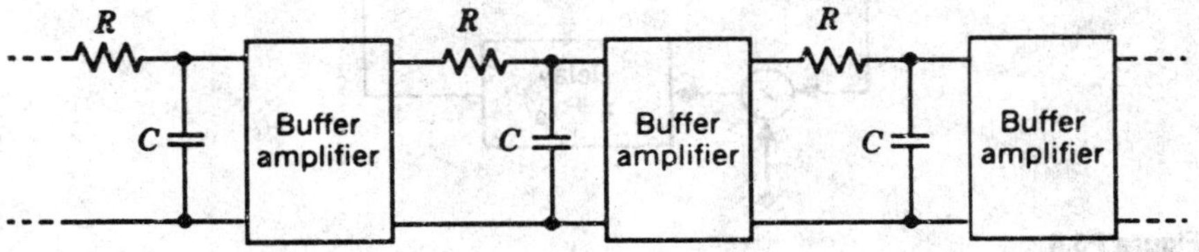

Figure P3.2

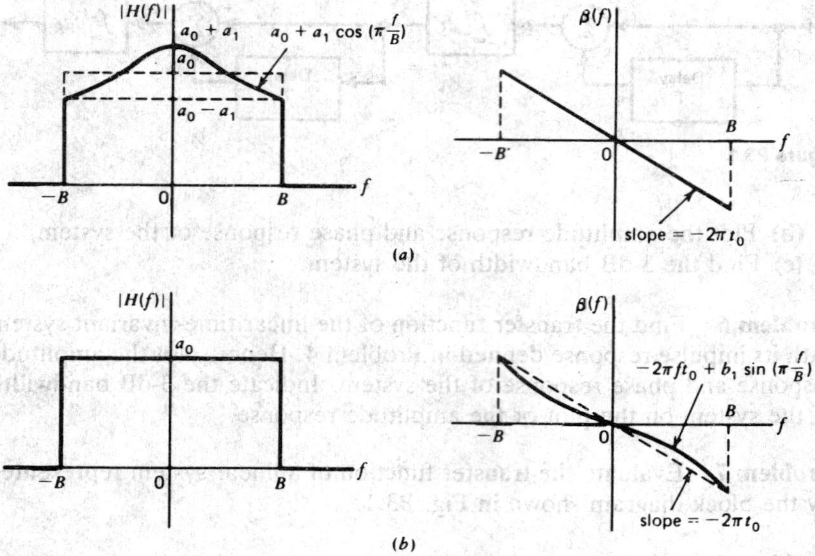

Figure P3.3

(b) Suppose that the system has a constant amplitude response but nonlinear phase, as in Fig. P3.3*b*. Determine the resulting output. Assume that the constant b_1 is small enough to justify using the approximation:

$$\exp\left[jb_1 \sin\left(\frac{\pi f}{B}\right)\right] \simeq 1 + jb_1 \sin\left(\frac{\pi f}{B}\right)$$

Problem 10 Figure P3.4 shows an idealized model of a radio channel. It consists of two paths. One path introduces a propagation delay t_0. The

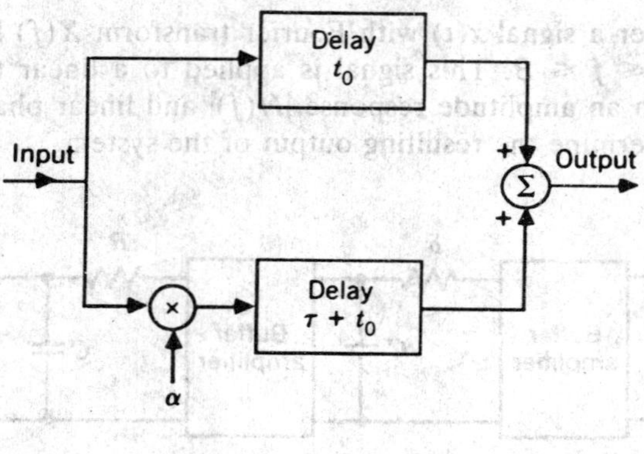

Figure P3.4

other path introduces an additional delay τ and an attenuation represented by the scaling factor α that is less than one. A channel so characterized is referred to as a *multipath channel*. To correct for the multipath distortion, a three-tap equalizer is connected in cascade with the channel. Given that $\alpha^2 \ll 1$, calculate the three tap coefficients of the tapped-delay-line equalizer.
Hint: Use the binomial expansion:

$$\frac{1}{1 + \alpha \exp(-j2\pi f\tau)} = 1 - \alpha \exp(-j2\pi f\tau) + \alpha^2 \exp(-j4\tau f\tau) - \cdots$$

P3.4 Ideal Low-Pass Filters

Problem 11 The transfer function of an ideal low-pass filter is defined by

$$H(f) = \begin{cases} K \exp(-j2\pi f t_0), & |f| < 1 \\ 0, & |f| > 1 \end{cases}$$

where t_0 is a constant. Find the impulse response of the system.

Problem 12 An ideal low-pass filter has zero time delay and bandwidth B. It is driven by a rectangular pulse of unit amplitude and duration T equal to $1/B$ and centered at $t = 0$.

(a) Show that the filter output at $t = 0$ is given by

$$y(0) = \frac{2}{\pi} \mathrm{Si}(\pi)$$

where $\mathrm{Si}(\pi)$ is the value of the sine integral for an argument equal to π.
(b) Show that the filter output at $t = T/2$ is given by

$$y\left(\frac{T}{2}\right) = \frac{1}{\pi} \mathrm{Si}(2\pi)$$

where $\mathrm{Si}(2\pi)$ is the value of the sine integral for an argument of 2π.
(c) Calculate these two values of the filter output and check them against the corresponding pulse response shown in Fig. 3.10*b*.

Note that $\mathrm{Si}(\pi) = 1.85$ and $\mathrm{Si}(2\pi) = 1.42$.

Problem 13 Suppose that, for a given signal $x(t)$, the integrated value of the signal over an interval T is required, as shown by

$$y(t) = \int_{t-T}^{t} x(\tau)\, d\tau$$

(a) Show that $y(t)$ can be obtained by transmitting the input signal $x(t)$ through a filter with its transfer function given by

$$H(f) = T \operatorname{sinc}(fT) \exp(-j\pi fT)$$

(b) An adequate approximation to this transfer function is obtained by using a low-pass filter with a bandwidth equal to $1/T$, passband amplitude response T, and delay $T/2$. Assuming this low-pass filter to be ideal, determine the filter output at time $t = T$ due to a unit step function applied to the filter and compare the result with the corresponding output of the ideal integrator.

Note that $\mathrm{Si}(\pi) = 1.85$ and $\mathrm{Si}(\infty) = \pi/2$.

P3.5 Band-Pass Transmission

Problem 14 An ideal band-pass filter has zero time delay and bandwidth B. An *RF* pulse of unit amplitude, duration $T = 1/2B$, and frequency f_c is applied to the filter; the pulse is centered at $t = 0$. Show that the filter output is given by

$$y(t) = \frac{1}{\pi} [\mathrm{Si}(2\pi Bt + \pi) - \mathrm{Si}(2\pi Bt - \pi)] \cos(2\pi f_c t)$$

where Si(•) is the sine integral. Sketch the waveform of $y(t)$.

Problem 15 Consider an ideal band-pass filter with center frequency f_c and bandwidth B, as defined in Fig. P3.5. The carrier wave $A \cos(2\pi f_0 t)$ is suddenly applied to this filter at time $t = 0$. Assuming that $|f_c - f_0|$ is large compared to the bandwidth B, determine the response of the filter.

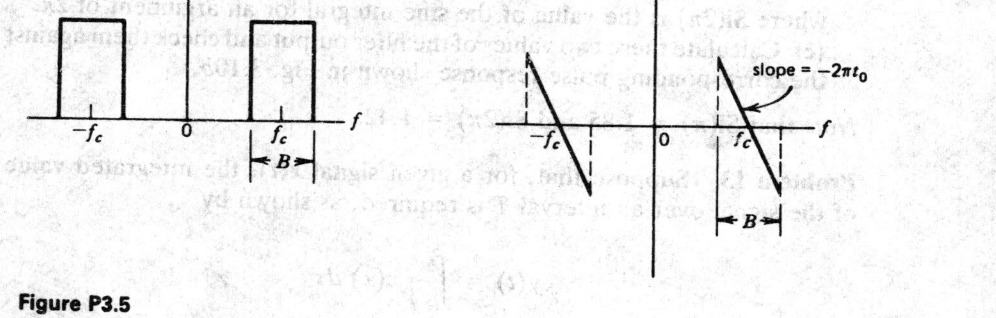

Figure P3.5

P3.6 Phase Delay and Group Delay

Problem 16 The impulse response of a linear time-invariant system is defined by

$$h(t) = \begin{cases} \exp(-t), & t > 0 \\ \dfrac{1}{2}, & t = 0 \\ 0, & t < 0 \end{cases}$$

(a) Determine the phase delay $\tau_p(f)$ and group delay $\tau_g(f)$ of the system.
(b) Plot both delays versus frequency f, and comment on your results.

P3.7 Nonlinear Distortion

Problem 17 Verify the frequencies and amplitudes of the intermodulation products listed in Table 3.1 for an input consisting of the sum of two sinusoidal waves of frequencies f_1 and f_2 and amplitudes A_1 and A_2.

CHAPTER FOUR

SPECTRAL DENSITY AND CORRELATION

In this Chapter we complete the characterization of signals and systems by focusing on the *energy* or *power* of a signal. In so doing, we introduce the notion of *spectral density,* which defines the distribution of energy or power per unit bandwidth as a function of frequency. When dealing with energy signals, it is natural to use *energy spectral density* as the parameter of interest. Likewise, when dealing with power signals, *power spectral density* is used to characterize the signal. In this chapter we also introduce another important parameter called *correlation,* which may be viewed as the time-domain counterpart of spectral density. Throughout the chapter, we deal with real-valued energy and power signals. We begin the discussion with energy spectral density.

4.1 ENERGY SPECTRAL DENSITY

Consider an *energy signal* $g(t)$ defined over the interval $-\infty < t < \infty$, and let its Fourier transform or spectrum be denoted by $G(f)$. The signal $g(t)$ is assumed to be *real valued*. The total *energy* of the signal is defined by (see Section 1.2)

$$E = \int_{-\infty}^{\infty} g^2(t)\, dt \tag{4.1}$$

Equation 4.1 is the standard formula for evaluating the energy E. Nevertheless, there is another method based on the amplitude spectrum $|G(f)|$, which may also be used to evaluate the energy E. To develop this alternative method, we start with the relation (see Exercise 10, Chapter 2):

$$\int_{-\infty}^{\infty} g_1(t)g_2(t)\, dt = \int_{-\infty}^{\infty} G_1(f)G_2(-f)\, df \tag{4.2}$$

where $g_1(t)$ and $g_2(t)$ are a pair of energy signals with Fourier transforms $G_1(f)$ and $G_2(f)$, respectively. Let

$$g_1(t) = g_2(t) = g(t)$$

Correspondingly, we may set

$$G_1(f) = G(f)$$

and for real-valued signals,

$$G_2(-f) = G^*(f)$$

Accordingly, we may simplify Eq. 4.2 as

$$\int_{-\infty}^{\infty} g^2(t)\, dt = \int_{-\infty}^{\infty} |G(f)|^2\, df \tag{4.3}$$

where $|G(f)|$ is the amplitude spectrum of the signal $g(t)$. Equation 4.3 is known as the *Rayleigh energy theorem.*

The Rayleigh energy theorem is important not only because it provides a useful method for evaluating energy, but also because it highlights $|G(f)|^2$ as the distribution of energy of the signal $g(t)$ in the frequency domain. It is for this reason that the squared amplitude spectrum $|G(f)|^2$ is called the *energy spectral density* or *energy density spectrum*. Using $\Psi_g(f)$ to denote this new parameter, we may thus write

$$\Psi_g(f) = |G(f)|^2 \tag{4.4}$$

EXAMPLE 1 SINC PULSE

Consider the sinc pulse defined by

$$g(t) = A \operatorname{sinc}(2Wt)$$

The energy of this pulse equals

$$E = A^2 \int_{-\infty}^{\infty} \operatorname{sinc}^2(2Wt)\, dt \tag{4.5}$$

The integral on the right side of Eq. 4.5 is difficult to evaluate. We may obtain the desired result indirectly by applying the Rayleigh energy theorem. We start with the Fourier transform pair (see Example 6, Chapter 2)

$$A \operatorname{sinc}(2Wt) \rightleftharpoons \frac{A}{2W} \operatorname{rect}\left(\frac{f}{2W}\right)$$

Hence, with the Fourier transform

$$G(f) = \frac{A}{2W} \operatorname{rect}\left(\frac{f}{2W}\right)$$

and $\operatorname{rect}^2(f/2W) = \operatorname{rect}(f/2W)$, the energy spectral density of the sinc pulse is given by

$$\Psi_g(f) = \left(\frac{A}{2W}\right)^2 \operatorname{rect}\left(\frac{f}{2W}\right) \tag{4.6}$$

Hence, the application of the Rayleigh energy theorem yields the result

$$E = \left(\frac{A}{2W}\right)^2 \int_{-\infty}^{\infty} \operatorname{rect}\left(\frac{f}{2W}\right) df$$

$$= \left(\frac{A}{2W}\right)^2 \int_{-W}^{W} df$$

$$= \frac{A^2}{2W} \tag{4.7}$$

EXERCISE 1 Show that the total area under the curve of $\text{sinc}^2(t)$ equals 1; that is,

$$\int_{-\infty}^{\infty} \text{sinc}^2(t)\, dt = 1 \tag{4.8}$$

PROPERTIES OF ENERGY SPECTRAL DENSITY

The energy spectral density $\Psi_g(f)$ has several properties that follow from the basic definition given in Eq. 4.4, which are formally described in the sequel.

PROPERTY 1

The energy spectral density of an energy signal $g(t)$ is a nonnegative real-valued function of frequency; that is,

$$\Psi_g(f) \geq 0, \qquad \text{for all } f \tag{4.9}$$

This property follows directly from the fact that the amplitude spectrum $|G(f)|$ of a signal $g(t)$ is a nonnegative real function of the frequency f.

PROPERTY 2

The energy spectral density of a real-valued energy signal $g(t)$ is an even function of frequency, that is,

$$\Psi_g(-f) = \Psi_g(f) \tag{4.10}$$

This property means that the energy spectral density of a real-valued signal is symmetric about zero frequency. It follows directly from the fact that the amplitude spectrum $|G(f)|$ of a real-valued signal $g(t)$ is an even function of the frequency f, as shown by

$$|G(-f)| = |G(f)|$$

PROPERTY 3

The total area under the curve of energy spectral density of an energy signal $g(t)$ equals the signal energy; that is,

$$E = \int_{-\infty}^{\infty} \Psi_g(f)\, df \tag{4.11}$$

Suppose that $g(t)$ denotes the voltage of a source connected across a 1-ohm load resistor. Then, the integral

$$\int_{-\infty}^{\infty} g^2(t)\, dt$$

equals the energy E delivered by the source to the load. From Rayleigh's energy theorem described by Eq. 4.3, the total area under the $\Psi_g(f)$ curve equals the energy E.

EXERCISE 2 Using the energy spectral density for the sinc pulse given by Eq. 4.6, derived in Example 1, demonstrate the validity of Properties 1 through 3.

PROPERTY 4

When an energy signal is transmitted through a linear time-invariant system, the energy spectral density of the output equals the energy spectral density of the input multiplied by the squared amplitude response of the system.

This property follows from the frequency-domain description of a linear time-invariant system given in Eq. 3.23. Specifically, with $X(f)$ denoting the Fourier transform of a signal $x(t)$ applied to the input of a linear time-invariant system of transfer function $H(f)$, the Fourier transform $Y(f)$ of the signal $y(t)$ produced at the output of the system is given by

$$Y(f) = H(f)X(f) \tag{4.12}$$

Taking the squared amplitude of both sides of this equation, we get

$$|Y(f)|^2 = |H(f)|^2\,|X(f)|^2 \tag{4.13}$$

Equivalently, we may write

$$\Psi_y(f) = |H(f)|^2\,\Psi_x(f) \tag{4.14}$$

where $\Psi_y(f) = |Y(f)|^2$ and $\Psi_x(f) = |X(f)|^2$. The quantities $\Psi_y(f)$ and $\Psi_x(f)$ denote the energy spectral densities of the output $y(t)$ and the input $x(t)$, respectively. Equation 4.14 is a mathematical statement of Property 4.

EXAMPLE 2

A rectangular pulse of unit amplitude and unit duration is passed through an ideal low-pass filter of bandwidth B, as illustrated in Fig. 4.1a. Part b of the figure depicts the waveform of the rectangular pulse. The amplitude response of the filter is defined by (see Fig. 4.1c)

$$|H(f)| = \begin{cases} 1, & -B \leq f \leq B \\ 0, & \text{otherwise} \end{cases}$$

The rectangular pulse constituting the filter input has unit energy. We wish to evaluate the effect of varying the bandwidth B on the energy of the filter output.

We start with the Fourier transform pair:

$$\text{rect}(t) \rightleftharpoons \text{sinc}(f)$$

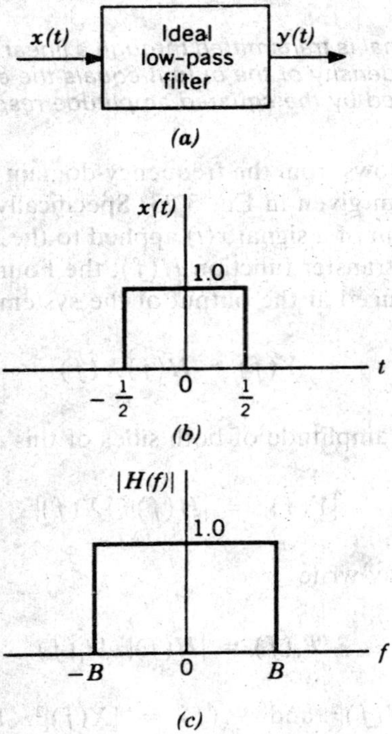

Figure 4.1
(a) *Ideal low-pass filter.* (b) *Filter input.* (c) *Amplitude response of the filter.*

This represents the *normalized* version of the Fourier transform pair given in Eq. 2.33. Hence, with the filter input defined by

$$x(t) = \text{rect}(t)$$

its Fourier transform equals

$$X(f) = \text{sinc}(f)$$

The energy spectral density of the filter input therefore equals

$$\begin{aligned}\Psi_x(f) &= |X(f)|^2 \\ &= \text{sinc}^2(f)\end{aligned} \tag{4.15}$$

This normalized energy spectral density is shown plotted in Fig. 4.2.

To evaluate the energy spectral density $\Psi_y(f)$ of the filter output $y(t)$, we use Eq. 4.14. We thus obtain

$$\begin{aligned}\Psi_y(f) &= |H(f)|^2\, \Psi_x(f) \\ &= \begin{cases}\Psi_x(f), & -B \le f \le B \\ 0, & \text{otherwise}\end{cases}\end{aligned} \tag{4.16}$$

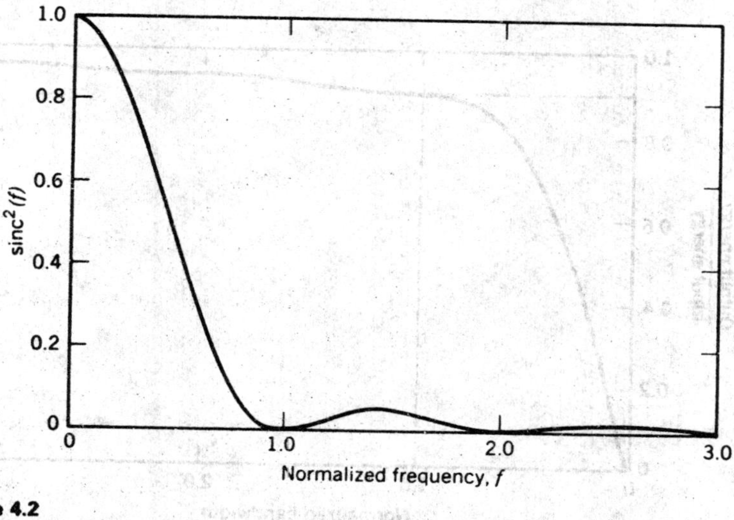

Figure 4.2
Normalized energy spectral density of filter input x(t).

The energy of the filter output therefore equals

$$\begin{aligned} E_y &= \int_{-\infty}^{\infty} \Psi_y(f)\, df \\ &= \int_{-B}^{B} \Psi_x(f)\, df \\ &= 2\int_{0}^{B} \Psi_x(f)\, df \end{aligned} \tag{4.17}$$

Substituting Eq. 4.15 in 4.17 yields

$$E_y = 2\int_0^B \operatorname{sinc}^2(f)\, df \tag{4.18}$$

Since the filter input is normalized to have unit energy, we may also view the result given in Eq. 4.18 as *the ratio of the energy of the filter ouput to that of the filter input* for the general case of a rectangular pulse of arbitrary amplitude and arbitrary duration, processed by an ideal band-pass filter of bandwidth B. Accordingly, we may also write

$$\begin{aligned} \rho &= \frac{\text{Energy of filter output}}{\text{Energy of filter input}} \\ &= 2\int_0^B \operatorname{sinc}^2(f)\, df \end{aligned} \tag{4.19}$$

According to Fig. 4.1*b*, the rectangular pulse applied to the filter input has

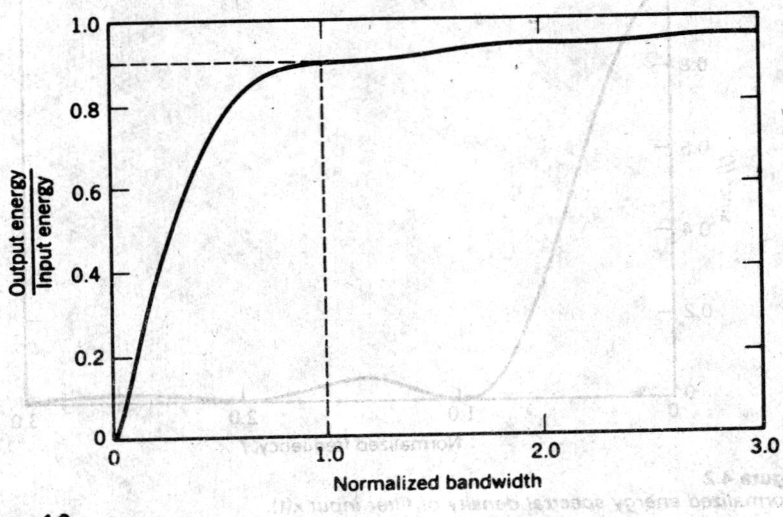

Figure 4.3
Output energy-to-input energy ratio versus normalized bandwidth.

unit duration; hence, the variable f in Eq. 4.19 represents a *normalized frequency*. Equation 4.19 is plotted in Fig. 4.3.

The graph of Fig. 4.3 shows that just over 90% of the total energy of a rectangular pulse lies inside the main spectral lobe of this pulse.

INTERPRETATION OF THE ENERGY SPECTRAL DENSITY

Equation 4.14 is important because it not only relates the output energy spectral density of a linear time-invariant system to the input energy spectral density but it also provides a basis for the physical interpretation of the concept of energy spectral density itself. To be specific, consider the arrangement shown in Fig. 4.4*a*, where an energy signal $x(t)$ is passed

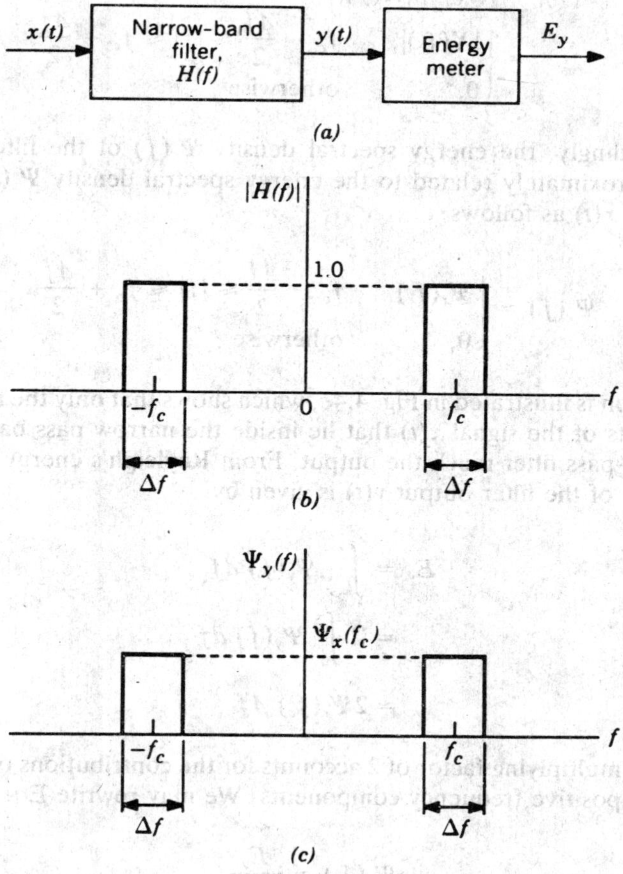

Figure 4.4
(a) Circuit arrangement for measuring energy spectral density. (b) Idealized amplitude response of the filter. (c) Energy spectral density of filter output.

through a narrow-band filter followed by an *energy meter.* Figure 4.4*b* shows the idealized amplitude response of the filter. That is, the amplitude response of the filter is defined by

$$|H(f)| = \begin{cases} 1, & f_c - \dfrac{\Delta f}{2} \leq |f| \leq f_c + \dfrac{\Delta f}{2} \\ 0, & \text{otherwise} \end{cases} \tag{4.20}$$

We assume that the filter bandwidth Δf is small enough for the amplitude response of the input signal $x(t)$ to be essentially constant over the frequency interval covered by the passband of the filter. Accordingly, we may express the amplitude spectrum of the filter output by the approximate formula:

$$\begin{aligned} |Y(f)| &= |H(f)|\,|X(f)| \\ &\simeq \begin{cases} |X(f_c)|, & f_c - \dfrac{\Delta f}{2} \leq |f| \leq f_c + \dfrac{\Delta f}{2} \\ 0, & \text{otherwise} \end{cases} \end{aligned} \tag{4.21}$$

Correspondingly, the energy spectral density $\Psi_y(f)$ of the filter output $y(t)$ is approximately related to the energy spectral density $\Psi_x(f)$ of the filter input $x(t)$ as follows

$$\Psi_y(f) \simeq \begin{cases} \Psi_x(f_c) & f_c - \dfrac{\Delta f}{2} \leq |f| \leq f_c + \dfrac{\Delta f}{2} \\ 0, & \text{otherwise} \end{cases} \tag{4.22}$$

This relation is illustrated in Fig. 4.4*c*, which shows that only the frequency components of the signal $x(t)$ that lie inside the narrow pass band of the ideal band-pass filter reach the output. From Rayleigh's energy theorem, the energy of the filter output $y(t)$ is given by

$$\begin{aligned} E_y &= \int_{-\infty}^{\infty} \Psi_y(f)\,df \\ &= 2\int_0^{\infty} \Psi_y(f)\,df \\ &\simeq 2\Psi_x(f_c)\,\Delta f \end{aligned} \tag{4.23}$$

where the multiplying factor of 2 accounts for the contributions of negative as well as positive frequency components. We may rewrite Eq. 4.23 as

$$\Psi_x(f_c) \simeq \frac{E_y}{2\,\Delta f} \tag{4.24}$$

Equation 4.24 states that the energy spectral density of the filter input at some frequency f_c equals the energy of the filter output divided by $2\,\Delta f$, where Δf is the filter bandwidth centered on f_c. We may therefore interpret the energy spectral density of an energy signal for any frequency f *as the energy per unit bandwidth, which is contributed by frequency components of the signal around the frequency* f.

The arrangement shown in the block diagram of Fig. 4.4*a* provides the basis for measuring the energy spectral density of an energy signal. Specifically, by using a *variable* band-pass filter to scan the frequency band of interest, and determining the energy of the filter output for each midband frequency setting of the filter, a plot of the energy spectral density versus frequency is obtained.

4.2 CORRELATION OF ENERGY SIGNALS

The energy spectral density is an important frequency-dependent parameter of an energy signal. With the interplay between time-domain and frequency-domain descriptions of a signal that we have become accustomed to, it is natural for us to seek the time-domain counterpart of energy spectral density. From the defining equation (4.4), we have

$$\Psi_g(f) = G(f)G^*(f) \tag{4.25}$$

where the signal $g(t)$ and its Fourier transform $G(f)$ are related by

$$g(t) \rightleftharpoons G(f)$$

Equation 4.25 states that the energy spectral density $\Psi_g(f)$ of an energy signal $g(t)$ equals the product of $G(f)$, the Fourier transform of $g(t)$, and its complex conjugate, $G^*(f)$. Hence, given $G(f)$, we need to perform two frequency-domain operations to get $\Psi_g(f)$, namely, complex conjugation and multiplication. This suggests that we may determine the inverse Fourier transform of $\Psi_g(f)$ by making use of two fundamental properties of the Fourier transform (see Section 2.3):

1. The complex conjugation property, according to which time reversal of a real-valued signal translates to complex conjugation of its Fourier transform.
2. The time-domain convolution property, according to which the convolution of two signals translates to the multiplication of their Fourier transforms.

Accordingly, we may formulate the following Fourier transform pair:

$$g(\tau) \star g(-\tau) \rightleftharpoons G(f)\,G^*(f) \tag{4.26}$$

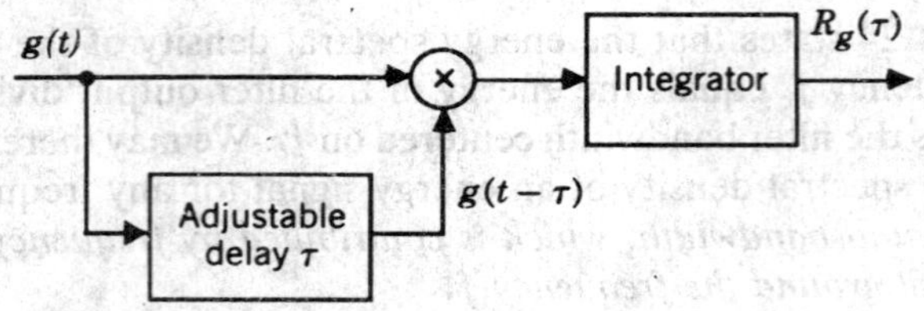

Figure 4.5
Circuit arrangement for measuring the autocorrelation function of an energy signal.

where $\star$ denotes convolution. For the convolution on the left side of Eq. 4.26, we have purposely used τ as the time variable of the energy signal of interest, because we wish to reserve the use of time t as the dummy variable of the integral describing the convolution there. Specifically, we define the time-domain convolution on the left side of Eq. 4.26 as

$$\begin{aligned} R_g(\tau) &= g(\tau) \star g(-\tau) \\ &= \int_{-\infty}^{\infty} g(t)g(t-\tau)\,dt \end{aligned} \tag{4.27}$$

The τ-dependent parameter $R_g(\tau)$ is called the *autocorrelation function* of the energy signal $g(t)$. In the defining equation (4.27), the time function $g(t-\tau)$ represents a *delayed version of the signal* $g(t)$, and $R_g(\tau)$ provides a measure of the similarity between the waveforms of the time functions $g(t)$ and $g(t-\tau)$. In particular, the *time lag* or *time delay* τ plays the role of a *scanning* or *searching parameter*. This role is highlighted in the block diagram of Fig. 4.5, which provides the basis for measuring the autocorrelation function $R_g(\tau)$.

EXERCISE 3 Show that the definition for the autocorrelation function $R_g(\tau)$ may also be formulated as

$$R_g(\tau) = \int_{-\infty}^{\infty} g(t+\tau)g(t)\,dt \tag{4.28}$$

PROPERTIES OF THE AUTOCORRELATION FUNCTION OF ENERGY SIGNALS

The autocorrelation function of a real-valued energy signal has several useful properties. They follow directly from the defining equation (4.27) or (4.26), from which the definition of autocorrelation function originated.

PROPERTY 1

The autocorrelation function of a real-valued energy signal g(t) is a real-valued even function, as shown by

$$R_g(-\tau) = R_g(\tau) \tag{4.29}$$

This property follows directly from Eq. 4.27. The implication of this property is that the autocorrelation function exhibits symmetry about the origin.

PROPERTY 2

The value of the autocorrelation function of an energy signal g(t) at the origin is equal to the energy of the signal; that is,

$$R_g(0) = E \tag{4.30}$$

This result is obtained by putting $\tau = 0$ in Eq. 4.27.

PROPERTY 3

The maximum value of the autocorrelation function of an energy signal g(t) occurs at the origin, as shown by

$$|R_g(\tau)| \leqslant R_g(0), \qquad \text{for all } \tau \tag{4.31}$$

To prove this property, we start with the observation that for any τ,

$$[g(t) \pm g(t - \tau)]^2 \geqslant 0$$

Equivalently, we may write

$$\pm 2g(t)g(t - \tau) \leqslant g^2(t) + g^2(t - \tau)$$

Integrating both sides of this relation with respect to time t from $-\infty$ to $+\infty$, and using Eqs. 4.27 and 4.30, we get the result given in Eq. 4.31.

According to Property 3, the degree of similarity between the signal $g(t)$ and its time-delayed version of $g(t - \tau)$ attains its maximum value at $\tau = 0$. This is intuitively satisfying.

PROPERTY 4

For an energy signal g(t), the autocorrelation function and energy spectral density form a Fourier transform pair; that is,

$$R_g(\tau) \rightleftharpoons \Psi_g(f) \tag{4.32}$$

This property follows directly from Eqs. 4.25 through 4.27.

EXERCISE 4 Write the formulas for the Fourier transform of $R_g(\tau)$ and the inverse Fourier transform of $\Psi_g(f)$. Hence, do the following:

(a) Show that the total area under $\Psi_g(f)$ equals the signal energy, by evaluating the formula for the Fourier transform of $R_g(\tau)$ for $\tau = 0$.
(b) Show that the total area under $R_g(\tau)$ equals $\Psi_g(0)$, by evaluating the inverse Fourier transform of $\Psi_g(f)$ for $f = 0$.

EXAMPLE 3 SINC PULSE (CONTINUED)

From Example 1, the energy spectral density of the sinc pulse $A\,\text{sinc}(2Wt)$ is given by (see Eq. 4.6)

$$\Psi_g(f) = \left(\frac{A}{2W}\right)^2 \text{rect}\left(\frac{f}{2W}\right)$$

Taking the inverse Fourier transform of $\Psi_g(f)$, we find that the autocorrelation function of the sinc pulse $A\,\text{sinc}(2Wt)$ is given by

$$R_g(\tau) = \frac{A^2}{2W}\,\text{sinc}(2W\tau) \tag{4.33}$$

which has a similar waveform to the sinc pulse itself.

CROSS-CORRELATION OF ENERGY SIGNALS

The autocorrelation function provides a measure of the similarity between a signal and its time-delayed version. In a similar way, we may use the *cross-correlation function* as a measure of the similarity between a signal and the time-delayed version of a second signal. Let $g_1(t)$ and $g_2(t)$ denote a pair of real-valued energy signals. The cross-correlation function of this pair of signals is defined by

$$R_{12}(\tau) = \int_{-\infty}^{\infty} g_1(t)g_2(t-\tau)\,dt \tag{4.34}$$

We see that if the two signals $g_1(t)$ and $g_2(t)$ are somewhat similar, then the cross-correlation function $R_{12}(\tau)$ will be finite over some range of τ, thereby providing a quantitative measure of the similarity, or coherence, between them. The energy signals $g_1(t)$ and $g_2(t)$ are said to be *orthogonal* over the entire time interval if $R_{12}(0)$ is zero, that is, if

$$\int_{-\infty}^{\infty} g_1(t)g_2(t)\,dt = 0 \tag{4.35}$$

Equation 4.34 defines one possible value for the cross-correlation function for a specified value of the delay variable τ. We may define a second cross-correlation function for the energy signals $g_1(t)$ and $g_2(t)$ as

$$R_{21}(\tau) = \int_{-\infty}^{\infty} g_2(t)g_1(t - \tau)\, dt \tag{4.36}$$

From the definitions of the cross-correlation functions $R_{12}(\tau)$ and $R_{21}(\tau)$ just given, we obtain the fundamental relationship

$$R_{12}(\tau) = R_{21}(-\tau) \tag{4.37}$$

Equation 4.37 indicates that unlike convolution, correlation is not in general commutative, that is, $R_{12}(\tau) \neq R_{21}(\tau)$.

Another important property of cross-correlation is shown by the Fourier transform pair

$$R_{12}(\tau) \rightleftharpoons G_1(f)\, G_2^*(f) \tag{4.38}$$

This relation is known as the *correlation theorem.* The correlation theorem states that *the cross-correlation of two energy signals corresponds to the multiplication of the Fourier transform of one signal by the complex conjugate of the Fourier transform of the other.*

EXERCISE 5 Prove the property of cross-correlation functions described in Eq. 4.37.

EXERCISE 6 Prove the correlation theorem described by Eq. 4.38.

4.3 POWER SPECTRAL DENSITY

Consider next the case of a *power signal* $g(t)$, which remains finite as time t approaches infinity. We assume $g(t)$ to be real valued. The *average power* of the signal is defined by (see Section 1.2)

$$P = \lim_{T\to\infty} \frac{1}{2T} \int_{-T}^{T} g^2(t)\, dt \tag{4.39}$$

To develop a frequency-domain description of power, we need to know the Fourier transform of the signal $g(t)$. However, this may pose a problem, because power signals have infinite energy and may therefore not be Fourier transformable. To overcome the problem, we consider a *truncated* version of the signal $g(t)$. In particular, we define

$$\begin{aligned} g_T(t) &= g(t)\,\mathrm{rect}\left(\frac{t}{2T}\right) \\ &= \begin{cases} g(t), & -T \leqslant t \leqslant T \\ 0, & \text{otherwise} \end{cases} \end{aligned} \tag{4.40}$$

As long as T is finite, the truncated signal $g_T(t)$ has finite energy; hence $g_T(t)$ is Fourier transformable. Let $G_T(f)$ denote the Fourier transform of $g_T(t)$; that is,

$$g_T(t) \rightleftharpoons G_T(f)$$

Using the definition of $g_T(t)$, we may rewrite Eq. 4.39 for the average power P in terms of $g_T(t)$ as

$$P = \lim_{T\to\infty} \frac{1}{2T} \int_{-\infty}^{\infty} g_T^2(t)\,dt \tag{4.41}$$

Since $g_T(t)$ has finite energy, we may use the Rayleigh energy theorem to express the energy of $g_T(t)$ in terms of its Fourier transform $G_T(f)$ as

$$\int_{-\infty}^{\infty} g_T^2(t)\,dt = \int_{-\infty}^{\infty} |G_T(f)|^2\,df \tag{4.42}$$

where $|G_T(f)|$ is the amplitude spectrum of $g_T(t)$. Accordingly, we may rewrite Eq. 4.41 in the equivalent form

$$P = \lim_{T\to\infty} \frac{1}{2T} \int_{-\infty}^{\infty} |G_T(f)|^2\,df \tag{4.43}$$

As T increases, the energy of $g_T(t)$ increases. Correspondingly, the energy spectral density $|G_T(f)|^2$ increases with T. Indeed as T approaches infinity, so will $|G_T(f)|^2$. However, for the average power P to be finite, $|G_T(f)|^2$ must approach infinity at the same rate as T. This requirement ensures the *convergence* of the integral on the right side of Eq. 4.43 in the limit as T approaches infinity. This convergence, in turn, permits us to interchange the order in which the limiting operation and integration in Eq. 4.43 are performed. We may thus rewrite this equation as

$$P = \int_{-\infty}^{\infty} \lim_{T\to\infty} \frac{1}{2T} |G_T(f)|^2\,df \tag{4.44}$$

Let the integrand be denoted by

$$S_g(f) = \lim_{T\to\infty} \frac{1}{2T} |G_T(f)|^2 \quad (4.45)$$

The frequency-dependent function $S_g(f)$ is called the *power spectral density* or *power spectrum* of a power signal, and $|G_T(f)|^2/2T$ is called the *periodogram*[1] of the signal.

EXAMPLE 4 MODULATED WAVE

Consider the *modulated wave*

$$x(t) = g(t)\cos(2\pi f_c t) \quad (4.46)$$

where $g(t)$ is a power signal band-limited to B hertz. We refer to $x(t)$ as a "modulated wave" in the sense that the amplitude of the sinusoidal "carrier" of frequency f_c is varied linearly with the signal $g(t)$. We wish to find the power spectral density of $x(t)$ in terms of that of $g(t)$, given that the frequency f_c is larger than the bandwidth B.

Adapting the formula of Eq. 4.45 to the situation at hand, we may define the power spectral density of the modulated wave $x(t)$ as

$$S_x(f) = \lim_{T\to\infty} \frac{1}{2T} |X_T(f)|^2 \quad (4.47)$$

where $X_T(f)$ is the Fourier transform of $x_T(t)$, the truncated version of $x(t)$. From Eq. 4.46, we have

$$x_T(t) = g_T(t)\cos(2\pi f_c t) \quad (4.48)$$

where the truncated signal $g_T(t)$ is itself defined in Eq. 4.40. Since

$$\cos(2\pi f_c t) = \frac{1}{2}[\exp(j2\pi f_c t) + \exp(-j2\pi f_c t)] \quad (4.49)$$

it follows from the frequency-shifting property of the Fourier transform that

$$X_T(f) = \frac{1}{2}[G_T(f - f_c) + G_T(f + f_c)] \quad (4.50)$$

where $G_T(f)$ is the Fourier transform of $g_T(t)$.

[1]The periodogram is a misnomer since it is a function of frequency not period. Nevertheless, the term has wide usage. It was first used by statisticians to look for periodicities such as seasonal trends in data.

Given that $f_c > B$, we find that $G_T(f - f_c)$ and $G_T(f + f_c)$ represent *nonoverlapping spectra;* their product is therefore zero. Accordingly, using Eq. 4.50 to evaluate the squared amplitude of $X_T(f)$, we get

$$|X_T(f)|^2 = \frac{1}{4}[|G_T(f - f_c)|^2 + |G_T(f + f_c)|^2], \qquad f_c > B \qquad (4.51)$$

Finally, substituting Eq. 4.51 in 4.47, and then using the definition of Eq. 4.45 for the power spectral density of the power signal $g(t)$, we get the desired result:

$$S_x(f) = \frac{1}{4}[S_g(f - f_c) + S_g(f + f_c)], \qquad f_c > B \qquad (4.52)$$

PROPERTIES OF POWER SPECTRAL DENSITY

The role of power spectral density for power signals is similar to that of energy density for energy signals. Indeed, the power spectral density has properties that parallel those of the energy spectral density. In the sequel, we present the properties of power spectral density without proof; these properties may be verified by using arguments similar to those used in Section 4.1 for verifying the properties of energy spectral density.

PROPERTY 1

The power spectral density of a power signal $g(t)$ is a nonnegative real-valued function of frequency; that is,

$$S_g(f) \geq 0, \qquad \text{for all } f \qquad (4.53)$$

PROPERTY 2

The power spectral density of a real-valued power signal $g(t)$ is an even function of frequency; that is,

$$S_g(-f) = S_g(f) \qquad (4.54)$$

PROPERTY 3

The total area under the curve of the power spectral density of a power signal $g(t)$ equals the average signal power; that is,

$$P = \int_{-\infty}^{\infty} S_g(f)\, df \qquad (4.55)$$

PROPERTY 4

When a power signal is transmitted through a linear time-invariant system, the power spectral density of the output equals the power spectral density of the input multiplied by the squared amplitude response of the system. That is, if $S_x(f)$ is the power spectral density of a power signal $x(t)$ applied to a linear time-invariant system of transfer function $H(f)$, the power spectral density $S_y(f)$ of the power signal $y(t)$ produced at the output of the system is defined by

$$S_y(f) = |H(f)|^2 S_x(f) \tag{4.56}$$

where $|H(f)|$ is the amplitude response of the system.

EXERCISE 7 Justify the validity of the input–output relation described in Eq. 4.56.

INTERPRETATION OF POWER SPECTRAL DENSITY

The input–output relation of Eq. 4.56 provides a basis for the physical interpretation of power spectral density, and therefore its measurement. Just as we did for the interpretation of energy spectral density, suppose a power signal $x(t)$ is applied to a band-pass filter followed by a power meter as in Fig. 4.6*a*. The filter has a narrow bandwidth Δf centered on some frequency f_c, as in Fig. 4.6*b*. Application of Eq. 4.56 yields the power spectral density of the resulting filter output $y(t)$ approximately as follows

$$S_y(f) \simeq \begin{cases} S_x(f_c) & f_c - \dfrac{\Delta f}{2} \leq |f| \leq f_c + \dfrac{\Delta f}{2} \\ 0, & \text{otherwise} \end{cases} \tag{4.57}$$

The average power of the filter output $y(t)$ is therefore approximately given by

$$P_y \simeq 2S_x(f_c)\,\Delta f \tag{4.58}$$

The evaluation of $S_x(f)$ is illustrated in Fig. 4.6*c*. Equivalently, we may write

$$S_x(f_c) \simeq \frac{P_y}{2\Delta f} \tag{4.59}$$

In other words, the power spectral density of the filter input $x(t)$ at some frequency f_c is equal to the average power of the filter output divided by $2\Delta f$, where Δf is the bandwidth of the filter centered on f_c. The factor of 2 accounts for the contributions of negative as well as positive frequency components.

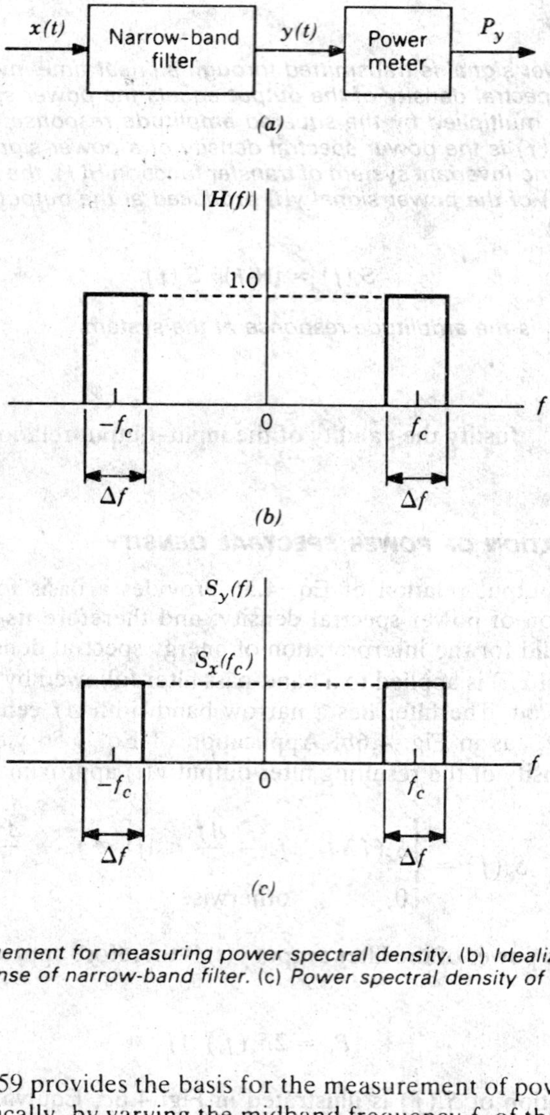

Figure 4.6
(a) *Circuit arrangement for measuring power spectral density.* (b) *Idealized amplitude response of narrow-band filter.* (c) *Power spectral density of the filter output.*

Equation 4.59 provides the basis for the measurement of power spectral density. Specifically, by varying the midband frequency f_c of the band-pass filter in Fig. 4.6*a*, and measuring the average power of the filter output for each setting of f_c, we may measure the power spectral density of a power signal (applied to the filter input) over a frequency band of interest.

4.4 CORRELATION OF POWER SIGNALS

We may develop a formula for the autocorrelation function of power signals by following a procedure similar to that described for the case of energy signals in Section 4.2. Specifically, we start with the defining equation (4.45) for the power spectral density $S_g(f)$ of a power signal $g(t)$, and rewrite it in the form

$$S_g(f) = \lim_{T\to\infty} \frac{1}{2T} G_T(f)G_T^*(f) \tag{4.60}$$

where $G_T(f)$ is the Fourier transform of the truncated version $g_T(t)$ of the power signal $g(t)$. Next, we use the Fourier transform pair

$$g_T(\tau) \star g_T(-\tau) \rightleftharpoons G_T(f)\, G_T^*(f) \tag{4.61}$$

Multiplying both members of this pair by the factor $1/2T$ and then taking the limit as T approaches infinity, we have

$$\lim_{T\to\infty} \frac{1}{2T} g_T(\tau) \star g_T(-\tau) \rightleftharpoons \lim_{T\to\infty} \frac{1}{2T} G_T(f)G_T^*(f) \tag{4.62}$$

The function on the right side of this pair is recognized as the power spectral density of the power signal $g(t)$. Accordingly, we adopt the function on the left side of Eq. 4.62 as the autocorrelation function of the power signal $g(t)$, and thus write

$$R_g(\tau) = \lim_{T\to\infty} \frac{1}{2T} \int_{-\infty}^{\infty} g_T(t)g_T(t-\tau)\,dt \tag{4.63}$$

We may formulate the autocorrelation function $R_g(\tau)$ in terms of the power signal $g(t)$ itself by using the definition given in Eq. 4.40 for the truncated signal $g_T(t)$. By so doing, we define the autocorrelation function of a power signal $g(t)$ as follows

$$R_g(\tau) = \lim_{T\to\infty} \frac{1}{2T} \int_{-T}^{T} g(t)g(t-\tau)\,dt \tag{4.64}$$

PROPERTIES OF THE AUTOCORRELATION FUNCTION OF POWER SIGNALS

The autocorrelation function of a power signal has properties that are similar to those of the autocorrelation function of energy signals. Indeed, by following arguments similar to those presented in Section 4.2, we may readily establish the following properties of the autocorrelation function of power signals, which are presented without proof.

PROPERTY 1

The autocorrelation function of a real-valued power signal g(t) is a real-valued even function, as shown by

$$R_g(-\tau) = R_g(\tau) \tag{4.65}$$

PROPERTY 2

The value of the autocorrelation function of a power signal $g(t)$ at the origin is equal to the average power of the signal; that is,

$$R_g(0) = P \tag{4.66}$$

PROPERTY 3

The maximum value of the autocorrelation function of a power signal $g(t)$ occurs at the origin, as shown by

$$|R_g(\tau)| \leq R_g(0) \tag{4.67}$$

PROPERTY 4

For a power signal $g(t)$, the autocorrelation function and power spectral density form a Fourier transform pair; that is,

$$R_g(\tau) \rightleftharpoons S_g(f) \tag{4.68}$$

Equation 4.68 states that the power spectral density $S_g(f)$ of a power signal $g(t)$ is the Fourier transform of the autocorrelation function $R_g(\tau)$ of the signal, as shown in the expanded form:

$$S_g(f) = \int_{-\infty}^{\infty} R_g(\tau) \exp(-j2\pi f\tau)\, d\tau \tag{4.69}$$

Equation 4.68 also states that the autocorrelation function $R_g(\tau)$ is the inverse Fourier transform of the power spectral density $S_g(f)$, as shown in the expanded form:

$$R_g(\tau) = \int_{-\infty}^{\infty} S_g(f) \exp(j2\pi f\tau)\, df \tag{4.70}$$

Equations 4.69 and 4.70 are known as the *Einstein–Wiener–Khintchine relations.*[2] Given the autocorrelation function $R_g(\tau)$, we may use Eq. 4.69 to compute the power spectral density $S_g(f)$. Conversely, given the power

[2]Traditionally, Eqs. 4.69 and 4.70 have been referred to in the literature as the Wiener–Khintchine relations in recognition of pioneering work done by Wiener and Khintchine; for their original papers, see Wiener (1930) and Khintchine (1934). A recent discovery of a forgotten paper by Albert Einstein on time-series analysis (delivered at the Swiss Physical Society's February 1914 meeting in Basel) reveals that Einstein had discussed the autocorrelation function and its relationship to the spectral content of a time series many years before Wiener and Khintchine. For this very brief paper, see Einstein (1914). An English translation of Einstein's paper is reproduced in the *IEEE Acoustics, Speech, and Signal Processing Magazine*, vol. 4, October 1987. This particular issue also contains articles by W. A. Gardner and A. M. Yaglom, which elaborate on Einstein's original work.

spectral density $S_g(f)$, we may use Eq. 4.70 to compute the autocorrelation function $R_g(\tau)$.

EXERCISE 8 Verify the properties of autocorrelation function of a power signal, described in Eqs. 4.65 through 4.68.

EXAMPLE 5

Consider again the modulated wave $x(t)$, defined in Eq. 4.46, reproduced here for convenience:

$$x(t) = g(t)\cos(2\pi f_c t)$$

The signal $g(t)$ is a power signal band-limited to B hertz, where $B < f_c$. In this example, we evaluate the autocorrelation function of $x(t)$ in terms of that of $g(t)$.

We do the evaluation by using Property 4 of the autocorrelation function, namely, the fact that autocorrelation function and power spectral density form a Fourier transform. From Eq. 4.52 of Example 4, we have

$$S_x(f) = \frac{1}{4}[S_g(f - f_c) + S_g(f + f_c)]$$

Therefore, taking the inverse Fourier transform of both sides of the equation, we get

$$\begin{aligned} R_x(\tau) &= \frac{1}{4}[R_g(\tau)\exp(j2\pi f_c\tau) + R_g(\tau)\exp(-j2\pi f_c\tau)] \\ &= \frac{1}{2}R_g(\tau)\cos(2\pi f_c\tau) \end{aligned} \tag{4.71}$$

which is the desired result.

EXERCISE 9 Using the relation of Eq. 4.71, show that the average power of the modulated signal $x(t)$ equals one-half the average power of the original signal $g(t)$.

CROSS-CORRELATION OF POWER SIGNALS

We complete the discussion of correlation of power signals by considering their cross-correlation. Let $g_1(t)$ and $g_2(t)$ denote a pair of power signals.

We define the *cross-correlation* between $g_1(t)$ and $g_2(t)$ as

$$R_{12}(\tau) = \lim_{T\to\infty} \frac{1}{2T} \int_{-T}^{T} g_1(t) g_2(t - \tau)\, dt \tag{4.72}$$

In a similar way we may define a second cross-correlation function $R_{21}(\tau)$.

The pair of power signals $g_1(t)$ and $g_2(t)$ are said to be *orthogonal* over the entire time interval if

$$\lim_{T\to\infty} \frac{1}{2T} \int_{-T}^{T} g_1(t) g_2(t)\, dt = 0 \tag{4.73}$$

4.5 FLOWCHART SUMMARIES

In this section we summarize the significance of the time-frequency relations derived for energy and power signals.

Given an energy signal $g(t)$ of Fourier transform $G(f)$, we may summarize this relationship and its interplay with the formula of Eq. 4.4 for the energy spectral density $\Psi_g(f)$ and that of Eq. 4.27 for the autocorrelation function $R_g(\tau)$ as in Fig. 4.7. This chart clearly shows that whatever operation or sequence of operations is used to obtain the autocorrelation function $R_g(\tau)$ or the energy spectral density $\Psi_g(f)$, that operation or sequence of operations is *irreversible*. The implication of this is that when a signal $g(t)$ is converted to $R_g(\tau)$ or $\Psi_g(f)$, in general, information is lost about the original signal $g(t)$ or its Fourier transform $G(f)$. In going from $g(t)$ to $R_g(\tau)$, dependence on the physical time t is destroyed. In going from $G(f)$ to $\Psi_g(f)$, information on the phase spectrum of the signal is destroyed. This means that if two (or more) different signals have the same amplitude spectrum but different phase spectra, then they will have the same energy spectral density or, equivalently, the same autocorrelation function. In other words, for a given energy signal $g(t)$, there is a unique energy spectral density $\Psi_g(f)$ or, equivalently, a unique autocorrelation function $R_g(\tau)$. The converse of this statement, however, is not true.

The flowchart of Fig. 4.7 shows that given the energy signal $g(t)$, we may compute the energy spectral density $\Psi_g(f)$ in one of two equivalent ways:

1. We compute the Fourier transform $G(f)$, and then use the definition of Eq. 4.4.
2. We compute the autocorrelation function $R_g(\tau)$ using Eq. 4.27, and then compute the Fourier transform of $R_g(\tau)$.

The interrelations between time-domain and frequency-domain descriptions of power signals are analogous to those for energy signals. In particular, for a power signal $g(t)$ we may draw a chart similar to that of Fig.

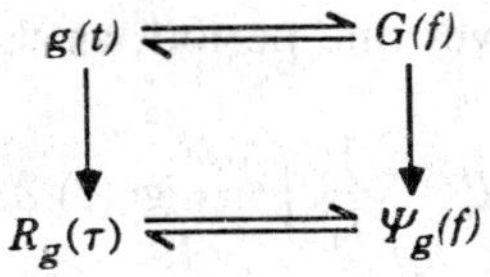

Figure 4.7
Flowchart summary of interrelations between time-domain and frequency-domain descriptions of energy signals.

4.7, except that the chart is now based on a truncated version $g_T(t)$ of the power signal. The point to note is that information is lost in the process of computing and retaining only the autocorrelation function $R_g(\tau)$ of the power signal or its power spectral density $\Psi_g(f)$. Moreover, given the signal $g(t)$, we may compute the power spectral density $S_g(f)$ using one of two equivalent procedures:

1. We compute the Fourier transform $G_T(f)$ of the power signal $g(t)$ for the interval $-T \leq t \leq T$ for large T, and then use Eq. 4.45 to compute the power spectral density $S_g(f)$.
2. We use Eq. 4.63 to compute the autocorrelation function $R_g(\tau)$, and then take the Fourier transform of $R_g(\tau)$.

EXERCISE 10 Given the energy signal $g(t)$, outline the two procedures that may be used to compute the autocorrelation function $R_g(\tau)$.

EXERCISE 11 Given the power signal $g(t)$, outline the two procedures that may be used to compute the autocorrelation function $R_g(\tau)$.

4.6 *SPECTRAL CHARACTERISTICS OF PERIODIC SIGNALS*

The definitions of power spectral density and autocorrelation function for power signals given in Eqs. 4.60 and 4.64 take on special forms for the case of periodic signals. These signals constitute an important class of power signals. Consider a periodic signal $g_p(t)$ of period T_0, represented in terms of its complex Fourier series as

$$g_p(t) = \sum_{n=-\infty}^{\infty} c_n \exp\left(\frac{j2\pi nt}{T_0}\right) \tag{4.74}$$

where the c_n are complex Fourier coefficients. For the situation at hand, the time average in the defining equation (4.39) for the average power of the signal may be taken over one period, as shown by

$$P = \frac{1}{T_0} \int_{-T_0/2}^{T_0/2} g_p^2(t)\, dt \tag{4.75}$$

Correspondingly, the formula for the power spectral density given in Eq. 4.60 takes on a discrete form defined in terms of the complex Fourier coefficients as

$$S_{g_p}(f) = \sum_{n=-\infty}^{\infty} |c_n|^2\, \delta\left(f - \frac{n}{T_0}\right) \tag{4.76}$$

Naturally, the power spectral density $S_{g_p}(f)$ has all the properties listed in Section 4.3. Moreover, it is *a discrete function of frequency,* which is a consequence of the periodic nature of the signal $g_p(t)$. Since the total area under a curve of power spectral density equals the average power, we may define the total average power of the periodic signal $g_p(t)$ in terms of its frequency-domain description as

$$P = \sum_{n=-\infty}^{\infty} |c_n|^2 \tag{4.77}$$

This relation is known as *Parseval's power theorem*. It states that the average power of a periodic signal $g_p(t)$ is equal to the sum of the squared amplitudes of all the harmonic components of the signal $g_p(t)$. Note that the Parseval power theorem, as with the Rayleigh energy theorem, requires knowledge of the amplitude spectrum only.

The power spectral density of Eq. 4.76 has a delta function at zero frequency, which is weighted by $|c_0|^2$. The presence of this delta function implies that the periodic signal $g_p(t)$ has *dc power,* given by

$$P_{dc} = |c_0|^2 \tag{4.78}$$

The coefficient c_0 equals the *mean* or *average value* of the periodic signal $g_p(t)$; that is,

$$c_0 = \frac{1}{T_0} \int_{-T_0/2}^{T_0/2} g_p(t)\, dt \tag{4.79}$$

The *ac power* of the periodic signal $g_p(t)$ is defined by the sum of the weights associated with the remaining delta functions in $S_{g_p}(f)$, as shown by

$$P_{ac} = \sum_{\substack{n=-\infty \\ n\neq 0}}^{\infty} |c_n|^2 \tag{4.80}$$

The square root of P_{ac} defines the *root mean square (rms) value* of the signal. Naturally. the sum of dc power P_{dc} and ac power P_{ac} equals the total average power P.

When the power signal of interest is periodic, the integrand in the defining equation (4.64) for the autocorrelation function $R_g(\tau)$ of the signal is likewise periodic. Hence, the time average in this formula may be taken over one period. Thus, we may express the autocorrelation function of a periodic signal $g_p(t)$ of period T_0 as

$$R_{g_p}(\tau) = \frac{1}{T_0}\int_{-T_0/2}^{T_0/2} g_p(t)g_p(t-\tau)\,dt \tag{4.81}$$

The autocorrelation function $R_{g_p}(\tau)$ exhibits all the properties listed in Section 4.4 for the autocorrelation function of power signals. In addition, *the autocorrelation function $R_{g_p}(\tau)$ is periodic with the same period as the periodic signal $g_p(t)$ itself;* that is

$$R_{g_p}(\tau) = R_{g_p}(\tau \pm nT_0), \qquad n = 1, 2, \ldots, \tag{4.82}$$

EXAMPLE 6 SINUSOIDAL WAVE

Consider the *sinusoidal wave*

$$g_p(t) = A\cos(2\pi f_c t + \theta) \tag{4.83}$$

which is plotted in Fig. 4.8*a*; the period $T_0 = 1/f_c$. The requirement is to evaluate the power spectral density, average power, and autocorrelation function of this sinusoidal wave.

To express the given sinusoidal wave as a complex Fourier series, we use the formula for a cosine function in terms of a pair of complex exponentials. We thus write

$$g_p(t) = c_1\exp(j2\pi f_c t) + c_{-1}\exp(-j2\pi f_c t)$$

where

$$c_1 = \frac{A}{2}\exp(j\theta)$$

and

$$c_{-1} = \frac{A}{2}\exp(-j\theta)$$

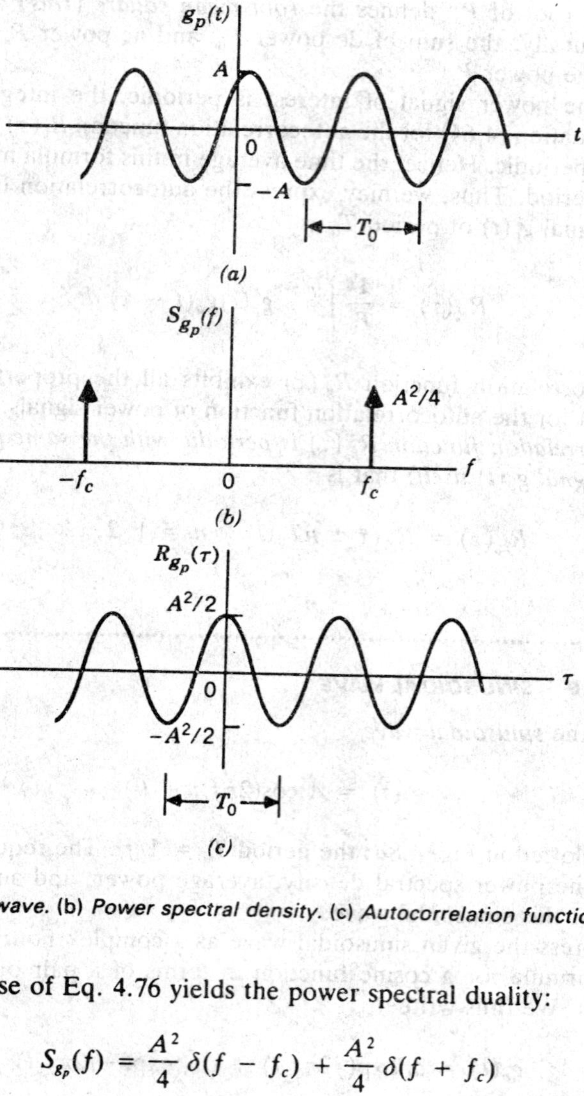

Figure 4.8
(a) *Sinusoidal wave.* (b) *Power spectral density.* (c) *Autocorrelation function.*

Hence, the use of Eq. 4.76 yields the power spectral duality:

$$S_{g_p}(f) = \frac{A^2}{4}\,\delta(f - f_c) + \frac{A^2}{4}\,\delta(f + f_c) \tag{4.84}$$

That is, the power spectral density of a sinusoidal wave consists of a pair of delta functions located at $f = \pm f_c$, both of which are weighted by the factor $A^2/4$, as depicted in Fig. 4.8*b*. Note that the power spectral density is independent of the phase θ of the sinusoidal wave.

By evaluating the total area under the power spectral density $S_{g_p}(f)$, we obtain the average power of the sinusoidal wave as

$$P = \frac{A^2}{2} \tag{4.85}$$

Finally, the use of Eq. 4.81 yields the autocorrelation function of the given sinusoidal wave as

$$R_{g_p}(\tau) = A^2 f_c \int_{-1/2f_c}^{1/2f_c} \cos(2\pi f_c t + \theta) \cos(2\pi f_c t - 2\pi f_c \tau + \theta)\, dt$$

$$= \frac{A^2 f_c}{2} \int_{-1/2f_c}^{1/2f_c} [\cos(2\pi f_c \tau) + \cos(4\pi f_c t - 2\pi f_c \tau + 2\theta)]\, dt$$

$$= \frac{A^2}{2} \cos(2\pi f_c \tau) \tag{4.86}$$

Equation 4.86 is plotted in Fig. 4.8*c*. It shows that the autocorrelation function of a sinusoidal wave is a sinusoidal function of τ, with the same period as the given sinusoidal wave. Moreover, putting $\tau = 0$ in Eq. 4.86, we find that $R_{g_p}(0) = P$, as expected.

EXERCISE 12 Show that the power spectral density and autocorrelation function of Eqs. 4.84 and 4.86 constitute a Fourier transform pair.

EXAMPLE 7 SQUARE WAVE

Consider next the square wave of Fig. 4.9*a*, one period of which is defined by

$$g_p(t) = \begin{cases} A, & -\frac{T_0}{4} \leq t \leq \frac{T_0}{4} \\ 0, & \text{for the remainder of the period} \end{cases} \tag{4.87}$$

The requirement is to determine the power spectral density and autocorrelation function of this square wave.

In Example 1, Chapter 2, we derived the formula for the complex Fourier coefficient c_n of a rectangular pulse train with arbitrary duty cycle; the result is given in Eq. 2.19. The square wave described here has a duty cycle of one-half. Hence, adapting Eq. 2.19 for this duty cycle, we find that the complex Fourier coefficient of the square wave of Fig. 4.9*a* is given by

$$c_n = \frac{A}{2} \operatorname{sinc}\left(\frac{n}{2}\right) \tag{4.88}$$

Substituting Eq. 4.88 in 4.76 yields the desired power spectral density of

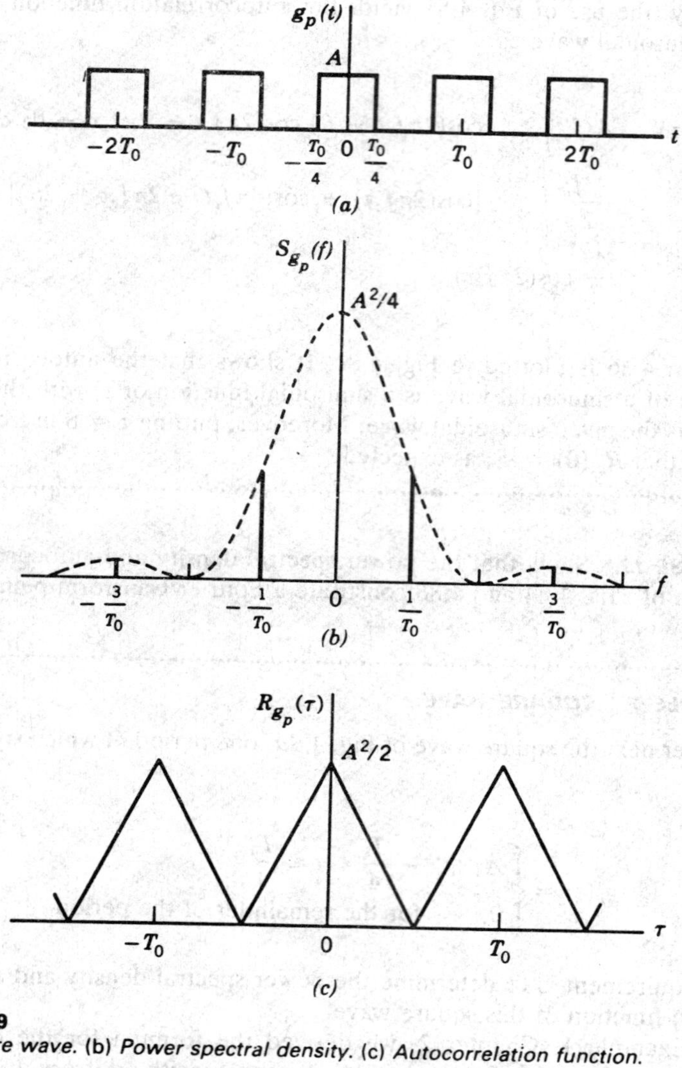

Figure 4.9
(a) *Square wave.* (b) *Power spectral density.* (c) *Autocorrelation function.*

the square wave as

$$S_{g_p}(f) = \frac{A^2}{4} \sum_{n=-\infty}^{\infty} \text{sinc}^2\left(\frac{n}{2}\right) \delta\left(f - \frac{n}{T_0}\right) \tag{4.89}$$

which is plotted in Fig. 4.9*b*.

The most expedient approach for obtaining the autocorrelation function is to use the formula of Eq. 4.81. Figure 4.10 presents a graphical portrayal of the steps involved in the application of this formula for

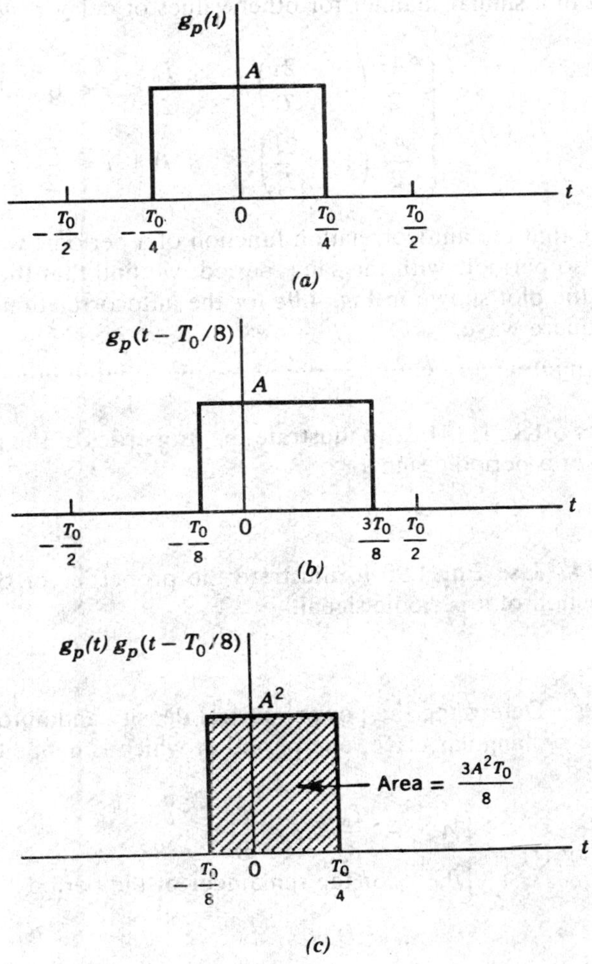

Figure 4.10
The computation of autocorrelation function $R_{g_p}(\tau)$ for lag $\tau = T_0/8$.

a delay $\tau = T_0/8$. Parts *a*, *b*, and *c* of the figure present plots of the square wave $g_p(t)$, its delayed version $g_p(t - T_0/8)$, and the product $g_p(t)g_p(t - T_0/8)$, respectively, for the period $-(T_0/2) \leq t \leq (T_0/2)$. The area under the product $g_p(t)g_p(t - T_0/8)$ for this period is shown shaded in Fig. 4.10*c*. Evaluating this area and scaling it by the factor $1/T_0$, we get

$$R_{g_p}\left(\frac{T_0}{8}\right) = \frac{3A^2}{8}$$

Proceeding in a similar manner for other values of delay τ, we obtain

$$R_{g_p}(\tau) = \begin{cases} \dfrac{A^2}{2}\left(1 + \dfrac{2\tau}{T_0}\right), & -\dfrac{T_0}{2} \leq \tau \leq 0 \\ \dfrac{A^2}{2}\left(1 - \dfrac{2\tau}{T_0}\right), & 0 \leq \tau \leq \dfrac{T_0}{2} \end{cases} \qquad (4.90)$$

Recognizing that the autocorrelation function of a periodic wave with period T_0 is also periodic with the same period, we find that the use of Eq. 4.90 yields the plot shown in Fig. 4.9*c* for the autocorrelation function of the given square wave.

EXERCISE 13 Use Eq. 4.89 to illustrate the properties of the power spectral density of a periodic signal.

EXERCISE 14 Use Eq. 4.90 to illustrate the properties of the autocorrelation function of a periodic signal.

EXERCISE 15 Determine the power spectral density and autocorrelation function of a rectangular wave, one period of which is defined by

$$g_p(t) = \begin{cases} A, & -\dfrac{T_0}{8} \leq t \leq \dfrac{T_0}{8} \\ 0, & \text{for the remainder of the period} \end{cases}$$

4.7 SPECTRAL CHARACTERISTICS OF RANDOM SIGNALS AND NOISE

Random signals constitute another important class of power signals. We say a signal is random if there is *uncertainty* about the signal before it actually occurs. Such a signal may be viewed as belonging to an *ensemble* of signals, the generation of which is governed by a mechanism that is *probabilistic* in nature. Hence, no two signals in the ensemble exhibit the same variation with time. Each waveform (signal) in the ensemble is referred to as a *sample function,* and the ensemble of all possible sample functions is referred to as a *random process*.

Let $x(t)$ denote a sample function of a random process $X(t)$. Figure 4.11*a* shows a plot of the waveform of $x(t)$. Since $x(t)$ is a power signal, its Fourier transform does not exist. This necessitates dealing with a trun-

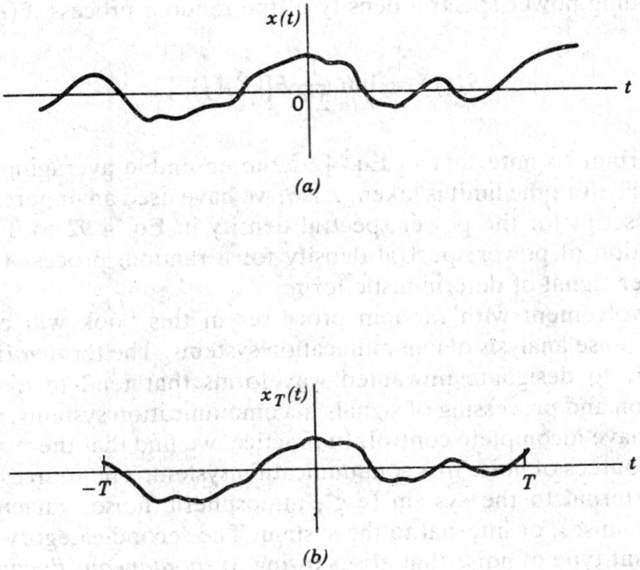

Figure 4.11
(a) *Sample function of a random process.* (b) *Truncated version of the sample function.*

cated version of the sample function, namely,

$$x_T(t) = \begin{cases} x(t), & -T \leq t \leq T \\ 0, & \text{otherwise} \end{cases} \tag{4.91}$$

Figure 4.11*b* depicts the truncated signal $x_T(t)$. From the discussion presented in Section 4.3, we note that the time average power spectral density of the sample function $x(t)$ over the interval $-T \leq t \leq T$ is $|X_T(f)|^2/2T$, where $X_T(f)$ is the Fourier transform of $x_T(t)$. This time-averaged power spectral density depends on the particular sample function $x(t)$ drawn from the random process $X(t)$. Accordingly, we must perform an *ensemble averaging* operation, and then take the limit as T approaches infinity. The value of frequency f is held fixed while averaging over the ensemble. The ensemble averaging operation requires using the probability distribution of the ensemble.[3] For the purpose of our present discussion, it is sufficient to acknowledge the ensemble averaging operation by using the operator E, commonly referred to as the *expectation operator*. We thus write the *ensemble-averaged* or *mean value* of $|X_T(f)|^2$ simply as $E[|X_T(f)|^2]$ and the

[3] The issue of ensemble averaging is considered in Chapter 8.

corresponding power spectral density of the random process $X(t)$ as

$$S_X(f) = \lim_{T\to\infty} \frac{1}{2T} E[|X_T(f)|^2] \tag{4.92}$$

It is important to note that in Eq. 4.92 the ensemble averaging must be performed before the limit is taken. Also, we have used an uppercase letter as the subscript for the power spectral density in Eq. 4.92 to distinguish this definition of power spectral density for a random process from that for a power signal of deterministic form.

Our involvement with random processes in this book will be in the context of noise analysis of communication systems. The term *noise* is used customarily to designate unwanted waveforms that tend to disturb the transmission and processing of signals in communication systems, and over which we have incomplete control. In practice, we find that there are many potential sources of noise in a communication system. The sources of noise may be external to the system (e.g., atmospheric noise, galactic noise, man-made noise), or internal to the system. The second category includes an important type of noise that arises owing to *spontaneous fluctuations* of current or voltage in electrical circuits. This type of noise, in one way or another, is present in every communication system and represents a basic limitation on the reliable transmission of information. It originates at the front end of the receiver part of the system; hence, it is commonly referred to as *receiver noise*.[4] It is also referred to as *channel noise*.

WHITE NOISE

The noise analysis of communication systems is customarily based on an idealized form of a noise process called *white noise,* the power spectral density of which is independent of frequency. The adjective *white* is used in the sense that white light contains equal amounts of all frequencies within the visible band of electromagnetic radiation. We denote the power spectral density of a white-noise process $W(t)$ as

$$S_W(f) = \frac{N_0}{2} \tag{4.93}$$

where the factor 1/2 has been included to indicate that half the power is associated with positive frequencies and half with negative frequencies, as illustrated in Fig. 4.12*a*. The dimensions of N_0 are in watts per hertz. The parameter N_0 is usually measured at the input stage of the receiver of a communication system.

[4]For a discussion of various types of noise encountered in communication systems, see Appendix C.

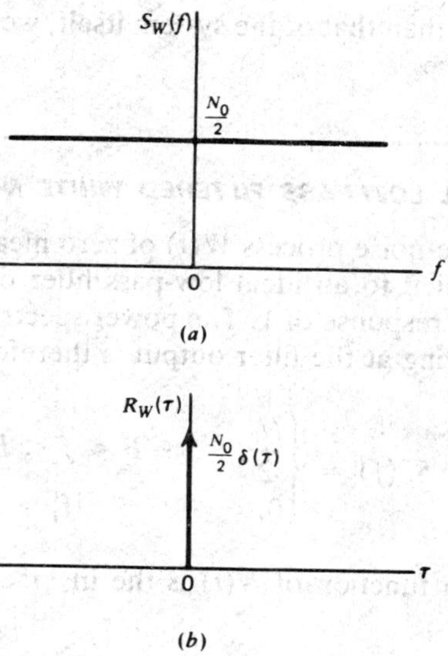

Figure 4.12
Characteristics of white noise. (a) *Power spectral density.* (b) *Autocorrelation function.*

The absence of a delta function in the power spectral density of Fig. 4.12*a* at the origin means that the white noise so described has no dc power. That is, its mean or average value is zero.

Since the autocorrelation function is the inverse Fourier transform of the power spectral density, it follows that for white noise

$$R_W(\tau) = \frac{N_0}{2}\delta(\tau) \tag{4.94}$$

That is, the autocorrelation function of white noise consists of a delta function weighted by the factor $N_0/2$ and occurring at $\tau = 0$, as in Fig. 4.12*b*. We note that $R_W(\tau)$ is zero for $\tau \neq 0$. Accordingly, any two different samples of white noise, no matter how close together in time they are taken, are uncorrelated.

Strictly speaking, white noise has infinite average power and, as such, it is not physically realizable. Nevertheless, white noise has convenient mathematical properties and therefore is useful in system analysis.

The utility of a white-noise process is parallel to that of an impulse function or delta function in the analysis of linear systems. The effect of an impulse is observed only after it has been passed through a system with finite bandwidth. Likewise, the effect of white noise is observed only after passing through a system with finite bandwidth. We may state, therefore, that as long as the bandwidth of a noise process at the input of a system is appreciably larger than that of the system itself, we may model the noise process as white noise.

EXAMPLE 8 IDEAL LOW-PASS FILTERED WHITE NOISE

Suppose that a white-noise process $W(t)$ of zero mean and power spectral density $N_0/2$ is applied to an ideal low-pass filter of bandwidth B and a passband amplitude response of 1. The power spectral density of the noise process $N(t)$ appearing at the filter output is therefore (see Fig. 4.13a)

$$S_N(f) = \begin{cases} \dfrac{N_0}{2}, & -B < f < B \\ 0, & |f| > B \end{cases} \tag{4.95}$$

The autocorrelation function of $N(t)$ is the inverse Fourier transform of

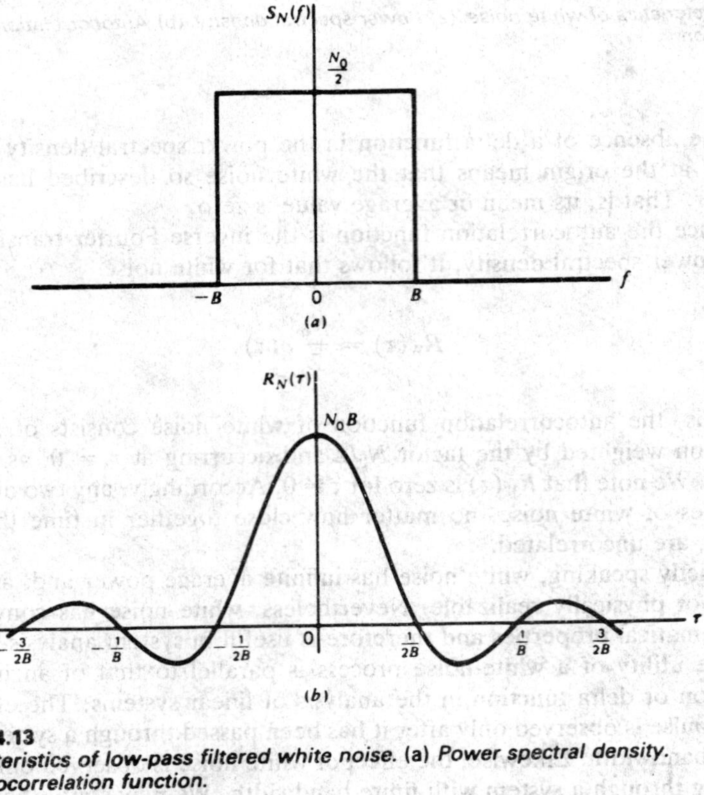

Figure 4.13
Characteristics of low-pass filtered white noise. (a) Power spectral density. (b) Autocorrelation function.

the power spectral density shown in Fig. 4.13*a*:

$$\begin{aligned} R_N(\tau) &= \int_{-B}^{B} \frac{N_0}{2} \exp(j2\pi f\tau)\, df \\ &= N_0 B \operatorname{sinc}(2B\tau) \end{aligned} \tag{4.96}$$

This autocorrelation function is plotted in Fig. 4.13*b*. We see that $R_N(\tau)$ has its maximum value of N_0B at the origin, and it passes through zero at $\tau = \pm n/2B$, where $n = 1, 2, 3, \ldots$.

EXAMPLE 9 RC LOW-PASS FILTERED WHITE NOISE

Consider next a white-noise process $W(t)$ of zero mean and power spectral density $N_0/2$ applied to a low-pass filter, as in Fig. 4.14*a*. The transfer function of the filter is

$$H(f) = \frac{1}{1 + j2\pi fRC} \tag{4.97}$$

The power spectral density of the noise $N(t)$ appearing at the low-pass RC filter output is therefore (see Fig. 4.14*b*)

$$S_N(f) = \frac{N_0/2}{1 + (2\pi fRC)^2} \tag{4.98}$$

From Example 4 of Chapter 2, we have

$$\exp(-|\tau|) \rightleftharpoons \frac{2}{1 + (2\pi f)^2} \tag{4.99}$$

Therefore, using the time-scaling property of the Fourier transform, we

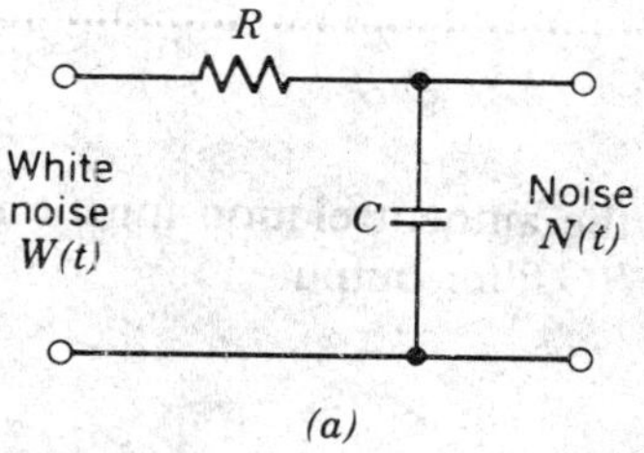

Figure 4.14
Characteristics of RC-filtered white noise. (a) Low-pass RC filter. (b) Power spectral density of filter output N(t). (c) Autocorrelation function of N(t).

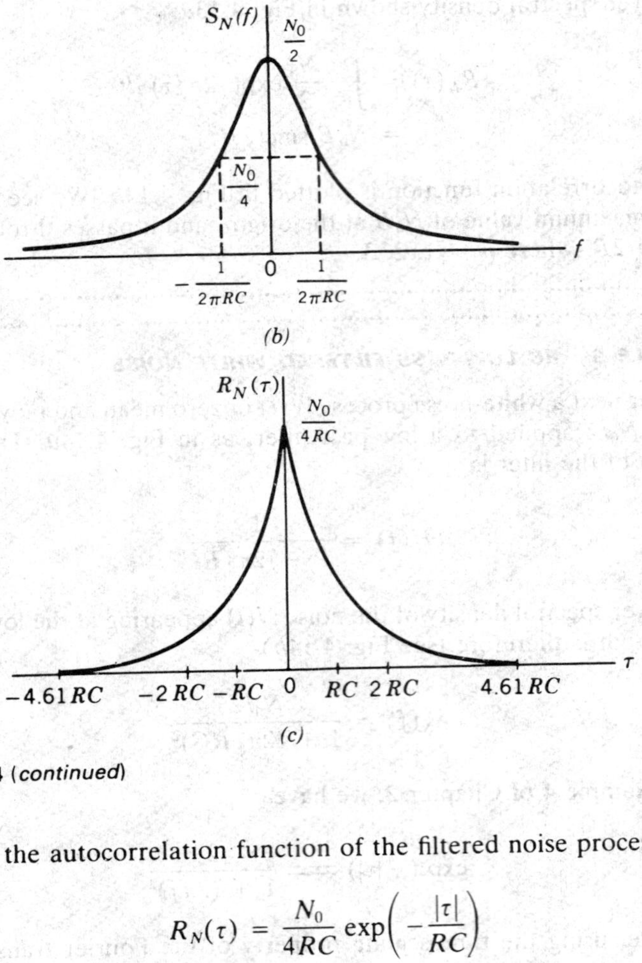

Figure 4.14 *(continued)*

find that the autocorrelation function of the filtered noise process $N(t)$ is

$$R_N(\tau) = \frac{N_0}{4RC} \exp\left(-\frac{|\tau|}{RC}\right) \tag{4.100}$$

which is plotted in Fig. 4.14*c*.

EXERCISE 16 Using the autocorrelation function of Eq. 4.100, find the average power of the *RC* filter output.

EXERCISE 17 Using the power spectral density of Eq. 4.98, find the average power of the *RC* filter output. Check this result against that of Exercise 16.

EXAMPLE 10 AUTOCORRELATION OF A SINUSOIDAL WAVE PLUS WHITE NOISE

Consider a random process $X(t)$ consisting of a sinusoidal wave component and a white-noise process of zero mean and power spectral density $N_0/2$. A sample function (i.e., single realization) of $X(t)$ is denoted by

$$x(t) = A\cos(2\pi f_c t + \theta) + w(t) \tag{4.101}$$

The phase θ of the sinusoidal component may lie anywhere inside the interval $-\pi \leqslant \theta \leqslant \pi$ with equal likelihood. The problem is to determine the autocorrelation function of the random process $X(t)$ represented by the sample function $x(t)$.

The two components of $x(t)$ originate from independent sources. Therefore, the autocorrelation function of $X(t)$ is the sum of the individual autocorrelation functions of the sinusoidal wave and white-noise components. In Example 6, we showed that the autocorrelation function of the sinusoidal component is equal to $(A^2/2)\cos(2\pi f_c\tau)$. The autocorrelation function of the white-noise component is equal to $(N_0/2)\delta(\tau)$. We may therefore write

$$R_X(\tau) = \frac{A^2}{2}\cos(2\pi f_c\tau) + \frac{N_0}{2}\delta(\tau) \tag{4.102}$$

which is plotted in Fig. 4.15. We thus see that for $|\tau| > 0$, the autocorrelation function of the random process $X(t)$ is the same as that of the sinusoidal wave component. This shows that by determining the autocorrelation function of $X(t)$ we can detect the presence of a periodic signal component that is corrupted by additive white noise.

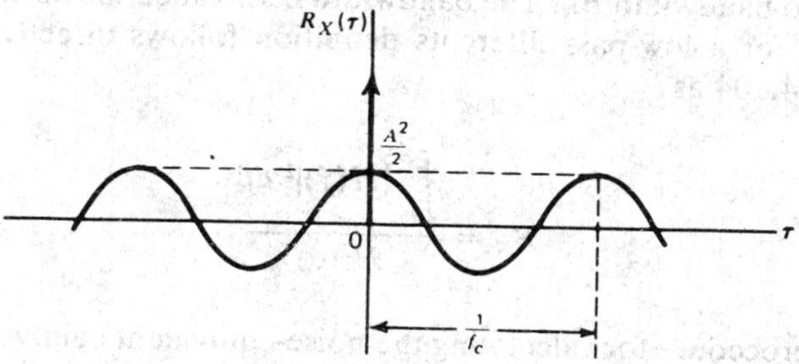

Figure 4.15
Autocorrelation function of sinusoidal wave plus white noise.

4.8 NOISE-EQUIVALENT BANDWIDTH

In Example 8 we observed that when a source of white noise of zero mean and power spectral density $N_0/2$ is connected across the input of an ideal low-pass filter of bandwidth B and passband amplitude response of one, the average output noise power [or equivalently $R_N(0)$] is equal to $N_0 B$. In Example 10 we observed that when such a similar noise source is connected to the input of the simple RC low-pass filter of Fig. 4.14*a*, the corresponding value of the average output noise power is equal to $N_0/(4RC)$. For this filter, the half-power or 3-dB bandwidth is equal to $1/(2\pi RC)$. We may therefore make two important observations. First, filtered white noise has *finite* average power. Second, the average power is proportional to bandwidth.

We may generalize these observations to include all kinds of low-pass filters by defining a noise-equivalent bandwidth as follows. Suppose that we have a source of white noise of zero mean and power spectral density $N_0/2$ connected to the input of an arbitrary low-pass filter of transfer function $H(f)$. The resulting average output noise power is therefore

$$\begin{aligned} P_N &= \frac{N_0}{2}\int_{-\infty}^{\infty} |H(f)|^2\, df \\ &= N_0 \int_0^{\infty} |H(f)|^2\, df \end{aligned} \tag{4.103}$$

where, in the last line, we have made use of the fact that the amplitude response $|H(f)|$ is an even function of frequency. Consider next the same source of white noise connected to the input of an ideal low-pass filter of zero-frequency response $H(0)$ and bandwidth B_N. In this case, the average output noise power is

$$P_N = N_0 B_N H^2(0) \tag{4.104}$$

Equation 4.104 shows that the filtered noise power P_N is finite and proportional to bandwidth B_N. The bandwidth B_N is called the *noise-equivalent bandwidth* for a low-pass filter; its definition follows directly from Eqs. 4.103 and 4.104 as

$$B_N = \frac{\int_0^{\infty} |H(f)|^2\, df}{H^2(0)} \tag{4.105}$$

Thus the procedure for calculating the noise-equivalent bandwidth consists of replacing the arbitrary low-pass filter of transfer function $H(f)$ by an equivalent ideal low-pass filter of zero-frequency response $H(0)$ and bandwidth B_N, as illustrated in Fig. 4.16.

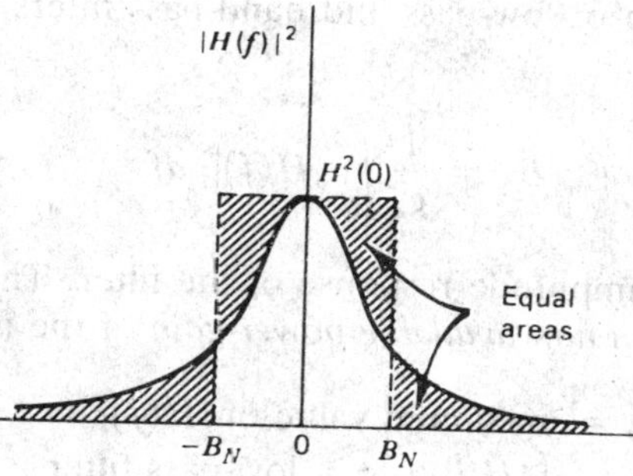

Figure 4.16
The definition of noise-equivalent bandwidth for low-pass filter.

In a similar way, we may define a noise-equivalent banc band-pass filter, as illustrated in Fig. 4.17; this figure depicts the amplitude response of the filter for positive frequencies only. Thu noise-equivalent bandwidth for a band-pass filter may be defined as

$$B_N = \frac{\int_0^\infty |H(f)|^2 \, df}{|H(f_c)|^2} \tag{4.106}$$

where $|H(f_c)|$ is the center-frequency amplitude response of the filter.

We may combine the definitions of Eqs. 4.105 and 4.106 for the noise-

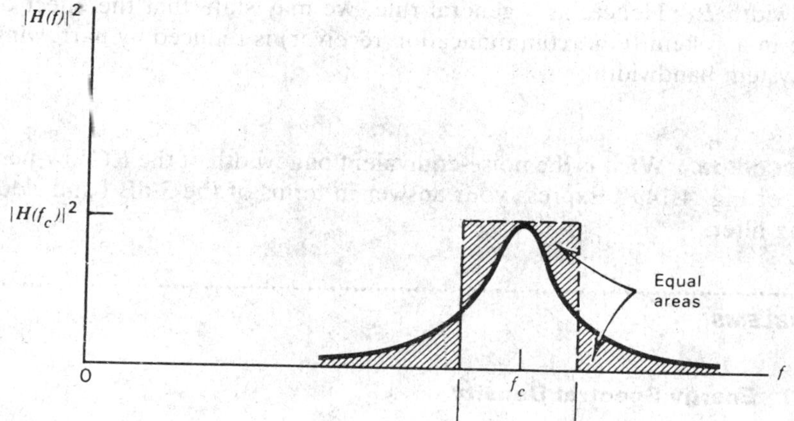

Figure 4.17
The definition of noise equivalent bandwidth for band-pass filters; only the response for positive frequencies is shown.

equivalent bandwidth of low-pass and band-pass filters into a single for-

$$B_N = \frac{1}{g_a}\int_0^{\infty} |H(f)|^2 \, df \tag{4.107}$$

$|H(f)|$ is the amplitude response of the filter. The parameter g_a in 107 is the *maximum available power gain* of the filter, defined by

$$\begin{aligned} g_a &= \text{maximum value of } |H(f)|^2 \\ &= \begin{cases} |H(0)|^2, & \text{low-pass filter} \\ |H(f_c)|^2, & \text{band-pass filter} \end{cases} \end{aligned} \tag{4.108}$$

Correspondingly, we may express the output noise power of a filter (for both positive and negative frequencies) as

$$P_N = N_0 g_a B_N \tag{4.109}$$

where $N_0/2$ is the noise power spectral density at the filter input. According to Eq. 4.109, the effect of passing white noise through a filter may be separated into two parts:

1. The maximum available power gain of the filter, g_a.
2. The noise-equivalent bandwidth B_N, representing *relative frequency selectivity* of the filter.

Eq. 4.109 also shows that, whether the filter of interest is low-pass or band-pass, the filtered noise power P_N is proportional to the noise-equivalent bandwidth B_N. Hence, as a general rule, we may state that the effect of noise in a system (e.g., communication receiver) is reduced by narrowing the system bandwidth.

EXERCISE 18 What is the noise-equivalent bandwidth of the RC low-pass filter of Fig. 4.14*a*? Express your answer in terms of the 3-dB bandwidth of the filter.

PROBLEMS

P4.1 Energy Spectral Density

Problem 1 Consider the decaying exponential pulse

$$g(t) = \begin{cases} \exp(-at), & t > 0 \\ \dfrac{1}{2}, & t = 0 \\ 0, & t < 0 \end{cases}$$

Find the percentage of the total energy of $g(t)$ contained inside the frequency band $-W \leq f \leq W$, where $W = a/2\pi$.

Problem 2 Show that the two different pulses defined in parts a and b of Fig. P4.1 have the same energy spectral density:

$$\psi_g(f) = \frac{4A^2T^2 \cos^2(\pi Tf)}{\tau^2(4T^2f^2 - 1)^2}$$

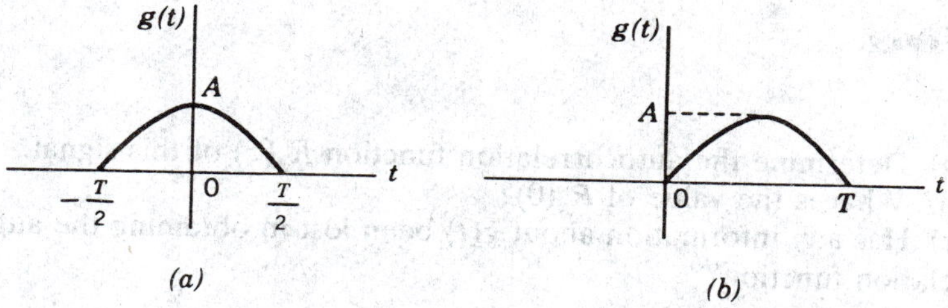

Figure P4.1

P4.2 Correlation of Energy Signals

Problem 3 Determine and sketch the autocorrelation functions of the following exponential pulses:

(a) $g(t) = \exp(-at)u(t)$
(b) $g(t) = \exp(-a|t|)$
(c) $g(t) = \exp(-at)u(t) - \exp(at)u(-t)$

where $u(t)$ is the unit step function, and $u(-t)$ is its time-reversed version.

Problem 4 Determine and sketch the autocorrelation function of a Gaussian pulse defined by

$$g(t) = \frac{1}{t_0} \exp\left(-\frac{\pi t^2}{t_0^2}\right)$$

Problem 5 The Fourier transform of a signal is defined by $|\text{sinc}(f)|$. Show that the autocorrelation function of this signal is triangular in form.

Problem 6 Specify two distinctly different pulse signals that have exactly the same autocorrelation function.

Problem 7 Consider a signal $g(t)$ defined by

$$g(t) = A_0 + A_1 \cos(2\pi f_1 t + \theta_1) + A_2 \cos(2\pi f_2 t + \theta_2)$$

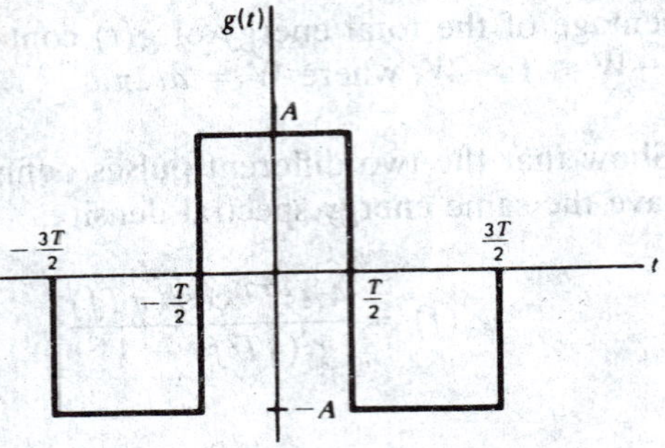

Figure P4.2

(a) Determine the autocorrelation function $R_g(\tau)$ of this signal.
(b) What is the value of $R_g(0)$?
(c) Has any information about $g(t)$ been lost in obtaining the autocorrelation function?

Problem 8 Determine the autocorrelation function of the triplet pulse shown in Fig. P4.2

Problem 9 Let $G(f)$ denote the Fourier transform of a real-valued energy signal $g(t)$, and $R_g(\tau)$ its autocorrelation function. Show that

$$\int_{-\infty}^{\infty} \left[\frac{dR_g(\tau)}{d\tau}\right]^2 d\tau = 4\pi^2 \int_{-\infty}^{\infty} f^2 \, |G(f)|^4 \, df$$

Problem 10 Determine the cross-correlation function $R_{12}(\tau)$ of the pair of rectangular pulses shown in Fig. P4.3, and sketch it. What is $R_{21}(\tau)$?

Problem 11 Determine the cross-correlation function $R_{12}(\tau)$ of the rectangular pulse $g_1(t)$ and triplet pulse $g_2(t)$ shown in Fig. P4.4, and sketch it. What is $R_{21}(\tau)$? Are these signals orthogonal? Why?

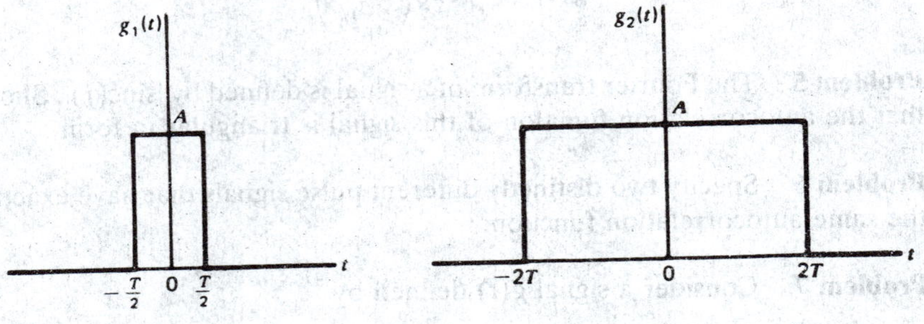

Figure P4.3

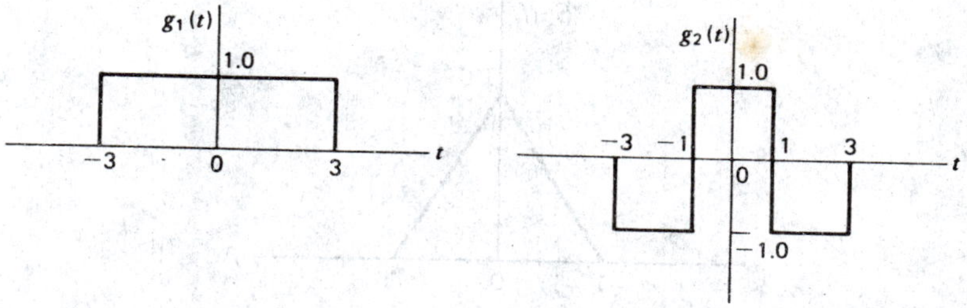

Figure P4.4

Problem 12 Consider two signals $g_1(t)$ and $g_2(t)$. These two signals are delayed by amounts equal to t_1 and t_2 seconds, respectively. Show that the time delays are additive in convolving the pair of delayed signals, whereas they are subtractive in cross-correlating them.

P4.3 Power Spectral Density

Problem 13 Consider the truncated version of a complex exponential, defined by

$$g_T(t) = A \exp(j2\pi f_c t)\,\mathrm{rect}\left(\frac{t}{2T}\right)$$

where rect$(t/2T)$ is a rectangular function of unit amplitude and duration $2T$. Find the power spectral density of $g_T(t)$ for finite T. What is the limiting value of this power spectral density as T approaches infinity?

Problem 14 Figure P4.5 shows the power spectral density of a power signal $g(t)$. Find the average power of the signal.

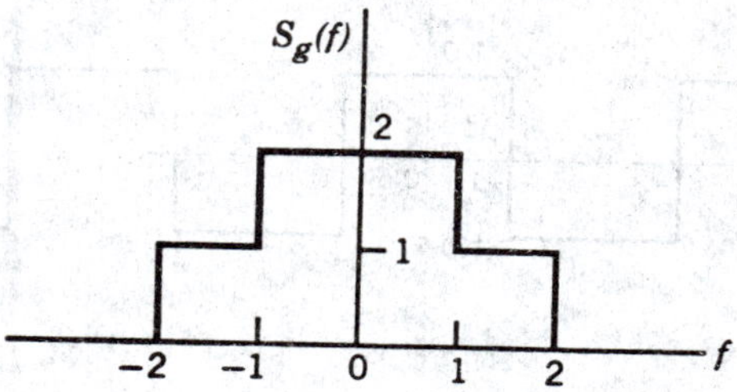

Figure P4.5

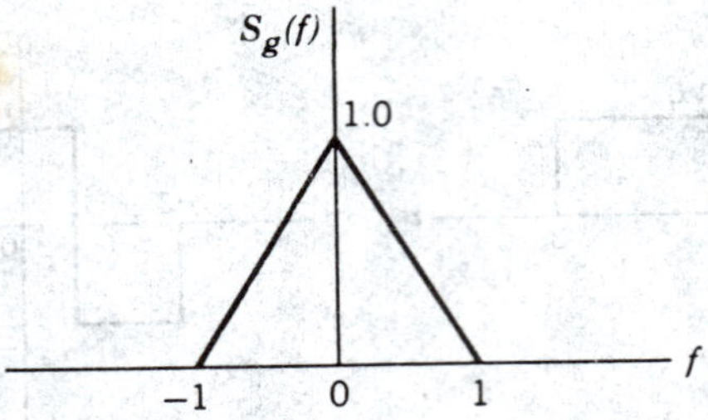

Figure P4.6

P4.4 Correlation of Power Signals

Problem 15 Find the autocorrelation function of the truncated version of a complex exponential, defined in Problem 13. What is the limiting value of this autocorrelation as T approaches infinity?

Problem 16 Find the autocorrelation function of a power signal $g(t)$ whose power spectral density is depicted in Fig. P4.6. What is the value of this autocorrelation function at the origin?

P4.6 Spectral Characteristics of Periodic Signals

Problem 17 Consider the square wave shown in Fig. P4.7. Find the power spectral density, average power, and autocorrelation function of this square wave. Does the wave have dc power? Explain your answer.

Problem 18 Consider two periodic signals $g_{p1}(t)$ and $g_{p2}(t)$ that have the following complex Fourier series representations:

$$g_{p1}(t) = \sum_{n=-\infty}^{\infty} c_{1,n} \exp\left(-\frac{j2\pi nt}{T_0}\right)$$

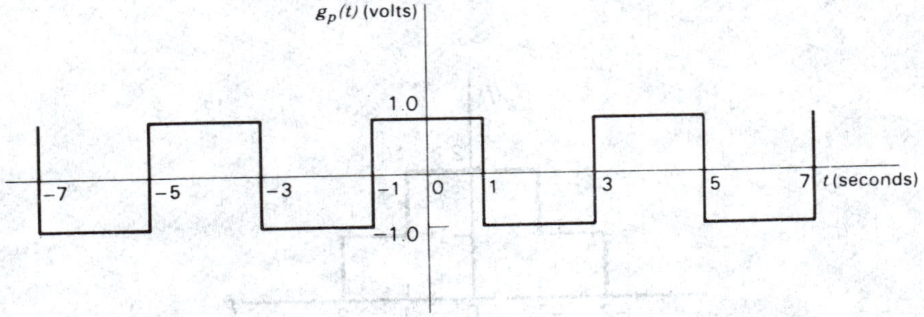

Figure P4.7

and

$$g_{p2}(t) = \sum_{n=-\infty}^{\infty} c_{2,n} \exp\left(-\frac{j2\pi nt}{T_0}\right)$$

The two signals have a common period equal to T_0.

Using the following definition of cross-correlation for a pair of periodic signals,

$$R_{12}(\tau) = \frac{1}{T_0}\int_{-T_0/2}^{T_0/2} g_{p1}(t)g_{p2}(t-\tau)\,dt$$

show that the prescribed pair of periodic signals satisfies the Fourier transform pair

$$R_{12}(\tau) \rightleftharpoons \sum_{n=-\infty}^{\infty} c_{1,n}c_{2,n}^{*}\,\delta\left(f - \frac{n}{T_0}\right)$$

P4.7 Spectral Characteristics of Random Signals and Noise

Problem 19 The power spectral density of a random process $X(t)$ is shown in Fig. P4.8.

(a) What is the dc power contained in this random process?
(b) What is the ac power contained in it?

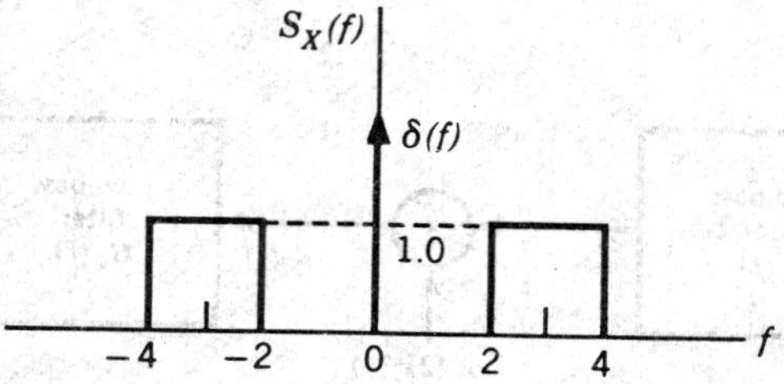

Figure P4.8

Problem 20 A white noise process of zero mean and power spectral density $N_0/2$ is applied to the low-pass RL filter shown in Fig. 4.9. Deter-

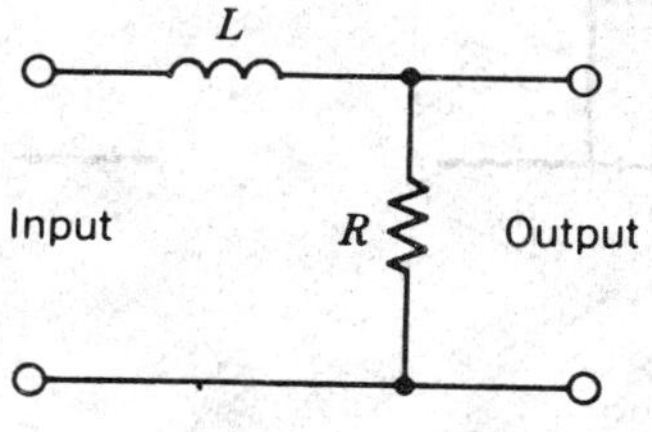

Figure P4.9

mine the power spectral density and autocorrelation function of the output.

Problem 21 Consider a white-noise process of zero mean and power spectral density $N_0/2$ applied to the input of the system shown in Fig. P4.10.

(a) Find the power spectral density of the random process at the output of the system.
(b) What is the average power of this output?

Hint: You may use Eq. 4.52, interpreted for a random process, to evaluate the power spectral density of the low-pass filter input.

P4.8 Noise-Equivalent Bandwidth

Problem 22 Find the noise-equivalent bandwidth for the low-pass *RL* filter shown in Fig. P4.9.

Problem 23 A white-noise process of power spectral density $N_0/2$ is applied to a *Butterworth* low-pass filter of order *n* with its amplitude response

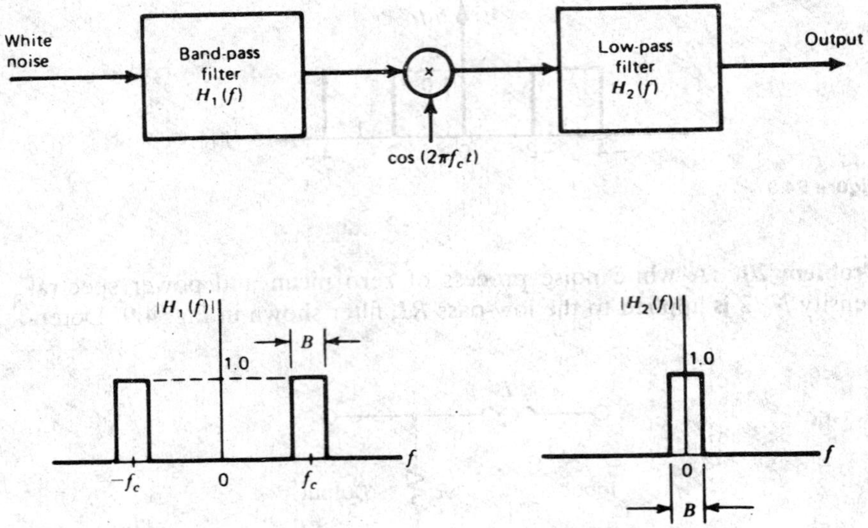

Figure P4.10

defined by

$$|H(f)| = \frac{1}{[1 + (f/f_0)^{2n}]^{1/2}}$$

(a) Determine the noise-equivalent bandwidth for this low-pass filter.
(b) What is the limiting value of the noise-equivalent bandwidth as n approaches infinity?

CHAPTER FIVE

DIGITAL CODING OF ANALOG WAVEFORMS

To transport an information-bearing signal from one point to another over a communication channel, we may use digital or analog techniques. As mentioned in Chapter 1, the use of digital communications offers several important advantages as compared to analog communications. In particular, a digital communication system offers the following highly attractive features:

1. *Ruggedness* to channel noise and external interference, unmatched by any analog communication system.
2. *Flexible* operation of the system.
3. *Integration* of diverse sources of information into a common format.
4. *Security* of information in the course of its transmission from source to destination.

For these reasons, digital communications have become the dominant form of communication technology in our society.

To handle the transmission of analog message signals (e.g., voice and video signals) by digital means, the signal has to undergo an *analog-to-digital conversion.* In the next section, we present an overview of three important methods of analog-to-digital conversion, which are known as *pulse-code modulation, differential pulse-code modulation,* and *delta modulation.* Their detailed descriptions are presented in subsequent sections of the chapter.

5.1 DIGITAL PULSE MODULATION

The process of analog-to-digital conversion is sometimes referred to as *digital pulse modulation.* The use of the terminology "pulse modulation" is justified by virtue of the fact that the first operation performed in the conversion of an analog signal into digital form involves the representation of the signal by a sequence of uniformly spaced *pulses,* the amplitude of which is *modulated* by the signal. Naturally, the pulse-repetition frequency must be chosen in accordance with the *sampling theorem.* In both pulse-code modulation and differential pulse-code modulation, the pulse-repetition frequency or the sampling rate is chosen to be *slightly greater* than the Nyquist rate (i.e., greater than twice the highest frequency component) of the analog signal. In delta modulation, on the other hand, the sampling rate is purposely chosen to be *much greater* than the Nyquist rate. The reason for such a choice in the latter case is to increase correlation between adjacent samples derived from the information-bearing analog signal and thereby to simplify the physical implementation of the delta modulation process. The distinguishing feature between pulse-code modulation and differential pulse-code modulation is that in the latter case, additional circuitry (designed to perform linear prediction) is used to exploit the correlation between adjacent samples of the analog signal so as to reduce the transmitted bit rate.

Figure 5.1 summarizes the comparison between delta modulation, pulse-code modulation, and differential pulse-code modulation in the context of two important system features: *circuit complexity* and *transmitted bit rate.* The bit rate refers to the rate at which *bits* (*bi*nary digi*ts*) constituting the digital version of an analog information-bearing signal are transmitted over the communication channel.

Pulse-code modulation is usually viewed as a *benchmark* against which other methods of digital pulse modulation are measured in performance and circuit complexity. It is therefore appropriate that we begin our study of digital pulse modulation by considering the operations involved in pulse-code modulation, which we do in the next section.

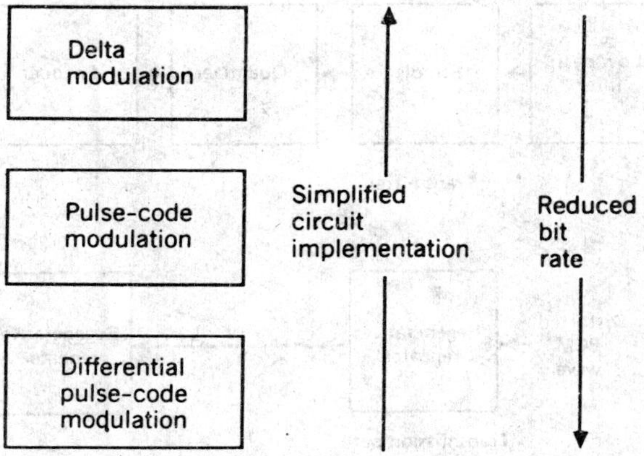

Figure 5.1
Diagrammatic comparison of the three basic forms of digital pulse modulation.

5.2 PULSE-CODE MODULATION

Pulse-code modulation[1] (PCM) is complex in the sense that the message signal is subjected to a great number of operations. The essential operations in the transmitter of a PCM system are *sampling, quantizing,* and *encoding,* as shown in Fig. 5.2. The quantizing and encoding operations are usually performed in the same circuit, which is called an *analog-to-digital converter.* The essential operations in the receiver are *regeneration* of impaired signals, *decoding,* and *demodulation* of the train of quantized samples. These operations are usually performed in the same circuit, which is called a *digital-to-analog converter.* At intermediate points along the transmission route from the transmitter to the receiver, *regenerative repeaters* are used to reconstruct (regenerate) the transmitted sequence of coded pulses in order to combat the accumulated effects of signal distortion and noise.

Quantizing refers to the use of a *finite set of amplitude levels* and the selection of a level nearest to a particular sample value of the message signal as the *representation* for it. This operation, combined with sampling, permits the use of *coded pulses* for representing the message signal. Indeed, it is the combined use of quantizing and coding that distinguishes pulse-code modulation from analog modulation techniques.

In the next three sections, we discuss the operations of sampling, quantizing, and coding, in that order.

[1]Pulse-code modulation is the oldest method for analog-to-digital conversion. It was invented by Reeves in 1937. For a historical account of this invention, see Reeves (1975).

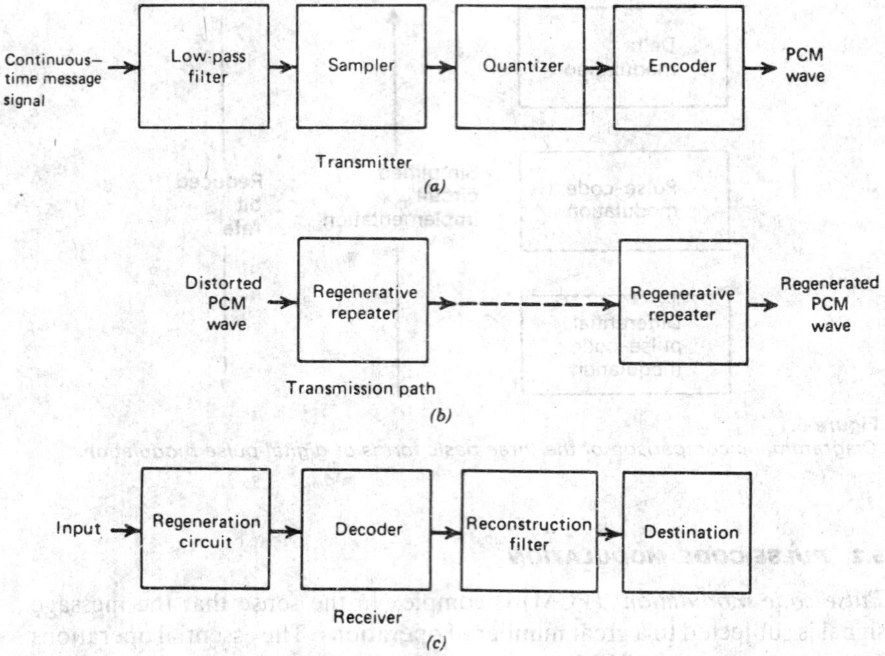

Figure 5.2
The basic elements of a PCM system. (a) *Transmitter.* (b) *Transmission path.* (c) *Receiver.*

5.3 SAMPLING

The sampling operation is performed in accordance with the *sampling theorem.* Specifically, we may state the sampling theorem for band-limited signals of finite energy in two equivalent parts (see Section 2.7):

1. *A band-limited signal of finite energy, which has no frequency components higher than W hertz, is completely described by specifying the values of the signal at instants of time separated by $1/2W$ seconds.*
2. *A band-limited signal of finite energy, which has no frequency components higher than W hertz, may be completely recovered from a knowledge of its samples taken at the rate of $2W$ per second.*

Part 1 of the sampling theorem is exploited in the transmitter; part 2 of the theorem is exploited in the receiver. The sampling rate $2W$ is called the *Nyquist rate,* and its reciprocal $1/2W$ is called the *Nyquist interval.*

The derivation of the sampling theorem, presented in Section 2.7, was based on the assumption that the message signal of interest is strictly band-limited. In practice, however, the amplitude spectrum of the signal approaches zero asymptotically as the frequency approaches infinity, as illustrated in Fig. 5.3*a*. This factor gives rise to an effect called *aliasing* or *fold-over,* which refers to a high-frequency component in the spectrum of the message signal apparently taking on the identity of a lower frequency in the spectrum of a sampled version of the signal.

The aliasing effect is illustrated in Fig. 5.3*b*. This figure shows the message spectrum and two frequency-shifted replicas of it; one replica is shifted to the right by the sampling rate $f_s = 2W$, and the other replica is shifted to the left by f_s. These replicas are manifestations of the periodic spectrum that results from sampling the message signal at the rate f_s; see Section 2.7. Inspection of the spectrum of the sampled signal, which is the sum of the message spectrum and its frequency-shifted replicas, shows that we are no longer able to recover the original message spectrum without distortion, owing to the presence of aliasing.

The presence of aliasing results in signal *distortion*. To combat the effects of aliasing in practice, we use two corrective measures:

1. Prior to sampling, a low-pass *pre-alias filter* is used to attenuate those high-frequency components of the signal that lie outside the band of interest.
2. The filtered signal is sampled at a rate higher than the Nyquist rate.

Figure 5.4 is the block diagram of a system for performing the sampling process. The low-pass pre-alias filter is included at the input of the sampling system, in accordance with point 1. The sampling rate is determined in accordance with point 2 by setting the pulse repetition frequency f_s of the

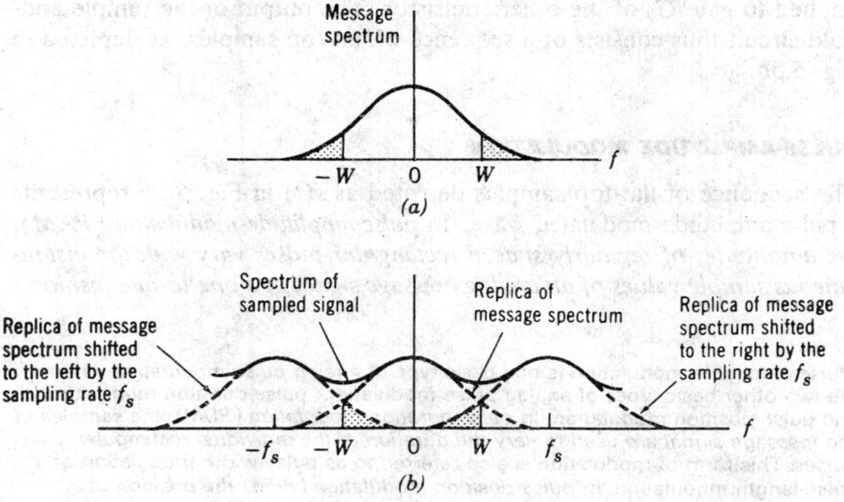

Figure 5.3
The aliasing effect. (a) *Message spectrum.* (b) *Spectrum of sampled signal.*

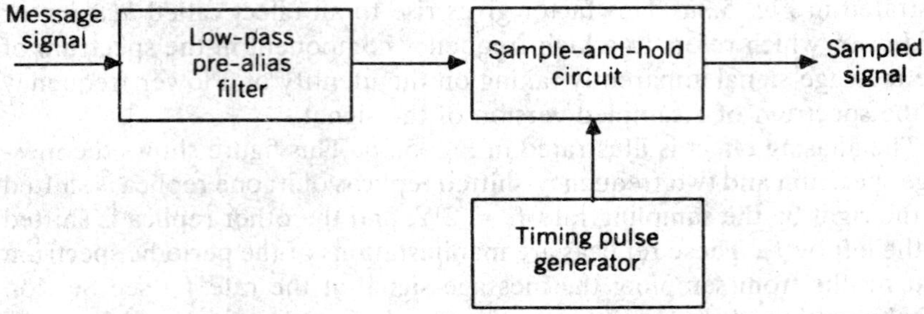

Figure 5.4
Practical sampling circuit arrangement.

timing pulse generator at a value greater than the Nyquist rate $2W$, where W is the pre-alias filter bandwidth.

SAMPLE-AND-HOLD CIRCUIT

The generation of samples is actually performed by a functional block termed the *sample-and-hold circuit* in Fig. 5.4. This circuit produces *flat-top samples* rather than the idealized instantaneous samples as postulated by the sampling theorem. Basically, the sample-and-hold circuit consists of two field-effect transistor (FET) switches and a capacitor connected together as in Fig. 5.5*a*. The "sampling switch" is closed briefly by a short pulse applied to gate G_1 of one transistor. The capacitor is thereby quickly charged up to a voltage equal to the instantaneous sample value of the incoming signal. It holds the sampled voltage until discharged by a pulse applied to gate G_2 of the other transistor. The output of the sample-and-hold circuit thus consists of a sequence of flat-top samples, as depicted in Fig. 5.5*b*.

PULSE-AMPLITUDE MODULATION

The sequence of flat-top samples depicted as $s(t)$ in Fig. 5.5*b* represents a pulse-amplitude modulated wave. In *pulse-amplitude modulation* (*PAM*), *the amplitudes of regularly spaced rectangular pulses vary with the instantaneous sample values of an analog message signal in a one-to-one fashion.*[2]

[2]Pulse-amplitude modulation is one basic type of *analog pulse modulation*. There are two other basic types of analog pulse modulation: pulse-duration modulation and pulse-position modulation. In *pulse-duration modulation* (*PDM*), *the samples of the message signal are used to vary the duration of the individual rectangular pulses*. This form of modulation is also referred to as pulse-width modulation or pulse-length modulation. In *pulse-position modulation* (*PPM*) *the position of a pulse relative to its unmodulated time of occurrence is varied in accordance with the message signal*. For a detailed discussion of analog pulse modulation techniques, see Carlson (1986, Chapter 10) or Black (1953).

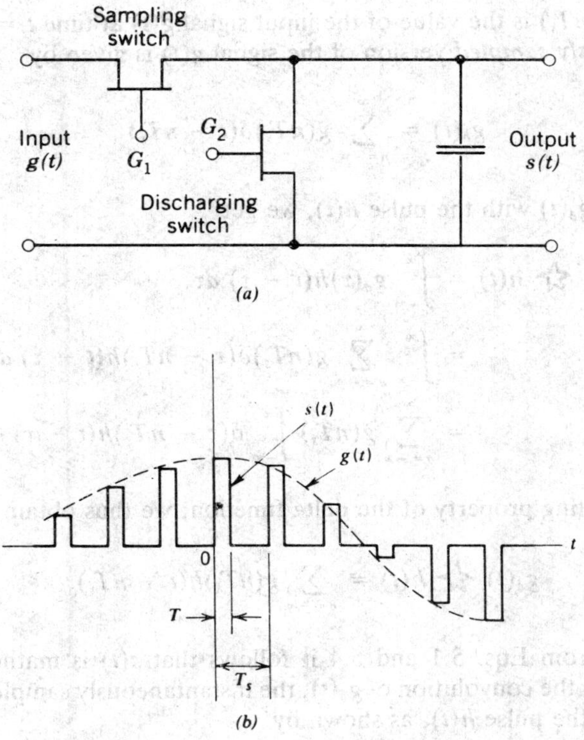

Figure 5.5
(a) *Sample-and-hold circuit.* (b) *Flat-top samples.*

The waveform denoted by $s(t)$ in Fig. 5.5*b* befits this definition exactly. Note that in PAM the carrier wave consists of a periodic train of rectangular pulses, and the carrier frequency (i.e., the pulse repetition frequency) is the same as the sampling rate.

For a mathematical representation of the PAM wave $s(t)$, we may write

$$s(t) = \sum_{n=-\infty}^{\infty} g(nT_s)h(t - nT_s) \qquad (5.1)$$

The term $h(t)$ is a rectangular pulse of unit amplitude and duration T, defined as follows (see Fig. 5.6*a*)

$$h(t) = \begin{cases} 1, & 0 < t < T \\ \frac{1}{2}, & t = 0,\ t = T \\ 0, & \text{otherwise} \end{cases} \qquad (5.2)$$

The term $g(nT_s)$ is the value of the input signal $g(t)$ at time $t = nT_s$. The *instantaneously sampled* version of the signal $g(t)$ is given by

$$g_\delta(t) = \sum_{n=-\infty}^{\infty} g(nT_s)\delta(t - nT_s) \tag{5.3}$$

Convolving $g_\delta(t)$ with the pulse $h(t)$, we get

$$\begin{aligned} g_\delta(t) \star h(t) &= \int_{-\infty}^{\infty} g_\delta(\tau)h(t - \tau)\, d\tau \\ &= \int_{-\infty}^{\infty} \sum_{n=-\infty}^{\infty} g(nT_s)\delta(\tau - nT_s)h(t - \tau)\, d\tau \\ &= \sum_{n=-\infty}^{\infty} g(nT_s) \int_{-\infty}^{\infty} \delta(\tau - nT_s)h(t - \tau)\, d\tau \end{aligned}$$

Using the sifting property of the delta function, we thus obtain

$$g_\delta(t) \star h(t) = \sum_{n=-\infty}^{\infty} g(nT_s)h(t - nT_s) \tag{5.4}$$

Therefore, from Eqs. 5.1 and 5.4 it follows that $s(t)$ is mathematically equivalent to the convolution of $g_\delta(t)$, the instantaneously sampled version of $g(t)$, and the pulse $h(t)$, as shown by

$$s(t) = g_\delta(t) \star h(t) \tag{5.5}$$

Taking the Fourier transform of both sides of Eq. 5.5 and recognizing that the convolution of two time functions is transformed into the multiplication of their respective Fourier transforms, we get

$$S(f) = G_\delta(f)H(f) \tag{5.6}$$

where $S(f) = F[s(t)]$, $G_\delta(f) = F[g_\delta(t)]$, and $H(f) = F[h(t)]$. In Section 2.7 we showed that instantaneous sampling of the time function $g(t)$ introduces periodicity into the spectrum, as described in Eq. 2.131. This equation is reproduced here in the form

$$G_\delta(f) = f_s \sum_{m=-\infty}^{\infty} G(f - mf_s) \tag{5.7}$$

where $f_s = 1/T_s$ is the sampling rate. Therefore, substitution of Eq. 5.7 into 5.6 yields

$$S(f) = f_s \sum_{m=-\infty}^{\infty} G(f - mf_s)H(f) \tag{5.8}$$

where $G(f) = F[g(t)]$.

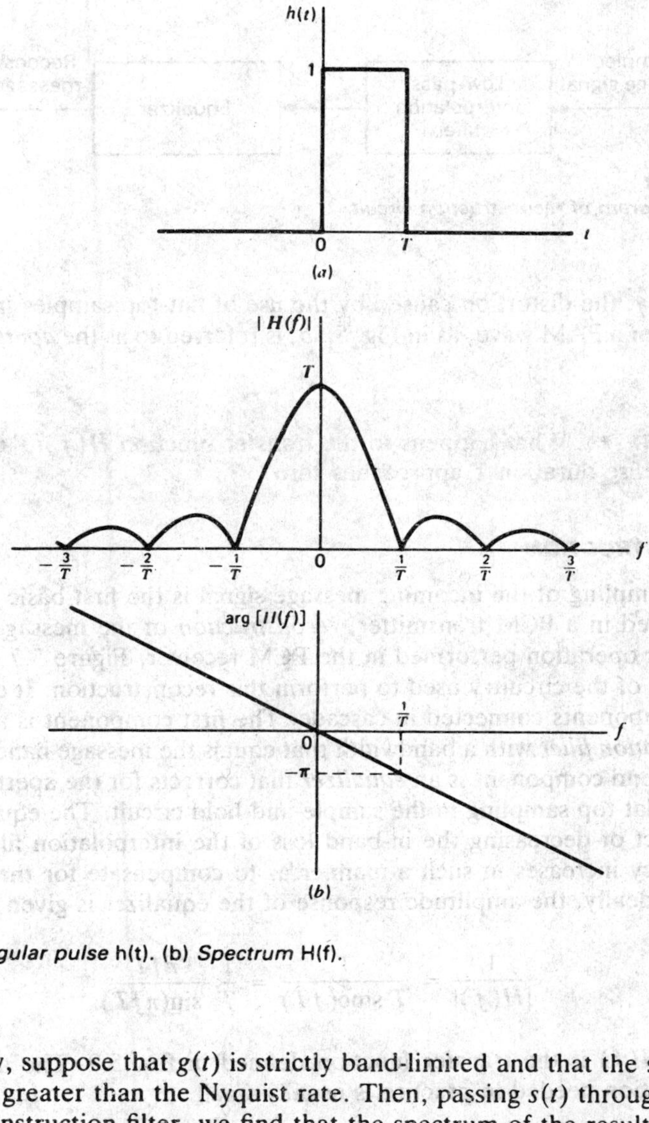

Figure 5.6
(a) *Rectangular pulse* h(t). (b) *Spectrum* H(f).

Finally, suppose that $g(t)$ is strictly band-limited and that the sampling rate f_s is greater than the Nyquist rate. Then, passing $s(t)$ through a low-pass reconstruction filter, we find that the spectrum of the resulting filter output is equal to $G(f)H(f)$. This is equivalent to passing the original analog signal $g(t)$ through a low-pass filter of transfer function $H(f)$.

From Eq. 5.2 we find that

$$H(f) = T \operatorname{sinc}(fT) \exp(-j\pi fT) \tag{5.9}$$

which is plotted in Fig. 5.6*b*. Hence, we see that by using pulse-amplitude modulation to represent an analog message signal we introduce *amplitude distortion* as well as a *delay* of $T/2$. This effect is similar to that caused by the finite size of the scanning aperture in television and facsimile. Ac-

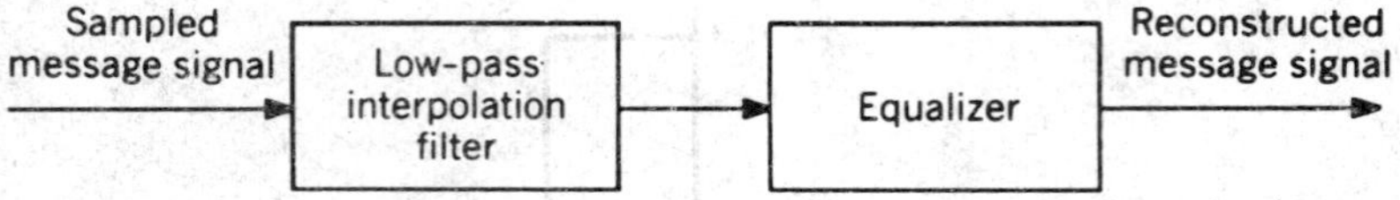

Figure 5.7
Block diagram of reconstruction circuit.

cordingly, the distortion caused by the use of flat-top samples in the generation of a PAM wave, as in Fig. 5.5*b*, is referred to as the *aperture effect.*

EXERCISE 1 What happens to the transfer function $H(f)/T$ of Eq. 5.9 as the pulse duration T approaches zero?

RECONSTRUCTION

Since sampling of the incoming message signal is the first basic operation performed in a PCM transmitter, *reconstruction* of the message signal is the final operation performed in the PCM receiver. Figure 5.7 is a block diagram of the circuitry used to perform this reconstruction. It consists of two components connected in cascade. The first component is a low-pass *interpolation filter* with a bandwidth that equals the message bandwidth W. The second component is an *equalizer* that corrects for the aperture effect due to flat-top sampling in the sample-and-hold circuit. The equalizer has the effect of decreasing the in-band loss of the interpolation filter as the frequency increases in such a manner as to compensate for the aperture effect. Ideally, the amplitude response of the equalizer is given by

$$\frac{1}{|H(f)|} = \frac{1}{T\,\mathrm{sinc}(fT)} = \frac{1}{T}\,\frac{\pi f T}{\sin(\pi f T)}$$

where $H(f)$ is the transfer function defined in Eq. 5.9. The amount of equalization needed in practice is usually small.

EXAMPLE 1

At $f = f_s/2$, which corresponds to the highest frequency component of the message signal for a sampling rate equal to the Nyquist rate, we find from Eq. 5.9 that the amplitude response of the equalizer normalized to that at zero frequency is equal to

$$\frac{1}{\mathrm{sinc}(0.5T/T_s)} = \frac{(\pi/2)(T/T_s)}{\sin[(\pi/2)T/T_s)]}$$

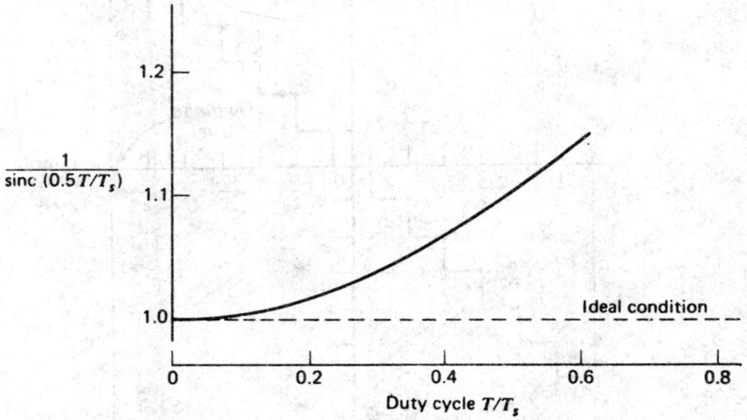

Figure 5.8
Normalized equalization (to compensate for aperture effect) plotted versus T/T_s.

where the ratio T/T_s is equal to the duty cycle of the sampling pulses. In Fig. 5.8 this result is plotted as a function of T/T_s. Ideally, it should be equal to 1 for all values of T/T_s. For a duty cycle of 10%, it is equal to 1.0041. It follows therefore that for duty cycles of less than 10% the aperture effect becomes negligible.

5.4 QUANTIZING

A continuous signal, such as voice, has a continuous range of amplitudes and therefore its samples have a continuous amplitude range. In other words, within the finite amplitude range of the signal we find an infinite number of amplitude levels. It is not necessary in fact to transmit the exact amplitudes of the samples. Any human sense (the ear or the eye), as ultimate receiver, can only detect finite intensity differences. This means that the original continuous signal may be approximated by a signal constructed of discrete amplitudes selected on a minimum error basis from an available set. The existence of a finite number of discrete amplitude levels is a basic condition of PCM. Clearly, if we assign the discrete amplitude levels with sufficiently close spacing, we may make the approximated signal practically indistinguishable from the original continuous signal.

The conversion of an analog (continuous) sample of the signal into a digital (discrete) form is called the *quantizing* process. Graphically, the quantizing process means that a straight line representing the relation between the input and output of a linear continuous system is replaced by a *staircase* characteristic, as in Fig. 5.9*a*. The difference between two adjacent discrete values is called a *quantum* or *step size*. Signals applied to a *quantizer*, with the input–output characteristic of Fig. 5.9*a*, are sorted into

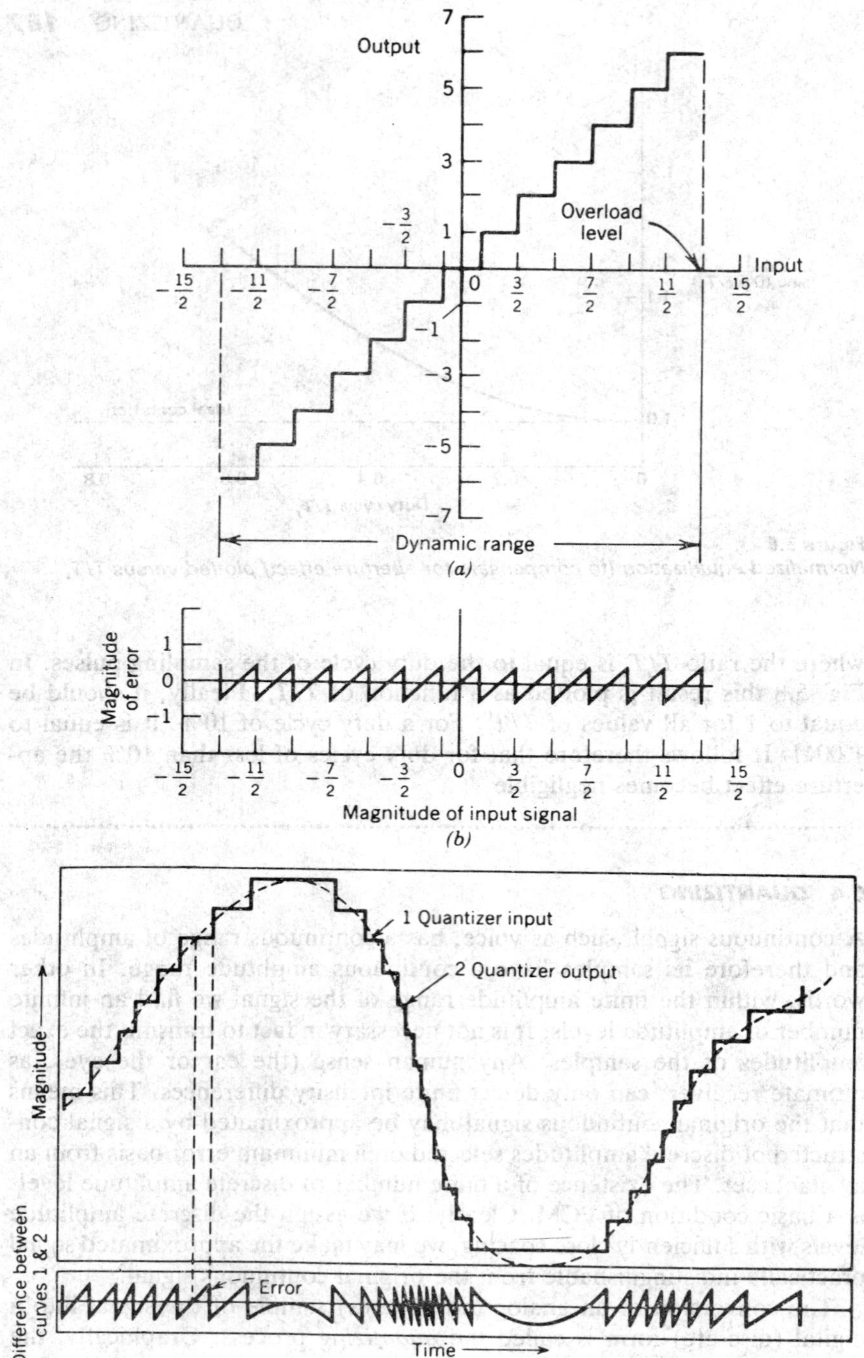

Figure 5.9
The quantizing principle. (a) *Quantizing characteristic.* (b) *Characteristic of errors in quantizing.* (c) *A quantized signal wave and the corresponding error curve. This figure is adapted from Bennett (1948) by permission of AT and T.*

amplitude slices (the treads of the staircase), and all input signals within plus or minus half a quantum step of the midvalue of a slice are replaced in the output by the midvalue in question.

The *quantizing error* consists of the difference between the input and output signals of the quantizer. It is apparent that the maximum instantaneous value of this error is half of one quantum step, and the total range of variation is from minus half a step to plus half a step. In part *b* of Fig. 5.9 the error is shown plotted as a function of the input signal. In part *c* of the figure typical variations of the quantizer input, the quantizer output, and the difference between them (i.e., the quantizing error) as functions of time are indicated.

A quantizer having the input-output amplitude characteristic of Fig. 5.9*a* is said to be of the *midtread* type, because the origin lies in the middle of a tread of the staircase-like graph. According to this characteristic, the quantizer output may be expressed as $i\Delta$, where $i = 0, \pm 1, \pm 2, \ldots, \pm K$. These discrete amplitude values of the quantizer output are called *representation levels*. A quantizer of the midtread type has an *odd* number of representation levels, as shown by

$$L = 2K + 1 \qquad (5.10)$$

The *dynamic range* or *peak-to-peak excursion* of the quantizer input is $L\Delta$. One half of this excursion defines the absolute value of the *overload level* of the quantizer. Clearly, the amplitude of the quantizer input must not exceed the overload level; otherwise, *overload distortion* results.

In the quantizer example illustrated in Fig. 5.9*a*, the step size Δ equals 1; the integer K is 6; and the number of representation levels L is 13. The corresponding absolute value of the overload level is 13/2.

QUANTIZING NOISE

Quantizing noise or *quantizing error* is produced in the transmitting end of a PCM system by *rounding off the sampled values of a continuous message signal to the nearest representation level.* We assume a quantizing process with a uniform step size denoted by Δ volts, so that the representation levels are at $0, \pm\Delta, \pm 2\Delta, \pm 3\Delta, \ldots$. Consider a particular sample at the quantizer input, with an amplitude that lies in the range $i\Delta - (\Delta/2)$ to $i\Delta + (\Delta/2)$, where i is an integer (positive or negative, including zero) and $i\Delta$ defines the corresponding quantizer output. We thus have a region of *uncertainty* of width Δ, centered about $i\Delta$, as illustrated in Fig. 5.10. Let q_e denote the value of the error produced by the quantizing process. Then the amplitude of the sample at the quantizer input is $i\Delta + q_e$. It is apparent that with a random input signal, the quantizing error q_e varies randomly within the interval $-\Delta/2 \leq q_e \leq \Delta/2$.

When the quantization is fine enough (say, the number of representation levels is greater than 64), the distortion produced by quantizing noise affects

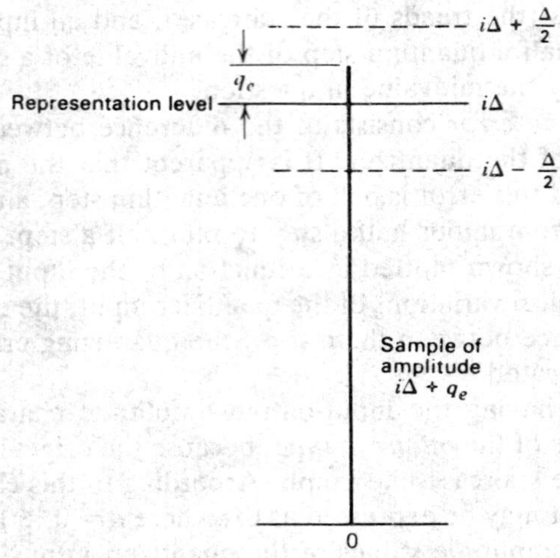

Figure 5.10
Illustrating the quantizing error q_e.

the performance of a PCM system as though it were an additive independent source of noise with zero mean and mean-square value determined by the quantizer step size Δ. The reason for this is that the power spectral density of the quantizing noise in the quantizer output is practically independent of that of the message signal over a wide range of input signal amplitudes. Furthermore, for a message signal of a root mean-square value that is large compared to a quantum step, it is found that the power spectral density of the quantizing noise has a large bandwidth compared with the signal bandwidth. Thus, with the quantizing noise uniformly distributed throughout the signal band, its interfering effect on a signal is similar to that of thermal noise. (Thermal noise is discussed in Appendix C.)

We say that the quantizing noise is *uniformly distributed* when the error may take on a sample value q_e anywhere inside the interval $(-\Delta/2, \Delta/2)$ with equal likelihood. Under this assumption, we may determine the *average power of the quantizing noise* by averaging q_e^2, the squared quantizing error, over all possible values of q_e. We may thus express the *average power of quantizing noise,* P_q, as follows

$$\begin{aligned} P_q &= \frac{1}{\Delta}\int_{-\Delta/2}^{\Delta/2} q_e^2 \, dq_e \\ &= \frac{\Delta^2}{12} \end{aligned} \qquad (5.11)$$

Thus, the average power of quantizing noise grows as the square of the step size Δ. This is perhaps the most often used result in quantization.

The step size Δ is under the designer's control. Hence, the signal distortion due to quantizing noise can be controlled by choosing the step size Δ small enough, as illustrated in the following example.

EXAMPLE 2 SIGNAL-TO-QUANTIZING NOISE RATIO FOR SINUSOIDAL MODULATION

Consider the special case of a *full-load* sinusoidal modulating wave of amplitude A_m, which uses all the representation levels provided. The average signal power is $A_m^2/2$. The peak-to-peak excursion of the quantizer input is $2A_m$, because the modulating wave swings between $-A_m$ and A_m. Assuming that the number of representation levels equals L, the quantizer step size is

$$\Delta = \frac{2A_m}{L} \tag{5.12}$$

Therefore Eq. 5.11 gives the average quantizing noise power as

$$P_q = \frac{A_m^2}{3L^2} \tag{5.13}$$

Thus the *output signal-to-quantizing noise ratio* of the PCM system, for a full-load test tone, is

$$(\mathrm{SNR})_0 = \frac{A_m^2/2}{A_m^2/3L^2} = \frac{3L^2}{2} \tag{5.14}$$

Expressing the signal-to-quantizing noise ratio in decibels, we get

$$10 \log_{10}(\mathrm{SNR})_0 = 1.8 + 20 \log_{10} L \tag{5.15}$$

Hence, the output signal-to-noise ratio of a PCM system in decibels, due to quantizing noise, increases logarithmically with the number of representation levels, L.

TABLE 5.1

Number of Representation Levels, L	Code Word Length n	Signal-to-Quantizing Noise Ratio, dB
32	5	31.8
64	6	37.8
128	7	43.8
256	8	49.8

For various values of L, the corresponding values of signal-to-quantizing noise ratio are as given in Table 5.1. The second column of this table corresponds to the binary code word length, an issue that is considered in Section 5.5.

EXERCISE 2 A sinusoidal signal is transmitted using PCM. The output signal-to-quantizing noise ratio is required to be 55.8 dB. Find the minimum number of representation levels L required to achieve this performance.

COMPANDING

The quantizing process based on Fig. 5.9*a* uses a *uniform separation* between the representation levels. In certain applications, however, it is preferable to use a variable separation between the representation levels. For example, the ratio of voltage levels covered by voice signals, from the peaks of loud talk to the weak passages of weak talk, is on the order of 1000 to 1. The excursions of the voice signal into the large amplitude ranges, which occur in practice relatively infrequently, can be taken care of by using a *nonuniform quantizer.* Such a quantizer is designed so that the step size increases as the separation from the origin of the input-output amplitude characteristic is increased. We thus find that the weak passages, which need more protection, are favored at the expense of the loud passages. In this way, a nearly uniform percentage precision is achieved throughout the amplitude range of the input signal, with the result that fewer steps are needed than would be the case if a uniform quantizer were used.

The use of a nonuniform quantizer is equivalent to passing the baseband signal through a *compressor* and then applying the compressed signal to a uniform quantizer. A particular form of compression law that is used in practice is the so-called *μ-law* defined by

$$|v_2| = \frac{\log(1 + \mu|v_1|)}{\log(1 + \mu)} \tag{5.16}$$

where v_1 and v_2 are normalized input and output voltages, and μ is a positive constant. Figure 5.11*a* plots the μ-law for varying μ. The case of uniform quantization corresponds to $\mu = 0$. For a given value of μ, the reciprocal slope of the compression curve, which defines the quantum steps, is

$$\frac{d|v_1|}{d|v_2|} = \frac{\log(1 + \mu)}{\mu}(1 + \mu|v_1|) \tag{5.17}$$

We see therefore that the μ-law is neither strictly linear nor strictly logarithmic, but it is approximately linear at low input levels corresponding to

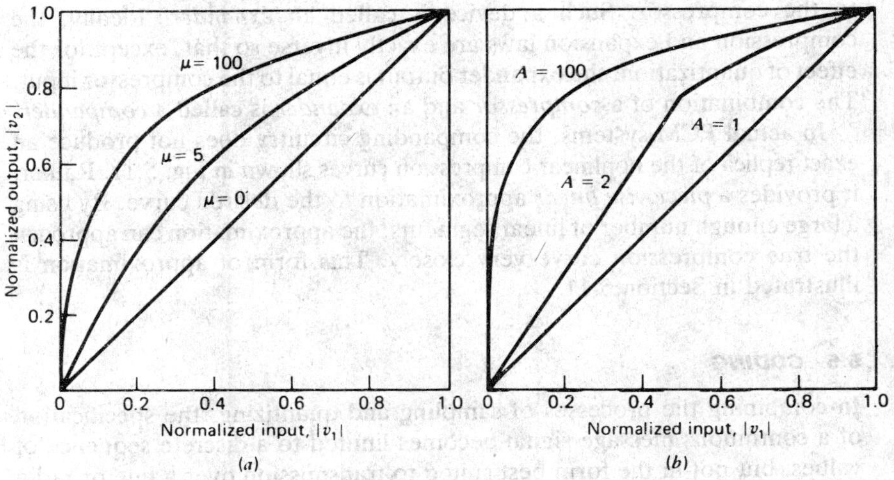

Figure 5.11
Compression laws. (a) μ-*law.* (b) A-*law.*

$\mu|v_1| \ll 1$, and approximately logarithmic at high input levels corresponding to $\mu|v_1| \gg 1$.

Another compression law that is used in practice is the so-called *A-law* defined by

$$|v_2| = \begin{cases} \dfrac{A|v_1|}{1 + \log A} & 0 \leqslant |v_1| \leqslant \dfrac{1}{A} \\ \dfrac{1 + \log(A|v_1|)}{1 + \log A} & \dfrac{1}{A} \leqslant |v_1| \leqslant 1 \end{cases} \tag{5.18}$$

which is shown plotted in Fig. 5.11*b*. Practical values of A (as of μ in the μ-law) tend to be in the vicinity of 100. The case of uniform quantization corresponds to $A = 1$. The reciprocal slope of this compression curve is

$$\frac{d|v_1|}{d|v_2|} = \begin{cases} \dfrac{1 + \log A}{A} & 0 \leqslant |v_1| \leqslant \dfrac{1}{A} \\ (1 + \log A)|v_1| & \dfrac{1}{A} \leqslant |v_1| \leqslant 1 \end{cases} \tag{5.19}$$

Thus the quantum steps over the central linear segment, which have a dominant effect on small signals, are diminished by the factor $A/(1 + \log A)$. This is typically about 25 dB in practice, as compared with uniform quantization.

To restore the signal samples to their correct relative levels, we must, of course, use a device in the receiver with a characteristic complementary to the compressor. Such a device is called an *expander*. Ideally, the compression and expansion laws are exactly inverse so that, except for the effect of quantization, the expander output is equal to the compressor input. The combination of a *compressor* and an *expander* is called a *compander*.

In actual PCM systems, the companding circuitry does not produce an exact replica of the nonlinear compression curves shown in Fig. 5.11. Rather, it provides a *piecewise linear* approximation to the desired curve. By using a large enough number of linear segments, the approximation can approach the true compression curve very closely. This form of approximation is illustrated in Section 5.11.

5.5 CODING

In combining the processes of sampling and quantizing, the specification of a continuous message signal becomes limited to a discrete sequence of values, but not in the form best suited to transmission over a line or radio path. To fully exploit the advantages of sampling and quantizing, we require the use of an *encoding process* to translate the discrete sequence of sample values to a more appropriate form of signal. Any plan for representing each element of this discrete set of values as a particular arrangement of discrete events is called a *code*. One of the discrete events in a code is called a *code element* or *symbol*. For example, the presence or absence of

TABLE 5.2

Ordinal Number of Representation Level	Level Number Expressed as Sum of Powers of 2	Binary Number
0		0000
1	2^0	0001
2	2^1	0010
3	$2^1 - 2^0$	0011
4	2^2	0100
5	$2^2 \quad - 2^0$	0101
6	$2^2 + 2^1$	0110
7	$2^2 + 2^1 - 2^0$	0111
8	2^3	1000
9	$2^3 \quad - 2^0$	1001
10	$2^3 \quad + 2^1$	1010
11	$2^3 \quad + 2^1 - 2^0$	1011
12	$2^3 + 2^2$	1100
13	$2^3 + 2^2 \quad - 2^0$	1101
14	$2^3 + 2^2 + 2^1$	1110
15	$2^3 + 2^2 + 2^1 - 2^0$	1111

a pulse is a symbol. A particular arrangement of symbols used in a code to represent a single value of the discrete set is called a *code word* or *character*.

In a *binary code*, each symbol may be either of two distinct values or kinds, such as the presence or absence of a pulse. The two symbols of a binary code are customarily denoted as 0 and 1. In a *ternary code*, each symbol may be one of three distinct values or kinds, and so on for other codes. However, *the maximum advantage over the effects of noise in a transmission medium is obtained by using a binary code, because a binary symbol withstands a relatively high level of noise and is easy to regenerate.*

Suppose that, in a binary code, each code word consists of *n bits*; the bit is an acronym for *bi*nary digi*t*. Then, using such a code, we may represent a total of 2^n distinct numbers. For example, a sample quantized into one of 128 levels may be represented by a 7-bit code word. There are several ways of establishing a one-to-one correspondence between representation levels and code words. A convenient one is to express the ordinal number of the representation level as a binary number. In the binary number system, each digit has a place-value that is a power of 2, as illustrated in Table 5.2 for the case of $n = 4$.

EXAMPLE 3 SIGNAL-TO-QUANTIZING NOISE RATIO FOR SINUSOIDAL MODULATION (CONTINUED)

In this example, we reformulate the output signal-to-quantizing noise ratio of Eq. 5.15 for a PCM system operating with a full-load test tone. This equation is reproduced here for convenience

$$10 \log_{10}(\text{SNR})_0 = 1.8 + 20 \log_{10} L, \text{ dB}$$

where L is the number of representation levels used in the system. Assuming the use of an n-bit binary code word, we may define L for a quantizer of the midtread type as

$$\begin{aligned} L &= 2^n - 1 \\ &\simeq 2^n, \qquad n \text{ large} \end{aligned} \tag{5.20}$$

Accordingly, we may redefine the output signal-to-quantizing noise ratio in terms of the code word length n as

$$10 \log_{10}(\text{SNR})_0 = (1.8 + 6n), \text{ dB} \tag{5.21}$$

For various values of n, the corresponding values of signal-to-quantizing noise ratio are as given in Table 5.1. The formula of Eq. 5.21 states that each bit in the code word of a PCM system contributes 6 dB to the output signal-to-quantizing noise ratio.

EXAMPLE 4 TRADEOFF BETWEEN CHANNEL BANDWIDTH AND SIGNAL-TO-QUANTIZING NOISE RATIO

We may develop further insight into the performance of a PCM system by examining the relationship between the signal-to-quantizing noise ratio and transmission bandwidth requirement of a binary PCM system. For the purpose of this evaluation, we will again consider the use of a sinusoidal modulating wave.

From our discussion of the sampling process, we have seen that a message signal of bandwidth W requires a minimum sampling rate of $2W$. With each signal sample represented by an n-bit code word, the bit duration T_b has a maximum value of $1/2nW$. In Section 6.4, it is shown that the channel bandwidth B required to transmit a pulse of this duration is given by

$$B = \kappa n W \tag{5.22}$$

where κ is a constant with a value between 1 and 2.

Expressing the output signal-to-quantizing noise ratio simply as a ratio, we have from Eqs. 5.14 and 5.20:

$$(\text{SNR})_0 = \frac{3}{2}(4^n) \tag{5.23}$$

Accordingly, using Eqs. 5.22 and 5.23, we get

$$(\text{SNR})_0 = \frac{3}{2}(4^{B/\kappa W}) \tag{5.24}$$

This relation shows that a PCM system is capable of improving the output signal-to-noise ratio *exponentially* with the bandwidth expansion ratio B/W.

EXERCISE 3 A *television (TV) signal* with a bandwidth of 4.2 MHz is transmitted using binary PCM. The number of representation levels is 512. Calculate the following parameters:

(a) The code word length.
(b) The final bit rate.
(c) The transmission bandwidth, assuming that $\kappa = 2$ in Eq. 5.22.

EXERCISE 4 The frequency content of a *studio-quality audio signal* that we like to hear extends from 20 Hz to 20 kHz. For professional use, the signal is sampled at the rate of 48×10^3 samples per second. The standard code word used for conversion into a PCM format is 16 bits per sample. What is the final bit rate for digital storage of the signal?

DIGITAL FORMATS

To send the encoded digital data over a channel, we require the use of a *format* or *waveform* for representing the data.[3] In this context, we have a number of formats available to us. Figure 5.12 illustrates some commonly used ones for the example of binary sequence 0110100011. Specifically, we have illustrated the following formats:

(a) Symbol 1 is represented by transmitting a pulse of constant amplitude for the duration of the symbol, and symbol 0 is represented by switching off the pulse, as in Fig. 5.12*a*. This type of format is referred to as *on–off* or *unipolar signaling.*
(b) Symbols 1 and 0 are represented by pulses of equal positive and negative amplitudes, as in Fig. 5.12*b*. This type of format is referred to as *polar signaling.*
(c) A rectangular pulse (half-symbol wide) is used for a 1 and no pulse for a 0, as in Fig. 5.12*c*. This type of format is called *return-to-zero* (RZ) *signaling*.
(d) Positive and negative pulses (of equal amplitude) are used alternately for symbol 1, and no pulse for symbol 0, as in Fig. 5.12*d*. This type of format is called *bipolar signaling*. A useful property of this method of signaling is that the power spectrum of the transmitted signal has no dc component and relatively insignificant low-frequency components.
(e) Symbol 1 is represented by a positive pulse followed by a negative pulse, with both pulses being of equal amplitude and half-symbol wide; for symbol 0, the polarities of these pulses are reversed, as in Fig. 5.12*e*. This type of format is called a *split-phase* or *Manchester code*. It also suppresses the dc component and has relatively insignificant low-frequency components.

Note that the polar signal waveform of Fig. 5.12*b* and the Manchester code of Fig. 5.12*e* are examples of *nonreturn-to-zero* (*NRZ*) *signaling*.

The binary code is a special case of *M-ary code.* In practice, we usually find that M, the number of symbols in the code, is an integer power of 2. Then, each code word in the M-ary code carries the equivalent information of $\log_2 M$ bits. Consider, for example, a *four-level* (*quarternary*) *code* (i.e., $M = 4$). In such a code, we may identify four distinct

[3]Digital formats (waveforms) are also referred to in the literature as *line* or *transmission codes.*

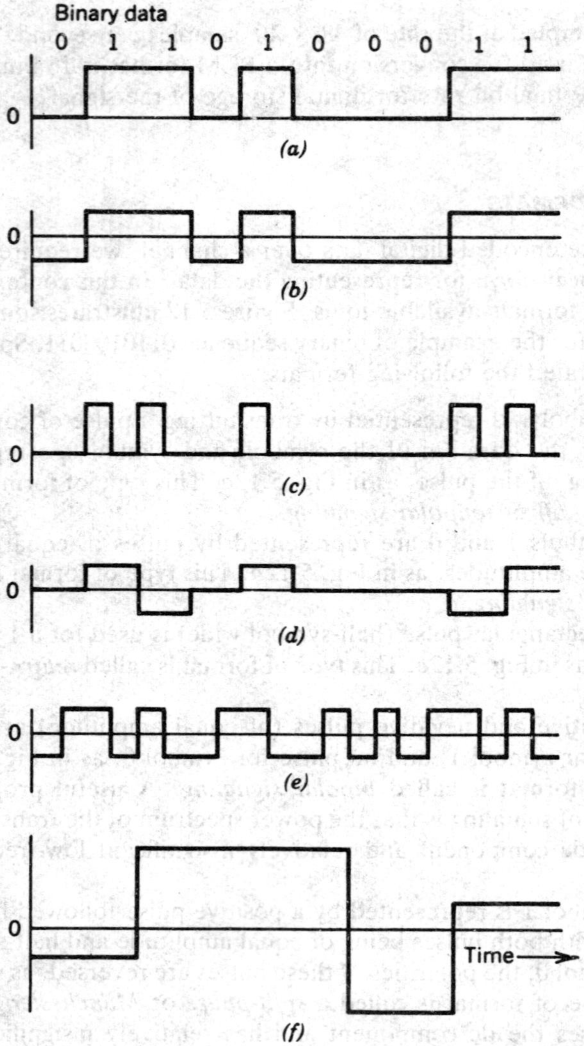

Figure 5.12
Electrical representations of binary data. (a) *On-off signaling.* (b) *Polar signaling.* (c) *Return-to-zero signaling.* (d) *Bipolar signaling.* (e) *Split-phase or Manchester code.* (f) *four-level Gray coding.*

dibits (pairs of bits). In Table 5.3*a*, we show two arrangements for the four possible dibits together with their individual electrical representations. In particular, the dibits are shown in both their *natural code* and *Gray code.* Using the notations of the Gray code in Table 5.3*a*, the binary sequence 0110100011 is thus represented by the waveform shown in Fig. 5.12*f*. To obtain this waveform, the given sequence is viewed as a new sequence of dibits, namely, 01,10,10,00,11, and each dibit is represented in accordance with the Gray code of Table 5.3*a*.

TABLE 5.3 Examples of Natural and Gray Codes

(a) Four-level code

Code Word Number	Natural Code	Gray Code	Electrical Representation
0	00	00	$-\frac{3}{2}$
1	01	01	$-\frac{1}{2}$
2	10	11	$+\frac{1}{2}$
3	11	10	$+\frac{3}{2}$

(b) Eight-level Code

Code Word Number	Natural Code	Gray Code	Electrical Representation
0	000	000	$-\frac{7}{2}$
1	001	001	$-\frac{5}{2}$
2	010	011	$-\frac{3}{2}$
3	011	010	$-\frac{1}{2}$
4	100	110	$+\frac{1}{2}$
5	101	111	$+\frac{3}{2}$
6	110	101	$+\frac{5}{2}$
7	111	100	$+\frac{7}{2}$

The distinguishing feature of a Gray code[4] is that there is a *one-bit change* as we move from one code word to another. This is well illustrated in the two Gray codes shown in Table 5.3 for $M = 4,8$. Note that in Table 5.3*b* for $M = 8$, for example, the rule of a one-bit change per transition applies not only to all the transitions for code word 0 to

[4]The origin of Gray codes goes back to the development of the rotary form of mechanical encoders known as *shaft encoders.* The use of Gray encoding makes it possible for electromechanical arrangements to change from one code word to another by changing the state of a single digit. With natural encoding, on the other hand, two or more digits may be required to change state *simultaneously,* which is difficult for electromechanical devices.

code word 1, from code word 1 to code word 2, and so on up to the transition from code word 6 to code word 7, but also to the "wrap-around" transition from code word 7 to code word 0. This wrap-around feature makes a Gray encoder *cyclic* in nature.

The choice of a particular digital waveform is influenced by the application of interest. Nevertheless, it is highly desirable for a digital waveform to have the following properties:

1. *Timing content.* The transmitted digital waveform should have adequate timing content to permit the extraction of *clock* information required for the purpose of synchronizing the receiver to the transmitter.
2. *Ruggedness.* The waveform should possess immunity to channel noise and interference for prescribed channel bandwidth and transmitted power.
3. *Error detection capability.* The waveform should permit the detection of errors that may occur in the course of transmission due to the presence of channel noise.
4. *Matched power spectrum.* The power spectral density of the transmitted digital waveform should match the frequency response of the channel as closely as possible so as to minimize signal distortion.
5. *Transparency.* The correct transmission of digital data over a channel should be transparent to the pattern of 1's and 0's contained in the data.

It is for these reasons that we find, for example, the bipolar format has become the standard for transmitting binary encoded PCM data over telephone channels.

EXERCISE 5 Rank the six digital waveforms depicted in Fig. 5.12 in increasing order of transmission bandwidth requirement.

DECODING

The first operation in the receiver is to regenerate (i.e., reshape and clean up) the received pulses. These clean pulses are then regrouped into code words and decoded (i.e., mapped back) into a quantized PAM signal. The *decoding* process involves generating a pulse the amplitude of which is the linear sum of all the pulses in the code word, with each pulse weighted by its place-value (2^0, 2^1, 2^2, 2^3, . . .) in the code.

It is noteworthy that every operation performed in the transmitter of a PCM system, except for the quantizing operation, is reversible. Specifically, the operations of sampling and encoding performed in the transmitter are reversed by performing decoding and interpolation (in that order) in the receiver. On the other hand, quantizing is an irreversible process that manifests itself by destroying information; once quantizing noise is introduced in the transmitter, there is nothing we can do in the receiver to make up for the loss of information thereby incurred.

5.6 REGENERATION

The most important feature of PCM systems lies in the ability to control the effects of distortion and noise produced by transmitting a PCM wave through a channel. This capability is accomplished by reconstructing the PCM wave by means of a chain of *regenerative repeaters* sufficiently close along the transmission route. As illustrated in Fig. 5.13, three basic functions are performed by a regenerative repeater: *equalization, timing,* and *decision making*. The equalizer shapes the received pulses so as to compensate for the effects of amplitude and phase distortions produced by the transmission characteristics of the channel. The timing circuitry provides a periodic pulse train, derived from the received pulses, for sampling the equalized pulses at the instants of time where the signal-to-noise ratio is a maximum. The decision device is enabled at the sampling times determined by the timing circuitry. It makes its decision based on whether or not the amplitude of the quantized pulse plus noise exceeds a predetermined voltage level. Thus, for example, in a PCM system with on–off signaling, the repeater makes a decision in each bit interval as to whether or not a pulse is present. If the decision is "yes" a clean new pulse is transmitted to the next repeater. If, on the other hand, the decision is "no," a clean base line is transmitted. In this way, the accumulation of distortion and noise in a repeater span is completely removed, provided that the disturbance is not too large to cause an error in the decision-making process. Ideally, except for delay, the regenerated signal is exactly the same as the signal originally transmitted. In practice, however, the regenerated signal departs from the original signal for two main reasons:

1. The presence of transmission noise and interference causes the repeater to make wrong decisions occasionally, thereby introducing *bit errors* into the regenerated signal; we will have more to say on this issue in Chapter 10.

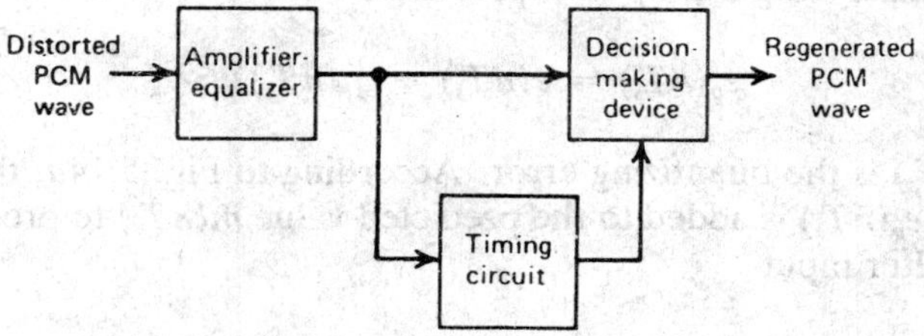

Figure 5.13
Block diagram of a regenerative repeater.

2. If the spacing between received pulses deviates from its assigned value, a *jitter* is introduced into the regenerated pulse position, thereby causing distortion.

5.7 DIFFERENTIAL PULSE-CODE MODULATION

When a voice or video signal is sampled at a rate slightly higher than the Nyquist rate, the resulting sampled signal is found to exhibit a high correlation between adjacent samples. The meaning of this high correlation is that, in an average sense, the signal does not change rapidly from one sample to the next. When these highly correlated samples are encoded, as in a standard PCM system, the resulting encoded signal contains *redundant information*. This means that symbols that are not absolutely essential to the transmission of information are generated as a result of the encoding process. By removing this redundancy before encoding, we obtain a more efficient coded signal.

Now, if we know a sufficient part of a redundant signal, we may infer the rest, or at least make the most probable estimate. In particular, if we know the past behavior of a signal up to a certain point in time, it is possible to make some inference about its future values; such a process is commonly called *prediction*. Suppose then a message signal $m(t)$ is sampled at the rate $1/T_s$ to produce a sequence of correlated samples T_s seconds apart; which is denoted by $\{m(nT_s)\}$. The fact that it is possible to predict future values of the signal $m(t)$ provides motivation for the *differential quantization* scheme shown in Fig. 5.14*a*. In this scheme, the input to the quantizer is

$$e(nT_s) = m(nT_s) - \hat{m}(nT_s) \tag{5.25}$$

which is the difference between the unquantized input sample $m(nT_s)$ and a prediction of it, denoted by $\hat{m}(nT_s)$. This predicted value is produced by using a *prediction filter* with an input, as we will see, that consists of a quantized version of the message sample $m(nT_s)$. The difference signal $e(nT_s)$ is called a *prediction error*, since it is the amount by which the prediction filter fails to predict the input exactly.

By encoding the quantizer output, as in Fig. 5.14*a*, we obtain a variation of PCM, known as *differential pulse-code modulation* (DPCM). It is this encoded signal that is used for transmission.

The quantizer output may be represented as

$$e_q(nT_s) = e(nT_s) + q_e(nT_s) \tag{5.26}$$

where $q_e(nT_s)$ is the quantizing error. According to Fig. 5.14*a*, the quantizer output $e_q(nT_s)$ is added to the predicted value $\hat{m}(nT_s)$ to produce the prediction-filter input

$$m_q(nT_s) = \hat{m}(nT_s) + e_q(nT_s) \tag{5.27}$$

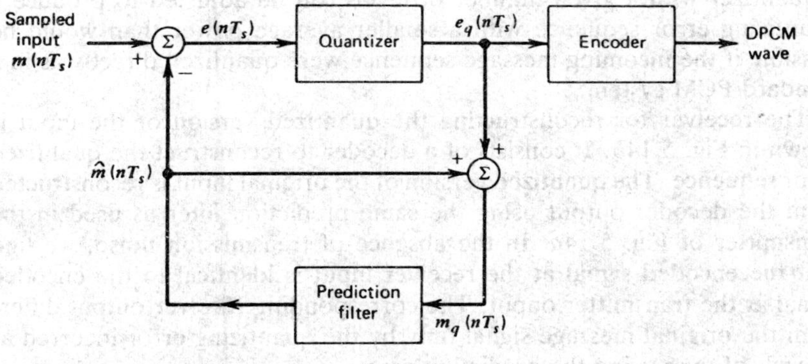

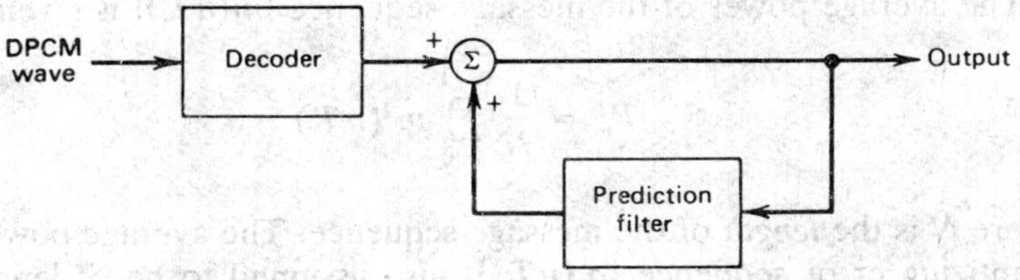

Figure 5.14
DPCM system. (a) *Transmitter.* (b) *Receiver.*

Substituting Eq. 5.26 in 5.27, we get

$$m_q(nT_s) = \hat{m}(nT_s) + e(nT_s) + q_e(nT_s) \tag{5.28}$$

However, from Eq. 5.25 we observe that $\hat{m}(nT_s)$ plus $e(nT_s)$ is equal to the incoming message sample $m(nT_s)$. Therefore, we may rewrite Eq. 5.28 as follows

$$m_q(nT_s) = m(nT_s) + q_e(nT_s) \tag{5.29}$$

which represents a quantized version of $m(nT_s)$. That is, irrespective of the properties of the prediction filter, the quantized sample, $m_q(nT_s)$, at the prediction filter input, differs from the sample $m(nT_s)$ of the original message signal $m(t)$ by the quantizing error $q_e(nT_s)$. Accordingly, if the prediction is good, the average power of the prediction error sequence $\{e(nT_s)\}$ will be smaller than that of the message sequence $\{m(nT_s)\}$. Hence, a quantizer with a given number of levels can be adjusted to produce a quantizing error sequence with a smaller average power than would be possible if the incoming message sequence were quantized directly as in a standard PCM system.

The receiver for reconstructing the quantized version of the input is shown in Fig. 5.14*b*. It consists of a decoder to reconstruct the quantized error sequence. The quantized version of the original input is reconstructed from the decoder output using the same prediction filter as used in the transmitter of Fig. 5.14*a*. In the absence of transmission noise, we find that the encoded signal at the receiver input is identical to the encoded signal at the transmitter ouput. The corresponding receiver output differs from the original message signal only by the quantizing error incurred as a result of quantizing the prediction error.

From the foregoing analysis we observe that, in a noise-free environment, the prediction filters in the transmitter and receiver operate on the same sequence of samples, $\{m_q(nT_s)\}$. It is with this purpose in mind that a feedback path is added to the quantizer in the transmitter, as shown in Fig. 5.14*a*.

The average power of the message sequence $\{m(nT_s)\}$ is given by

$$P_m = \frac{1}{N}\sum_{n=0}^{N-1} m^2(nT_s) \tag{5.30}$$

where N is the *length* of the message sequence. The average power of the quantizing error sequence $\{q_e(nT_s)\}$, also assumed to be of length N, is given by

$$P_q = \frac{1}{N}\sum_{n=0}^{N-1} q_e^2(nT_s) \tag{5.31}$$

We may thus define the *output signal-to-quantizing noise ratio* of a DPCM system as

$$(\text{SNR})_0 = \frac{P_m}{P_q} \tag{5.32}$$

It is clear that we may rewrite Eq. 5.32 as

$$(\text{SNR})_0 = \frac{P_m}{P_e}\frac{P_e}{P_q} = G_p(\text{SNR})_Q$$

where $(\text{SNR})_Q$ is the *signal-to-quantizing noise ratio* defined by

$$(\text{SNR})_Q = \frac{P_e}{P_q} \tag{5.33}$$

and G_p is the *prediction gain* produced by the differential quantization scheme, defined by

$$G_p = \frac{P_m}{P_e} \tag{5.34}$$

The quantity G_p, when greater than unity, represents the gain in signal-to-noise ratio that is due to the differential quantization scheme of Fig. 5.14. Now for a given message signal, the average power P_m is fixed, so that G_p is maximized by minimizing the average prediction error power P_e. Accordingly, our objective should be to design the prediction filter so as to minimize P_e, while the signal-to-quantizing noise ratio is maintained constant.

THE PREDICTION FILTER

One approach to specify the nature of the prediction filters in the transmitter and the receiver of the DPCM system shown in Fig. 5.14 is to use a *tapped-delay-line filter* as the basis of the design. An advantage of this approach is that it leads to tractable mathematics, and it is simple to implement. Thus the predicted value $\hat{m}(nT_s)$ is modeled as a linear combination of past values of the quantized input as shown by (see Fig. 5.15)

$$\hat{m}(nT_s) = \sum_{k=1}^{p} w_k m_q(nT_s - kT_s) \tag{5.35}$$

where the tap weights $w_1, w_2, \ldots, w_p$ define the prediction filter coefficients, and p is the *order* of the prediction filter. Substitution of Eq. 5.35 in 5.25 yields the prediction error

$$e(nT_s) = m(nT_s) - \sum_{k=1}^{p} w_k m_q(nT_s - kT_s) \tag{5.36}$$

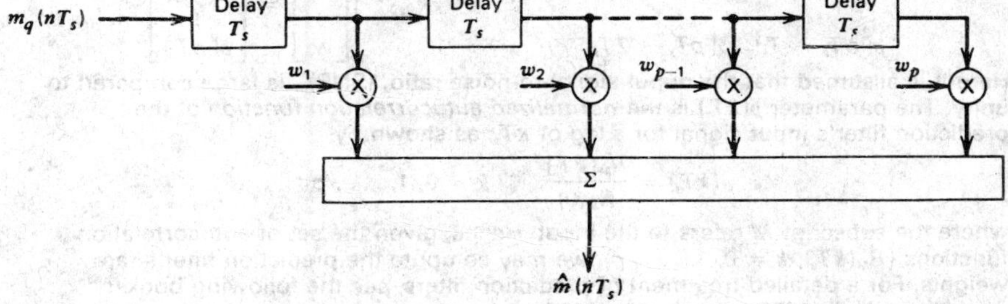

Figure 5.15
Tapped-delay-line filter used as a prediction filter.

The mathematical basis for the design of the prediction filter is that of *minimizing the average prediction-error power with respect to the tap weights of the filter.*[5]

5.8 DELTA MODULATION

The exploitation of signal correlations in DPCM suggests the further possibility of oversampling a message signal (i.e., at a rate much higher than the Nyquist rate) to purposely increase the correlation between adjacent samples of the signal. This would permit the use of a simple quantizing strategy for constructing the encoded signal. *Delta modulation* (DM), which is the one-bit (or two-level) version of DPCM, is precisely such a scheme.

In its simple form, DM provides a staircase approximation to the oversampled version of an incoming message signal, as illustrated in Fig. 5.16*a*. The difference between the input and the approximation is quantized into only two representation levels, namely, $\pm\delta$, corresponding to positive and negative differences. Thus, if the approximation falls below the signal at any sampling epoch, it is increased by δ. If, on the other hand, the approximation lies above the signal, it is diminished by δ. Provided that the signal does not change too rapidly from sample to sample, we find that the staircase approximation remains within $\pm\delta$ of the input signal.

Denoting the input signal as $m(t)$ and the staircase approximation as $m_q(t)$, the basic principle of delta modulation may be formalized in the following set of discrete-time relations:

$$e(nT_s) = m(nT_s) - m_q(nT_s - T_s) \tag{5.37}$$

$$e_q(nT_s) = \delta \operatorname{sgn}[e(nT_s)] \tag{5.38}$$

and

$$m_q(nT_s) = m_q(nT_s - T_s) + e_q(nT_s) \tag{5.39}$$

[5]The result of this minimization is a set of simultaneous equations, expressed in matrix form as follows

$$\begin{bmatrix} 1 & \rho(T_s) & \cdots & \rho(pT_s - T_s) \\ \rho(T_s) & 1 & \cdots & \rho(pT_s - 2T_s) \\ \vdots & \vdots & & \vdots \\ \rho(pT_s - T_s) & \rho(pT_s - 2T_s) & \cdots & 1 \end{bmatrix} \begin{bmatrix} w_1 \\ w_2 \\ \vdots \\ w_p \end{bmatrix} = \begin{bmatrix} \rho(T_s) \\ \rho(2T_s) \\ \vdots \\ \rho(pT_s) \end{bmatrix}$$

Here it is assumed that the ouput signal-to-noise ratio, $(\text{SNR})_0$, is large compared to unity. The parameter $\rho(kT_s)$ is the *normalized autocorrelation function* of the prediction filter's input signal for a lag of kT_s, as shown by

$$\rho(kT_s) = \frac{R_M(kT_s)}{R_M(0)}, \qquad k = 0, 1, \ldots, p$$

where the subscript M refers to the input. Hence, given the set of autocorrelation functions $\{R_M(kT_s), k = 0, 1, \ldots, p\}$, we may compute the prediction filter's tap weights. For a detailed treatment of prediction filters, see the following books: Jayant and Noll (1984) and Haykin (1986).

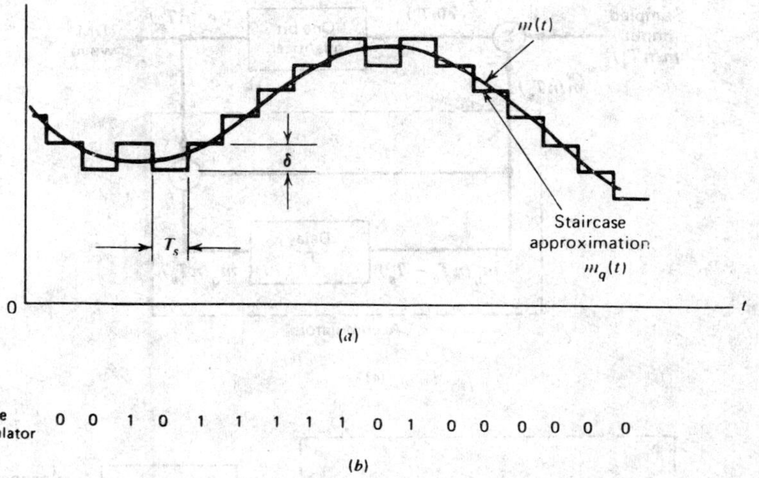

Figure 5.16
Delta modulation

where T_s is the sampling period; $e(nT_s)$ is an error signal representing the difference between the present sample value $m(nT_s)$ of the input signal and the latest approximation to it, namely, $\hat{m}(nT_s) = m_q(nT_s - T_s)$; and $e_q(nT_s)$ is the quantized version of $e(nT_s)$. The quantizer output $e_q(nT_s)$ is the desired DM wave for varying n.

Part *a* of Fig. 5.16 illustrates the way in which the staircase approximation $m_q(t)$ follows variations in the input signal $m(t)$ in accordance with Eqs. 5.37 through 5.39, and part *b* of the figure displays the corresponding binary sequence at the delta modulator output. It is apparent that in a delta modulation system the rate of information transmission is simply equal to the sampling rate $1/T_s$.

The principal virtue of delta modulation is its simplicity. It may be generated by applying the sampled version of the incoming message signal to a modulator that involves a summer, quantizer, and accumulator interconnected as shown in Fig. 5.17*a*. Details of the modulator follow directly from Eqs. 5.37 and 5.39. In particular, the quantizer consists of a *hard limiter* with an input–output relation defined by Eq. 5.38, which is depicted in Fig. 5.18. The quantizer output is applied to an *accumulator*, producing the result

$$
\begin{aligned}
m_q(nT_s) &= \delta \sum_{i=1}^{n} \operatorname{sgn}[e(iT_s)] \\
&= \sum_{i=1}^{n} e_q(iT_s) \qquad (5.40)
\end{aligned}
$$

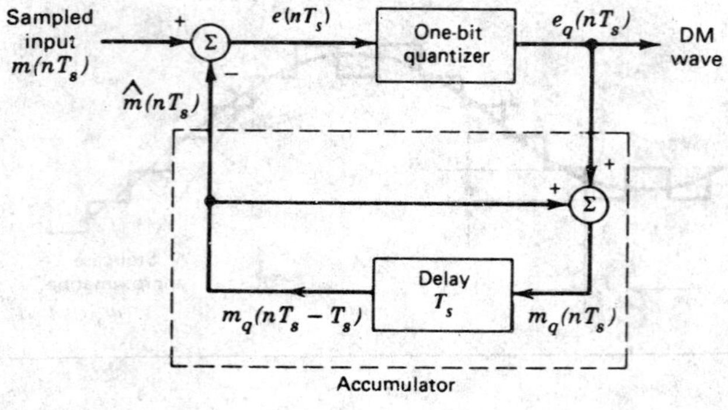

(a)

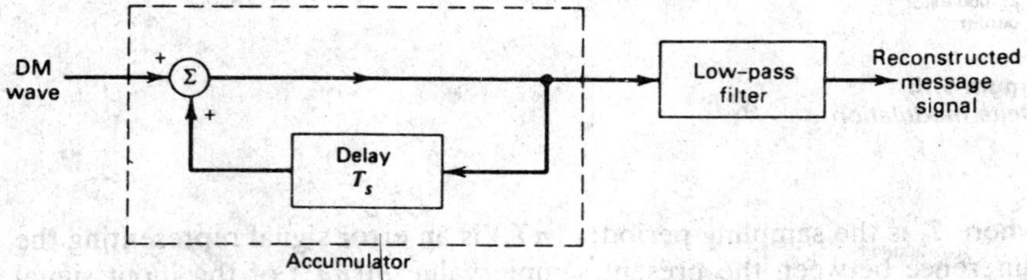

(b)

Figure 5.17
DM system. (a) *Transmitter.* (b) *Receiver.*

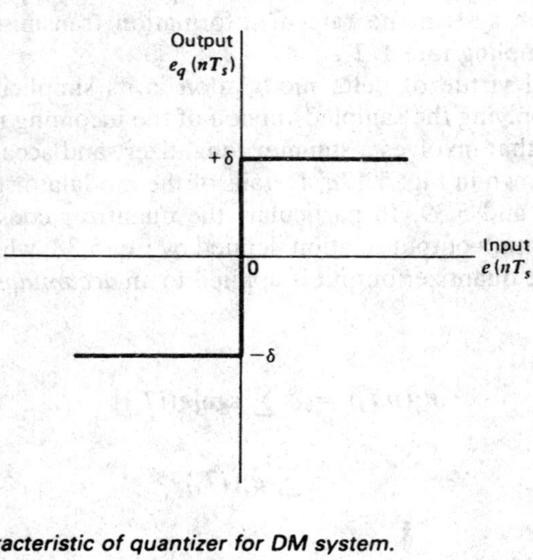

Figure 5.18
Input–output characteristic of quantizer for DM system.

which is obtained by solving Eqs. 5.38 and 5.39 for $m_q(nT_s)$. Thus, at the sampling instant nT_s, the accumulator increments the approximation by an amount equal to δ in a positive or negative direction, depending on the algebraic sign of the error signal $e(nT_s)$. If the input signal $m(nT_s)$ is greater than the most recent approximation $\hat{m}(nT_s)$, a positive increment $+\delta$ is applied to the approximation. If, on the other hand, the input signal is smaller, a negative increment $-\delta$ is applied to the approximation. In this way, the accumulator does the best it can to track the input samples by one step at a time. In the receiver, shown in Fig. 5.17*b*, the staircase approximation $m_q(t)$ is reconstructed by passing the sequence of positive and negative pulses, produced at the decoder output, through an accumulator in a manner similar to that used in the transmitter. The out-of-band quantizing noise in the high-frequency staircase waveform $m_q(t)$ is rejected by passing it through a low-pass filter with a bandwidth equal to the original message bandwidth.

In comparing the DPCM and DM networks of Fig. 5.14 and 5.17, we note that they are similar, except for two important differences, namely, the use of a one-bit (two-level) quantizer in the delta modulator and the replacement of the prediction filter by a single delay element.

QUANTIZING NOISE

Delta modulation systems are subject to two types of quantizing error: (1) slope overload distortion, and (2) granular noise. We first discuss the cause of slope overload distortion, and then granular noise.

We observe that Eq. 5.40 is the digital equivalent of integration in the sense that it represents the accumulation of positive and negative increments of magnitude δ. Also, denoting the quantizing error by $q_e(nT_s)$, as shown by,

$$m_q(nT_s) = m(nT_s) + q_e(nT_s) \tag{5.41}$$

we observe from Eq. 5.37 that the input to the quantizer is

$$e(nT_s) = m(nT_s) - m(nT_s - T_s) - q_e(nT_s - T_s) \tag{5.42}$$

Thus, except for the quantizing error $q_e(nT_s - T_s)$, the quantizer input is a *first backward difference* of the input signal, which may be viewed as a digital approximation to the derivative of the input signal or, equivalently, as the inverse of the digital integration process. If we consider the maximum slope of the original input waveform $m(t)$, it is clear that in order for the quantized sequence $\{m_q(nT_s)\}$ to increase as fast as the message sequence $\{m(nT_s)\}$ in a region of maximum slope of $m(t)$, we require that the condition

$$\frac{\delta}{T_s} \geq \max\left|\frac{d\,m(t)}{dt}\right| \tag{5.43}$$

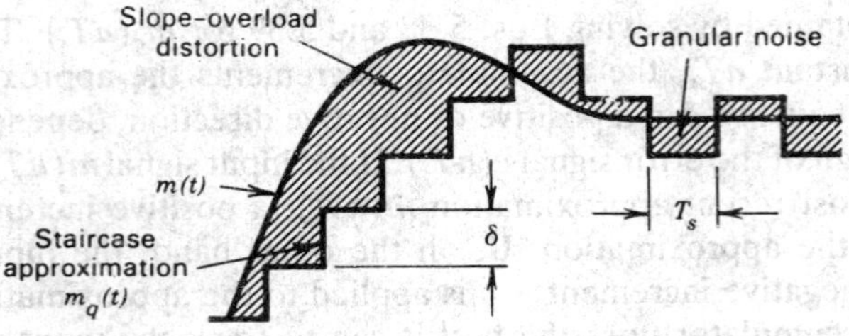

Figure 5.19
Quantizing error in delta modulation.

be satisfied. Otherwise, we find that the absolute value of the representation level δ is too small for the staircase approximation $m_q(t)$ to follow a steep segment of the input waveform $m(t)$, with the result that $m_q(t)$ falls behind $m(t)$, as illustrated in Fig. 5.19. This condition is called *slope overload*, and the resulting quantizing error is called *slope-overload distortion* (noise). Note that since the maximum slope of the staircase approximation $m_q(t)$ is fixed by the value of δ, increases and decreases in $m_q(t)$ tend to occur along straight lines. For this reason, a delta modulator using a fixed δ is often referred to as a *linear delta modulator*.

In contrast to slope-overload distortion, *granular noise* occurs when δ is too large relative to the local slope characteristics of the input waveform $m(t)$, thereby causing the staircase approximation $m_q(t)$ to hunt around a relatively flat segment of the input waveform; this phenomenon is also illustrated in Fig. 5.19. The granular noise is analogous to quantizing noise in a PCM system.

We thus see that there is a need to have a large δ so as to accommodate a wide dynamic range, whereas a small δ is required for the accurate representation of relatively low-level signals. It is therefore clear that the choice of the optimum δ that minimizes the mean-square value of the quantizing error in linear delta modulation will be the result of a compromise between slope overload distortion and granular noise.

EXERCISE 6 From Fig. 5.18 we see that the *step size* of a linear delta modulator is

$$\Delta = 2\delta$$

What is the average power of the granular noise expressed in terms of δ?

5.9 DISCUSSION

In this section we discuss the advantages and disadvantages of DPCM and DM, compared with standard PCM, for the encoding of voice and television signals.

The "relative" behavior of standard PCM and DPCM systems is much the same with either uniform or logarithmic quantizing, because the repertoire of signals consists of waveforms similar in character, but differing in mean level. In the case of voice signals, it is found that the signal-to-quantizing noise advantage of DPCM over standard PCM is in the neighborhood of 4–11 dB. The greatest improvement occurs in going from no prediction to first-order prediction, with some additional gain resulting from increasing the order of the prediction filter up to 4 or 5, after which little additional gain is obtained. Since 6 dB of quantizing noise is equivalent to 1 bit per sample, by virtue of Eq. 5.21, the advantage of DPCM may also be expressed in terms of bit rate. For a constant signal-to-quantizing noise ratio, and assuming a sampling rate of 8 kHz, the use of DPCM may provide a saving of about 8–16 kilobits per second (1 to 2 bits per sample) over standard PCM.

In the case of television signals, DPCM provides more of an advantage for high-resolution television systems than for low-resolution systems. For monochrome entertainment television, DPCM provides a signal-to-quantizing noise ratio of approximately 12 dB higher than standard PCM. For a constant signal-to-quantizing noise ratio, and assuming a sampling rate of 9 MHz, this represents a saving of about 18 megabits per second (2 bits per sample) by DPCM over PCM.

Turning next to delta modulation, subjective voice tests and noise measurements have shown that a DM system operating at 40 kilobits per second is equivalent to a standard PCM system operating with a sampling rate of 8 kHz and 5 bits per sample. At lower bit rates, DM is better than the standard PCM (the latter still using 8-kHz sampling and a reduced number of bits per sample), but at higher bit rates PCM is superior to DM. The quality of 5-bit PCM is low for most purposes in telephony. For telephone quality voice signals, it has become conventional to use 8-bit PCM. Equivalent voice quality with DM can be obtained only by using bit rates much higher than 64 kilobits per second.

Also, in a delta modulation system, operating on voice signals under optimum conditions, the SNR is increased by 9 dB by doubling the bit rate. By comparison, in the case of standard PCM, we achieve a 6 dB increase in SNR for each *added* bit. For example, by doubling the bit rate from 40 to 80 kilobits per second, the SNR is increased by 9 dB using DM. On the other hand, if PCM is employed and the bit rate is similarly doubled by increasing the number of bits per sample from 5 to 10 (keeping the sampling rate fixed at 8 kHz), the SNR is improved by 30 dB. Thus the increase of SNR with bit rate is much more dramatic for PCM than for DM.

The use of delta modulation is therefore recommended only in certain special circumstances: (1) if it is necessary to reduce the bit rate below 40 kilobits per second and limited voice quality is tolerable; or (2) if extreme circuit simplicity is of overriding importance and the accompanying use of a high-bit rate is acceptable. Note that since delta modulation uses a high sampling rate, there is no need for employing a pre-alias filter prior to sampling in the transmitter.

ADAPTIVE DIGITAL CODING OF WAVEFORMS

From the discussion presented on PCM using a uniform quantizer with a fixed step size, we see that we have a dilemma in quantizing speech signals. On the one hand, we wish to choose the quantization step size large enough to accommodate the maximum peak-to-peak range of the input signal with the lowest possible number of representation levels. On the other hand, we would like to make the quantization step size small enough to minimize the average power of the quantizing noise. This issue is further compounded by the fact that the amplitude of the speech signal can vary over a wide range, depending on the speaker, the communication environment, and within a given utterance, from *voiced* to *unvoiced sounds*.[6] One approach to accommodating these conflicting requirements is to use a nonuniform quantizer; this approach is commonly used in PCM systems for telephony as described in Section 5.4. An alternative approach is to use an *adaptive quantizer,* wherein the step size is varied automatically so as to match the average power of the input speech signal; this second approach is commonly used in *adaptive DPCM (ADPCM) systems.*

In ADPCM systems used in telephony, the prediction filter is also adaptive. An *adaptive prediction filter* is responsive to changing level and spectrum of the input speech signal. The variation of performance with speakers and speech material, together with variations in signal level inherent in the speech communication process, make the combined use of adaptive quantization and adaptive prediction necessary to achieve best performance over a wide range of speakers and speaking situations.

It is of interest to note that improvements in circuit design and technology have made it possible for ADPCM to provide toll quality speech coding at 32 kb/s[7]; this corresponds to a sampling rate of 8 kHz and 4 bits per sample. By "toll quality" we mean the quality of commercial telephone service. This performance is comparable to that of 64 kb/s PCM incorporating the use of μ-law (logarithmic) companding with $\mu = 255$. However, unlike log-PCM, the performance of the ADPCM system is very signal dependent.

Finally, we should mention that a delta modulator may also be made adaptive, wherein the variable step size increases during a steep segment

[6] *Voiced sounds* are produced by forcing air through the glottis with the tension of the vocal cords adjusted so that they vibrate in a relaxation oscillation, thereby producing quasiperiodic pulses of air that excite the vocal tract. *Fricative* or *unvoiced sounds* are generated by forming a constriction at some point in the vocal tract (usually toward the mouth end) and forcing air through the constriction at a high enough velocity to produce turbulence. This creates a broad-spectrum noise source to excite the vocal tract.

[7] Jayant and Noll, 1984, pp. 309–311.

of the input signal and decreases when the modulator is quantizing an input signal with a slowly varying segment. In this way the step size is adapted to the level of the input signal. The resulting system is called an *adaptive delta modulator* (ADM).

5.10 TIME-DIVISION MULTIPLEXING

The sampling theorem enables us to transmit the complete information contained in a band-limited message signal by using samples of the message signal taken uniformly at a rate that is usually slightly higher than the Nyquist rate. An important feature of the sampling process is a conservation of time. That is, the transmission of the message samples engages the transmission channel for only a fraction of the sampling interval on a periodic basis, and in this way some of the time interval between adjacent samples is cleared for use by other independent message sources on a time-shared basis. We thereby obtain a *time-division multiplex system* (TDM), which enables the joint use of a common transmission channel by a plurality of independent message sources without mutual interference.

The concept of TDM is illustrated by the block diagram shown in Fig. 5.20. Each input message signal is first restricted in bandwidth by a low-pass filter to remove the frequencies that are nonessential to an adequate signal representation. The low-pass filter outputs are then applied to a *commutator* that is usually implemented using electronic switching circuitry. The function of the commutator is two-fold: (1) to take a narrow sample of each of the N input messages at a rate $1/T_s$ that is slightly higher than $2W$, where W is the cutoff frequency of the low-pass input filter, and (2) to sequentially interleave these N samples inside a sampling interval T_s. Indeed, this latter function is the essence of the time-division multiplexing operation. Following the commutation process, the multiplexed signal is applied to a *pulse modulator,* (e.g., pulse-amplitude modulator), the purpose of which is to transform the multiplexed signal into a form suitable for transmission over the common channel. It is clear that the use of time-division multiplexing introduces a *bandwidth expansion factor* N, because the scheme must squeeze N samples derived from N independent message sources into a time slot equal to one sampling interval. At the receiving end of the system, the received signal is applied to a *pulse demodulator,* which performs the inverse operation of the pulse modulator. The narrow samples produced at the pulse demodulator output are distributed to the appropriate low-pass reconstruction filters by means of a decommutator, which operates in *synchronism* with the commutator in the transmitter. This synchronization is essential for the satisfactory operation of the system.

The TDM system is highly sensitive to dispersion in the common transmission channel, that is, to variations of amplitude with frequency or non-linear phase response. Accordingly, accurate equalization of both the amplitude and phase responses of the channel is necessary to ensure a satisfactory

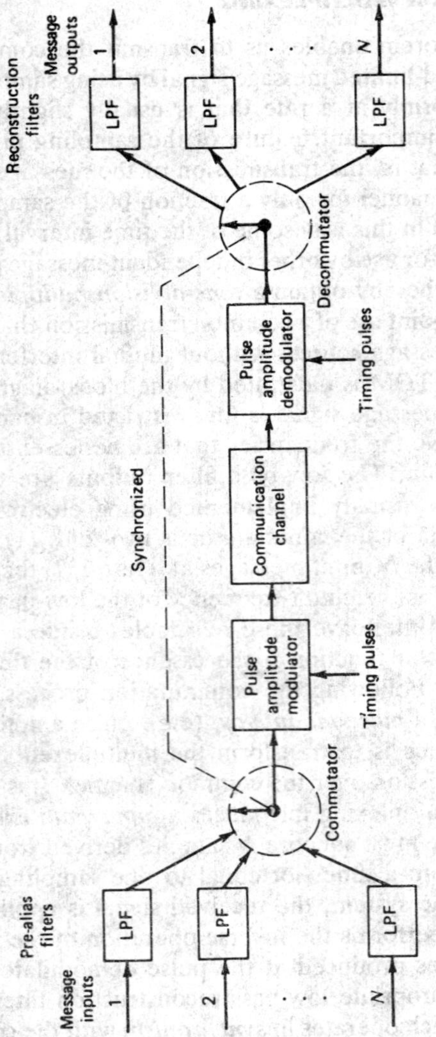

Figure 5.20
Block diagram of TDM system.

operation of the system. This issue is discussed in Chapter 6. To a first approximation, however, TDM is immune to amplitude nonlinearities in the channel as a source of crosstalk, because the different message signals are not simultaneously impressed on the channel.

5.11 *APPLICATION: DIGITAL MULTIPLEXERS FOR TELEPHONY*

In the previous section we introduced the idea of time-division multiplexing whereby a group of analog signals (e.g., voice signals) are sampled sequentially in time at a *common* sampling rate and then multiplexed for transmission over a common line. In this section we consider the *multiplexing of digital signals*[8] at different bit rates. This enables us to combine several digital signals, such as computer outputs, digitized voice signals, and digitized facsimile and television signals, into a single data stream (at a considerably higher bit rate than any of the inputs). Figure 5.21 shows a conceptual diagram of the digital multiplexing–demultiplexing operation.

The multiplexing of digital signals may be accomplished by using a *bit-by-bit interleaving procedure* with a selector switch that sequentially takes a bit from each incoming line and then applies it to the high-speed common line. At the receiving end of the system the output of this common line is separated out into its individual low-speed components and then delivered to their respective destinations.

Two major groups of digital multiplexers are used in practice:

1. One group of multiplexers is designed to combine relatively low-speed digital signals, up to a maximum rate of 4800 bits per second, into a higher speed multiplexed signal with a rate of up to 9600 bits per second. These multiplexers are used primarily to transmit data over voice-grade channels of a telephone network. Their implementation requires the use of *modems* in order to convert the digital format into an analog format suitable for transmission over telephone channels. The operation of a modem (modulator–demodulator) is covered in Section 7.14.

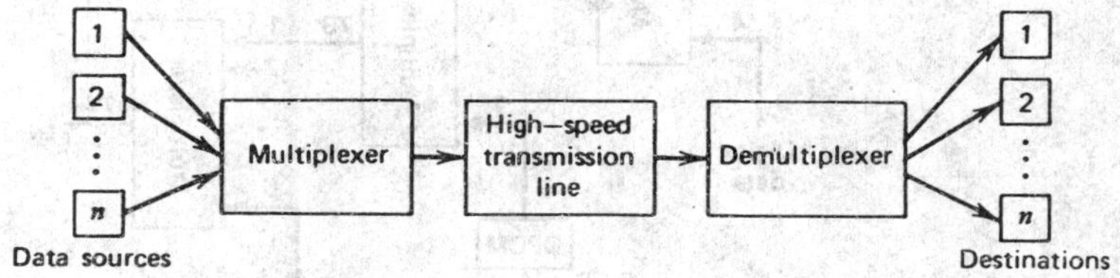

Figure 5.21
Conceptual diagram of multiplexing–demultiplexing.

[8]For more detailed information on the multiplexing of digital signals, see Bell Telephone Laboratories (1970).

2. The second group of multiplexers, designed to operate at much higher bit rates, forms part of the data transmission service generally provided by communication carriers. For example, Fig. 5.22 is a block diagram of the digital hierarchy based on the T1 carrier, which has been developed by the Bell System. The T1 carrier, described later on, is designed to operate at 1.544 megabits per second, the T2 at 6.312 megabits per second, the T3 at 44.736 megabits per second, and the T4 at 274.176 megabits per second. The system is thus made up of various combinations of lower order T-carrier subsystems. These subsystems are designed to accommodate the transmission of voice signals, Picturephone® service, and television signals using PCM, as well as (direct) digital signals from data terminal equipment.

There are some basic problems involved in the design of a digital multiplexer, irrespective of its grouping:

1. Digital signals cannot be directly interleaved into a format that allows for their eventual separation unless their bit rates are locked to a common clock. Accordingly, provision has to be made for *synchronization* of the incoming digital signals, so that they can be properly interleaved.
2. The multiplexed signal must include some form of *framing,* so that its individual components can be identified at the receiver.
3. The multiplexer has to handle small variations in the bit rates of the incoming digital signals. For example, a 1000-kilometer coaxial cable carrying 3×10^8 pulses per second will have about 1 million pulses in transit, with each pulse occupying about 1 meter of the cable. A 0.01%

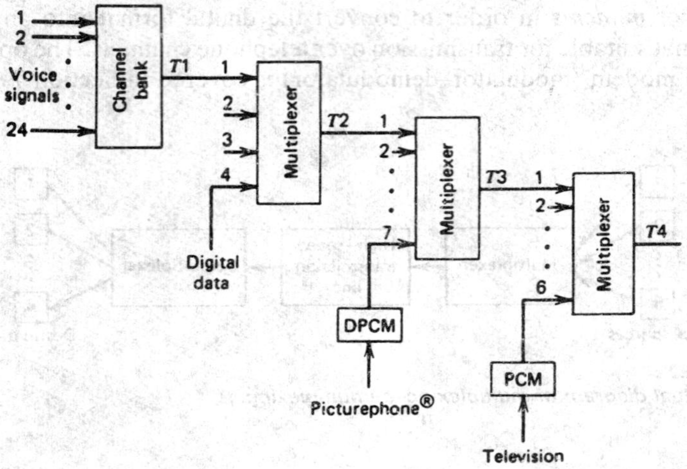

Figure 5.22
Digital hierarchy, Bell system.

variation in the propagation delay, produced by a 1°F decrease in temperature, will result in 100 fewer pulses in the cable. Clearly, these pulses must be absorbed by the multiplexer.

To cater to the requirements of synchronization and rate adjustment to accommodate small variations in the input data rates, we may use a technique known as *bit stuffing*. The idea here is to have the outgoing bit rate of the multiplexer slightly higher than the sum of the maximum expected bit rates of the input channels. This is achieved by stuffing in additional non-information-carrying pulses. All incoming digital signals are stuffed with a number of bits sufficient to raise their respective bit rates to that of a locally generated clock. To accomplish bit stuffing, each incoming digital signal or bit stream is fed into an *elastic store* at the multiplexer. The elastic store is a device that stores a bit stream in such a manner that the stream may be read out at a rate different from the rate at which it is read in. At the demultiplexer, the stuffed bits must obviously be removed from the multiplexed signal. This requires some method of identifying the stuffed bits. To illustrate one such method, and also to show one method of providing frame synchronization, we describe the signal format of the Bell System *M12 multiplexer,* which is designed to combine four T1 bit streams into one T2 bit stream. We begin the description by considering the T1 system.

T1 SYSTEM

The *T1 carrier system* is designed to accommodate 24 voice channels, primarily for short-distance, heavy usage in metropolitan areas. The T1 system was pioneered by the Bell System in the United States in the early 1960s, and with its introduction the shift to digital communication facilities started. The T1 system has been adopted for use throughout the United States, Canada, and Japan. It forms the basis for a complete hierarchy of higher-order multiplexed systems that are used for either long-distance transmission or transmission in heavily populated urban centers.

A voice signal (male or female) is essentially limited to a band from 300 to 3400 Hz in the sense that frequencies outside this band do not contribute much to articulation efficiency. Indeed, telephone circuits that respond to this range of frequencies give quite satisfactory service. Accordingly, it is customary to pass the voice signal through a low-pass filter with a cutoff frequency of about 3.4 kHz prior to sampling. Hence, the nominal value of the Nyquist rate is 6.8 kHz. The filtered voice signal is usually sampled at a slightly higher rate, namely, 8 kHz, which is the *standard* sampling rate in digital telephony.

For companding, the T1 system uses a *piecewise-linear* characteristic (consisting of 15 linear segments) to approximate the logarithmic μ-law of Eq. 5.16 with the constant $\mu = 255$. This approximation is constructed in such a way that the segment end points lie on the compression curve computed from Eq. 5.16, and their projections onto the vertical axis are spaced uniformly. Table 5.4 gives the projections of the segment end points onto the horizontal axis and the step sizes of the individual segments. The table is normalized to 8159, so that all values are represented as integer numbers. Segment 0 of the approximation is a linear segment, passing through the origin; it contains a total of 31 uniform representation levels. Linear segments $1a$, $2a$, . . . , $7a$ lie above the horizontal axis, whereas linear segments $1b$, $2b$, . . . , $7b$ lie below the horizontal axis; each of these 14 segments contains 16 uniform representation levels. For colinear segment 0 the representation levels at the compressor input are ± 1, ± 3, . . . , ± 31, and the corresponding compressor output levels are 0, ± 1, . . . , ± 15. For linear segments $1a$ and $1b$ the representation levels at the compressor input are ± 33, ± 35, . . . , ± 95, and the corresponding compressor output levels are ± 16, ± 17, . . . , ± 31, and so on for the other linear segments.

There are a total of $31 + 14 \times 16 = 255$ output levels associated with the 15-segment companding characteristic described herein. To accommodate this number of output levels, each of the 24 voice channels uses a binary code with an 8-bit word. The first bit indicates whether the input voice sample is positive or negative. The next three bits of the code word identify the particular segment inside which the amplitude of the input voice sample lies, and the last four bits identify the actual quantizing step inside that segment.

With a sampling rate of 8 kHz, each frame of the multiplexed signal occupies a period of 125 μs. In particular, it consists of twenty-four 8-bit words, plus a single bit that is added at the end of the frame for the purpose of synchronization. Hence, each frame consists of a total of $24 \times 8 + 1 = 193$ bits. Correspondingly, the duration of each bit equals 0.647 μs, and the resultant transmission rate is 1.544 megabits per second.

In addition to the voice signal, a telephone system must also pass special

TABLE 5.4 The 15-Segment μ-law Companding Characteristic ($\mu = 255$)

Linear Segment Number	Step Size	Projections of Segment End Point onto the Horizontal Axis
0	2	± 31
$1a$, $1b$	4	± 95
$2a$, $2b$	8	± 223
$3a$, $3b$	16	± 479
$4a$, $4b$	32	± 991
$5a$, $5b$	64	± 2015
$6a$, $6b$	128	± 4063
$7a$, $7b$	256	± 8159

supervisory signals to the far end. This *signaling information* is needed to transmit dial pulses, as well as telephone off-hook/on-hook signals. In the T1 system this requirement is accomplished as follows. Every sixth frame, the least significant (i.e., the eighth) bit of each voice channel is deleted and a *signaling bit* is inserted in its place, thereby yielding an average $7\frac{5}{6}$-bit operation for each voice input. The sequence of signaling bits is thus transmitted at a rate equal to the sampling rate divided by six, that is, 1.333 kilobits per second.

M12 MULTIPLEXER

Figure 5.23 illustrates the signal format of the M12 multiplexer. Each frame is subdivided into four subframes. The first subframe (first line in Fig. 5.23) is transmitted, then the second, the third, and the fourth, in that order.

Bit-by-bit interleaving of the incoming four T1 bit streams is used to accumulate a total of 48 bits, 12 from each input. A *control bit* is then inserted by the multiplexer. Each frame contains a total of 24 control bits, separated by sequences of 48 data bits. Three types of control bits are used in the M12 multiplexer to provide synchronization and frame indication, and to identify which of the four input signals has been stuffed. These control bits are labeled as F, M, and C in Fig. 5.23. Their functions are:

1. The F-control bits, two per subframe, constitute the *main* framing pulses. The subscripts on the F-control bits denote the actual bit (0 or 1) transmitted. Thus the main framing sequence is $F_0F_1F_0F_1F_0F_1F_0F_1$ or 01010101.
2. The M-control bits, one per subframe, form *secondary* framing pulses to identify the four subframes. Here again the subscripts on the M-control bits denote the actual bit (0 or 1) transmitted. Thus the secondary framing sequence is $M_0M_1M_1M_1$ or 0111.
3. The C-control bits, three per subframe are *stuffing indicators.* In particular, C_I refers to input channel I, C_{II} refers to input channel II, and so forth. For example, the three C-control bits following M_0 in the first subframe are stuffing indicators for the first T1 signal. The insertion of a stuffed bit in this T1 signal is indicated by setting all three C-control bits to 1. To indicate no stuffing, all three are set to 0. If the three C-control bits indicate stuffing, the stuffed bit is located in the position of

M_0	[48]	C_I	[48]	F_0	[48]	C_I	[48]	C_I	[48]	F_1	[48]
M_1	[48]	C_{II}	[48]	F_0	[48]	C_{II}	[48]	C_{II}	[48]	F_1	[48]
M_1	[48]	C_{III}	[48]	F_0	[48]	C_{III}	[48]	C_{III}	[48]	F_1	[48]
M_1	[48]	C_{IV}	[48]	F_0	[48]	C_{IV}	[48]	C_{IV}	[48]	F_1	[48]

Figure 5.23
Signal format of Bell system M12 multiplexer.

the first information bit associated with the first T1 signal that follows the F_1-control bit in the same subframe. In a similar way, the second, third, and fourth T1 signals may be stuffed, as required. By using *majority logic decoding* in the receiver, a single error in any of the three C-control bits can be detected. This form of decoding means simply that the majority of the C-control bits determine whether an all-one or all-zero sequence was transmitted. Thus three 1's or combinations of two 1's and a 0 indicate that a stuffed bit is present in the information sequence, following the control bit F_1 in the pertinent subframe. On the other hand, three 0's or combinations of two 0's and a 1 indicate that no stuffing is used.

The demultiplexer at the receiving M12 unit first searches for the main framing sequence $F_0F_1F_0F_1F_0F_1F_0F_1$. This establishes identity for the four input T1 signals and also for the M- and C-control bits. From the $M_0M_1M_1M_1$ sequence, the correct framing of the C-control bits is verified. Finally, the four T1 signals are properly demultiplexed and destuffed.

This signal format has two safeguards:

1. It is possible, although unlikely, that with just the $F_0F_1F_0F_1F_0F_1F_0F_1$ sequence, one of the incoming T1 signals may contain a similar sequence. This could then cause the receiver to lock onto the wrong sequence. The presence of the $M_0M_1M_1M_1$ sequence provides verification of the genuine $F_0F_1F_0F_1F_0F_1F_0F_1$ sequence, thereby ensuring that the four T1 signals are properly demultiplexed.
2. The single-error correction capability built into the C-control bits ensures that the four T1 signals are properly destuffed.

EXAMPLE 5: CAPACITY OF M12 MULTIPLEXER

The capacity of the M12 multiplexer to accommodate small variations in the input data rates can be calculated from the format of Fig. 5.23. In each M frame, defined as the interval containing one cycle of $M_0M_1M_1M_1$ bits, one bit can be stuffed into each of four input T1 signals. Each such signal has $12 \times 6 \times 4 = 288$ positions in each M frame. Also the T1 signal has a bit rate equal to 1.544 megabits per second. Hence, each input can be incremented by

$$1.544 \times 10^6 \times \frac{1}{288} = 5.4 \text{ kilobits/s}$$

This result is much larger than the expected change in the bit rate of the incoming T1 signal. It follows therefore that the use of only one stuffed bit per input channel in each frame is sufficient to accommodate expected variations in the input signal rate.

The local clock that determines the outgoing bit rate also determines the nominal *stuffing rate S*, defined as the average number of bits stuffed per channel in any frame. The M12 multiplexer is designed for $S = 1/3$. Accordingly, the nominal bit rate of the T2 line is

$$1.544 \times 4 \times \frac{49}{48} \times \frac{288}{288\text{-}S} = 6.312 \text{ megabits/s}$$

This also ensures that the nominal T2 clock frequency is a multiple of 8 kHz (the nominal sampling rate of a voice signal), which is a desirable feature.

EXERCISE 7 Given that the data rate for one Picturephone® service is 6:312 megabits per second, and that for one television service is 44.736 megabits per second, determine the capacity of each Bell Telephone system level measured in terms of the number of (a) voice, (b) picturephone, or (c) television channels that it can accommodate.

PROBLEMS

P5.3 Sampling

Problem 1 Figure P5.1 depicts the spectrum of a message signal $m(t)$. The signal is *undersampled* at a rate of 1.5 Hz.

(a) Sketch the spectrum of the sampled version of this signal.
(b) The sampled signal is passed through an idealized low-pass interpolation filter of bandwidth 1 Hz. Sketch the spectrum of the resulting filter output.

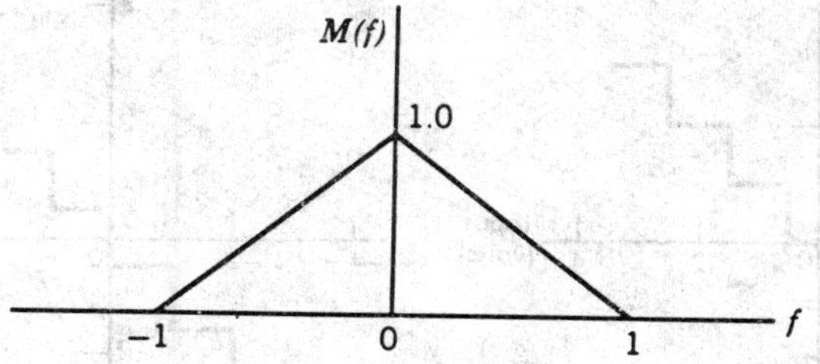

Figure P5.1

Problem 2 Consider the operation of a sample-and-hold circuit with the following parameters:

Message bandwidth, $W = 1$ Hz
Sampling period, $T_s = 0.4$ s
Pulse duration, $T = 0.2$ s

(a) Calculate the amplitude distortion produced by the aperture effect (arising from the use of flat-top samples) at the highest frequency component of the message signal.
(b) Find the amplitude response of the equalizer required to compensate for the aperture effect.

P5.4 Quantizing

Problem 3

(a) A sinusoidal signal, with an amplitude of 3.25 V is applied to a uniform quantizer of the *midtread* type with output values of 0, ±1, ±2, ±3 V, as in Fig. P5.2*a*. Sketch the waveform of the resulting quantizer output for one complete cycle of the input.
(b) Repeat this evaluation for the case when the quantizer is of the *midriser* type with output values ±0.5, ±1.5, ±2.5, ±3.5 V, as in Fig. P5.2*b*.

Problem 4 The signal

$$m(t) = 6 \sin(2\pi t) \text{ volts}$$

is transmitted using a 4-bit binary PCM system. The quantizer is of the midriser type, with a step size of 1 V. Sketch the resulting sequence of quantized samples for one complete cycle of the input. Assume a sampling

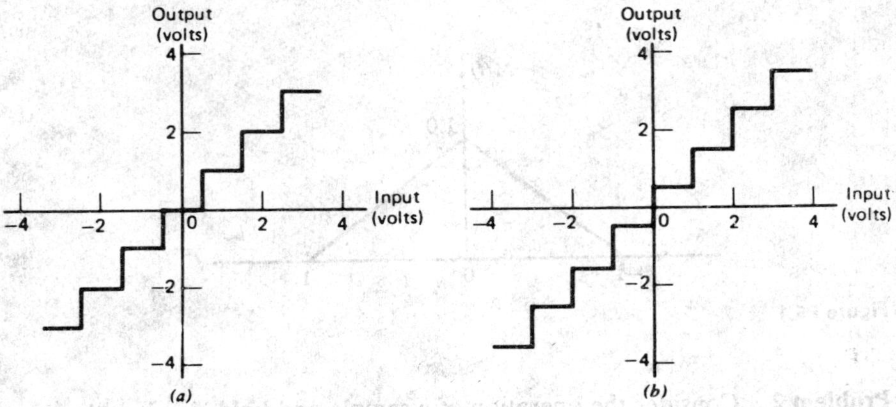

Figure P5.2

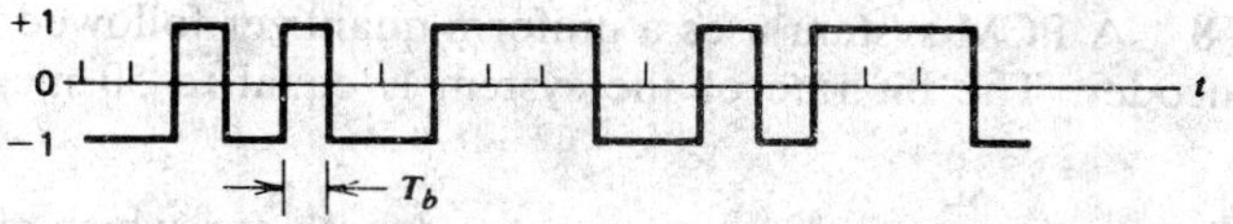

Figure P5.3

rate of four samples per second, with samples taken at $t = \pm 1/8, \pm 3/8, \pm 5/8, \ldots$, seconds.

P5.5 Coding

Problem 5 Consider the following binary sequences:

(a) An alternating sequence of 1's and 0's.
(b) A long sequence of 1's followed by a long sequence of 0's.
(c) A long sequence of 1's followed by a single 0 and than a long sequence of 1's.

Sketch the waveform for each of these sequences using the following methods of representing symbols 1 and 0:

(a) On–off signaling.
(b) Polar signaling.
(c) Return-to-zero signaling.
(d) Bipolar signaling.
(e) Manchester code.

Problem 6 Figure P5.3 shows a PCM wave in which the amplitude levels of +1 V and −1 V are used to represent binary symbols 1 and 0, respectively. The code word used consists of three bits. Find the sampled version of an analog signal from which this PCM wave is derived.

Problem 7 The bipolar waveform of Fig. 5.12*d*, representing the binary sequence 0110100011, is transmitted over a noisy channel. The received waveform is shown in Fig. P5.4, which contains a single error. Locate the position of this error, giving reasons for your answer.

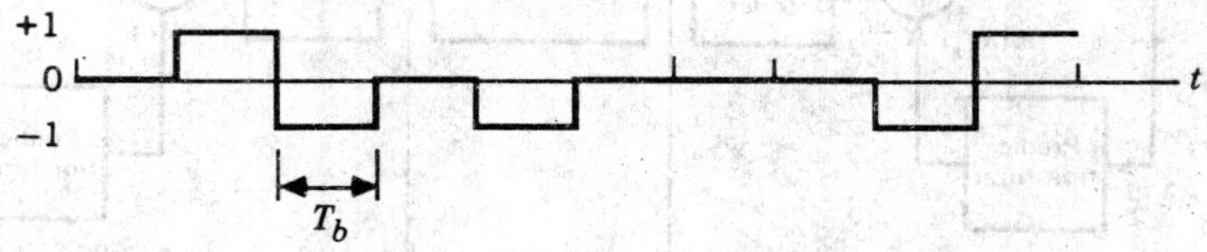

Figure P5.4

Problem 8 A PCM system uses a uniform quantizer followed by a 7-bit binary encoder. The bit rate of the system is equal to 50 megabits per second.

(a) What is the maximum message bandwidth for which the system operates satisfactorily?
(b) Determine the output signal-to-quantizing noise ratio when a full-load sinusoidal modulating wave of frequency 1 MHz is applied to the input.

P5.7 Differential Pulse-Code Modulation

Problem 9 In the DPCM system depicted in Fig. P5.5, show that in the absence of noise in the channel, the transmitting and receiving prediction filters operate on slightly different input signals.

P5.8 Delta Modulation

Problem 10 Consider a sine wave of frequency f_m and amplitude A_m, applied to a delta modulator with representation levels $\pm\delta$. Show that slope-overload distortion will occur if

$$A_m > \frac{\delta}{2\pi f_m T_s}$$

where T_s is the sampling period. What is the maximum power that may be transmitted without slope-overload distortion?

Problem 11 The ramp signal $m(t) = at$ is applied to a delta modulator that operates with a sampling period T_s and representation levels $\pm\delta$.

(a) Show that slope-overload distortion occurs if $\delta < aT_s$.
(b) Sketch the modulator output for the following three values of δ:

(i) $\delta = 0.75\, aT_s$
(ii) $\delta = aT_s$
(iii) $\delta = 1.25\, aT_s$

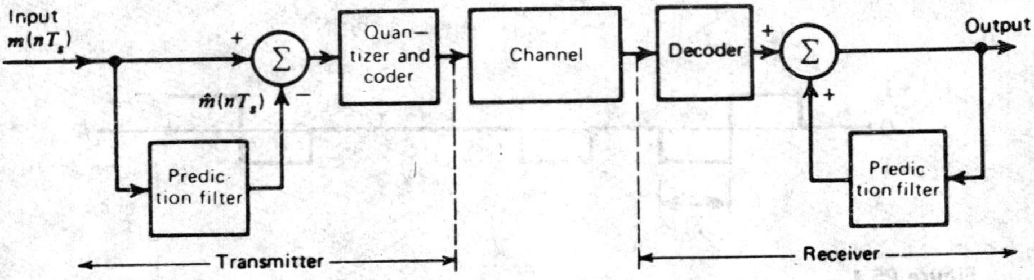

Figure P5.5

Problem 12 Consider a speech signal with maximum frequency of 3.4 kHz and maximum amplitude of 1 V. This speech signal is applied to a delta modulator with its bit rate set at 20 kilobits per second. Discuss the choice of an appropriate step size for the modulator.

P5.10 Time-Division Multiplexing

Problem 13 Six independent message sources of bandwidths W, W, $2W$, $2W$, $3W$, and $3W$ hertz are to be transmitted on a time-division multiplexed basis using a common communication channel.

(a) Set up a scheme for accomplishing this multiplexing requirement, with each message signal sampled at its Nyquist rate.
(b) Determine the minimum transmission bandwidth of the channel.

Problem 14 Twenty-four voice signals are sampled uniformly and then time-division multiplexed. The sampling operation uses flat-top samples with 1 μs duration. The multiplexing operation includes provision for synchronization by adding an extra pulse of sufficient amplitude and also 1 μs duration. The highest frequency component of each voice signal is 3.4 kHz.

(a) Assuming a sampling rate of 8 kHz, calculate the spacing between successive pulses of the multiplexed signal.
(b) Repeat your calculation assuming the use of Nyquist rate sampling.

CHAPTER SIX

INTERSYMBOL INTERFERENCE AND ITS CURES

When digital data (of whatever origin) is transmitted over a *band-limited channel,* dispersion in the channel gives rise to a troublesome form of interference called intersymbol interference. As the name implies, *intersymbol interference* refers to interference caused by the time response of the channel *spilling over* from one symbol into another. This has the effect of introducing deviations (errors) between the data sequence reconstructed at the receiver output and the original data sequence applied to the transmitter input. Hence, unless corrective measures are taken, intersymbol interference may impose a limit on the attainable rate of data transmission that is far below the physical capability of the channel.

In this chapter, we study the *intersymbol interference problem* and the use of *baseband pulse shaping* as the solution to the problem. The term "baseband" is used to designate the band of frequencies representing the original signal as delivered by a source of information.

6.1 BASEBAND TRANSMISSION OF BINARY DATA

For the *baseband transmission of digital data,* the use of *discrete pulse-amplitude modulation* (PAM) provides the most efficient form of discrete pulse modulation in terms of power and bandwidth use. In discrete PAM, *the amplitude of the transmitted pulses is varied in a discrete manner in accordance with the given digital data.*

The basic elements of a *baseband binary* PAM *system* are shown in Fig. 6.1. The signal applied to the input of the system consists of a binary data sequence $\{b_k\}$ with a *bit duration* of T_b seconds; b_k is in the form of 1 or 0. This signal is applied to a pulse generator, producing the pulse waveform

$$x(t) = \sum_{k=-\infty}^{\infty} A_k g(t - kT_b) \tag{6.1}$$

where $g(t)$ denotes a *shaping pulse* with its value at time $t = 0$ defined by

$$g(0) = 1$$

The amplitude A_k depends on the identity of the input bit b_k; specifically, we assume that

$$A_k = \begin{cases} +a, & \text{if the input bit } b_k \text{ is symbol 1} \\ -a, & \text{if the input bit } b_k \text{ is symbol 0} \end{cases} \tag{6.2}$$

The PAM signal $x(t)$ passes through a *transmitting filter* of transfer function $H_T(f)$. The resulting filter output defines the transmitted signal, which is modified in a deterministic fashion as a result of transmission through the channel of transfer function $H_C(f)$. The signal at the receiver input is passed through a *receiving filter* of transfer function $H_R(f)$. This filter output is sampled *synchronously* with the transmitter, with the sampling instants being determined by a *clock* or *timing signal* that is usually extracted from the receiving filter output. Finally, the sequence of samples thus obtained is used to reconstruct the original data sequence by means of a *decision device.* The amplitude of each sample is compared to a *threshold.* If the threshold is exceeded, a decision is made in favor of symbol 1 (say). If the threshold is not exceeded, a decision is made in favor of symbol 0. If the sample amplitude equals the threshold exactly, the symbol may be chosen as 0 or 1 without affecting overall performance. In such an event, we will choose symbol 0.

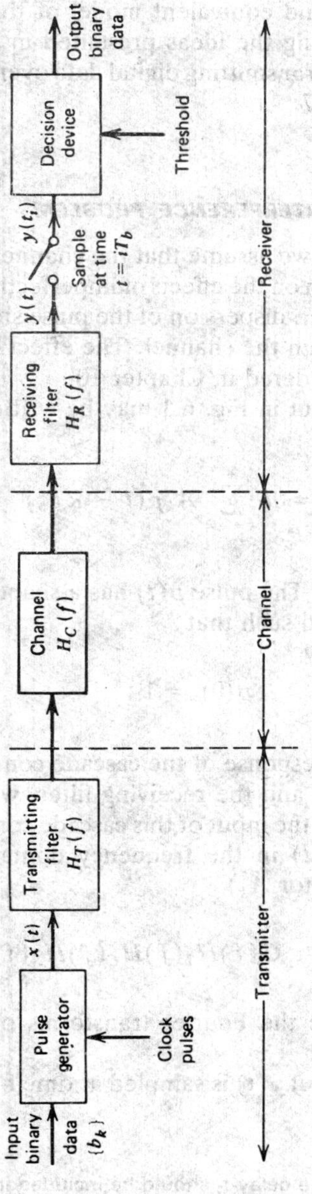

Figure 6.1
Baseband binary data transmission system.

The model shown in Fig. 6.1 represents not only a data transmission system inherently baseband in nature (e.g., data transmission over a coaxial cable) but also the baseband *equivalent* of a linear modulation system used to transmit data over a band-pass channel (e.g., telephone channel). In the latter case, the baseband equivalent model of the data transmission system is developed by using the ideas presented in Section 3.5. Linear modulation techniques for transmitting digital data over band-pass channels are considered in Chapter 7.

6.2 THE INTERSYMBOL INTERFERENCE PROBLEM

For the present discussion, we assume that the channel is *noiseless*. We do so in order to focus attention on the effects of imperfections in the frequency response of the channel (i.e., dispersion of the pulse shape by the channel) on data transmission through the channel. The effect of channel noise on the receiver output is considered in Chapter 10.

The receiving filter output in Fig. 6.1 may be written as[1]

$$y(t) = \mu \sum_{k=-\infty}^{\infty} A_k p(t - kT_b) \tag{6.3}$$

where μ is a scaling factor. The pulse $p(t)$ has a shape different from that of $g(t)$, but it is normalized such that

$$p(0) = 1$$

The pulse $\mu A_k p(t)$ is the response of the cascade connection of the transmitting filter, the channel, and the receiving filter, which is produced by the pulse $A_k g(t)$ applied to the input of this cascade connection. Therefore, we may relate $p(t)$ to $g(t)$ in the frequency domain as follows (after cancelling the common factor A_k)

$$\mu P(f) = G(f)H_T(f)H_C(f)H_R(f) \tag{6.4}$$

where $P(f)$ and $G(f)$ are the Fourier transforms of $p(t)$ and $g(t)$, respectively.

The receiving filter output $y(t)$ is sampled at time $t_i = iT_b$ (with i taking

[1]To be precise, an arbitrary time delay t_0 should be included in the argument of the pulse $p(t - kT_b)$ in Eq. 6.3 to represent the effect of transmission delay through the system. For convenience, we have put this delay equal to zero in Eq. 6.3.

on integer values), yielding

$$y(t_i) = \mu \sum_{k=-\infty}^{\infty} A_k p[(i-k)T_b]$$

$$= \mu A_i + \mu \sum_{\substack{k=-\infty \\ k\neq i}}^{\infty} A_k p[(i-k)T_b] \qquad i = 0, \pm 1, \pm 2, \ldots \tag{6.5}$$

In Eq. 6.5 the first term μA_i represents the contribution of the ith transmitted bit. The second term represents the residual effect of all other transmitted bits on the decoding of the ith received bit; this residual effect is called *intersymbol interference* (ISI).

In the absence of ISI (and, of course, channel noise), we observe from Eq. 6.5 that

$$y(t_i) = \mu A_i \tag{6.6}$$

which shows that, under these conditions, the ith transmitted bit can be decoded correctly. The unavoidable presence of ISI in the system, however, introduces errors in the decision device at the receiver output. Therefore, in the design of the transmitting and receiving filters, the objective is to minimize the effects of ISI, and thereby deliver the digital data to its destination with the smallest error rate possible.

Typically, the channel transfer function $H_C(f)$ and the pulse spectrum $G(f)$ are specified, and the problem is to determine the transfer functions of the transmitting and receiving filters, $H_T(f)$ and $H_R(f)$, so as to enable the receiver to correctly decode the received sequence of sample values $\{y(t_i)\}$ in accordance with Eq. 6.6. Deviation from this ideal condition is caused by the presence of intersymbol interference that arises owing to dispersion of the pulse shape by the channel. To solve the problem, we have to exercise control over intersymbol interference, an issue that is discussed next.

6.3 IDEAL SOLUTION

Control of intersymbol interference in the system is achieved in the time domain by controlling the function $p(t)$, or in the frequency domain by controlling $P(f)$. One signal waveform that produces *zero* intersymbol interference is defined by the *sinc function:*

$$p(t) = \frac{\sin(2\pi B_0 t)}{2\pi B_0 t} = \text{sinc}(2B_0 t) \tag{6.7}$$

where

$$B_0 = \frac{1}{2T_b} \tag{6.8}$$

The parameter B_0 is called the *Nyquist bandwidth;* it defines the minimum transmission bandwidth for zero intersymbol interference. According to Eq. 6.8, the Nyquist bandwidth B_0 is equal to one half of the *bit rate* $1/T_b$. Note the analogy between this relation and the sampling theorem for strictly band-limited signals. (The sampling theorem was discussed in Sections 2.7 and 5.3).

The frequency response $P(f)$, representing the Fourier transform of the pulse $p(t)$ of Eq. 6.7, is defined by

$$P(f) = \begin{cases} \dfrac{1}{2B_0}, & 0 \leq |f| < B_0 \\ 0 & B_0 < |f| \end{cases} \tag{6.9}$$

This means that no frequencies of absolute value exceeding half the bit rate are needed. The function $p(t)$ can be regarded as the impulse response of an ideal low-pass filter with an amplitude response of $1/(2B_0)$ in the passband and a bandwidth B_0. The function $p(t)$ has its peak value at the origin and goes through zero at integer multiples of the bit duration T_b. It is apparent that if the received waveform $y(t)$ is sampled at the instants of time $t = 0, \pm T_b, \pm 2T_b, \ldots$, then the pulses defined by $A_i p(t - iT_b)$ with arbitrary amplitude A_i and $i = 0, \pm 1, \pm 2, \ldots$, will not interfere with each other.

Although this ideal choice of pulse shape for $p(t)$ achieves economy in bandwidth in that it solves the problem of zero intersymbol interference with the minimum bandwidth possible, there are two difficulties that make its use for system design impractical:

1. It requires that frequency response $P(f)$ be flat from $-B_0$ to B_0, and zero elsewhere. This is physically unrealizable, and very difficult to approximate in practice because of the abrupt transitions at $\pm B_0$.
2. The time function $p(t)$ decreases as $1/|t|$ for large $|t|$, resulting in a slow rate of decay. This is caused by the discontinuity of $P(f)$ at $\pm B_0$. Accordingly, there is practically no margin of error in sampling times in the receiver.

To evaluate the effect of this *timing error,* consider the sample of $y(t)$ at $t = \Delta t$, where Δt is the timing error. To simplify the analysis, we have put the correct sampling time t_i equal to zero. We thus obtain, in the absence of noise:

$$\begin{aligned} y(\Delta t) &= \mu \sum_k A_k p(\Delta t - kT_b) \\ &= \mu \sum_k A_k \operatorname{sinc}[2B_0(\Delta t - kT_b)] \end{aligned}$$

Since $2B_0T_b = 1$, we may rewrite this relation as

$$\begin{aligned} y(\Delta t) &= \mu \sum_k A_k \operatorname{sinc}(2B_0\, \Delta t - k) \\ &= \mu A_0 \operatorname{sinc}(2B_0\, \Delta t) + \mu \frac{\sin(2\pi B_0\, \Delta t)}{\pi} \sum_{\substack{k \\ k \neq 0}} \frac{(-1)^k A_k}{2B_0\, \Delta t - k} \end{aligned} \tag{6.10}$$

The first term on the right side of Eq. 6.10 defines the desired symbol, whereas the remaining series represents the intersymbol interference caused by the timing error Δt in sampling the signal $y(t)$. In certain cases, it is possible for this series to diverge, thereby causing erroneous decisions in the receiver.

We therefore have to look to other pulse shapes not only to combat the intersymbol interference problem but also to do so in a feasible way. In the next section, we present one such solution that is a natural extension of the minimum-bandwidth (ideal) solution just described.

6.4 RAISED COSINE SPECTRUM

The solution we have in mind differs from the ideal solution in one important respect: the overall frequency response $P(f)$ decreases toward zero gradually rather than abruptly. In particular, $P(f)$ consists of a *flat* portion and a *rolloff* portion that has the form of a raised-cosine function, as follows[2]

$$P(f) = \begin{cases} \dfrac{1}{2B_0}, & 0 \leq |f| < f_1 \\ \dfrac{1}{4B_0}\left\{1 + \cos\left[\dfrac{\pi(|f| - f_1)}{2B_0 - 2f_1}\right]\right\}, & f_1 < |f| < 2B_0 - f_1 \\ 0, & 2B_0 - f_1 < |f| \end{cases} \tag{6.11}$$

[2]The solution described in Eq. 6.11 was first proposed by Nyquist (1928) in his studies of telegraph transmission theory.

The frequency f_1 and the Nyquist bandwidth B_0 are related by

$$\alpha = 1 - \frac{f_1}{B_0} \tag{6.12}$$

which is called the *rolloff factor*. For $\alpha = 0$, i.e., $f_1 = B_0$, we get the minimum bandwidth solution described in Section 6.3.

The frequency response $P(f)$, normalized by multiplying it by $2B_0$, is plotted in Fig. 6.2*a* for three values of α, namely, 0, 0.5, and 1. We see that for $\alpha = 0.5$ or 1, the function $P(f)$ cuts off gradually as compared with an ideal low-pass filter (corresponding to $\alpha = 0$), and it is therefore easier to realize in practice. Also the function $P(f)$ exhibits odd symmetry about the cutoff frequency B_0 of the ideal low-pass filter.

The time response $p(t)$, that is, the inverse Fourier transform of $P(f)$, is defined by

$$p(t) = \text{sinc}(2B_0t)\frac{\cos(2\pi\alpha B_0 t)}{1 - 16\alpha^2 B_0^2 t^2} \tag{6.13}$$

This function consists of the product of two factors: the factor $\text{sinc}(2B_0t)$ associated with the ideal solution, and a second factor that decreases as $1/|t|^2$ for large $|t|$. The first factor ensures zero crossings of $p(t)$ at the desired sampling instants of time $t = iT$ with i an integer (positive and negative). The second factor reduces the tails of the pulse considerably below that obtained from the ideal low-pass filter, so that the transmission of binary waves using such pulses is relatively insensitive to sampling time errors. In fact, the amount of intersymbol interference resulting from this timing error decreases as the rolloff factor α is increased from zero to unity.

The time response $p(t)$ is plotted in Fig. 6.2*b* for $\alpha = 0$, 0.5 and 1. For the special case of $\alpha = 1$, the function $p(t)$ simplifies as

$$p(t) = \frac{\text{sinc}(4B_0t)}{1 - 16B_0^2t^2} \tag{6.14}$$

This time response exhibits two interesting properties:

1. At $t = \pm T_b/2 = \pm 1/4B_0$, we have $p(t) = 0.5$; that is, the pulse width measured at half amplitude is exactly equal to the bit duration T_b.
2. There are zero crossings at $t = \pm 3T_b/2, \pm 5T_b/2, \ldots$ in addition to the usual zero crossings at the sampling times $t = \pm T_b, \pm 2T_b, \ldots$.

These two properties are particularly useful in generating a timing signal from the received signal for the purpose of synchronization.

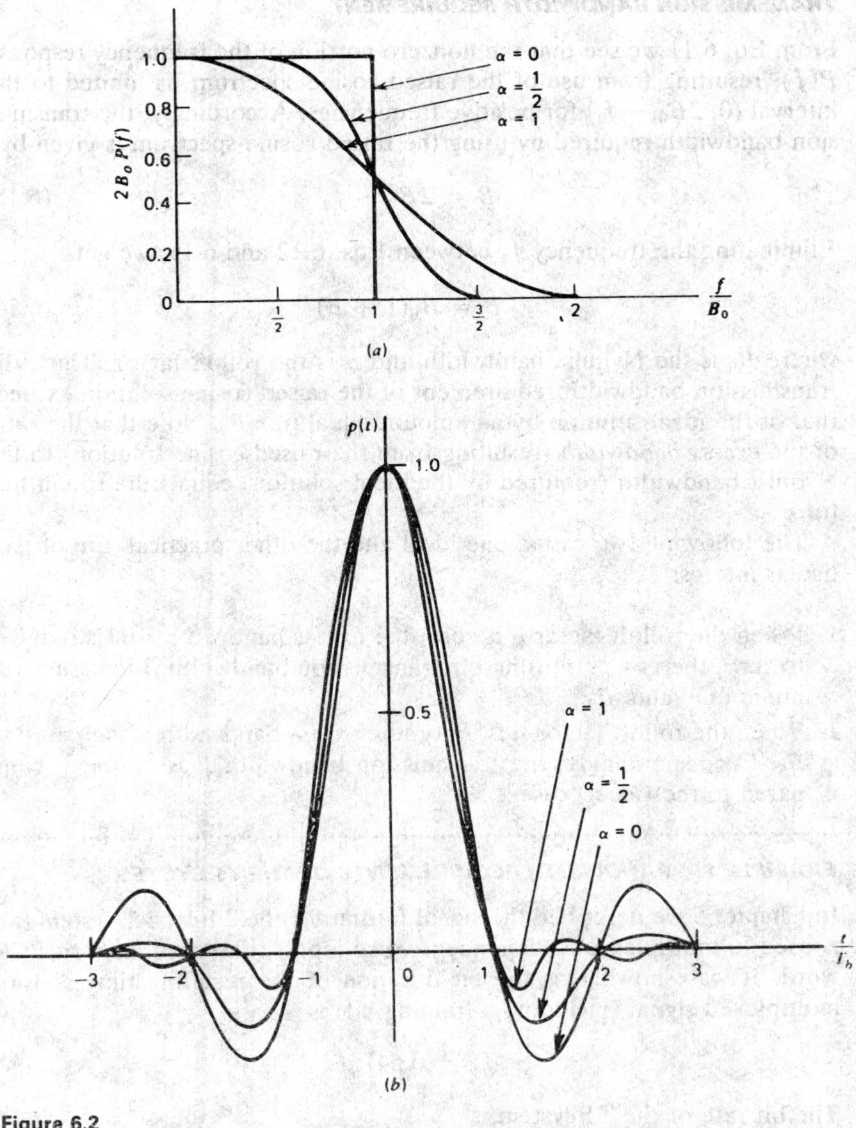

Figure 6.2
Responses for different rolloff factors. (a) Frequency response. (b) Time response.

EXERCISE 1 Given the frequency response $P(f)$ defined in Eq. 6.11, show that the inverse Fourier transform $p(t)$ is as given in Eq. 6.13.

TRANSMISSION BANDWIDTH REQUIREMENT

From Eq. 6.11 we see that the nonzero portion of the frequency response $P(f)$, resulting from use of the raised cosine spectrum, is limited to the interval $(0, 2B_0 - f_1)$ for positive frequencies. Accordingly, the transmission bandwidth required by using the raised cosine spectrum is given by

$$B = 2B_0 - f_1 \tag{6.15}$$

Eliminating the frequency f_1 between Eqs. 6.12 and 6.15, we get

$$B = B_0(1 + \alpha) \tag{6.16}$$

where B_0 is the Nyquist bandwidth and α is the rolloff factor. Thus, the transmission bandwidth requirement of the raised cosine solution exceeds that of the ideal solution by an amount equal to αB_0. Note that the ratio of the *excess bandwidth* (resulting from the raised cosine solution) to the Nyquist bandwidth (required by the ideal solution) equals the rolloff factor α.

The following two cases, one ideal and the other practical, are of particular interest:

1. When the rolloff factor α is zero, the excess bandwidth αB_0 is reduced to zero, thereby permitting the transmission bandwidth B to assume its minimum value B_0.
2. When the rolloff factor α is unity, the excess bandwidth is increased to B_0. Correspondingly, the transmission bandwidth B is doubled, compared to the (ideal) case 1.

EXAMPLE 1 BANDWIDTH REQUIREMENTS OF THE T1 SYSTEM

In Chapter 5 we described the signal format for the T1 carrier system that is used to multiplex 24 independent voice inputs, based on an 8-bit PCM word. It was shown that the bit duration of the resulting time-division multiplexed signal (including a framing bit) is

$$T_b = 0.647\ \mu\text{s}$$

The bit rate of the T1 system is

$$R_b = \frac{1}{T_b} = 1.544 \text{ Mb/s}$$

Assuming an ideal low-pass characteristic for the channel, it follows that the Nyquist bandwidth of the T1 system is

$$B_0 = \frac{1}{2T_b} = 772 \text{ kHz}$$

This is the minimum transmission bandwidth of the T1 system for zero intersymbol interference. However, a more realistic value for the transmission bandwidth B is obtained by using a raised cosine spectrum with $\alpha = 1$. In this case, we find that

$$B = \frac{1}{T_b} = 1.544 \text{ MHz}$$

EXERCISE 2 Calculate the transmission bandwidth requirement of the M12 multiplexer described in Section 5.11. Assume the use of a raised cosine spectrum with rolloff factor $\alpha = 1$ for the baseband pulse shaping.

6.5 CORRELATIVE CODING

Thus far we have treated intersymbol interference as an undesirable phenomenon that produces a degradation in system performance. Indeed, its very name connotes a nuisance effect. Nevertheless, by adding intersymbol interference to the transmitted signal in a controlled manner, it is possible to achieve a signaling rate of $2B_0$ symbols per second in a channel of bandwidth B_0 hertz. Such schemes are called *correlative coding* or *partial-response signaling* schemes.[3] The design of these schemes is based on the premise that since the intersymbol interference that is introduced into the transmitted signal is known, its effect can be accounted for at the receiver. Thus correlative coding may be regarded as a practical means of achieving the theoretical maximum signaling rate of $2B_0$ symbols per second in a bandwidth of B_0 hertz, using realizable and perturbation-tolerant filters.

In this section, we illustrate the basic idea of correlative coding by considering two specific examples: *duobinary signaling* and *modified duobinary signaling*. Duobinary signaling employs a correlation span of one binary digit, whereas modified duobinary signaling employs a correlation span of two binary digits; the use of "duo" is intended to imply doubling of the transmission capacity of a straight binary system.

DUOBINARY SIGNALING

Consider a binary input sequence $\{b_k\}$ consisting of uncorrelated binary digits each having duration T_b seconds, with symbol 1 represented by a

[3]Correlative coding and partial response signaling are synonomous; both terms are used in the literature. The idea of correlative coding was originated by Lender (1963). For an overview on correlative coding, see Pasupathy (1977).

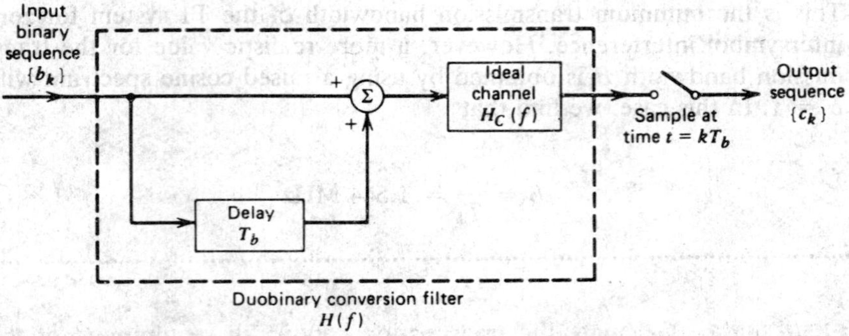

Figure 6.3
Duobinary signaling scheme.

pulse of amplitude $+1$ V, and symbol 0 by a pulse amplitude -1 V. When this sequence is applied to a *duobinary encoder,* it is converted into a *three-level output,* namely, -2, 0, and $+2$ V. To produce this transformation, we may express the digit c_k at the duobinary coder output as the sum of the present binary digit b_k and its previous value b_{k-1}, as shown by

$$c_k = b_k + b_{k-1} \tag{6.17}$$

One of the effects of the transformation described by Eq. 6.17 is to change the input sequence $\{b_k\}$ of uncorrelated binary digits into a sequence $\{c_k\}$ of correlated digits. This correlation between the adjacent transmitted levels may be viewed as introducing intersymbol interference into the transmitted signal in an artificial manner. However, this intersymbol interference is under the designer's control; this is the basis of correlative coding.

Figure 6.3 depicts the block diagram of a *duobinary encoder,* including a band-limited channel assumed to be ideal. The binary sequence $\{b_k\}$ is first passed through a simple filter consisting of the parallel combination of a direct path and an ideal element producing a *delay* of T_b seconds, where T_b is the bit duration. For every unit impulse applied to the input of this filter, we get two unit impulses spaced T_b seconds apart at the filter output. The output of this filter in response to the incoming binary sequence $\{b_k\}$ is then passed through the channel of transfer function $H_C(f)$. A continuous waveform is therefore produced at the channel output. The resulting waveform is sampled uniformly every T_b seconds, thereby producing the duobinary encoded sequence $\{c_k\}$. Note that the effect of the channel is included in this encoding operation.

The cascade connection of the delay-line filter and the channel is called a *duobinary conversion filter*. In Fig. 6.3, we have enclosed this filter inside a dashed rectangle. The response of the filter may be characterized in terms of an overall transfer function $H(f)$, which is evaluated next.

An ideal delay element, producing a delay of T_b seconds, has the transfer function $\exp(-j2\pi f T_b)$, so that the transfer function of the delay-line filter shown in Fig. 6.3 is $1 + \exp(-j2\pi f T_b)$. Hence, the overall transfer function of this filter connected in cascade with the ideal channel $H_C(f)$ is

$$\begin{aligned} H(f) &= H_C(f)[1 + \exp(-j2\pi f T_b)] \\ &= H_C(f)[\exp(j\pi f T_b) + \exp(-j\pi f T_b)] \exp(-j\pi f T_b) \\ &= 2H_C(f)\cos(\pi f T_b)\exp(-j\pi f T_b) \end{aligned} \tag{6.18}$$

For an ideal channel of bandwidth $B_0 = 1/2T_b$, we have

$$H_C(f) = \begin{cases} 1, & |f| \leq 1/2T_b \\ 0, & \text{otherwise} \end{cases} \tag{6.19}$$

Thus the overall frequency response has the form of a half-cycle cosine function, as shown by

$$H(f) = \begin{cases} 2\cos(\pi f T_b)\exp(-j\pi f T_b), & |f| \leq 1/2T_b \\ 0, & \text{otherwise} \end{cases} \tag{6.20}$$

for which the amplitude response and phase response are as shown in parts *a* and *b* of Fig. 6.4, respectively. An advantage of this frequency response is that it can be easily approximated in practice.

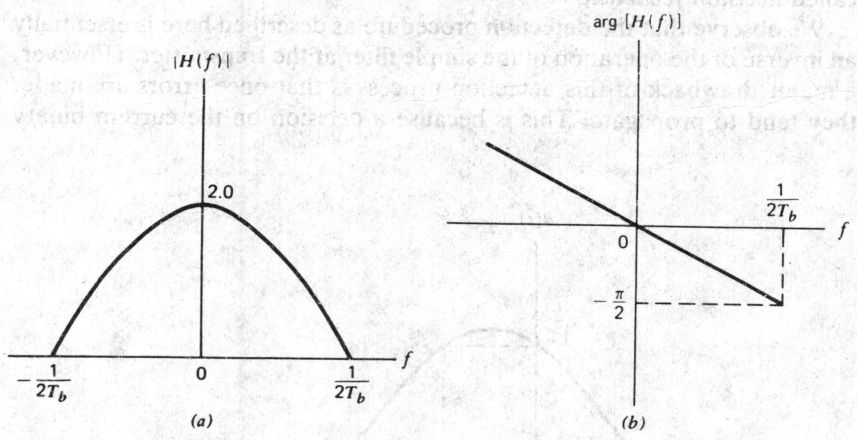

Figure 6.4
Frequency response of the duobinary conversion filter. (a) Amplitude response. (b) Phase response.

The corresponding value of the impulse response consists of two sinc pulses, time-displaced by T_b seconds, as shown by (except for a scaling factor)

$$\begin{aligned} h(t) &= \frac{\sin(\pi t/T_b)}{\pi t/T_b} + \frac{\sin[\pi(t - T_b)/T_b]}{\pi(t - T_b)/T_b} \\ &= \frac{\sin(\pi t/T_b)}{\pi t/T_b} - \frac{\sin(\pi t/T_b)}{\pi(t - T_b)/T_b} \\ &= \frac{T_b^2 \sin(\pi t/T_b)}{\pi t(T_b - t)} \end{aligned} \tag{6.21}$$

which is shown plotted in Fig. 6.5. We see that the overall impulse response $h(t)$ has only *two* distinguishable values at the sampling instants.

The original data $\{b_k\}$ may be detected from the duobinary-coded sequence $\{c_k\}$ by subtracting the previous decoded binary digit from the currently received digit c_k in accordance with Eq. 6.17. Specifically, letting $\hat{b}_k$ represent the *estimate* of the original binary digit b_k as conceived by the receiver at time $t = kT_b$, we have

$$\hat{b}_k = c_k - \hat{b}_{k-1} \tag{6.22}$$

It is apparent that if c_k is received without error and if the previous estimate $\hat{b}_{k-1}$ at time $t = (k - 1)T_b$ also corresponds to a correct decision, then the current estimate $\hat{b}_k$ will be correct too. The technique of using a stored estimate of the previous symbol in the estimation of the current symbol is called *decision feedback*.

We observe that the detection procedure as described here is essentially an inverse of the operation of the simple filter at the transmitter. However, a major drawback of this detection process is that once errors are made, they tend to propagate. This is because a decision on the current binary

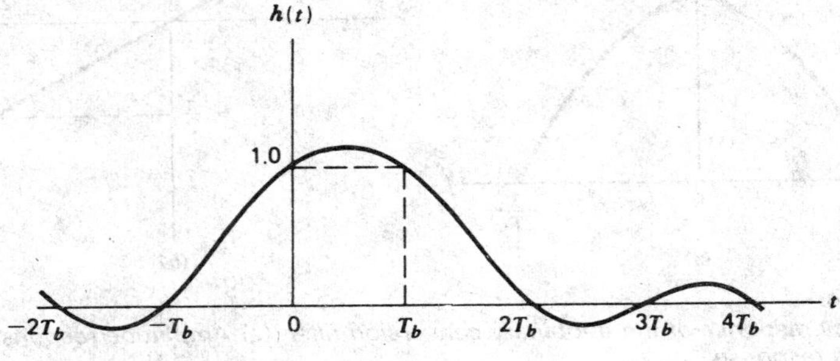

Figure 6.5
Impulse response of duobinary conversion filter.

digit b_k depends on the correctness of the decision made on the previous binary digit b_{k-1}.

A practical means of avoiding this error propagation is to use *precoding* before the duobinary coding, as shown in Fig. 6.6. The precoding operation performed on the input binary sequence $\{b_k\}$ converts it into another binary sequence $\{a_k\}$ defined by

$$a_k = b_k \oplus a_{k-1} \tag{6.23}$$

where the symbol $\oplus$ denotes *modulo-two addition* of the binary digits b_k and a_{k-1}. This addition is equivalent to the EXCLUSIVE OR operation. An EXCLUSIVE OR gate operates as follows. The output of a two-input EXCLUSIVE OR gate is a 1 if exactly one input is a 1; otherwise, the output remains a 0. The resulting precoder output $\{a_k\}$ is next applied to the duobinary coder, thereby producing the sequence $\{c_k\}$ that is related to $\{a_k\}$ as follows

$$c_k = a_k + a_{k-1} \tag{6.24}$$

Note that unlike the linear operation of duobinary coding, precoding is a nonlinear operation.

We assume that symbol 1 at the precoder output in Fig. 6.6 is represented by +1 V and symbol 0 by −1 V. Therefore, from Eqs. 6.23 and 6.24, we find that

$$c_k = \begin{cases} \pm 2\text{ V}, & \text{if } b_k \text{ is represented by symbol 0} \\ 0\text{ V} & \text{if } b_k \text{ is represented by symbol 1} \end{cases} \tag{6.25}$$

which is illustrated in Example 2. From Eq. 6.25 we deduce the following decision rule for constructing the decoded binary sequence $\{\hat{b}_k\}$ at the

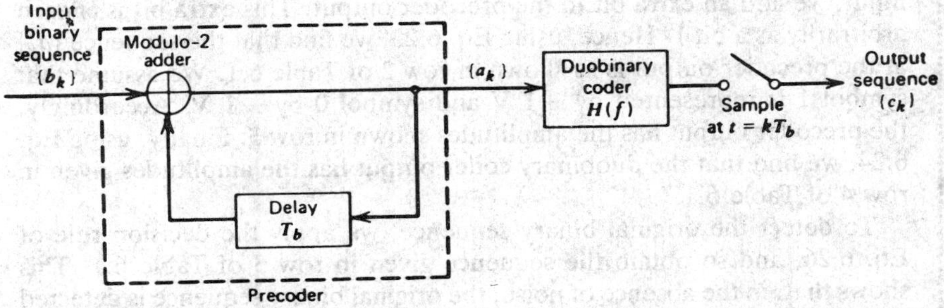

Figure 6.6
A precoded duobinary scheme. Details of the duobinary coder are given in Fig. 6.3.

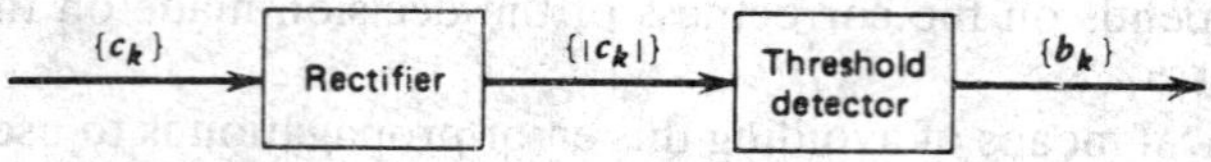

Figure 6.7
Detector for recovering original binary sequence from the precoded duobinary coder output.

receiver output:

$$\hat{b}_k = \begin{cases} \text{symbol 0}, & \text{if } |c_k| > 1 \text{ V} \\ \text{symbol 1}, & \text{if } |c_k| \leq 1 \text{ V} \end{cases} \tag{6.26}$$

According to Eq. 6.26, the detector (decoder) consists of a rectifier, the output of which is compared to a threshold of 1 V, and the original binary sequence $\{b_k\}$ is thereby detected. A block diagram of the detector is shown in Fig. 6.7. A useful feature of this detector is that no knowledge of any input sample other than the present one is required. Hence, error propagation cannot occur in the detector of Fig. 6.7.

Moreover, we may note the following two points:

1. In the absence of channel noise, the decoded sequence $\{\hat{b}_k\}$ derived from Eq. 6.26 is exactly the same as the original binary sequence $\{b_k\}$ at the transmitter input.
2. The use of Eq. 6.23 requires the addition of an extra bit to the precoded sequence $\{a_k\}$. The decoded sequence $\{\hat{b}_k\}$ is invariant to the use of a 1 or a 0 for this extra bit.

EXAMPLE 2

Consider the input binary sequence 0010110. To proceed with the precoding of this sequence, which involves feeding the precoder output back to the input, we add an extra bit to the precoder output. This extra bit is chosen arbitrarily as a bit 1. Hence, using Eq. 6.23, we find that the sequence $\{a_k\}$ at the precoder output is as shown in row 2 of Table 6.1. We assume that symbol 1 is represented by +1 V and symbol 0 by −1 V. Accordingly, the precoder output has the amplitudes shown in row 3. Finally, using Eq. 6.24, we find that the duobinary coder output has the amplitudes given in row 4 of Table 6.1.

To detect the original binary sequence, we apply the decision rule of Eq. 6.26, and so obtain the sequence given in row 5 of Table 6.1. This shows that, in the absence of noise, the original binary sequence is detected correctly.

TABLE 6.1

Input binary sequence $\{b_k\}$		0	0	1	0	1	1	0
Precoded binary sequence $\{a_k\}$	1	1	1	0	0	1	0	0
Polar representation of sequence $\{a_k\}$	+1	+1	+1	−1	−1	+1	−1	−1
Duobinary coder output, $\{c_k\}$		2	2	0	−2	0	0	−2
Decoded binary sequence $\{\hat{b}_k\}$		0	0	1	0	1	1	0

EXERCISE 3 Repeat the calculations of Table 6.1, assuming that the extra bit at the beginning of the precoded sequence $\{a_k\}$ is a 0. Hence, show that the decoded sequence $\{\hat{b}_k\}$ is unaffected by this change (compared to the initial bit used in Example 2).

EXERCISE 4 The duobinary, ternary, and bipolar signaling techniques have one common feature: They all employ three amplitude levels. In what way does the duobinary technique differ from the other two?

MODIFIED DUOBINARY SIGNALING

In the duobinary signaling technique just described, the transfer function $H(f)$, and consequently the power spectral density of the transmitted pulse, is nonzero at the origin. In some applications, this is an undesirable feature. We may correct for this drawback by using the *modified duobinary signaling* technique, which involves a correlation span of two binary digits. This is achieved by subtracting input binary digits spaced $2T_b$ seconds apart. Specifically, the output of the modified duobinary conversion filter is related to the sequence $\{a_k\}$ at its input as follows

$$c_k = a_k - a_{k-2} \tag{6.27}$$

Here, again, we find that a three-level signal is generated. If $a_k = \pm 1$ V, as assumed previously, c_k takes on one of three values: 2, 0, and −2 V.

Figure 6.8 depicts the complete block diagram of a *modified duobinary encoder*, incorporating an appropriate precoder and a band-limited channel assumed to be ideal. Here again the channel is included as an integral part of the encoding operation. The incoming binary sequence $\{b_k\}$ produces a continuous waveform at the channel output. This waveform is therefore

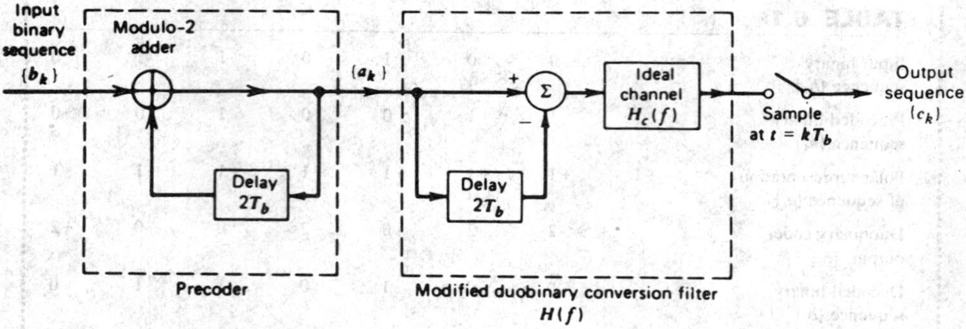

Figure 6.8
Modified duobinary signaling scheme.

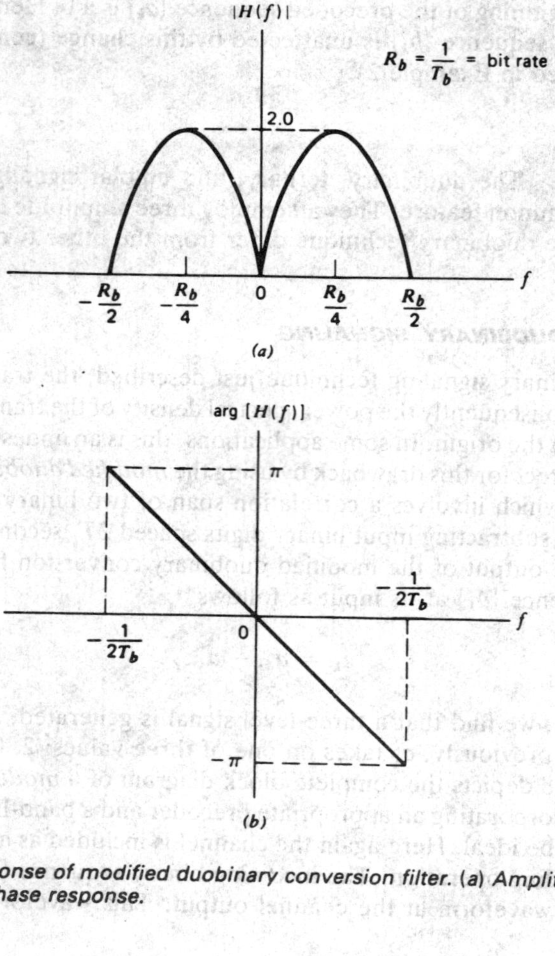

Figure 6.9
Frequency response of modified duobinary conversion filter. (a) Amplitude response. (b) Phase response.

sampled uniformly every T_b seconds to produce the modified duobinary encoded sequence $\{c_k\}$.

Let $H(f)$ denote the overall transfer function of the *modified duobinary conversion filter* that consists of the cascade connection of the delay-line-filter and the channel; this filter is enclosed inside the second dashed rectangle in Fig. 6.8. Hence, we may write

$$\begin{aligned} H(f) &= H_C(f)[1 - \exp(-j4\pi f T_b)] \\ &= 2jH_C(f)\sin(2\pi f T_b)\exp(-j2\pi f T_b) \end{aligned} \tag{6.28}$$

where $H_C(f)$ is defined in Eq. 6.19. We, therefore, have an overall frequency response in the form of a half-cycle sine function, as shown by

$$H(f) = \begin{cases} 2j\sin(2\pi f T_b)\exp(-j2\pi f T_b), & |f| \leq 1/2T_b \\ 0, & \text{elsewhere} \end{cases} \tag{6.29}$$

The corresponding amplitude response and phase response of the modified duobinary-coder are as shown in parts *a* and *b* of Fig. 6.9, respectively. Note that the phase response depicted in Fig. 6.9*b* does not include the constant 90°-phase shift due to the multiplying factor j in Eq. 6.29. A useful feature of the modified duobinary coder is the fact that its output has no dc component. This property is important since, in practice, many communication channels cannot transmit a dc component.

The impulse response of the modified duobinary coder consists of two sinc pulses that are time-displaced by $2T_b$ seconds, as shown by (except for a scaling factor)

$$\begin{aligned} h(t) &= \frac{\sin(\pi t/T_b)}{\pi t/T_b} - \frac{\sin[\pi(t - 2T_b)/T_b]}{\pi(t - 2T_b)/T_b} \\ &= \frac{\sin(\pi t/T_b)}{\pi t/T_b} - \frac{\sin(\pi t/T_b)}{\pi(t - 2T_b)/T_b} \\ &= \frac{2T_b^2\sin(\pi t/T_b)}{\pi t(2T_b - t)} \end{aligned} \tag{6.30}$$

This impulse response is plotted in Fig. 6.10, which shows that it has *three* distinguishable levels at the sampling instants.

To eliminate the possibility of error propagation in the modified duobinary system, we use a precoding procedure similar to that used for the duobinary case. Specifically, prior to the generation of the modified duobinary signal, a modulo-two logical addition is used on signals $2T_b$ seconds apart, as shown by (see Fig. 6.8)

$$a_k = b_k \oplus a_{k-2} \tag{6.31}$$

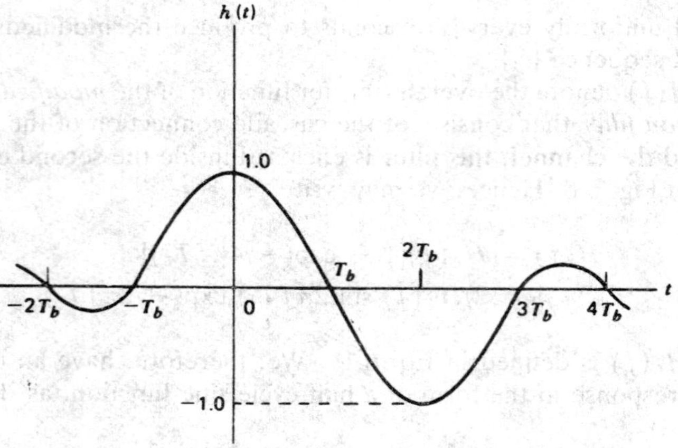

Figure 6.10
Impulse response of the modified duobinary conversion filter.

where $\{b_k\}$ is the input binary sequence and $\{a_k\}$ is the sequence at the precoder output. The sequence $\{a_k\}$ thus produced is then applied to the modified duobinary conversion filter.

In Fig. 6.8, the output digit c_k equals 0, +2, or −2V, assuming the use of a polar representation for the precoded sequence $\{a_k\}$. Also we find that the decoded (detected) digit $\hat{b}_k$ at the receiver output may be extracted from c_k by disregarding the polarity of c_k. Specifically, we may write

$$\hat{b}_k = \begin{cases} \text{symbol 1 if } |c_k| > 1 \text{ V} \\ \text{symbol 0 if } |c_k| \leq 1 \text{ V} \end{cases} \tag{6.32}$$

As with the duobinary signaling, we may note the following:

1. In the absence of channel noise, the decoded binary sequence $\{\hat{b}_k\}$ is exactly the same as the original binary sequence $\{b_k\}$ at the transmitter input.
2. The use of Eq. 6.31 requires the addition of two extra bits to the precoded sequence $\{a_k\}$. The composition of the decoded sequence $\{\hat{b}_k\}$ using Eq. 6.32 is invariant to the selection made for these two bits.

EXERCISE 5 Consider again the binary sequence 0010110 used to illustrate the operation of the duobinary signaling scheme in Example 2. Using this sequence as the input $\{b_k\}$, calculate the following sequences for the modified duobinary signaling scheme of Fig. 6.8:

(a) The sequence $\{a_k\}$ at the precoder output.

(b) The polar representation of $\{a_k\}$.

(c) The sequence $\{c_k\}$ at the modified duobinary conversion filter output, assuming the addition of bits 11 at the beginning of the precoded sequence $\{a_k\}$.

(d) The decoded sequence $\{\hat{b}_k\}$ at the receiver output. Compare this sequence with the original binary sequence $\{b_k\}$.

EXERCISE 6 Repeat the calculations of Exercise 5, assuming that the bits added at the beginning of the precoded sequence $\{a_k\}$ are 00. Hence, show that the decoded sequence $\{\hat{b}_k\}$ is unaffected by this choice of initial bits for the sequence $\{a_k\}$.

6.6 BASEBAND TRANSMISSION OF M-ARY DATA

In the baseband binary PAM system of Fig. 6.1, the output of the pulse generator consists of binary pulses, that is, pulses with one of two possible amplitude levels. On the other hand, in a *baseband M-ary* version of the system, the output of the pulse generator takes on one of M possible amplitude levels with $M > 2$; the digital waveform of a *quaternary system* (that is, $M = 4$) is illustrated in Fig. 5.12f. In an M-ary system, the information source emits a sequence of symbols from an alphabet that consists of M symbols. Each amplitude level at the pulse generator output corresponds to a distinct symbol, so that there are M distinct amplitude levels to be transmitted.

Consider then an M-ary PAM system with a signal alphabet that contains M symbols, with the *symbol duration* denoted by T seconds. We refer to $1/T$ as the *signaling rate* of the system, which is expressed in *symbols per second* or *bauds*. It is informative to relate the signaling rate of this system to that of an equivalent binary PAM system for which the value of M is 2 and the bit duration is T_b seconds. The binary PAM system transmits data at the rate of $1/T_b$ *bits per second.* We also observe that in the case of a *quaternary* PAM system, for example, the four possible symbols may be identified with the dibits 00, 10, 11, and 01. We thus see that each symbol represents 2 bits of data and 1 baud is equal to 2 bits per second. We may generalize this result by stating that in an M-ary PAM system, 1 baud is equal to $\log_2 M$ bits per second, and the symbol duration T of the M-ary PAM system is related to the bit duration T_b of the equivalent binary PAM system as follows:

$$T = T_b \log_2 M \tag{6.33}$$

Therefore, in a given channel bandwidth, we find that by using an M-ary PAM system we are able to transmit data at a rate that is $\log_2 M$ faster than the corresponding binary PAM system.

However, this improvement in bandwidth use is attained at a price.

Specifically, the transmitted power must be increased by a factor equal to $M^2/\log_2 M$, compared to a binary PAM system, if we are to realize the same performance in the presence of channel noise.[4] Also, system complexity is increased.

EXERCISE 7 An M-ary PAM system uses a raised cosine spectrum with rolloff factor α. Show that the signaling rate of the system is given by

$$\frac{1}{T} = \frac{2\log_2 M}{1+\alpha} B$$

where B is the channel bandwidth.

6.7 EYE PATTERN

One way to study intersymbol interference in a PCM or data transmission system experimentally is to apply the received wave to the vertical deflection plates of an oscilloscope and to apply a sawtooth wave at the transmitted symbol rate $1/T$ to the horizontal deflection plates. The waveforms in successive symbol intervals are thereby translated into one interval on the oscilloscope display, as illustrated in Fig. 6.11 for the case of a binary wave for which $T = T_b$. The resulting display is called an *eye pattern* because of its resemblance to the human eye for binary waves. The interior region of the eye pattern is called the *eye opening*.

An eye pattern provides a great deal of information about the performance of the pertinent system, as described here (see Fig. 6.12):

1. The width of the eye opening defines the time interval over which the received wave can be sampled without error from intersymbol interference. It is apparent that the preferred time for sampling is the instant of time at which the eye is open widest.
2. The sensitivity of the system to timing error is determined by the rate of closure of the eye as the sampling time is varied.
3. The height of the eye opening, at a specified sampling time, defines the margin over channel noise.

[4]The performance of a data transmission system in the presence of channel noise is usually measured in terms of the average probability of symbol error. When M is much larger than 2 and the average probability of symbol error is small compared to unity, an M-ary PAM system requires a transmitted power larger than in a binary PAM by a factor of $M^2/\log_2 M$. For a proof of this result, see Haykin (1988), pp. 78–80.

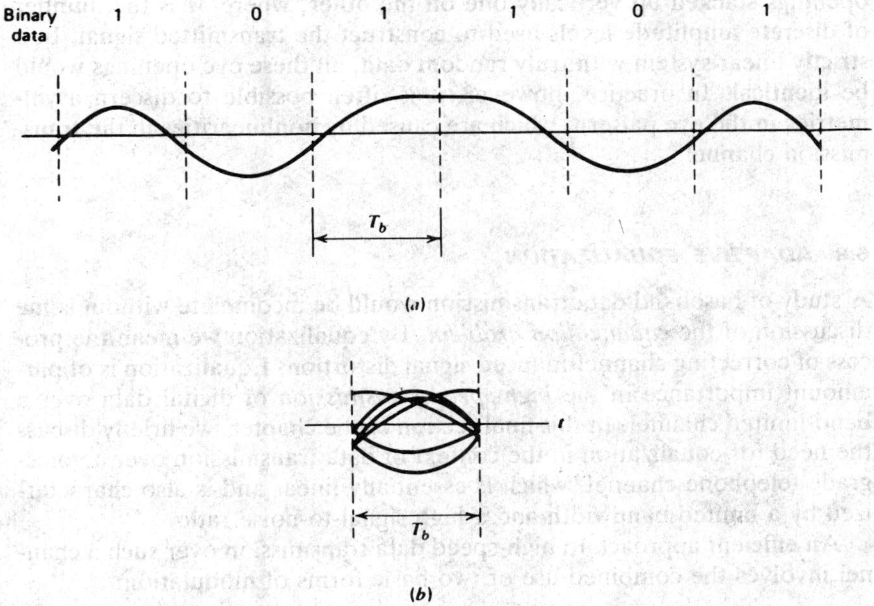

Figure 6.11
(a) Distorted binary wave. (b) Eye pattern.

When the effect of intersymbol interference is severe, traces from the upper portion of the eye pattern cross traces from the lower portion, with the result that the eye is completely closed. In such a situation, it is impossible to avoid errors due to the combined presence of intersymbol interference and channel noise in the system.

In the case of an M-ary system, the eye pattern contains $(M - 1)$ eye

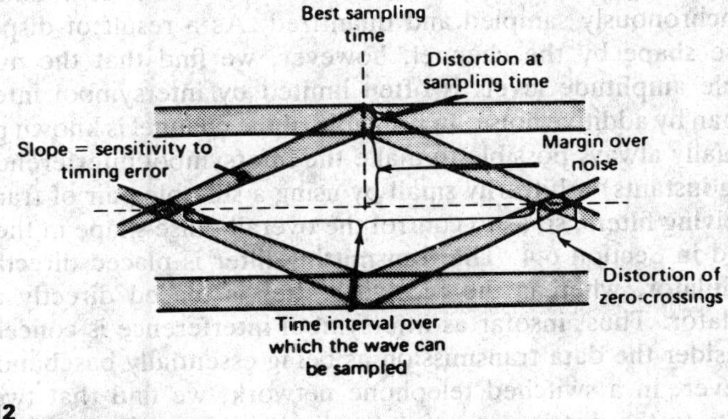

Figure 6.12
Interpretation of the eye pattern.

openings stacked up vertically one on the other, where M is the number of discrete amplitude levels used to construct the transmitted signal. In a strictly linear system with truly random data, all these eye openings would be identical. In practice, however, it is often possible to discern asymmetries in the eye pattern, which are caused by nonlinearities in the transmission channel.

6.8 ADAPTIVE EQUALIZATION

A study of baseband data transmission would be incomplete without some discussion of the *equalization problem*. By equalization we mean the process of correcting channel-induced signal distortion. Equalization is of paramount importance in the *high-speed transmission* of digital data over a band-limited channel. In this final section of the chapter, we briefly discuss the need for equalization in the context of data transmission over a voice-grade telephone channel, which is essentially linear and is also characterized by a limited bandwidth and a high signal-to-noise ratio.

An efficient approach to high-speed data transmission over such a channel involves the combined use of two basic forms of modulation:

1. *Discrete pulse-amplitude modulation* (PAM): In this operation, the amplitudes of successive pulses in a periodic train (acting as a carrier) are varied in a discrete fashion in accordance with the incoming data stream.
2. *Linear modulation:* In this second operation, the amplitude or phase of a sinusoidal carrier is varied in accordance with the discrete PAM signal resulting from the first stage of modulation. The selection of a specific type of linear modulation is made with the aim of conserving channel bandwidth. Linear modulation schemes for data transmission are considered in Sections 7.15 and 10.7.

At the receiving end of the system, the received wave is demodulated, and then synchronously sampled and quantized. As a result of dispersion of the pulse shape by the channel, however, we find that the number of detectable amplitude levels is often limited by intersymbol interference rather than by additive noise. In principle, if the channel is known precisely, it is virtually always possible to make the intersymbol interference (at the sampling instants) arbitrarily small by using a suitable pair of transmitting and receiving filters, so as to control the overall pulse shape in the manner described in Section 6.4. The transmitting filter is placed directly before the modulator, whereas the receiving filter is placed directly after the demodulator. Thus, insofar as intersymbol interference is concerned, we may consider the data transmission as being essentially baseband.

However, in a switched telephone network, we find that two factors contribute to the distribution of pulse distortion on different link connections: (1) differences in the transmission characteristics of the individual links that may be switched together, and (2) differences in the number of links in a connection. The result is that the telephone channel is random in the sense of being one of an ensemble of possible channels. Consequently, the use of a fixed pair of transmitting and receiving filters designed on the basis of average channel characteristics may not adequately reduce intersymbol interference. To realize the full transmission capability of a telephone channel, there is need for *adaptive equalization.* An *equalizer* is a filter that compensates for the dispersive effects of a channel. The process of equalization is said to be *adaptive* when the equalizer is capable of adjusting its coefficients continuously during the transmission of data; it does so by operating on the received signal (channel output) in accordance with some *algorithm*.

Among the philosophies for adaptive equalization of data transmission systems, we have *prechannel equalization* at the transmitter and *postchannel equalization* at the receiver. Because the first approach requires a feedback channel, we consider only adaptive equalization at the receiving end of the system. This equalization can be achieved, prior to data transmission, by training the filter with the guidance of a suitable *training sequence* transmitted through the channel so as to adjust the filter parameters to optimum values. The typical telephone channel changes little during an average data call, so that precall equalization with a training sequence is sufficient in most cases encountered in practice. The equalizer is positioned after the receiving filter in the receiver.

Figure 6.13 shows a popular structure used to design adaptive equalizers. The structure is a tapped-delay-line filter that consists of a set of delay elements, a set of multipliers connected to the delay-line taps, a corresponding set of adjustable tap weights, and a summer for adding the multiplier outputs. Let the sequence $\{x(nT)\}$, appearing at the output of the receiving filter, be applied to the input of this tapped-delay-line filter, producing the output (see Fig. 6.13).

$$y(nT) = \sum_{i=0}^{M-1} w_i x(nT - iT) \tag{6.34}$$

where w_i is the weight at the ith tap, and M is the total number of taps. These M tap weights constitute the adaptive filter coefficients. The tap spacing is chosen equal to the symbol duration T of the transmitted signal or the reciprocal of the signaling rate.

The adaptation of the filter may be achieved by proceeding as follows:

1. A *known sequence* $\{d(nT)\}$ is transmitted, and in the receiver the resulting *response sequence* $\{y(nT)\}$ is obtained by measuring the filter output at the sampling instants.
2. Viewing the known transmitted sequence $\{d(nT)\}$ as the *desired response*, the differences between it and the response sequence $\{y(nT)\}$

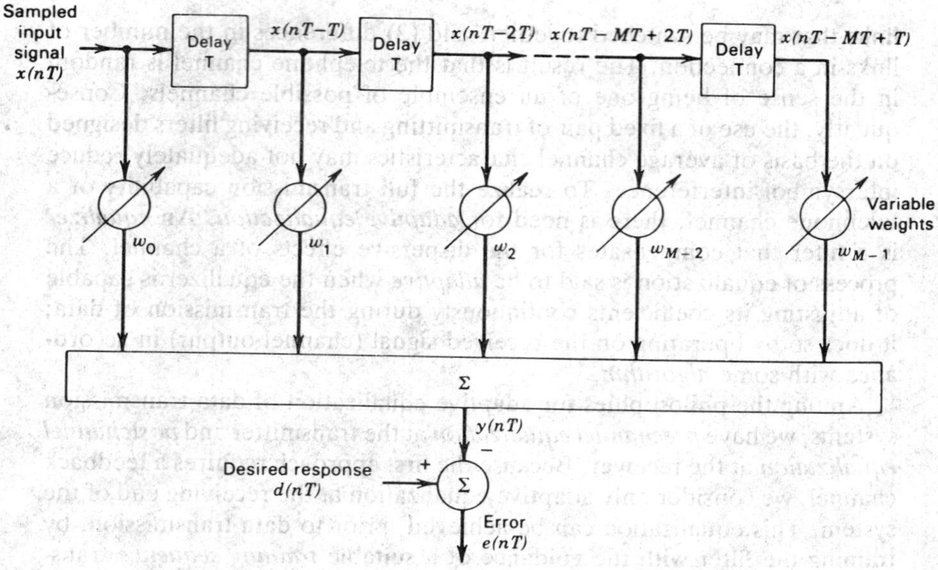

Figure 6.13
Elements of an adaptive filter.

is computed. This difference is called the *error sequence,* denoted by $\{e(nT)\}$; thus,

$$e(nT) = d(nT) - y(nT), \qquad n = 0, 1, \ldots N - 1 \quad (6.35)$$

where N is the total *length* of the sequence.

3. The error sequence $\{e(nT)\}$ is used to estimate the direction in which the weights $\{w_i\}$ of the filter are changed so as to make them approach their optimum settings $\{w_{oi}\}$.

We assume that all sequences (signals) of interest are real valued. A criterion appropriate for optimization is the *total error energy* defined by

$$\mathcal{E} = \sum_{n=0}^{N-1} e^2(nT) \qquad (6.36)$$

The *optimum values* of the tap weights, namely, $w_{o0}, w_{o1}, \ldots w_{o,M-1}$ result when the total error energy $\mathcal{E}$ is minimized.

The solution to this optimization problem may be developed in the form of an *algorithm* that adjusts the tap weights of the filter in a *recursive* manner, which means that the tapped-delay-line filter assumes a *time-varying* form. In particular, the present estimate of each tap weight is updated by incrementing it by a *correction term* proportional to the error signal at that time. Thus, starting from some arbitrary *initial condition,* the algorithm *learns* (about the operating channel conditions) from the incoming data, sample by sample, and thereby automatically adjusts the tap weights toward the optimum solution.

A simple and yet effective solution to this adaptation procedure is provided by the *least-mean-square* (*LMS*) *algorithm.*[5] According to the LMS algorithm, the tap weights are adapted as follows:

$$\hat{w}_i(nT + T) = \hat{w}_i(nT) + \mu e(nT)x(nT - iT) \tag{6.37}$$

where $i = 0, 1, \ldots, M - 1$, and $\hat{w}_i(nT)$ is the *present estimate* of the optimum weight w_{oi} for tap i at time nT, and $\hat{w}_i(nT + T)$ is the *updated estimate*. The parameter μ in Eq. 6.37 is called the *adaptation constant.* In particular, it controls the amount of *correction* applied to the old estimate $\hat{w}_i(nT)$ to produce the updated estimate $\hat{w}_i(nT + T)$. In addition to the parameter μ, the correction depends on the filter input $x(nT - iT)$ and the error signal $e(nT)$, both measured at time nT. Thus, by a proper choice of the adaptation constant μ, the use of the recursive equation (6.37) helps the adjustment of the tap weights move toward their optimum settings in a step-by-step fashion. Typically, for the starting condition, all the tap weights of the equalizer are set equal to zero.

The LMS algorithm requires knowledge of the desired response $d(nT)$ and the filter response $y(nT)$ to form the error signal $e(nT)$ in accordance with Eq. 6.35. For $y(nT)$, we may use Eq. 6.34 with $\hat{w}_i(nT)$ substituted for w_i. However, by the very nature of data communications, the desired response (providing a frame of reference for the adaptation process) originates at the channel input, which is separated physically from the receiver where the adaptive equalization is preformed. There are two methods in which a *replica* of the desired response $d(nT)$ may be obtained, as illustrated in Fig. 6.14. These two methods and their applicability are described in the following paragraphs.

In the first method, a replica of the desired response is *stored* in the receiver. Naturally, the generator of this stored reference has to be *synchronized* with the known transmitted sequence. The use of a stored reference is well suited for the *initial training* of the equalizer. This operation of the equalizer corresponds to position 1 of the switch in Fig. 6.14. (In Section 8.9 we describe a pseudo-random sequence known as a *linear maximal sequence* that may be used for this purpose.)

In the second method, the output from a decision device in the receiver is used. Under normal operating conditions, the decisions made by the receiver are correct with high probability. This means that the *error esti-*

[5]For a detailed discussion of the LMS and other adaptive filtering algorithms, see the following references: Haykin (1986), and Widrow and Stearns (1985).

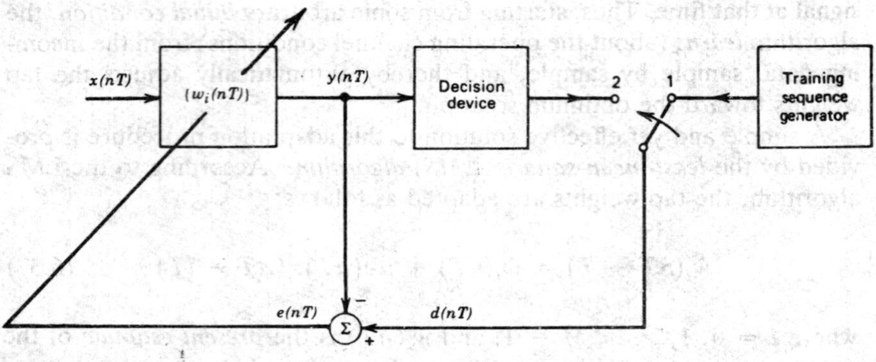

Figure 6.14
Illustrating the two modes of operation of an adaptive equalizer: Position 1 of the switch corresponds to the training mode. Position 2 corresponds to the decision-directed mode.

mates thus obtained are correct most of the time, thereby permitting the adaptive equalizer to operate satisfactorily. This second method of operation is referred to as the *decision-directed mode* of the adaptive equalizer; it corresponds to position 2 of the switch in Fig. 6.14. It is well suited for *tracking* relatively slow variations in channel characteristics during the course of transmission.

The adaptive equalizer depicted in Fig. 6.14 represents a *closed-loop feedback system,* irrespective of its mode of operation. As such, there is a tendency for the adaptive equalizer to become unstable. To ensure *stability,* care has to be exercised in the value assigned to the adaptation constant μ in the time update of Eq. 6.37. On the one hand, μ must be large enough to ensure a reasonably fast rate of convergence of the LMS algorithm. On the other hand, it must be small enough to make it possible for the LMS algorithm to track slow statistical variations in the channel.

PROBLEMS

P6.3 Ideal Solution

Problem 1 The pulse shape $p(t)$ of a baseband binary PAM system is defined by

$$p(t) = \text{sinc}\left(\frac{t}{T_b}\right)$$

where T_b is the bit duration of the input binary data. The amplitude levels at the pulse generator output are $+1$ V or -1 V, depending on whether the binary symbol at the input is 1 or 0, respectively. Sketch the waveform at the output of the receiving filter in response to the input data 001101001.

P6.4 Raised Cosine Spectrum

Problem 2 An analog signal is sampled, quantized, and encoded into a binary PCM wave. The specifications of the PCM system include the following:

Sampling rate = 8 kHz
Number of representation levels = 64

The PCM wave is transmitted over a baseband channel using discrete pulse-amplitude modulation. Determine the minimum bandwidth required for transmitting the PCM wave if each pulse is allowed to take on the following number of amplitude levels:

(a) 2
(b) 4
(c) 8

Problem 3 The raised cosine pulse spectrum for a rolloff factor of unity is given by

$$P(f) = \begin{cases} \dfrac{1}{2B_0}\cos^2\left(\dfrac{\pi f}{4B_0}\right), & 0 \leq |f| < 2B_0 \\ 0, & 2B_0 \leq |f| \end{cases}$$

Show that the time response $p(t)$, the inverse Fourier transform of $P(f)$, is

$$p(t) = \frac{\text{sinc}(4B_0 t)}{1 - 16B_0^2 t^2}$$

Problem 4 A computer puts out binary data at the rate of 56 kilobits per second. The computer output is transmitted using a baseband binary PAM system that is designed to have a raised cosine pulse spectrum. Determine the transmission bandwidth required for each of the following rolloff factors:

(a) $\alpha = 0.25$
(b) $\alpha = 0.5$
(c) $\alpha = 0.75$
(d) $\alpha = 1.0$

Problem 5 A binary PAM wave is to be transmitted over a low-pass channel with an absolute maximum bandwidth of 75 kHz. The bit duration is 10 μs. Find a raised cosine spectrum that satisfies these requirements.

P6.5 Correlative Coding

Problem 6 The binary data 001101001 is applied to the input of a duobinary system.

(a) Construct the duobinary coder output and corresponding receiver output, without a precoder.
(b) Suppose that owing to error during transmission, the level at the receiver input produced by the second input digit is reduced to zero. Construct the new receiver output.

Problem 7 Repeat Problem 6, assuming the use of a precoder in the transmitter.

Problem 8 The binary data 011100101 is applied to the input of a modified duobinary system.

(a) Construct the modified duobinary coder output and corresponding receiver output, without a precoder.
(b) Suppose that owing to error during transmission, the level produced by the third input digit is zero. Construct the new receiver output.

Problem 9 Repeat Problem 8, assuming the use of a precoder in the transmitter.

Problem 10 Using conventional analog filter design methods, it is difficult to approximate the frequency response of the modified duobinary system defined by Eq. 6.29. To get around this problem, we may use the arrangement shown in Fig. P6.1. Justify the validity of this scheme.

P6.6 Baseband Transmission of *M*-ary Data

Problem 11 Repeat Problem 4, given that each set of three successive binary digits in the computer output are coded into one of eight possible

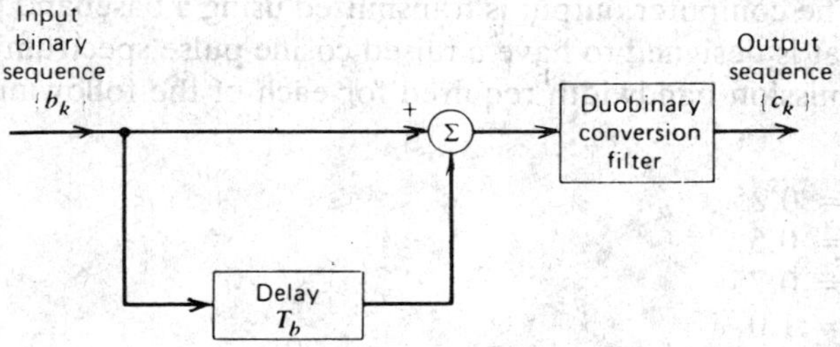

Figure P6.1

amplitude levels, and the resulting signal is transmitted by using an 8-level PAM system designed to have a raised cosine pulse spectrum.

Problem 12 An analog signal is sampled, quantized, and encoded into a binary PCM wave. The number of representation levels used is 128. A synchronizing pulse is added at the end of each code word representing a sample of the analog signal. The resulting PCM wave is transmitted over a channel of bandwidth 12 kHz using a *quaternary* PAM system with a raised cosine pulse spectrum. The rolloff factor is unity.

(a) Find the rate (in bits per second) at which information is transmitted through the channel.
(b) Find the rate at which the analog signal is sampled. What is the maximum possible value for the highest frequency component of the analog signal?

P6.7 Eye Pattern

Problem 13 A binary wave using polar signaling is generated by representing symbol 1 by a pulse of amplitude +1 V and symbol 0 by a pulse of amplitude −1 V; in both cases the pulse duration equals the bit duration. This signal is applied to a low-pass *RC* filter with transfer function:

$$H(f) = \frac{1}{1 + jf/f_0}$$

Construct the eye pattern for the filter output for the following sequences:

(a) Alternating 1's and 0's.
(b) A long sequence of 1's followed by a long sequence of 0's.
(c) A long sequence of 1's followed by a single 0 and then a long sequence of 1's.

Assume a bit rate of $2f_0$ bits per second.

Problem 14 The binary sequence 011010 is transmitted through a channel having a raised cosine characteristic with a rolloff factor of unity. Assume the use of polar signaling, with symbols 1 and 0 represented by +1 and −1 V, respectively.

(a) Construct, to scale, the received wave, and indicate the best sampling times for regeneration.
(b) Construct the eye pattern for this received wave and show that it is completely open.
(c) Determine the zero crossings of the received wave.

MODULATION TECHNIQUES

Ordinarily, the transmission of a message signal (be it in analog or digital form) over a *band-pass communication channel* (e.g., telephone line, satellite channel) requires a shift of the range of frequencies contained in the signal into other frequency ranges suitable for transmission, and a corresponding shift back to the original frequency range after reception. For example, a radio system must operate with frequencies of 30 kHz and upward, whereas the message signal usually contains frequencies in the audio frequency range, so some form of frequency-band shifting must be used for the system to operate satisfactorily. A shift of the range of frequencies in a signal is accomplished by using *modulation*, defined as *the process by which some characteristic of a carrier is varied in accordance with a modulating wave.*[1] The message signal is referred to as the *modulating wave*, and the result of the modulation process is referred to as the *modulated wave*. At the receiving end of the communication system, we usually require the message signal to be recovered. This is accomplished by using a process known as *demodulation*, or *detection*, which is the inverse of the modulation process.

In this chapter we study modulation techniques for both analog and digital forms of message (information-bearing) signals. The chapter is a long one, which is the result of integrating a variety of modulation techniques, side-by-side. The chapter is organized as follows:

1. In Sections 7.1 through 7.8, we study the various types of amplitude modulation that constitute the first family of analog modulation techniques. In *amplitude modulation* the amplitude of a sinusoidal carrier wave is varied in accordance with the information-bearing signal. The applications of amplitude modulation in broadcasting are considered in Section 7.9.
2. In Sections 7.10 through 7.13, we study the second family of analog modulation techniques known collectively as *angle modulation*. In this method of modulation the phase or frequency of a sinusoidal carrier wave is varied in accordance with the information-bearing signal. The application of frequency modulation, an important type of angle modulation, in broadcasting is considered in Section 7.14.
3. Finally, in Section 7.15 we describe digital modulation techniques. The discussion is completed in Section 7.16 with a description of digital satellite communications.

7.1 AMPLITUDE MODULATION

Consider a sinusoidal *carrier wave* $c(t)$ defined by

$$c(t) = A_c \cos(2\pi f_c t) \tag{7.1}$$

where the peak value A_c is called the *carrier amplitude* and f_c is called the *carrier frequency*. For convenience, we have assumed that the phase of the carrier wave is zero in Eq. 7.1. We are justified in making this assumption since the carrier source is always independent of the message source. Let $m(t)$ denote the baseband signal that carries specification of the message. From here on, we refer to $m(t)$ as the *message signal*. *Amplitude modulation is defined as a process in which the amplitude of the carrier wave $c(t)$ is varied linearly with the message signal $m(t)$*. This definition is general enough

[1] *IEEE Standard Dictionary of Electrical and Electronics Terms*, p. 351 (Wiley-Interscience, 1972).

to permit different interpretations of the linearity. Correspondingly, amplitude modulation may take on different forms, depending on *the frequency content of the modulated wave.* In the following section we consider the *standard* form of amplitude modulation.

TIME-DOMAIN DESCRIPTION

The standard form of an amplitude-modulated (AM) wave is defined by

$$s(t) = A_c[1 + k_a m(t)] \cos(2\pi f_c t) \tag{7.2}$$

where k_a is a constant called the *amplitude sensitivity* of the modulator. The modulated wave so defined is said to be a "standard" AM wave, because (as we will see presently) its frequency content is *fully* representative of amplitude modulation.

The amplitude of the time function multiplying $\cos(2\pi f_c t)$ in Eq. 7.2 is called the *envelope* of the AM wave $s(t)$. Using $a(t)$ to denote this envelope, we may thus write

$$a(t) = A_c|1 + k_a m(t)| \tag{7.3}$$

Two cases of particular interest arise, depending on the magnitude of $k_a m(t)$, compared to unity. For *case 1,* we have

$$|k_a m(t)| \leqslant 1, \qquad \text{for all } t \tag{7.4}$$

Under this condition, the term $1 + k_a m(t)$ is always nonnegative. We may therefore simplify the expression for the envelope of the AM wave by writing

$$a(t) = A_c[1 + k_a m(t)], \qquad \text{for all } t \tag{7.5}$$

For *case 2,* on the other hand, we have

$$|k_a m(t)| > 1, \qquad \text{for some } t \tag{7.6}$$

Under this condition, we must use Eq. 7.3 for evaluating the envelope of the AM wave.

The maximum absolute value of $k_a m(t)$ multiplied by 100 is referred to as the *percentage modulation.* Accordingly, case 1 corresponds to a percentage modulation less than or equal to 100%, whereas case 2 corresponds to a percentage modulation in excess of 100%.

The waveforms of Fig. 7.1 illustrate the amplitude modulation process. Part *a* of the figure depicts the waveform of a message signal $m(t)$. Part *b* of the figure depicts an AM wave produced by this message signal for a value of k_a for which the percentage modulation is 66.7% (i.e., case 1).

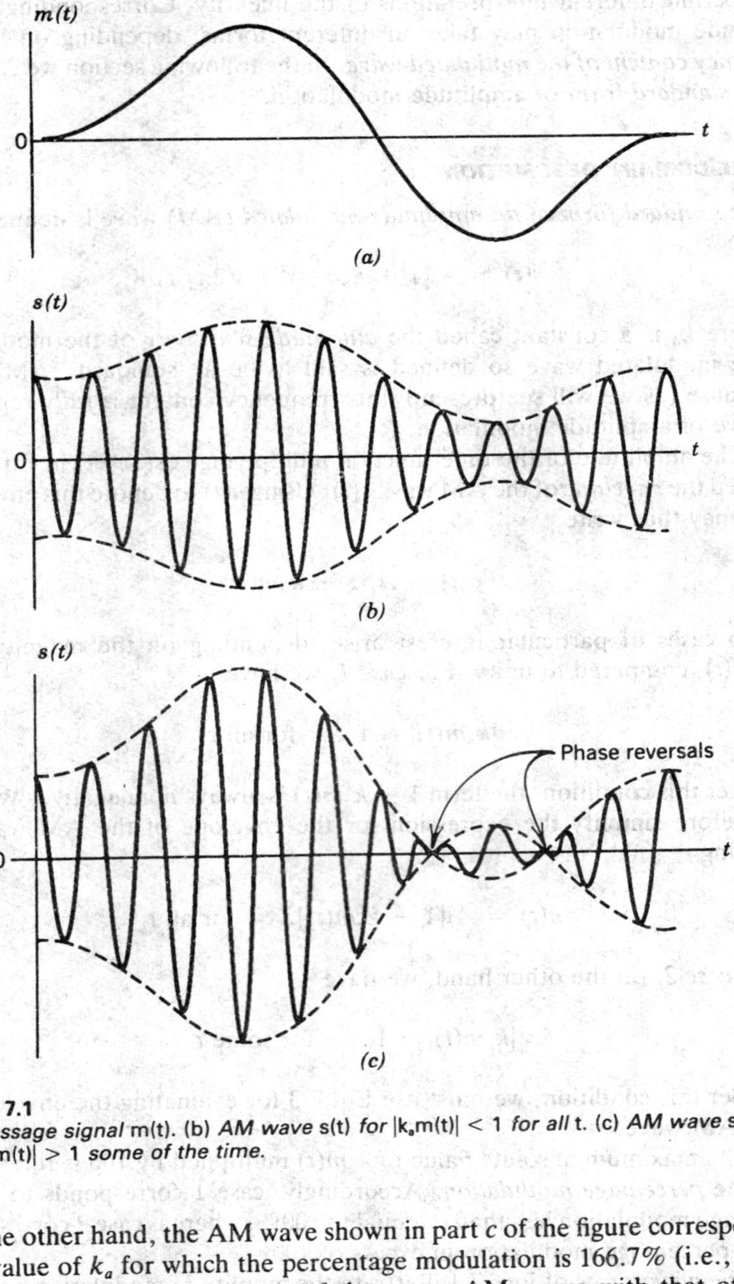

Figure 7.1
(a) *Message signal* m(t). (b) *AM wave* s(t) *for* $|k_a m(t)| < 1$ *for all* t. (c) *AM wave* s(t) *for* $|k_a m(t)| > 1$ *some of the time.*

On the other hand, the AM wave shown in part *c* of the figure corresponds to a value of k_a for which the percentage modulation is 166.7% (i.e., case 2). Comparing the waveforms of these two AM waves with that of the message signal, we draw an important conclusion. Specifically, the envelope of the AM wave has a waveform that bears a *one-to-one correspondence* with that of the message signal if and only if the percentage modulation is less than or equal to 100%. This correspondence is destroyed if the percentage modulation exceeds 100%. In the latter case, the modulated wave is said to suffer from *envelope distortion,* and the wave itself is said to be *overmodulated.*

The complexity of the *detector* (i.e., the demodulation circuit used to recover the message signal from the incoming AM wave at the receiver) is greatly simplified if the transmitter is designed to produce an envelope $a(t)$ that has the same shape as the message signal $m(t)$. For this requirement to be realized, we must satisfy two conditions:

1. The percentage modulation is less than 100%, so as to avoid envelope distortion.
2. The *message bandwidth,* W, is small compared to the carrier frequency f_c, so that the envelope $a(t)$ may be visualized satisfactorily. Here, it is assumed that the spectral content of the message signal is negligible for frequencies outside the interval $-W \leqslant f \leqslant W$.

EXERCISE 1 Demonstrate that the percentage modulation for the AM wave shown in Fig. 7.1*b* equals 66.7%, whereas for the AM wave shown in Fig. 7.1*c* it equals 166.7%.

FREQUENCY-DOMAIN DESCRIPTION

Equation 7.2 defines the standard AM wave $s(t)$ as a function of time. To develop the frequency description of this AM wave, we take the Fourier transform of both sides of Eq. 7.2. Let $S(f)$ denote the Fourier transform of $s(t)$, and $M(f)$ denote the Fourier transform of the message signal $m(t)$; we refer to $M(f)$ as the *message spectrum.* Accordingly, using the Fourier transform of the cosine function $A_c \cos(2\pi f_c t)$ and the frequency-shifting property of the Fourier transform (see Sections 2.3 and 2.5), we may write

$$S(f) = \frac{A_c}{2}[\delta(f - f_c) + \delta(f + f_c)] + \frac{k_a A_c}{2}[M(f - f_c) + M(f + f_c)] \tag{7.7}$$

Let the message signal $m(t)$ be band-limited to the interval $-W \leqslant f \leqslant W$, as in Fig. 7.2*a*. The shape of the spectrum shown in this figure is intended for the purpose of illustration only. We find from Eq. 7.7 that the spectrum $S(f)$ of the AM wave is as shown in Fig. 7.2*b* for the case when $f_c > W$. This spectrum consists of two delta functions weighted by the factor $A_c/2$ and occurring at $\pm f_c$ and two versions of the baseband spectrum translated

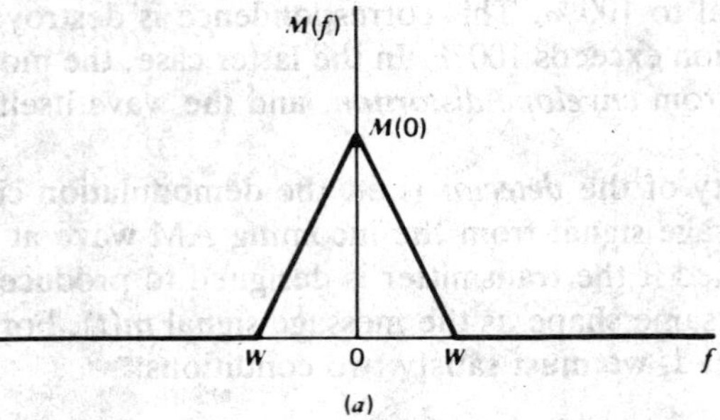

(a)

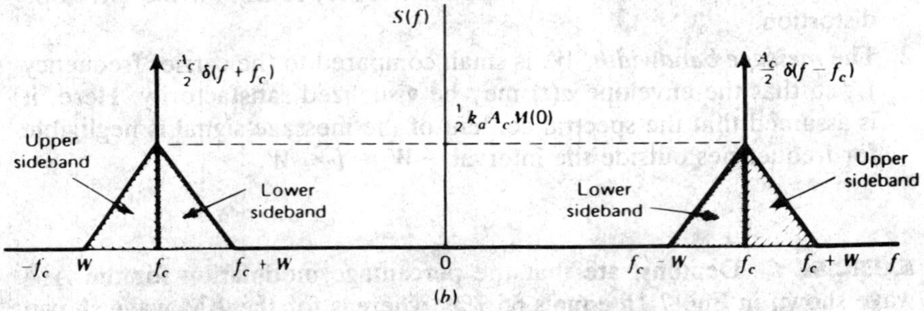

(b)

Figure 7.2
(a) *Spectrum of messages signal.* (b) *Spectrum of AM wave.*

in frequency by $\pm f_c$ and scaled in amplitude by $k_a A_c/2$. The spectrum of Fig. 7.2*b*, may be described as follows:

1. For positive frequencies, the portion of the spectrum of the modulated wave lying above the carrier frequency f_c is called the *upper sideband,* whereas the symmetric portion below f_c is called the *lower sideband.* For negative frequencies, the image of the upper sideband is represented by the portion of the spectrum below $-f_c$ and the image of the lower sideband by the portion above $-f_c$. The condition $f_c > W$ ensures that the sidebands do not overlap. Otherwise, the modulated wave exhibits *spectral overlap* and, therefore, frequency distortion.
2. For positive frequencies, the highest frequency component of the AM wave is $f_c + W$, and the lowest frequency component is $f_c - W$. The difference between these two frequencies defines the *transmission bandwidth B* for an AM wave, which is exactly twice the message bandwidth W; that is,

$$B = 2W \tag{7.8}$$

The spectrum of the AM wave as depicted in Fig. 7.2*b* is *full* in that the carrier, the upper sideband, and the lower sideband are all completely represented. It is for this reason that we treat this form of amplitude modulation as the "standard" against which other forms of amplitude modulation are compared.

EXAMPLE 1 SINGLE-TONE MODULATION

Consider a modulating wave $m(t)$ that consists of a single tone or frequency component, that is,

$$m(t) = A_m \cos(2\pi f_m t) \tag{7.9}$$

where A_m is the amplitude of the modulating wave and f_m is its frequency (see Fig. 7.3*a*). The sinusoidal earrier wave $c(t)$ has amplitude A_c and frequency f_c (see Fig. 7.3*b*). The requirement is to evaluate the time-domain and frequency-domain characteristics of the resulting AM wave.

The AM wave is described by

$$s(t) = A_c[1 + \mu \cos(2\pi f_m t)] \cos(2\pi f_c t) \tag{7.10}$$

where

$$\mu = k_a A_m \tag{7.11}$$

The dimensionless constant μ is the *modulation factor*, or the percentage modulation when it is expressed numerically as a percentage. To avoid envelope distortion due to overmodulation, the modulation factor μ must be kept below unity.

Figure 7.3*c* is a sketch of $s(t)$ for μ less than unity. Let $A_{\max}$ and $A_{\min}$ denote the maximum and minimum values of the envelope of the modulated wave. Then, from Eq. 7.10 we get

$$\frac{A_{\max}}{A_{\min}} = \frac{A_c(1 + \mu)}{A_c(1 - \mu)}$$

That is,

$$\mu = \frac{A_{\max} - A_{\min}}{A_{\max} + A_{\min}} \tag{7.12}$$

Expressing the product of the two cosines in Eq. 7.10 as the sum of two sinusoidal waves, one having frequency $f_c + f_m$ and the other having frequency $f_c - f_m$, we get

$$s(t) = A_c \cos(2\pi f_c t) + \tfrac{1}{2}\mu A_c \cos[2\pi(f_c + f_m)t] + \tfrac{1}{2}\mu A_c \cos[2\pi(f_c - f_m)t] \tag{7.13}$$

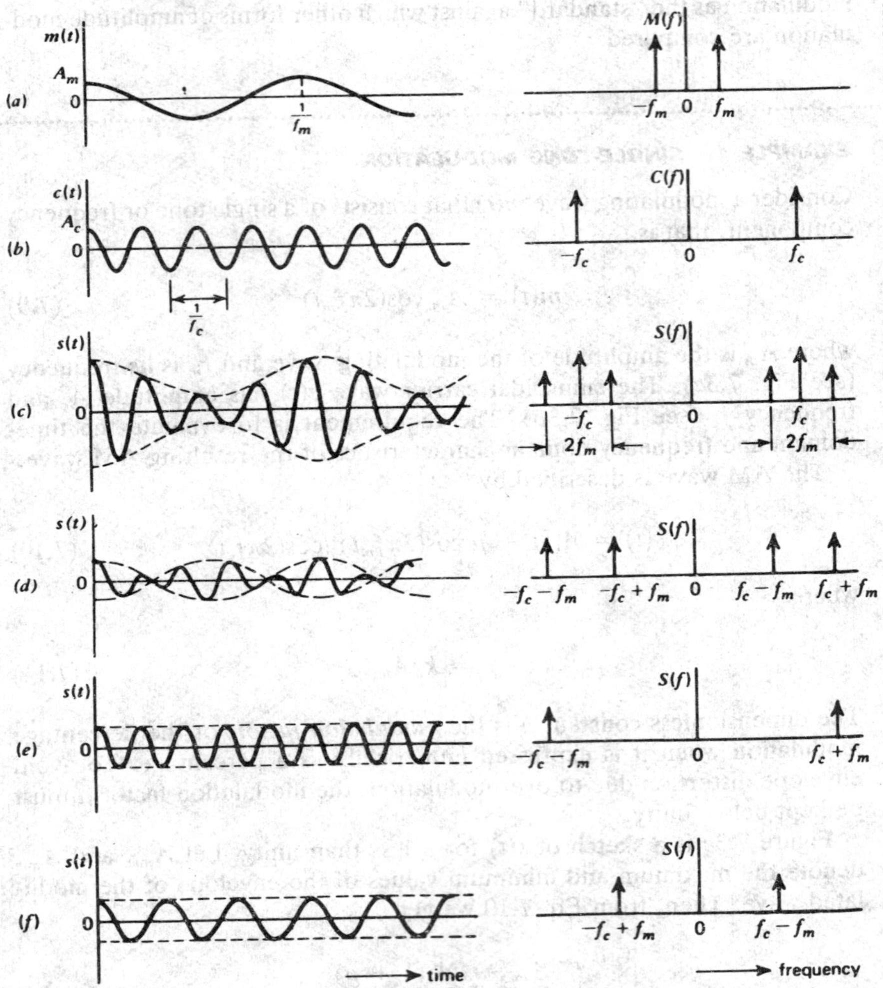

Figure 7.3
The time-domain and frequency-domain characteristics of different modulated waves produced by a single tone.

The Fourier transform of $s(t)$ is therefore

$$\begin{aligned} S(f) = {} & \tfrac{1}{2}A_c[\delta(f - f_c) + \delta(f + f_c)] \\ & + \tfrac{1}{4}\mu A_c[\delta(f - f_c - f_m) + \delta(f + f_c + f_m)] \\ & + \tfrac{1}{4}\mu A_c[\delta(f - f_c + f_m) + \delta(f + f_c - f_m)] \end{aligned} \tag{7.14}$$

Thus the spectrum of an AM wave, for the special case of sinusoidal modulation, consists of delta functions at $\pm f_c$, $f_c \pm f_m$, and $-f_c \pm f_m$, as in Fig. 7.3*c*.

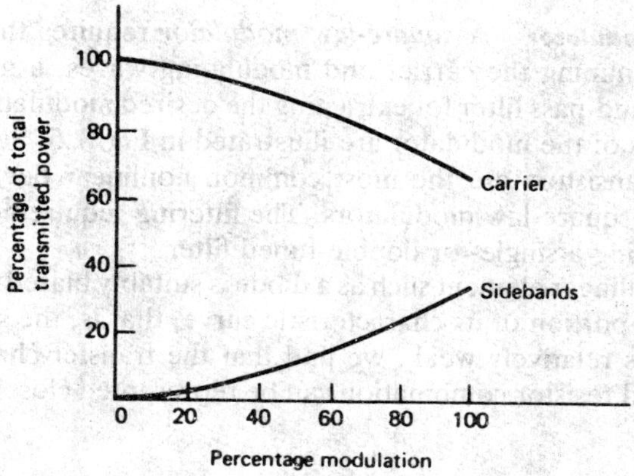

Figure 7.4
Variations of carrier power and total sideband power with percentage modulation.

In practice, the AM wave $s(t)$ is a voltage or current wave. In either case, the average power delivered to a 1-ohm load resistor by $s(t)$ is comprised of three components:

$$\text{Carrier power} = \tfrac{1}{2}A_c^2$$
$$\text{Upper side-frequency power} = \tfrac{1}{8}\mu^2 A_c^2$$
$$\text{Lower side-frequency power} = \tfrac{1}{8}\mu^2 A_c^2$$

The ratio of the total sideband power to the total power in the modulated wave is therefore equal to $\mu^2/(2 + \mu^2)$, which depends only on the modulation factor μ. If $\mu = 1$, that is, 100% modulation is used, the total power in the two side-frequencies of the resulting AM wave is only one third of the total power in the modulated wave.

Figure 7.4 shows the percentage of total power in both side-frequencies and in the carrier plotted versus the percentage modulation. Note that when the percentage modulation is less than 20%, the power in one side-frequency is less than 1% of the total power in the AM wave.

GENERATION OF AM WAVES

Having familiarized ourselves with the characteristics of a standard AM wave, we may go on to describe devices for its generation. Specifically, we describe the square-law modulator and the switching modulator, both of which require the use of a nonlinear element for their implementation. These two devices are well-suited for low-power modulation purposes.

Square-Law Modulator A *square-law modulator* requires three features: a means of summing the carrier and modulating waves, a nonlinear element, and a band-pass filter for extracting the desired modulation products. These features of the modulator are illustrated in Fig. 7.5. Semiconductor diodes and transistors are the most common nonlinear devices used for implementing square-law modulators. The filtering requirement is usually satisfied by using a single- or double-tuned filter.

When a nonlinear element such as a diode is suitably biased and operated in a restricted portion of its characteristic curve, that is, the signal applied to the diode is relatively weak, we find that the transfer characteristic of the diode–load resistor combination can be represented closely by a *square law*:

$$v_2(t) = a_1 v_1(t) + a_2 v_1^2(t) \tag{7.15}$$

where a_1 and a_2 are constants. The input voltage $v_1(t)$ consists of the carrier wave plus the modulating wave, that is,

$$v_1(t) = A_c \cos(2\pi f_c t) + m(t) \tag{7.16}$$

Therefore, substituting Eq. 7.16 in 7.15, the resulting voltage developed across the primary winding of the output transformer is given by

$$v_2(t) = \underbrace{a_1 A_c \left[1 + \frac{2a_2}{a_1} m(t)\right] \cos(2\pi f_c t)}_{\text{AM wave}} + \underbrace{a_1 m(t) + a_2 m^2(t) + a_2 A_c^2 \cos^2(2\pi f_c t)}_{\text{Unwanted terms}} \tag{7.17}$$

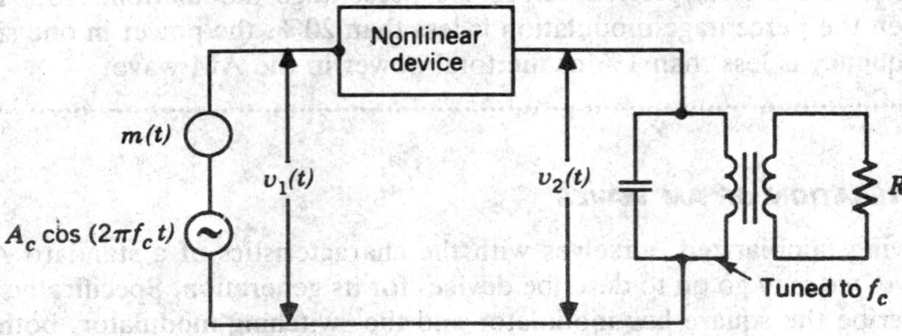

Figure 7.5
Square-law modulator.

The first term in Eq. 7.17 is the desired AM wave with amplitude sensitivity $k_a = 2a_2/a_1$. The remaining three terms are unwanted terms; they are removed by appropriate filtering.

EXERCISE 2 Show that the unwanted terms in Eq. 7.17 are removed by the tuned (band-pass) filter at the modulator output of Fig. 7.5 provided that it satisfies the following specifications:

$$\text{Midband frequency} = f_c$$

$$\text{Bandwidth} = 2W$$

$$f_c > 3W$$

Switching Modulator A *switching modulator* is shown in Fig. 7.6*a*, where it is assumed that the carrier wave $c(t)$ applied to the diode is large in amplitude, so that it swings right across the characteristic curve of the diode. We assume that the diode acts as an *ideal switch*; that is, it presents zero impedance when it is forward-biased [corresponding to $c(t) > 0$] and infinite impedance when it is reverse-biased [corresponding to $c(t) < 0$].

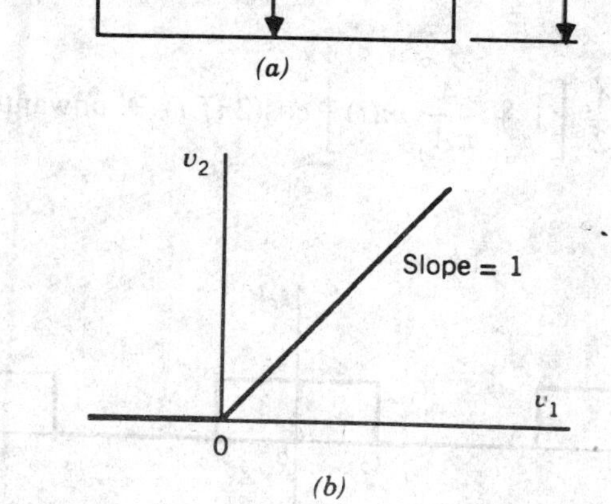

Figure 7.6
Switching modulator. (a) *Circuit diagram.* (b) *Idealized input–output relation.*

We may thus approximate the transfer characteristic of the diode–load resistor combination by a *piecewise-linear* characteristic, as shown in Fig. 7.6*b*. Accordingly, for an input voltage $v_1(t)$ given by

$$v_1(t) = A_c \cos(2\pi f_c t) + m(t) \tag{7.18}$$

where $|m(t)| \ll A_c$, the resulting load voltage $v_2(t)$ is

$$v_2(t) \simeq \begin{cases} v_1(t), & c(t) > 0 \\ 0, & c(t) < 0 \end{cases} \tag{7.19}$$

That is, the load voltage $v_2(t)$ varies periodically between the values $v_1(t)$ and zero at a rate equal to the carrier frequency f_c. In this way, by assuming a modulating wave that is weak compared with the carrier wave, we have effectively replaced the nonlinear behavior of the diode by an approximately equivalent linear time-varying operation.

We may express Eq. 7.19 mathematically as

$$v_2(t) \simeq [A_c \cos(2\pi f_c t) + m(t)]g_p(t) \tag{7.20}$$

where $g_p(t)$ is a periodic pulse train of duty cycle equal to one half and period $T_0 = 1/f_c$, as in Fig. 7.7. Representing this $g_p(t)$ by its Fourier series, we have

$$\begin{aligned} g_p(t) &= \frac{1}{2} + \frac{2}{\pi}\sum_{n=1}^{\infty} \frac{(-1)^{n-1}}{2n-1}\cos[2\pi f_c t(2n-1)] \\ &= \frac{1}{2} + \frac{2}{\pi}\cos(2\pi f_c t) + \text{odd harmonic components} \end{aligned} \tag{7.21}$$

Therefore substituting Eq. 7.21 in 7.20, we find that the load voltage $v_2(t)$ is as follows:

$$v_2(t) = \frac{A_c}{2}\left[1 + \frac{4}{\pi A_c} m(t)\right]\cos(2\pi f_c t) + \text{unwanted terms} \tag{7.22}$$

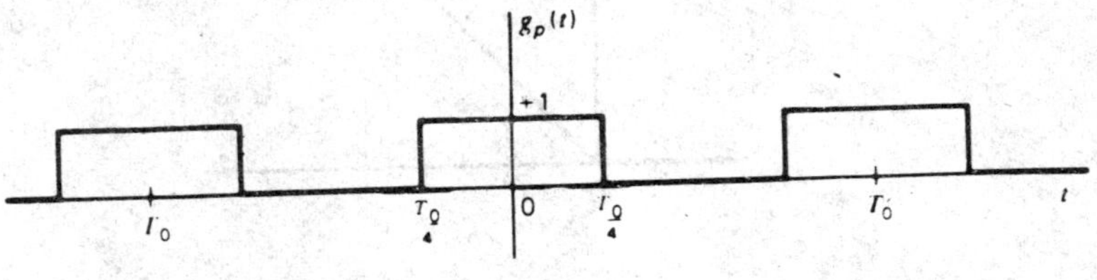

Figure 7.7
Periodic pulse train.

The first term of Eq. 7.22 is the desired AM wave with amplitude sensitivity $k_a = 4/\pi A_c$. The unwanted terms are removed from the load voltage $v_2(t)$ by means of a band-pass filter.

EXERCISE 3 Show that removal of the unwanted terms in Eq. 7.22 is accomplished if the band-pass filter satisfies the following specifications:

$$\text{Midband frequency} = f_c$$

$$\text{Bandwidth} = 2W$$

$$f_c > 2W$$

DETECTION OF AM WAVES

The process of *detection* or *demodulation* provides a means of recovering the message signal from an incoming modulated wave. In effect, detection is the inverse of modulation. In the sequel, we describe two devices for the detection of AM waves, namely, the square-law detector and the envelope detector.

Square-Law Detector A *square-law detector* is essentially obtained by using a square-law modulator for the purpose of detection. Consider Eq. 7.15 defining the transfer characteristic of a nonlinear device, which is reproduced here for convenience:

$$v_2(t) = a_1 v_1(t) + a_2 v_1^2(t) \tag{7.23}$$

where $v_1(t)$ and $v_2(t)$ are the input and output voltages, respectively, and a_1 and a_2 are constants. When such a device is used for the demodulation of an AM wave, we have for the input

$$v_1(t) = A_c[1 + k_a m(t)] \cos(2\pi f_c t) \tag{7.24}$$

Therefore, substituting Eq. 7.24 in 7.23, we get

$$\begin{aligned} v_2(t) = {} & a_1 A_c[1 + k_a m(t)] \cos(2\pi f_c t) \\ & + \tfrac{1}{2} a_2 A_c^2[1 + 2k_a m(t) + k_a^2 m^2(t)][1 + \cos(4\pi f_c t)] \end{aligned} \tag{7.25}$$

The desired signal, namely, $a_2 A_c^2 k_a m(t)$, is due to the $a_2 v_1^2(t)$ term—hence, the description "square-law detector." This component can be extracted by means of a low-pass filter. This is not the only contribution within the baseband spectrum, however, because the term $\frac{1}{2} a_2 A_c^2 k_a^2 m^2(t)$ will give rise to a plurality of similar frequency components. The ratio of wanted signal to distortion is equal to $2/k_a m(t)$. To make this ratio large we limit the percentage modulation, that is, we choose $|k_a m(t)|$ small compared with unity for all t. We conclude therefore that distortionless recovery of the baseband signal $m(t)$ is possible only if the applied AM wave is weak (so as to justify the use of a square-law input–output relation as in Eq. 7.23) and if the percentage modulation is very small.

Envelope Detector An *envelope detector* is a simple and yet highly effective device that is well-suited for the demodulation of a narrow-band AM wave (i.e., the carrier frequency is large compared with the message bandwidth), for which the percentage modulation is less than 100%. Ideally, an envelope detector produces an output signal that follows the envelope of the input signal waveform exactly; hence, the name. Some version of this circuit is used in almost all commercial AM radio receivers.

Figure 7.8*a* shows the circuit diagram of an envelope detector that consists of a diode and a resistor-capacitor filter. The operation of this envelope detector is as follows. On the positive half-cycle of the input signal, the diode is forward-biased and the capacitor C charges up rapidly to the peak value of the input signal. When the input signal falls below this value, the diode becomes reverse-biased and the capacitor C discharges slowly through the load resistor R_l. The discharging process continues until the next positive half-cycle. When the input signal becomes greater than the voltage across the capacitor, the diode conducts again and the process is repeated. We assume that the diode is ideal, presenting zero impedance to current flow in the forward-biased region, and infinite impedance in the reverse-biased region. We further assume that the AM wave applied to the envelope detector is supplied by a voltage source of internal impedance R_s. The charging time constant $R_s C$ must be short compared with the carrier period $1/f_c$, that is,

$$R_s C \ll \frac{1}{f_c} \tag{7.26}$$

Hence, the capacitor C charges rapidly and thereby follows the applied voltage up to the positive peak when the diode is conducting. On the other hand, the discharging time constant $R_l C$ must be long enough to ensure that the capacitor discharges slowly through the load resistor R_l between positive peaks of the carrier wave, but not so long that the capacitor voltage will not discharge at the maximum rate of change of the modulating wave, that is,

$$\frac{1}{f_c} \ll R_l C \ll \frac{1}{W} \tag{7.27}$$

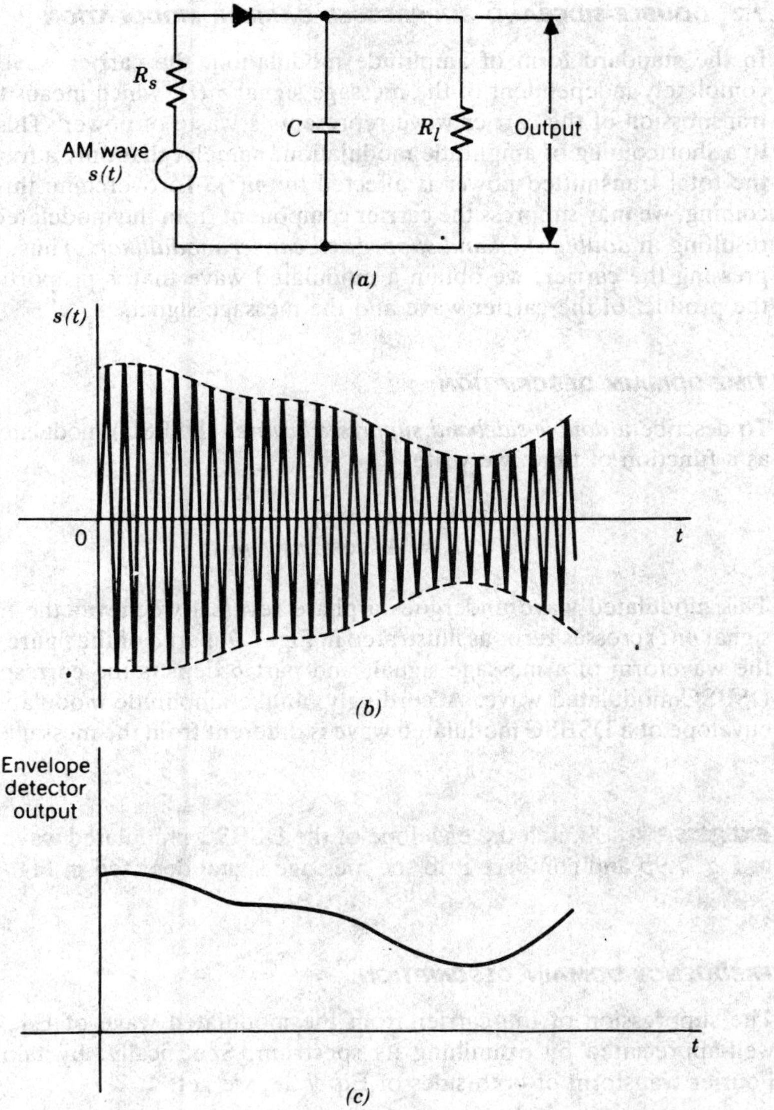

Figure 7.8
Envelope detector. (a) *Circuit diagram.* (b) *AM wave input.* (c) *Envelope detector output.*

where W is the message bandwidth. The result is that the capacitor voltage or detector output is very nearly the same as the envelope of the AM wave, as illustrated in Figs. 7.8*b* and *c*. The detector output usually has a small ripple (not shown in Fig. 7.8*c*) at the carrier frequency; this ripple is easily removed by low-pass filtering.

7.2 DOUBLE-SIDEBAND SUPPRESSED-CARRIER MODULATION

In the standard form of amplitude modulation, the carrier wave $c(t)$ is completely independent of the message signal $m(t)$, which means that the transmission of the carrier wave represents a waste of power. This points to a shortcoming of amplitude modulation; namely, that only a fraction of the total transmitted power is affected by $m(t)$. To overcome this shortcoming, we may suppress the carrier component from the modulated wave,. resulting in *double-sideband suppressed carrier modulation*. Thus, by suppressing the carrier, we obtain a modulated wave that is proportional to the product of the carrier wave and the message signal.

TIME-DOMAIN DESCRIPTION

To describe a *double-sideband suppressed-carrier* (DSBSC) modulated wave as a function of time, we write

$$\begin{aligned} s(t) &= c(t)m(t) \\ &= A_c \cos(2\pi f_c t)m(t) \end{aligned} \tag{7.28}$$

This modulated wave undergoes a phase reversal whenever the message signal $m(t)$ crosses zero, as illustrated in Fig. 7.9; part *a* of the figure depicts the waveform of a message signal, and part *b* depicts the corresponding DSBSC-modulated wave. Accordingly, unlike amplitude modulation, the envelope of a DSBSC modulated wave is different from the message signal.

EXERCISE 4 Sketch the envelope of the DSBSC modulated wave shown in Fig. 7.9*b* and compare it to the message signal depicted in Fig. 7.9*a*.

FREQUENCY-DOMAIN DESCRIPTION

The suppression of the carrier from the modulated wave of Eq. 7.28 is well-appreciated by examining its spectrum. Specifically, by taking the Fourier transform of both sides of Eq. 7.28, we get

$$S(f) = \tfrac{1}{2}A_c[M(f - f_c) + M(f + f_c)] \tag{7.29}$$

where, as before, $S(f)$ is the Fourier transform of the modulated wave $s(t)$, and $M(f)$ is the Fourier transform of the message signal $m(t)$. When the message signal $m(t)$ is limited to the interval $-W \leq f \leq W$, as in Fig. 7.10*a*, we find that the spectrum $S(f)$ is as illustrated in part *b* of the figure. Except for a change in scale factor, the modulation process simply translates the spectrum of the baseband signal by $\pm f_c$. Of course, the transmission bandwidth required by DSBSC modulation is the same

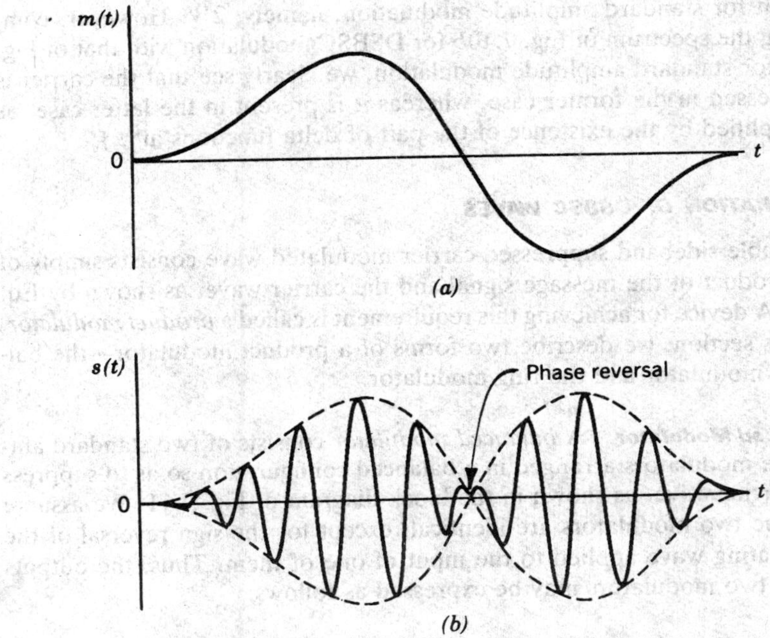

Figure 7.9
(a) *Message signal.* (b) *DSBSC-modulated wave* s(t).

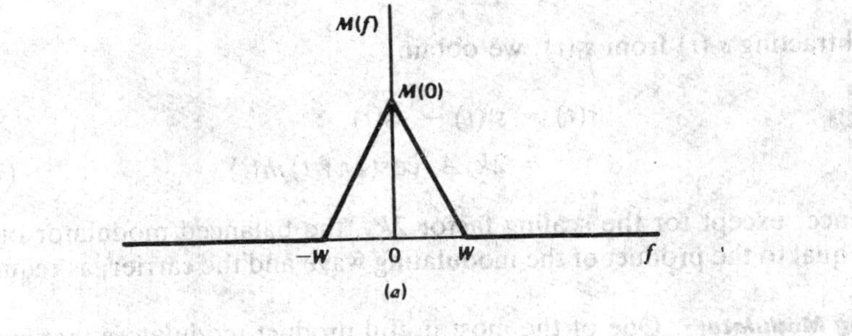

Figure 7.10
(a) *Spectrum of message signal.* (b) *Spectrum of DSBSC modulated wave.*

as that for standard amplitude modulation, namely, $2W$. However, comparing the spectrum of Fig. 7.10*b* for DSBSC modulation with that of Fig. 7.2*b* for standard amplitude modulation, we clearly see that the carrier is suppressed in the former case, whereas it is present in the latter case, as exemplified by the existence of the pair of delta functions at $\pm f_c$.

GENERATION OF DSBSC WAVES

A double-sideband suppressed-carrier modulated wave consists simply of the product of the message signal and the carrier wave, as shown by Eq. 7.28. A device for achieving this requirement is called a *product modulator.* In this section, we describe two forms of a product modulator—the balanced modulator and the ring modulator.

Balanced Modulator A *balanced modulator* consists of two standard amplitude modulators arranged in a balanced configuration so as to suppress the carrier wave, as shown in the block diagram of Fig. 7.11. We assume that the two modulators are identical, except for the sign reversal of the modulating wave applied to the input of one of them. Thus, the outputs of the two modulators may be expressed as follows:

$$s_1(t) = A_c[1 + k_a m(t)] \cos(2\pi f_c t)$$

and

$$s_2(t) = A_c[1 - k_a m(t)] \cos(2\pi f_c t)$$

Subtracting $s_2(t)$ from $s_1(t)$, we obtain

$$\begin{aligned} s(t) &= s_1(t) - s_2(t) \\ &= 2k_a A_c \cos(2\pi f_c t)\, m(t) \end{aligned} \tag{7.30}$$

Hence, except for the scaling factor $2k_a$, the balanced modulator output is equal to the product of the modulating wave and the carrier, as required.

Ring Modulator One of the most useful product modulators that is well-suited for generating a DSBSC modulated wave is the *ring modulator* shown in Fig. 7.12*a*; it is also known as a *lattice* or *double-balanced modulator.* The four diodes in Fig. 7.12*a* form a ring in which they all point in the same way. The diodes are controlled by a square-wave carrier $c(t)$ of frequency f_c, which is applied by means of two center-tapped transformers. We assume that the diodes are ideal and the transformers are perfectly balanced. When the carrier supply is positive, the outer diodes are switched on, presenting zero impedance, whereas the inner diodes are switched off, presenting infinite impedance, as in Fig. 7.12*b*, so that the modulator

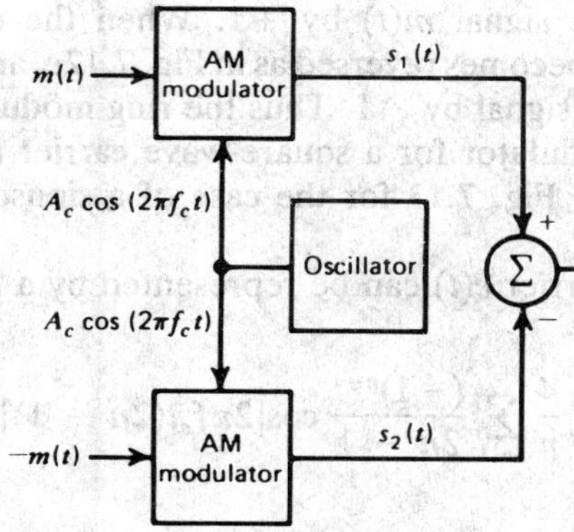

Figure 7.11
Balanced modulator.

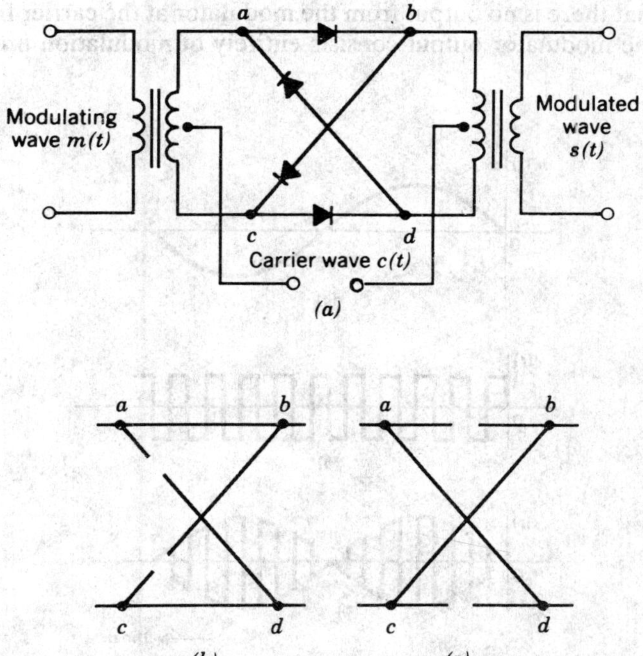

Figure 7.12
Ring modulator. (a) *Circuit diagram.* (b) *The condition when the outer diodes are switched on and the inner diodes are switched off.* (c) *The condition when the outer diodes are switched off and the inner diodes are switched on.*

multiplies the message signal $m(t)$ by $+1$. When the carrier supply is negative, the situation becomes reversed as in Fig. 7.12*c*, and the modulator multiplies the message signal by -1. Thus the ring modulator, in its ideal form, is a product modulator for a square-wave carrier and the message signal, as illustrated in Fig. 7.13 for the case of a sinusoidal modulating wave.

The square-wave carrier $c(t)$ can be represented by a Fourier series as

$$c(t) = \frac{4}{\pi} \sum_{n=1}^{\infty} \frac{(-1)^{n-1}}{2n-1} \cos[2\pi f_c t(2n-1)] \tag{7.31}$$

The ring modulator output is therefore

$$\begin{aligned} s(t) &= c(t)m(t) \\ &= \frac{4}{\pi} \sum_{n=1}^{\infty} \frac{(-1)^{n-1}}{2n-1} \cos[2\pi f_c t(2n-1)]m(t) \end{aligned} \tag{7.32}$$

We see that there is no output from the modulator at the carrier frequency; that is, the modulator output consists entirely of modulation products.

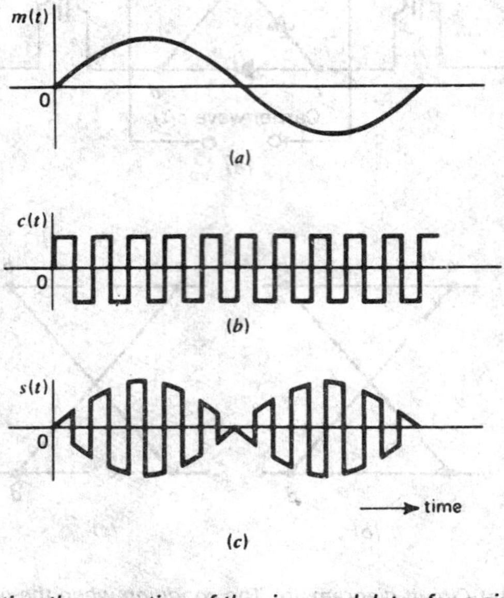

Figure 7.13
Waveforms illustrating the operation of the ring modulator for a sinusoidal modulating wave. (a) *Modulating wave.* (b) *Square-wave carrier.* (c) *Modulated wave.*

EXERCISE 5 The spectrum of the ring modulator output $s(t)$, defined by Eq. 7.32, consists of sidebands around the fundamental frequency of the square wave $c(t)$ and its odd harmonics. Suppose that the message signal $m(t)$ is band-limited to the interval $-W \leq f \leq W$. Hence, show that the DSBSC modulated wave $4\cos(2\pi f_c t)m(t)/\pi$ may be selected by using a band-pass filter with the following specifications:

$$\text{Midband frequency} = f_c$$
$$\text{Bandwidth} = 2W$$
$$f_c > W$$

COHERENT DETECTION OF DSBSC MODULATED WAVES

The message signal $m(t)$ is recovered from a DSBSC wave $s(t)$ by first multiplying $s(t)$ with a locally generated sinusoidal wave and then low-pass filtering the product, as in Fig. 7.14. It is assumed that the local oscillator output is exactly coherent or synchronized, in both frequency and phase, with the carrier wave $c(t)$ used in the product modulator to generate $s(t)$. This method of demodulation is known as *coherent detection* or *synchronous detection*.

It is instructive to derive coherent detection as a special case of the more general demodulation process using a local oscillator signal of the same frequency but arbitrary phase difference ϕ, measured with respect to the carrier wave $c(t)$. Thus, denoting the local oscillator signal by $\cos(2\pi f_c t + \phi)$, assumed to be of unit amplitude for convenience, and using Eq. 7.28 for the DSBSC modulated wave $s(t)$, we find that the product modulator output in Fig. 7.14 is given by

$$\begin{aligned} v(t) &= \cos(2\pi f_c t + \phi)s(t) \\ &= A_c \cos(2\pi f_c t)\cos(2\pi f_c t + \phi)m(t) \\ &= \underbrace{\tfrac{1}{2}A_c \cos\phi\, m(t)}_{\text{Scaled version of message signal}} + \underbrace{\tfrac{1}{2}A_c \cos(4\pi f_c t + \phi)m(t)}_{\text{Unwanted term}} \end{aligned} \tag{7.33}$$

The low-pass filter in Fig. 7.14 removes the unwanted term in the product modulator output of Eq. 7.33. The overall output $v_o(t)$ is therefore given by

$$v_o(t) = \tfrac{1}{2}A_c \cos\phi\, m(t) \tag{7.34}$$

The demodulated signal $v_o(t)$ is therefore proportional to $m(t)$ when the phase error ϕ is a constant. The amplitude of this demodulated signal is

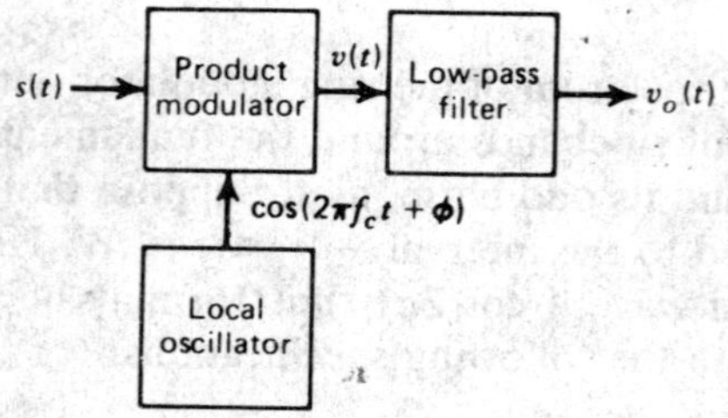

Figure 7.14
Coherent detection of DSBSC modulated wave.

maximum when $\phi = 0$, and is minimum (zero) when $\phi = \pm\pi/2$. The zero demodulated signal, which occurs for $\phi = \pm\pi/2$, represents the *quadrature null effect* of the coherent detector. Thus the phase error ϕ in the local oscillator causes the detector output to be attenuated by a factor equal to $\cos\phi$. As long as the phase error ϕ is constant, the detector output provides an undistorted version of the original message signal $m(t)$. In practice, however, we usually find that the phase error ϕ varies randomly with time, owing to random variations in the communication channel. The result is that at the detector output, the multiplying factor $\cos\phi$ also varies randomly with time, which is obviously undesirable. Therefore, circuitry must be provided in the receiver to maintain the local oscillator in perfect synchronism, in both frequency and phase, with the carrier wave used to generate the DSBSC modulated wave in the transmitter. The resulting increase in receiver complexity is the price that must be paid for suppressing the carrier wave to save transmitter power.

EXERCISE 6 Suppose that the message signal $m(t)$ is band-limited to the interval $-W \leq f \leq W$. Hence, show that the low-pass filter in Fig. 7.14 removes the unwanted term in the product modulator output of Eq. 7.33, provided that it satisfies the following specifications:

$$\text{Midband frequency} = f_c$$

$$\text{Bandwidth} = 2W$$

$$f_c > W$$

EXAMPLE 2 SINGLE-TONE MODULATION (CONTINUED)

Consider again the sinusoidal modulating signal

$$m(t) = A_m \cos(2\pi f_m t)$$

The corresponding DSBSC modulated wave is given by

$$\begin{aligned} s(t) &= A_c A_m \cos(2\pi f_c t) \cos(2\pi f_m t) \\ &= \tfrac{1}{2} A_c A_m \cos[2\pi(f_c + f_m)t] + \tfrac{1}{2} A_c A_m \cos[2\pi(f_c - f_m)t] \end{aligned} \tag{7.35}$$

Figure 7.3*d* is a sketch of this modulated wave.

The Fourier transform of $s(t)$ is therefore

$$\begin{aligned} S(f) = \tfrac{1}{4} A_c A_m [&\delta(f - f_c - f_m) + \delta(f + f_c + f_m) \\ &+ \delta(f - f_c + f_m) + \delta(f + f_c - f_m)] \end{aligned} \tag{7.36}$$

Thus the spectrum of the DSBSC modulated wave, for the case of a sinusoidal modulating wave, consists of delta functions located at $f_c \pm f_m$ and $-f_c \pm f_m$, as in Fig. 7.3*d*.

Assuming perfect synchronism between the local oscillator in Fig. 7.14 and the carrier wave, we find that the product modulator output is

$$\begin{aligned} v(t) &= \cos(2\pi f_c t)\{\tfrac{1}{2} A_c A_m \cos[2\pi(f_c - f_m)t] \\ &\quad + \tfrac{1}{2} A_c A_m \cos[2\pi(f_c + f_m)t]\} \\ &= \tfrac{1}{4} A_c A_m \cos[2\pi(2f_c - f_m)t] + \tfrac{1}{4} A_c A_m \cos(2\pi f_m t) \\ &\quad + \tfrac{1}{4} A_c A_m \cos[2\pi(2f_c + f_m)t] + \tfrac{1}{4} A_c A_m \cos(2\pi f_m t) \end{aligned} \tag{7.37}$$

where the first two terms are produced by the lower side-frequency, and the last two terms are produced by the upper side-frequency. The first and third terms, of frequencies $2f_c - f_m$ and $2f_c + f_m$, respectively, are removed by the low-pass filter in Fig. 7.14. The coherent detector output thus reproduces the original modulating wave. Note, however, that this detector output appears as two equal terms, one derived from the upper side-frequency and the other from the lower side-frequency. We conclude, therefore, that for the transmission of information, only one side-frequency is necessary. We will have more to say about this issue in Section 7.4.

COSTAS LOOP

One method of obtaining a practical synchronous receiving system, suitable for use with DSBSC modulated waves, is to use the *Costas loop*[2] shown in Fig. 7.15. This receiver consists of two coherent detectors supplied with the same input signal, namely, the incoming DSBSC modulated wave $A_c \cos(2\pi f_c t)m(t)$, but with individual local oscillator signals that are in phase quadrature to each other. The frequency of the local oscillator is

The Costas loop is named in honor of its inventor; see Costas (1956).

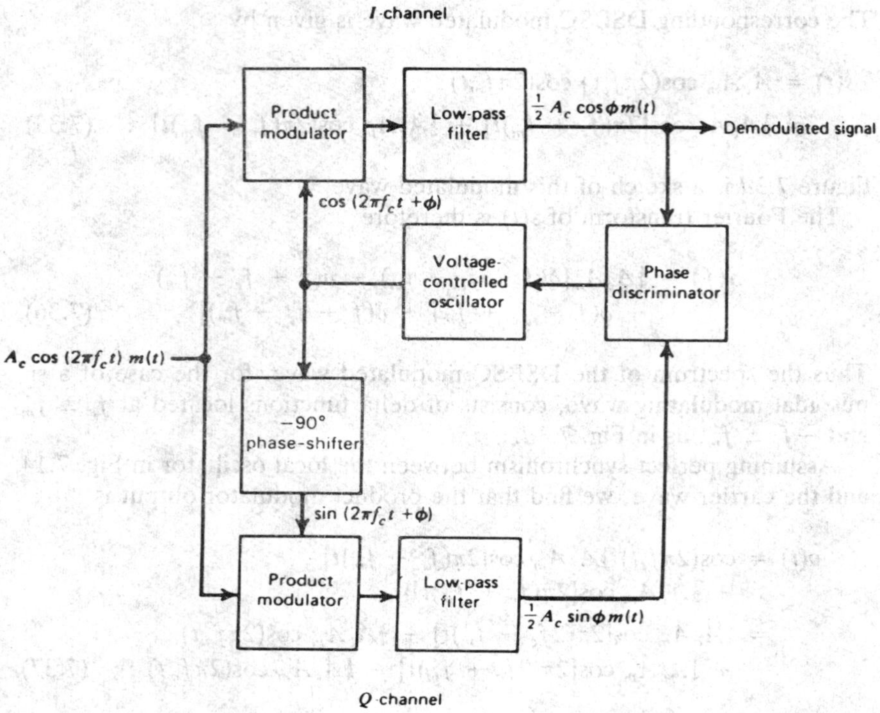

Figure 7.15
Costas loop.

adjusted to be the same as the carrier frequency f_c, which is assumed known a priori. The detector in the upper path is referred to as the *in-phase coherent detector* or *I-channel,* and that in the lower path is referred to as the *quadrature-phase coherent detector* or *Q-channel.* These two detectors are coupled to form a negative feedback system designed in such a way as to maintain the local oscillator synchronous with the carrier wave. To understand the operation of this receiver, suppose that the local oscillator signal is of the same phase as the carrier wave $A_c\cos(2\pi f_c t)$ used to generate the incoming DSBSC wave. Under these conditions, we find that the I-channel output contains the desired demodulated signal $m(t)$, whereas the Q-channel output is zero owing to the quadrature null effect of the Q-channel. Suppose next the local oscillator phase drifts from its proper value by a small amount ϕ radians. The I-channel output will remain essentially unchanged, but there will now be some signal appearing at the Q-channel output, which is proportional to $\sin\phi \simeq \phi$. This Q-channel output will have the same polarity as the I-channel output for one direction of local oscillator phase drift and opposite polarity for the opposite direction of local oscillator phase drift. The I- and Q-channel outputs are combined in a *phase dis-*

criminator (which consists of a multiplier followed by a low-pass filter). A dc *control signal* proportional to the phase error ϕ is obtained at the discriminator output. Hence, the receiver automatically corrects for local oscillator phase errors.

It is apparent that phase control in the Costas loop ceases with modulation, and that phase-lock has to be re-established with the reappearance of modulation. This is not a serious problem when receiving voice transmission, because the lock-up process normally occurs so rapidly that no perceptible distortion is observed.

EXERCISE 7 Show that the phase discriminator output in the receiver of Fig. 7.15 is proportional to $\alpha\phi$, where α is the average value of $m^2(t)$ and ϕ is the phase error (assumed small).

7.3 QUADRATURE-CARRIER MULTIPLEXING

A *quadrature-carrier multiplexing* or *quadrature-amplitude modulation* (QAM) scheme enables two DSBSC modulated waves (resulting from the application of two *independent* message signals) to occupy the same transmission bandwidth, and yet it allows for the separation of the two message signals at the receiver output. It is therefore a *bandwidth-conservation scheme*.

Figure 7.16 is a block diagram of the quadrature-carrier multiplexing system. The transmitter of the system, shown in part *a* of the figure, involves the use of two separate product modulators that are supplied with two carrier waves of the same frequency but differing in phase by $-90°$. The multiplexed signal $s(t)$ consists of the sum of these two product modulator outputs, as shown by

$$s(t) = A_c m_1(t) \cos(2\pi f_c t) + A_c m_2(t) \sin(2\pi f_c t) \tag{7.38}$$

where $m_1(t)$ and $m_2(t)$ denote the two different message signals applied to the product modulators. Thus, the multiplexed signal $s(t)$ occupies a transmission bandwidth of $2W$, centered at the carrier frequency f_c, where W is the message bandwidth of $m_1(t)$ or $m_2(t)$, whichever is largest.

The receiver of the system is shown in Fig. 7.16*b*. The multiplexed signal $s(t)$ is applied simultaneously to two separate coherent detectors that are supplied with two local carriers of the same frequency, but differing in phase by $-90°$. The output of the top detector is $\frac{1}{2}A_c m_1(t)$, whereas the output of the bottom detector is $\frac{1}{2}A_c m_2(t)$.

For the quadrature-carrier multiplexing system to operate satisfactorily, it is important to maintain the correct phase and frequency relationships between the local oscillators used in the transmitter and receiver parts of

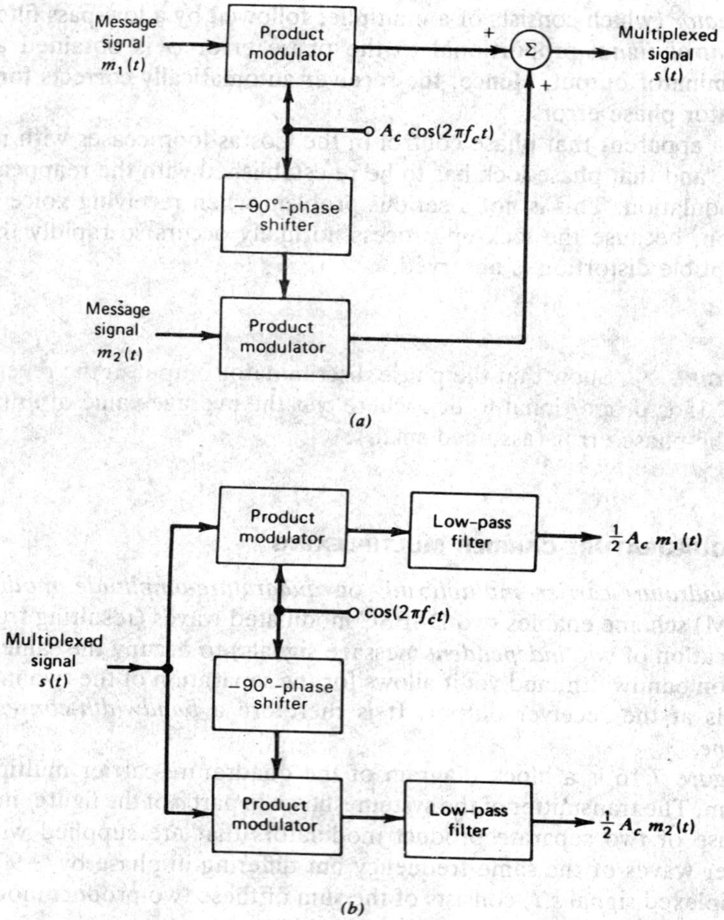

Figure 7.16
Quadrature-carrier multiplexing system. (a) *Transmitter.* (b) *Receiver.*

the system. This requirement may be satisfied, for example, by using a Costas loop; see Section 7.2.

7.4 SINGLE-SIDEBAND MODULATION

Standard amplitude modulation and double-sideband suppressed-carrier modulation are wasteful of bandwidth because they both require a transmission bandwidth equal to twice the message bandwidth. In either case, one half the transmission bandwidth is occupied by the upper sideband of the modulated wave, whereas the other half is occupied by the lower sideband. However, the upper and lower sidebands are uniquely related to each other by virtue of their symmetry about the carrier frequency; that is, given the amplitude and phase spectra of either sideband, we can uniquely determine the other. This means that insofar as the transmission of information is concerned, only one sideband is necessary, and if both the carrier and the other sideband are suppressed at the transmitter, no information is lost. Thus the channel needs to provide only the same bandwidth as the message signal, a conclusion that is intuitively satisfying. When only one sideband is transmitted, the modulation is referred to as *single-sideband modulation*.

In the study of standard amplitude modulation and double sideband-suppressed carrier modulation, pursued in Sections 7.1 and 7.2, we first formulated a time-domain description of the modulated wave and then moved on to its frequency-domain description. In the study of single-sideband modulation, we find it easier in conceptual terms to reverse the order in which these two descriptions are presented.

FREQUENCY-DOMAIN DESCRIPTION

The precise frequency-domain description of a *single-sideband (SSB) modulated* wave depends on which sideband is transmitted. Consider a message signal $m(t)$ with a spectrum $M(f)$ limited to the band $-W \leq f \leq W$, as in Fig. 7.17*a*. The spectrum of the DSBSC modulated wave, obtained by multiplying $m(t)$ by the carrier wave $A_c \cos(2\pi f_c t)$, is as shown in Fig. 7.17*b*. The upper sideband is represented in duplicate by the frequencies above f_c and those below $-f_c$; and when only the upper sideband is transmitted, the resulting SSB modulated wave has the spectrum shown in Fig. 7.17*c*. Likewise, the lower sideband is represented in duplicate by the frequencies below f_c (for positive frequencies) and those above $-f_c$ (for negative frequencies); and when only the lower sideband is transmitted, the spectrum of the corresponding SSB modulated wave is as shown in Fig. 7.17*d*. Thus the essential function of SSB modulation is to translate the spectrum of the modulating wave, either with or without inversion, to

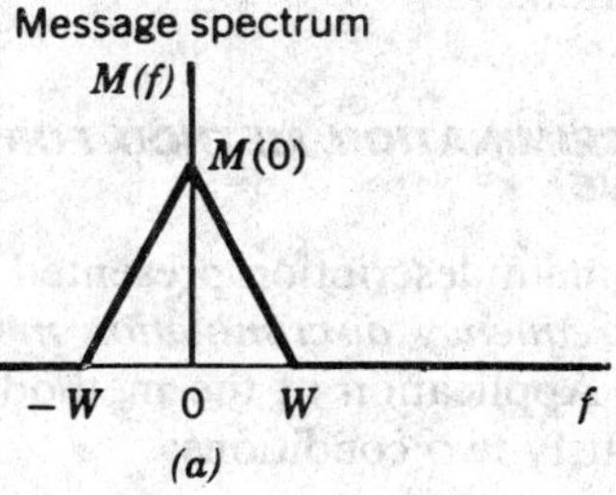

Figure 7.17
(a) *Spectrum of message signal.* (b) *Spectrum of DSBSC modulated wave.* (c) *Spectrum of SSB modulated wave with the upper sideband transmitted.* (d) *Spectrum of SSB modulated wave with the lower sideband transmitted.*

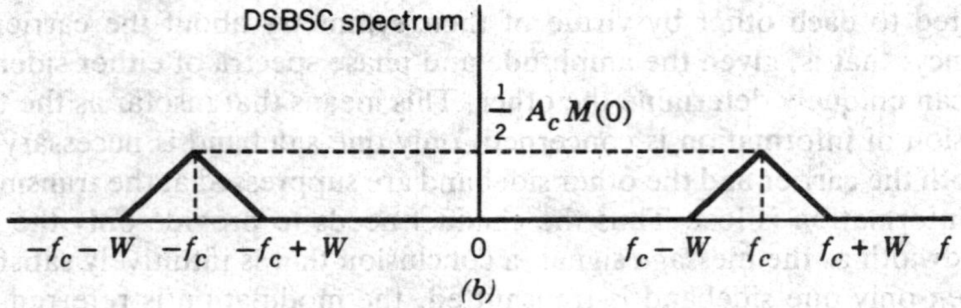

(b)

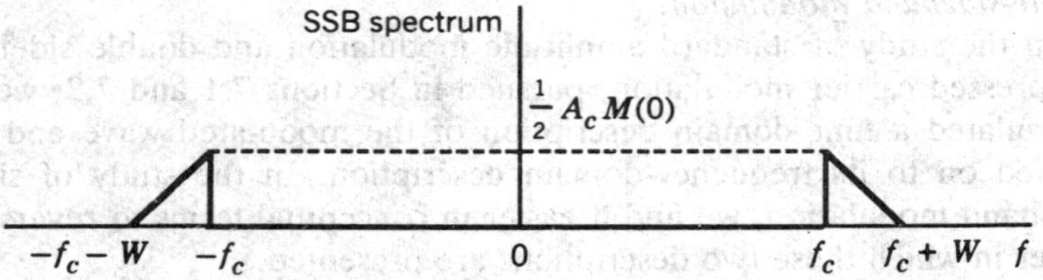

(c)

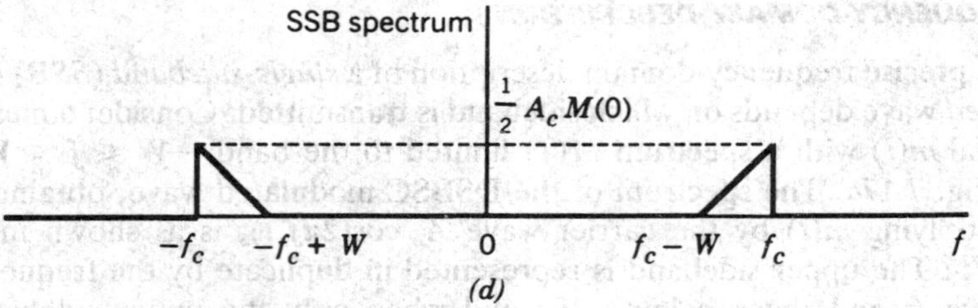

(d)

Figure 7.17 (*continued*)

a new location in the frequency domain. Moreover, the transmission bandwidth requirement of an SSB modulation system is one half that of a standard AM or DSBSC modulation system. The benefit of using SSB modulation is therefore derived principally from the reduced bandwidth requirement and the elimination of the high-power carrier wave. The principal disadvantage of SSB modulation, however, is the cost and complexity of its implementation.

FREQUENCY DISCRIMINATION METHOD FOR GENERATING AN SSB MODULATED WAVE

The frequency-domain description presented for SSB modulation leads us naturally to the *frequency discrimination method* for generating an SSB modulated wave. Application of the method, however, requires that the message signal satisfy two conditions:

1. The message signal $m(t)$ has little or no low-frequency content; that is, the message spectrum $M(f)$ has "holes" at zero frequency. An important type of message signal with such a property is an audio signal

(speech or music). In telephony, for example, the useful frequency content of a speech signal is restricted to the band 0.3–3.4 kHz, thereby creating an *energy gap* from zero to 300 Hz.

2. The highest frequency component W of the message signal $m(t)$ is much less than the carrier frequency f_c.

Then, under these conditions, the desired sideband will appear in a non-overlapping interval in the spectrum in such a way that it may be selected by an appropriate filter. Thus an SSB modulator based on frequency discrimination consists basically of a product modulator and a filter designed to pass the desired sideband of the DSBSC modulated wave at the product modulator output and reject the other sideband. A block diagram of this modulator is shown in Fig. 7.18*a*. The most severe requirement of this method of SSB generation usually arises from the unwanted sideband, the nearest frequency component of which is separated from the desired sideband by twice the lowest frequency component of the message signal.

In designing the band-pass filter in the SSB modulation scheme of Fig. 7.18*a*, we must therefore satisfy two basic requirements:

1. The passband of the filter occupies the same frequency range as the spectrum of the desired SSB modulated wave.
2. The width of the guardband of the filter, separating the passband from the stopband where the unwanted sideband of the filter input lies, is twice the lowest frequency component of the message signal.

We usually find that this kind of frequency discrimination can be satisfied only by using highly selective filters, which can be realized using crystal resonators with a Q factor per resonator in the range of 1000 to 2000.

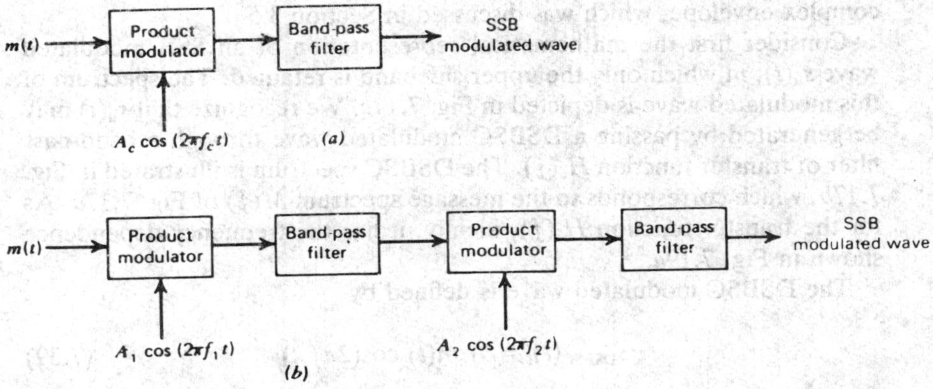

Figure 7.18
(a) *Block diagram of the frequency discrimination method (single stage) for generating SSB modulated waves.* (b) *Block diagram of a two-stage SSB modulator.*

When it is necessary to generate an SSB modulated wave occupying a frequency band that is much higher than that of the message signal (e.g., translating a voice signal to the high-frequency region of the radio spectrum), it becomes very difficult to design an appropriate filter that will pass the desired sideband and reject the other using the simple arrangement of Fig. 7.18*a*. In such a situation it is necessary to resort to a multiple-modulation process so as to ease the filtering requirement. This approach is illustrated in Fig. 7.18*b* involving two stages of modulation. The SSB modulated wave at the first filter output is used as the modulating wave for the second product modulator, which produces a DSBSC modulated wave with a spectrum that is symmetrically spaced about the second carrier frequency f_2. The frequency separation between the sidebands of this DSBSC modulated wave is effectively twice the first carrier frequency f_1, thereby permitting the second filter to remove the unwanted sideband.

TIME-DOMAIN DESCRIPTION

The spectra shown in Fig. 7.17 clearly display the frequency-domain description of SSB modulated waves; also, they highlight the relation between this frequency-domain description and that of the message signal. It is interesting to observe that we were able to relate the spectral content of SSB modulated waves to that of the message signal without having to resort to the use of mathematics. But how do we define an SSB modulated wave in the time domain? The answer to this question is desired not only because it completes the description of SSB modulated waves but also it provides the mathematical basis of another method for their generation. Unfortunately, the task of developing the time-domain description of SSB modulated waves is mathematically more difficult than that of standard AM or DSBSC modulated waves. To solve the problem, we use the idea of a complex envelope, which was discussed in Section 3.5.

Consider first the mathematical representation of an SSB modulated wave $s_u(t)$, in which only the upper sideband is retained. The spectrum of this modulated wave is depicted in Fig. 7.17*c*. We recognize that $s_u(t)$ may be generated by passing a DSBSC modulated wave through a band-pass filter of transfer function $H_u(f)$. The DSBSC spectrum is illustrated in Fig. 7.17*b*, which corresponds to the message spectrum $M(f)$ of Fig. 7.17*a*. As for the transfer function $H_u(f)$, ideally, it has the frequency dependence shown in Fig. 7.19*a*.

The DSBSC modulated wave is defined by

$$s_{\text{DSBSC}}(t) = A_c m(t) \cos(2\pi f_c t) \tag{7.39}$$

where $m(t)$ is the message signal and $A_c \cos(2\pi f_c t)$ is the carrier wave. Naturally, it is a band-pass signal with an in-phase component only. The

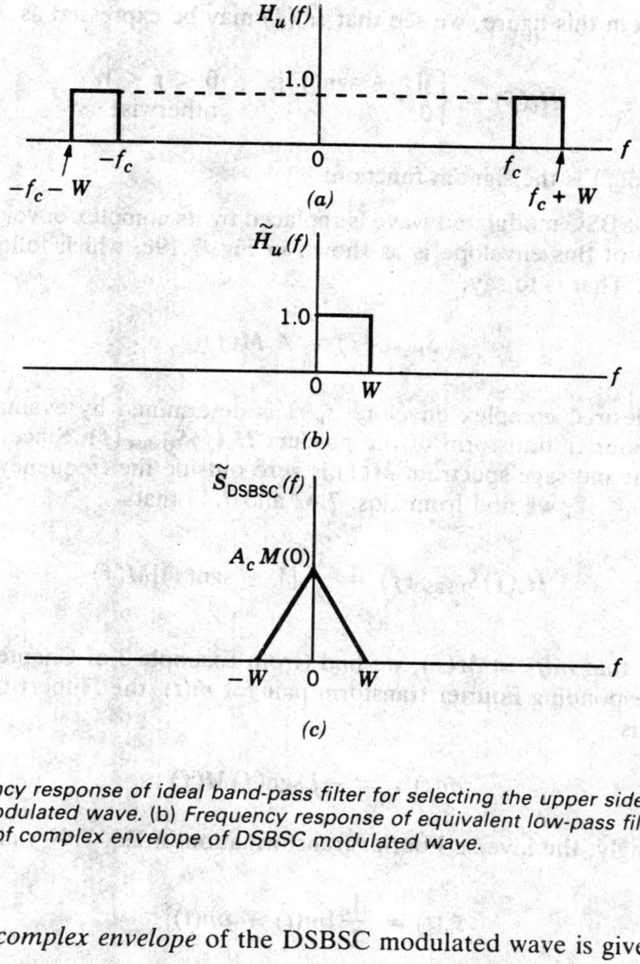

Figure 7.19
(a) *Frequency response of ideal band-pass filter for selecting the upper sideband of a DSBC modulated wave.* (b) *Frequency response of equivalent low-pass filter.* (c) *Spectrum of complex envelope of DSBSC modulated wave.*

low-pass *complex envelope* of the DSBSC modulated wave is given by

$$\tilde{s}_{DSBSC}(t) = A_c m(t) \tag{7.40}$$

The SSB modulated wave $s_u(t)$ is also a band-pass signal. However, unlike the DSBSC modulated wave, it has a quadrature as well as an in-phase component. Let the low-pass signal $\tilde{s}_u(t)$ denote the complex envelope of $s_u(t)$. We may then write

$$s_u(t) = \text{Re}[\tilde{s}_u(t) \exp(j2\pi f_c t)] \tag{7.41}$$

To determine $\tilde{s}_u(t)$, we proceed as follows (see Section 3.5):

1. The band-pass filter of transfer function $H_u(f)$ is replaced by an equivalent low-pass filter of transfer function $\tilde{H}_u(f)$, which is as shown in Fig.

7.19*b*. From this figure, we see that $\tilde{H}_u(f)$ may be expressed as

$$\tilde{H}_u(f) = \begin{cases} \frac{1}{2}[1 + \text{sgn}(f)], & 0 < f < W \\ 0, & \text{otherwise} \end{cases} \tag{7.42}$$

where sgn(f) is the signum function.

2. The DSBSC modulated wave is replaced by its complex envelope. The spectrum of this envelope is as shown in Fig. 7.19*c*, which follows from Eq. 7.40. That is to say,

$$\tilde{S}_{\text{DSBSC}}(f) = A_c M(f) \tag{7.43}$$

3. The desired complex envelope $\tilde{s}_u(t)$ is determined by evaluating the inverse Fourier transform of the product $\tilde{H}_u(f)\tilde{S}_{\text{DSBSC}}(f)$. Since, by definition, the message spectrum $M(f)$ is zero outside the frequency interval $-W < f < W$, we find from Eqs. 7.42 and 7.43 that

$$\tilde{H}_u(f)\tilde{S}_{\text{DSBSC}}(f) = \frac{A_c}{2}[1 + \text{sgn}(f)]M(f) \tag{7.44}$$

Given that $m(t) \rightleftharpoons M(f)$, we find (from Example 3 of Chapter 3) that the corresponding Fourier transform pair for $\hat{m}(t)$, the Hilbert transform of $m(t)$, is

$$\hat{m}(t) \rightleftharpoons -j\,\text{sgn}(f)M(f) \tag{7.45}$$

Accordingly, the inverse Fourier transformation of Eq. 7.44 yields

$$\tilde{s}_u(t) = \frac{A_c}{2}[m(t) + j\hat{m}(t)] \tag{7.46}$$

which is the desired result.

Having determined $\tilde{s}_u(t)$, we are now ready to formulate the mathematical description of the SSB modulated wave $s_u(t)$. Specifically, placing Eq. 7.46 in Eq. 7.41, we get

$$s_u(t) = \frac{A_c}{2}[m(t)\cos(2\pi f_c t) - \hat{m}(t)\sin(2\pi f_c t)] \tag{7.47}$$

This equation reveals that, except for a scaling factor, a modulated wave containing only an upper sideband has an in-phase component equal to the message signal $m(t)$ and a quadrature component equal to $\hat{m}(t)$, the Hilbert transform of $m(t)$.

EXERCISE 8 Let $s_l(t)$ denote an SSB modulated wave in which only the lower sideband is retained. To determine $s_l(t)$, proceed as follows:

1. Identify the transfer function $H_l(f)$ of a band-pass filter the output of which equals $s_l(t)$ in response to a DSBSC modulated wave.
2. Determine the transfer function $\tilde{H}_l(f)$ of the equivalent low-pass filter corresponding to $H_l(f)$.
3. Hence, using the results in parts (1) and (2), show that $s_l(t)$ is given by

$$s_l(t) = \frac{A_c}{2}[m(t)\cos(2\pi f_c t) + \hat{m}(t)\sin(2\pi f_c t)] \tag{7.48}$$

What are the in-phase and quadrature components of $s_l(t)$?

DISCUSSION

Equations 7.47 and 7.48 are *canonical* representations of upper and lower sidebands modulated on a carrier of frequency f_c. These two equations clearly demonstrate how the upper and lower sidebands can be isolated from each other by subtracting or adding the outputs of two product modulators. The modulators differ from each other by the insertion of $-90°$ phase shifts between the modulating waves as well as between the carrier waves at their inputs; we will have more to say on this issue when we revisit the generation of SSB modulated waves. The mathematical complexity of Eqs. 7.47 and 7.48, involving not only the message signal $m(t)$ but also its Hilbert transform $\hat{m}(t)$, makes it difficult for us to sketch the waveforms of SSB modulated waves, in general. We therefore have to resort to the use of single-tone modulation in order to infer time-domain properties of SSB modulation.

EXAMPLE 3 SINGLE-TONE MODULATION (CONTINUED)

Consider again the sinusoidal modulating wave

$$m(t) = A_m \cos(2\pi f_m t) \tag{7.49}$$

The Hilbert transform of this signal is obtained by passing it through a $-90°$ phase shifter, which yields

$$\hat{m}(t) = A_m \sin(2\pi f_m t) \tag{7.50}$$

Therefore, substituting Eqs. 7.49 and 7.50 in 7.47, we find that the SSB wave, obtained by transmitting only the upper side-frequency, is defined by

$$
\begin{aligned}
s(t) &= \tfrac{1}{2}A_cA_m[\cos(2\pi f_m t)\cos(2\pi f_c t) - \sin(2\pi f_m t)\sin(2\pi f_c t)] \\
&= \tfrac{1}{2}A_cA_m \cos[2\pi(f_c + f_m)t]
\end{aligned}
\tag{7.51}
$$

This is exactly the same as the result obtained by suppressing the lower side-frequency $f_c - f_m$ of the corresponding DSBSC wave of Eq. 7.35. The SSB wave of Eq. 7.51 and its spectrum are illustrated in Fig. 7.3*e*.

Next, using Eq. 7.48, we find that the SSB wave, obtained by transmitting only the lower side-frequency, is defined by

$$
\begin{aligned}
s(t) &= \tfrac{1}{2}A_cA_m[\cos(2\pi f_m t)\cos(2\pi f_c t) + \sin(2\pi f_c t)\sin(2\pi f_m t)] \\
&= \tfrac{1}{2}A_cA_m \cos[2\pi(f_c - f_m)t]
\end{aligned}
\tag{7.52}
$$

which is exactly the same as the result obtained by suppressing the upper side-frequency $f_c + f_m$ of the DSBSC wave of Eq. 7.35. The SSB wave of Eq. 7.52 and its spectrum are illustrated in Fig. 7.3*f*.

PHASE DISCRIMINATION METHOD FOR GENERATING AN SSB MODULATED WAVE

The *phase discrimination method* of generating an SSB modulated wave involves two separate simultaneous modulation processes and subsequent combination of the resulting modulation products, as shown in Fig. 7.20. The derivation of this system follows directly from Eq. 7.47 or 7.48, which defines the canonical representation of SSB modulated waves in the time-domain. The system uses two product modulators, I and Q, supplied with carrier waves in phase quadrature to each other. The incoming baseband signal $m(t)$ is applied to product modulator I, producing a modulated DSBSC wave that contains *reference phase* sidebands symmetrically spaced about carrier frequency f_c. The Hilbert transform $\hat{m}(t)$ of $m(t)$ is applied to product modulator Q, producing a DSBSC modulated wave that contains sidebands having identical amplitude spectra to those of modulator I, but with phase spectra such that vector addition or subtraction of the two modulator outputs results in cancellation of one set of sidebands and reinforcement of the other set. The use of a plus sign at the summing junction yields an SSB wave with only the lower sideband, whereas the use of a minus sign yields an SSB wave with only the upper sideband. In this way the desired SSB modulated wave is produced. The SSB modulator of Fig. 7.20 is also known as the *Hartley modulator*.

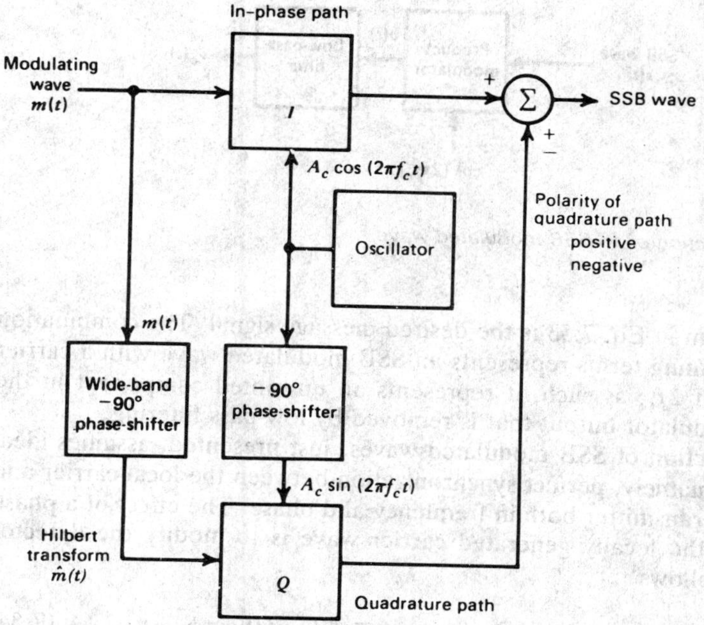

Figure 7.20
Block diagram of the phase discrimination method for generating SSB modulated waves.

DEMODULATION OF SSB WAVES

To recover the baseband signal $m(t)$ from the SSB wave $s(t)$, equal to $s_u(t)$ or $s_l(t)$, we have to shift the spectrum in Fig. 7.17*c* or *d* by the amounts $\pm f_c$ so as to convert the transmitted sideband back into the baseband signal. This can be accomplished using coherent detection, which involves applying the SSB wave $s(t)$, together with a locally generated carrier $\cos(2\pi f_c t)$, assumed to be of unit amplitude for convenience, to a product modulator and then low-pass filtering the modulator output, as in Fig. 7.21. Thus, using Eq. 7.47 or 7.48, we find that the product modulator output is given by

$$\begin{aligned} v(t) &= \cos(2\pi f_c t)s(t) \\ &= \tfrac{1}{2}A_c \cos(2\pi f_c t)[m(t)\cos(2\pi f_c t) \pm \hat{m}(t)\sin(2\pi f_c t)] \\ &= \underbrace{\tfrac{1}{4}A_c m(t)}_{\substack{\text{Scaled} \\ \text{message} \\ \text{signal}}} + \underbrace{\tfrac{1}{4}A_c[m(t)\cos(4\pi f_c t) \pm \hat{m}(t)\sin(4\pi f_c t)]}_{\text{Unwanted component}} \end{aligned} \tag{7.53}$$

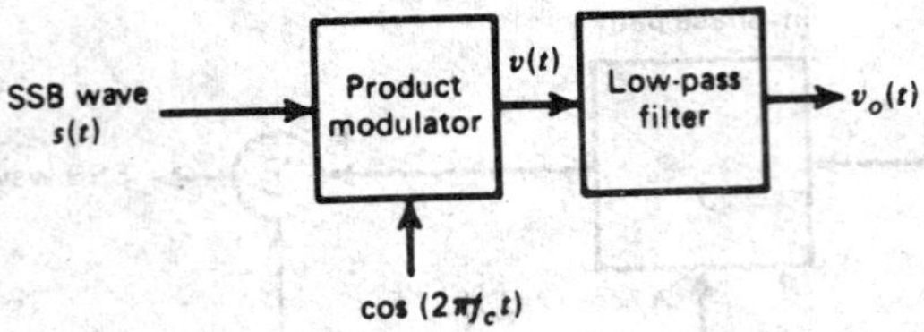

Figure 7.21
Coherent detection of an SSB modulated wave.

The first term in Eq. 7.53 is the desired message signal. The combination of the remaining terms represents an SSB modulated wave with a carrier frequency of $2f_c$; as such, it represents an unwanted component in the product modulator output that is removed by low-pass filtering.

The detection of SSB modulated waves, just presented, assumes ideal conditions, namely, perfect synchronization between the local carrier and that in the transmitter both in frequency and phase. The effect of a phase error ϕ in the locally generated carrier wave is to modify the detector output as follows[3]

$$v_o(t) = \tfrac{1}{4}A_c m(t)\cos\phi \mp \tfrac{1}{4}A_c \hat{m}(t)\sin\phi \tag{7.54}$$

where the plus sign applies to an incoming SSB modulated wave containing only the upper sideband (i.e., the modulated wave of Eq. 7.47), and the minus sign applies to one containing only the lower sideband (i.e., the modulated wave of Eq. 7.48). Owing to the phase error ϕ, the detector output $v_o(t)$ contains not only the message signal $m(t)$ but also its Hilbert transform $\hat{m}(t)$. Consequently, the detector output suffers from *phase distortion*. This phase distortion is usually not serious with voice communications because the human ear is relatively insensitive to phase distortion. The presence of phase distortion gives rise to what is called the Donald Duck voice effect. In the transmission of music and video signals, on the other hand, phase distortion in the form of a constant phase difference in all components can be intolerable.

EXERCISE 9 Show that the low-pass filter in the coherent detector of Fig. 7.21 only passes the message signal component of the product modulator output, provided it satisfies the following conditions:

(a) Bandwidth $= W$
(b) Width of guardband $\leq 2f_c - aW$, where $a = 1$ for an SSB mod-

[3]For a more complete discussion of the effects of carrier phase and frequency errors in single-sideband modulation, see Haykin (1983, pp. 146–149).

ulated wave containing only the upper sideband, and $a = 2$ for an SSB modulated wave containing only the lower sideband.

EXERCISE 10 Let $\cos(2\pi f_c t + \phi)$ denote the local carrier applied to the product modulator in Fig. 7.21. Show that the effect of the phase error ϕ is to modify the detector output $v_o(t)$ as in Eq. 7.54.

7.5 VESTIGIAL SIDEBAND MODULATION

Single-sideband modulation is well-suited for the transmission of voice because of the energy gap that exists in the spectrum of voice signals between zero and a few hundred hertz. When the message signal contains significant components at extremely low frequencies (as in the case of television signals and wideband data), the upper and lower sidebands meet at the carrier frequency. This means that the use of SSB modulation is inappropriate for the transmission of such message signals owing to the difficulty of isolating one sideband. This difficulty suggests another scheme known as *vestigial sideband* modulation (VSB), which is a compromise between SSB and DSBSC modulation. In this modulation scheme, one sideband is passed almost completely whereas just a trace, or *vestige*, of the other sideband is retained.

FREQUENCY-DOMAIN DESCRIPTION

Figure 7.22 illustrates the spectrum of a *vestigial sideband (VSB) modulated wave* $s(t)$ in relation to that of the message signal $m(t)$, assuming that the lower sideband is modified into the vestigial sideband. Specifically, the transmitted vestige of the lower sideband compensates for the amount removed from the upper sideband. The *transmission* bandwidth required by the VSB modulated wave is therefore given by

$$B = W + f_v \qquad (7.55)$$

where W is the message bandwidth and f_v is the width of the vestigial sideband.

Vestigial sideband modulation has the virtue of conserving bandwidth almost as efficiently as single-sideband modulation, while retaining the excellent low-frequency baseband characteristics of double-sideband modulation. Thus VSB modulation has become standard for the transmission of television and similar signals where good phase characteristics and transmission of low-frequency components are important, but the bandwidth required for double-sideband transmission is unavailable or uneconomical.

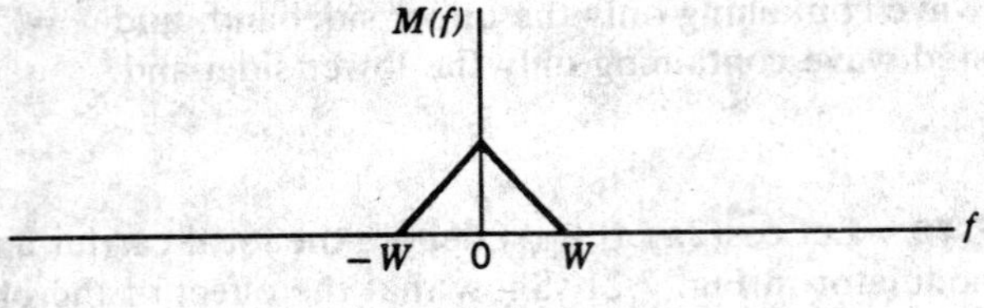

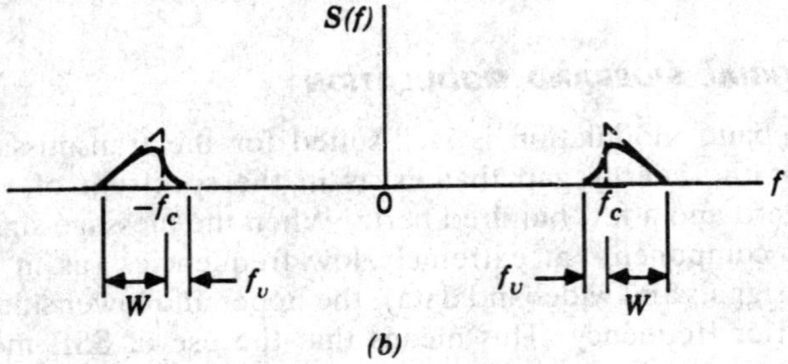

Figure 7.22
(a) *Spectrum of message signal.* (b) *Spectrum of VSB modulated wave containing a vestige of the lower sideband.*

GENERATION OF VSB MODULATED WAVE

To generate a VSB modulated wave, we pass a DSBSC modulated wave through a *sideband shaping filter,* as in Fig. 7.23*a*. The exact design of this filter depends on the desired spectrum of the VSB modulated wave. The relation between the transfer function $H(f)$ of the filter and the spectrum $S(f)$ of the VSB modulated wave $s(t)$ is defined by

$$S(f) = \frac{A_c}{2}[M(f - f_c) + M(f + f_c)]H(f) \tag{7.56}$$

where $M(f)$ is the message spectrum. We wish to determine the specification of the filter transfer function $H(f)$, so that $S(f)$ defines the spectrum of the desired VSB wave $s(t)$. This can be established by passing $s(t)$ through a coherent detector and then determining the necessary condition for the detector output to provide an undistorted version of the original message signal $m(t)$. Thus, multiplying $s(t)$ by a locally generated sine-wave $\cos(2\pi f_c t)$, which is synchronous with the carrier wave $A_c \cos(2\pi f_c t)$ in both frequency and phase, as in Fig. 7.23*b*, we get

$$v(t) = \cos(2\pi f_c t)s(t) \tag{7.57}$$

Transforming this relation into the frequency domain gives the Fourier

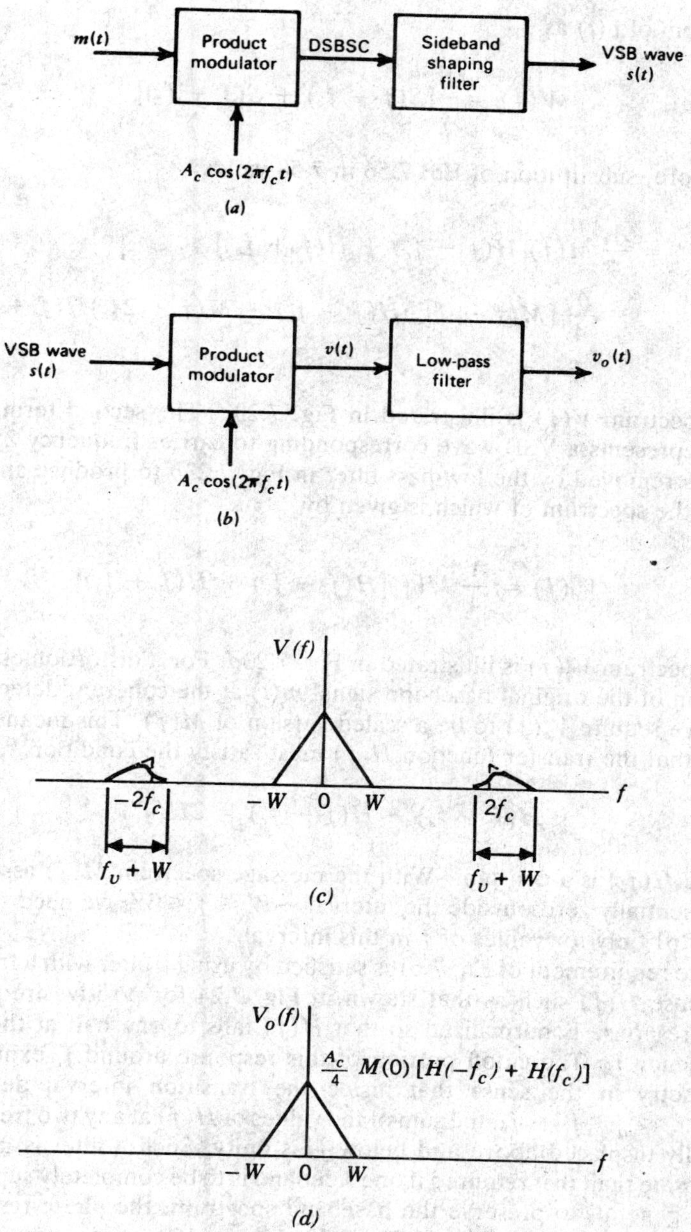

Figure 7.23
Scheme for the generation and demodulation of a VSB modulated wave. (a) Block diagram of VSB modulator. (b) Block diagram of VSB demodulator. (c) Spectrum of the product modulator output v(t) in the demodulation scheme. (d) Spectrum of the demodulated signal $v_o(t)$.

transform of $v(t)$ as

$$V(f) = \frac{1}{2}[S(f - f_c) + S(f + f_c)] \tag{7.58}$$

Therefore, substitution of Eq. 7.56 in 7.58 yields

$$V(f) = \frac{A_c}{4} M(f)[H(f - f_c) + H(f + f_c)] + \frac{A_c}{4}[M(f - 2f_c)H(f - f_c) + M(f + 2f_c)H(f + f_c)] \tag{7.59}$$

The spectrum $V(f)$ is illustrated in Fig. 7.23*c*. The second term in Eq. 7.59 represents a VSB wave corresponding to carrier frequency $2f_c$. This term is removed by the low-pass filter in Fig. 7.23*b* to produce an output $v_o(t)$, the spectrum of which is given by

$$V_o(f) = \frac{A_c}{4} M(f)[H(f - f_c) + H(f + f_c)] \tag{7.60}$$

The spectrum $V_o(f)$ is illustrated in Fig. 7.23*d*. For a distortionless reproduction of the original baseband signal $m(t)$ at the coherent detector output, we require $V_o(f)$ to be a scaled version of $M(f)$. This means, therefore, that the transfer function $H(f)$ must satisfy the condition

$$H(f - f_c) + H(f + f_c) = 2H(f_c) \tag{7.61}$$

where $H(f_c)$ is a constant. With the message spectrum $M(f)$ assumed to be essentially zero outside the interval $-W \leqslant f \leqslant W$, we need to satisfy Eq. 7.61 only for values of f in this interval.

The requirement of Eq. 7.61 is satisfied by using a filter with a frequency response $H(f)$ such as that shown in Fig. 7.24 for positive frequencies. This response is normalized so that $H(f)$ falls to one half at the carrier frequency f_c. The cutoff portion of this response around f_c exhibits odd symmetry in the sense that inside the transition interval defined by $f_c - f_v \leqslant f \leqslant f_c + f_v$, the sum of the values of $H(f)$ at any two frequencies equally displaced above and below f_c is unity. Such a filter is much less elaborate than that required if one sideband is to be completely suppressed.

In general, to preserve the baseband spectrum, the phase response of the sideband shaping filter in Fig. 7.23*a* must exhibit odd symmetry about the carrier frequency f_c. Specifically, it must be linear over the frequency intervals $f_c - f_v \leqslant |f| \leqslant f_c + W$, and its value at the frequency f_c has to equal zero or an integer multiple of 2π radians. The effect of this linear phase characteristic is merely to introduce a constant delay in the recovery of the message signal $m(t)$ at the receiver output.

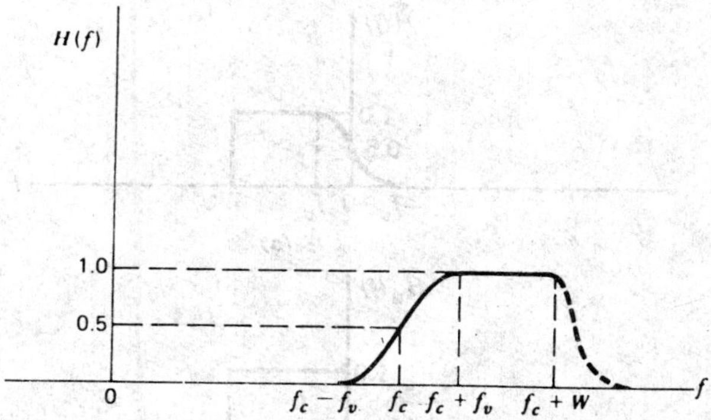

Figure 7.24
Frequency response of sideband shaping filter for a VSB modulated wave containing a vestige of lower sideband; only the positive-frequency portion is shown.

The frequency response of Fig. 7.24 pertains to a VSB modulated wave containing a vestige of the lower sideband. In the situation depicted here, control over the frequency response of the sideband shaping filter need only be exercised over the band $f_c - f_v \leq |f| \leq f_c + W$. This is the reason for showing the frequency response of the sideband shaping filter in Fig. 7.24 for $f > f_c + W$ as a dashed line.

EXERCISE 11 Construct the positive-frequency portion of the frequency response of a sideband shaping filter for a VSB modulated wave that contains a vestige of the upper sideband.

TIME-DOMAIN DESCRIPTION

Our next task is to determine the time-domain description of a VSB modulated wave. To do this, we follow a procedure similar to that used for SSB modulated waves in Section 7.4.

Let $s(t)$ denote a VSB modulated wave containing a vestige of the lower sideband. This modulated wave may be viewed as the output of a sideband shaping filter produced in response to a DSBSC modulated wave defined in Eq. 7.39. The filter has a transfer function $H(f)$ as illustrated in Fig. 7.24. Using the band-pass to low-pass transformation technique of Section 3.5, we may replace the sideband shaping filter by an equivalent complex low-pass filter of transfer function $H(f)$, which is depicted in Fig. 7.25*a*. (For convenience of presentation, we have ignored the dashed portion of $H(f)$ in Fig. 7.24 as it is not pertinent to our present discussion.) Clearly,

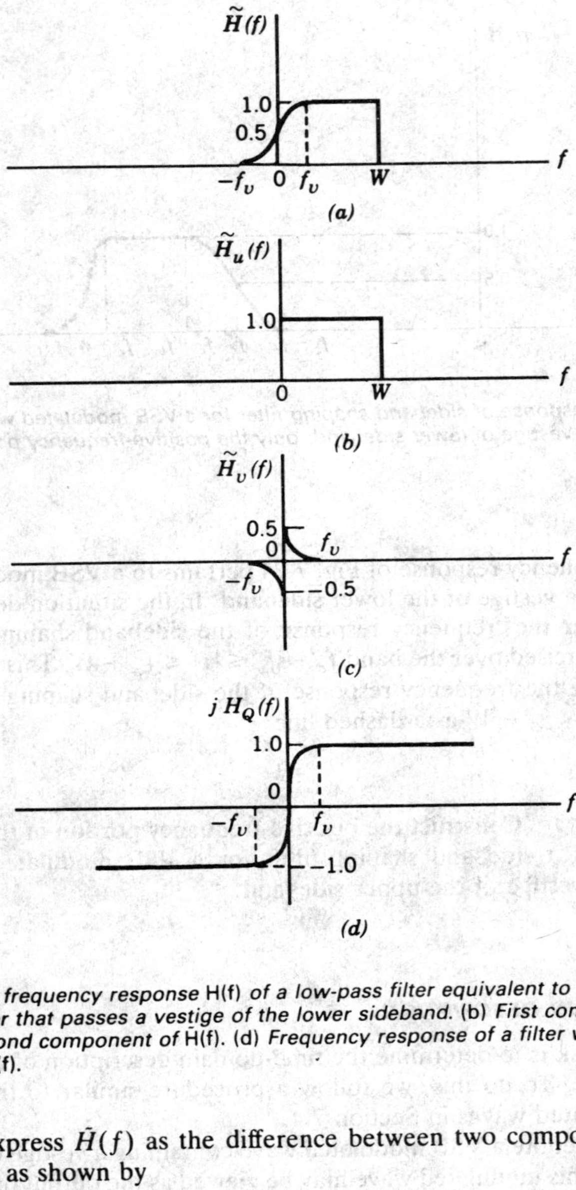

Figure 7.25
(a) *Idealized frequency response* H(f) *of a low-pass filter equivalent to the sideband shaping filter that passes a vestige of the lower sideband.* (b) *First component of* H̃(f). (c) *Second component of* H̃(f). (d) *Frequency response of a filter with transfer function* jH$_Q$(f).

we may express $\tilde{H}(f)$ as the difference between two components $\tilde{H}_u(f)$ and $\tilde{H}_v(f)$ as shown by

$$\tilde{H}(f) = \tilde{H}_u(f) - \tilde{H}_v(f) \tag{7.62}$$

These two components are described individually as follows:

1. The transfer function $\tilde{H}_u(f)$, shown in Fig. 7.25*b*, pertains to a complex low-pass filter equivalent to a band-pass filter designed to reject the lower sideband completely; it is defined in Eq. 7.42.

2. The transfer function $\tilde{H}_v(f)$, shown in Fig. 7.25*c*, accounts for both the generation of a vestige of the lower sideband and the removal of a corresponding portion from the upper sideband.

Thus, substituting Eq. 7.42 in 7.62, we may redefine the transfer function $\tilde{H}(f)$ as

$$\tilde{H}(f) = \begin{cases} \dfrac{1}{2}[1 + \text{sgn}(f) - 2\tilde{H}_v(f)], & -f_v < f < W \\ 0, & \text{otherwise} \end{cases} \tag{7.63}$$

The signum function sgn(f) and the transfer function $\tilde{H}_v(f)$ are both *odd* functions of the frequency f. Hence, they both have *purely imaginary* inverse Fourier transforms. Accordingly, we may introduce a new transfer function

$$H_Q(f) = \frac{1}{j}[\text{sgn}(f) - 2\tilde{H}_v(f)] \tag{7.64}$$

that has a *purely real* inverse Fourier transform. Let $h_Q(t)$ denote the inverse Fourier transform of $H_Q(f)$; that is,

$$h_Q(t) \rightleftharpoons H_Q(f) \tag{7.65}$$

Figure 7.25*d* shows a plot of $jH_Q(f)$ as a function of frequency in accordance with both Eq. 7.64 and Fig. 7.25*c*. To go on with our task, we rewrite Eq. 7.63 in terms of $H_Q(f)$ as

$$\tilde{H}(f) = \begin{cases} \dfrac{1}{2}[1 + jH_Q(f)], & -f_v < f < W \\ 0, & \text{elsewhere} \end{cases} \tag{7.66}$$

We are now ready to determine the VSB modulated wave $s(t)$. First, we write

$$s(t) = \text{Re}[\tilde{s}(t)\exp(j2\pi f_c t)] \tag{7.67}$$

where $\tilde{s}(t)$ is the complex envelope of $s(t)$. Since $\tilde{s}(t)$ is the output of the complex low-pass filter of transfer function $\tilde{H}(f)$, which is produced in response to the complex envelope of the DSBSC modulated wave, we may express the spectrum of $\tilde{s}(t)$ as

$$\tilde{S}(f) = \tilde{H}(f)\tilde{S}_{\text{DSBSC}}(f) \tag{7.68}$$

where $\tilde{S}_{\text{DSBSC}}(f)$ is defined in Eq. 7.43. Hence, substituting Eqs. 7.43 and

7.66 in 7.68, we get

$$\tilde{S}(f) = \frac{A_c}{2}[1 + j\tilde{H}_Q(f)]M(f) \tag{7.69}$$

Taking the inverse Fourier transform of $\tilde{S}(f)$, we thus obtain

$$\tilde{s}(t) = \frac{A_c}{2}[m(t) + jm_Q(t)] \tag{7.70}$$

where $m_Q(t)$ is the response produced by passing the message signal $m(t)$ through a low-pass filter of impulse response $h_Q(t)$. Finally, substituting Eq. 7.70 in 7.67, we get

$$s(t) = \frac{A_c}{2} m(t)\cos(2\pi f_c t) - \frac{A_c}{2} m_Q(t)\sin(2\pi f_c t) \tag{7.71}$$

This is the desired representation for a VSB modulated wave containing a vestige of the lower sideband.[4] The component $\frac{1}{2}A_c m(t)$ constitutes the in-phase component of this VSB modulated wave, and $\frac{1}{2}A_c m_Q(t)$ constitutes the quadrature component.

The DSBSC and SSB waves may be regarded as special cases of the VSB modulated wave defined by Eq. 7.71. If the vestigial sideband is increased to the width of a full sideband, the resulting wave becomes a DSBSC wave with the result that $m_Q(t)$ vanishes. If, on the other hand, the width of the vestigial sideband is reduced to zero, the resulting wave becomes an SSB wave containing the upper sideband, with the result that $m_Q(t) = \hat{m}(t)$, where $\hat{m}(t)$ is the Hilbert transform of $m(t)$.

EXERCISE 12 Show that a VSB modulated wave $s(t)$, containing a vestige of the upper sideband, is defined by

$$s(t) = \frac{1}{2} A_c m(t)\cos(2\pi f_c t) + \frac{1}{2} A_c m_Q(t)\sin(2\pi f_c t) \tag{7.72}$$

where $m(t)$ is the message signal, and $m_Q(t)$ is defined by Eqs. 7.64 and 7.65.

[4] Another time-domain representation of a VSB modulated signal consists of the product of a narrow-band "envelope" function and an SSB modulated signal. For details of this representation, see Hill (1974).

EXERCISE 13 How would you modify the block diagram of Fig. 7.20 so that it may be used to generate VSB modulated waves?

ENVELOPE DETECTION OF A VSB WAVE PLUS CARRIER

In commercial television broadcasting, a sizable carrier is transmitted together with the modulated wave. This makes it possible to demodulate the incoming modulated wave by an envelope detector in the receiver. It is, therefore, of interest to determine the distortion introduced by the envelope detector. Adding the carrier component $A_c \cos(2\pi f_c t)$ to Eq. 7.71, scaled by a factor k_a, modifies the modulated wave applied to the envelope detector input as

$$s(t) = A_c[1 + \tfrac{1}{2}k_a m(t)] \cos(2\pi f_c t) - \tfrac{1}{2}k_a A_c m_Q(t) \sin(2\pi f_c t) \tag{7.73}$$

where the constant k_a determines the percentage modulation. The envelope detector output, denoted by $a(t)$, is therefore

$$\begin{aligned} a(t) &= A_c\{[1 + \tfrac{1}{2}k_a m(t)]^2 + [\tfrac{1}{2}k_a m_Q(t)]^2\}^{1/2} \\ &= A_c[1 + \tfrac{1}{2}k_a m(t)] \left\{1 + \left[\frac{\tfrac{1}{2}k_a m_Q(t)}{1 + \tfrac{1}{2}k_a m(t)}\right]^2\right\}^{1/2} \end{aligned} \tag{7.74}$$

Equation 7.74 indicates that the distortion is contributed by the quadrature component $m_Q(t)$ of the incoming VSB wave. This distortion can be reduced using two methods: (1) reducing the percentage modulation to reduce k_a and (2) increasing the width of the vestigial sideband to reduce $m_Q(t)$. Both methods are used in practice. In commercial television broadcasting, the vestigial sideband occupies a width of about 1.25 MHz, or about one-quarter of a full sideband. This has been determined empirically as the width of vestigial sideband required to keep the distortion due to $m_Q(t)$ within tolerable limits when the percentage modulation is nearly 100.

7.6 COMPARISON OF AMPLITUDE MODULATION TECHNIQUES

Having studied the characteristics of the different forms of amplitude modulation, we are now in a position to compare their practical merits:

1. In standard AM systems the sidebands are transmitted in full, accompanied by the carrier. Accordingly, demodulation is accomplished simply by using an envelope detector or square-law detector. On the other

hand, in suppressed-carrier systems the receiver is more complex because additional circuitry must be provided for the purpose of carrier recovery. It is for this reason we find that in commercial AM radio *broadcast* systems, which involve one transmitter and numerous receivers, standard AM is used in preference to DSBSC or SSB modulation.

2. Suppressed-carrier modulation systems have an advantage over standard AM systems in that they require much less power to transmit the same amount of information, which makes the transmitters for such systems less expensive than those required for standard AM. Suppressed-carrier systems are therefore well-suited for *point-to-point communication* involving one transmitter and one receiver, which would justify the use of increased receiver complexity.
3. Single-sideband modulation requires the minimum transmitter power and minimum transmission bandwidth possible for conveying a message signal from one point to another. We thus find that single-sideband modulation is the preferred method of modulation for long-distance transmission of voice signals over metallic circuits, because it permits longer spacing between the *repeaters,* which is a more important consideration here than simple terminal equipment. A repeater is simply a wideband amplifier that is used at intermediate points along the transmission path so as to make up for the attenuation incurred during the course of transmission.
4. Vestigial-sideband modulation requires a transmission bandwidth that is intermediate between that required for SSB or DSBSC modulation, and the saving can be significant if modulating waves with large bandwidths are being handled, as in the case of television signals and wideband data.
5. Double-sideband suppressed-carrier modulation, single-sideband modulation, and vestigial-sideband modulation are all examples of *linear modulation.* The output of a linear modulator can be expressed in the *canonical form*

$$s(t) = s_I(t)\cos(2\pi f_c t) - s_Q(t)\sin(2\pi f_c t) \tag{7.75}$$

The in-phase component $s_I(t)$ is a scaled version of the incoming message signal $m(t)$. The quadrature component $s_Q(t)$ is derived from $m(t)$ by some linear filtering operation. Accordingly, the principle of superposition can be used to calculate the modulator output $s(t)$ as the sum of responses of the modulator to individual components of $m(t)$. In Table 7.1 we have summarized the definitions for $s_I(t)$ and $s_Q(t)$ in terms of $m(t)$ for DSBSC, SSB, and VSB modulated waves, assuming a carrier of unit amplitude. In a strict sense, ordinary amplitude modulation fails to meet the definition of a linear modulator with respect to the message signal. If $s_1(t)$ is the AM wave produced by a message

Table 7.1 Different Forms of Linear Modulation

Type of Modulation	In-phase component $s_I(t)$	Quadrature component $s_Q(t)$	Comments
DSBSC	$m(t)$	0	$m(t)$ = message signal
SSB			
1. Upper sideband transmitted	$\frac{1}{2}m(t)$	$\frac{1}{2}\hat{m}(t)$	$\hat{m}(t)$ = Hilbert transform of $m(t)$
2. Lower sideband transmitted	$\frac{1}{2}m(t)$	$-\frac{1}{2}\hat{m}(t)$	
VSB			
1. Vestige of lower sideband transmitted	$\frac{1}{2}m(t)$	$\frac{1}{2}m_Q(t)$	$m_Q(t)$ = output of filter of transfer function $H_Q(f)$, produced by $m(t)$
2. Vestige of upper sideband transmitted	$\frac{1}{2}m(t)$	$-\frac{1}{2}m_Q(t)$	For the definition of $H_Q(f)$, see Eq. 7.69

signal $m_1(t)$ and $s_2(t)$ is the AM wave produced by a second message signal $m_2(t)$, then the AM wave produced by $m_1(t)$ plus $m_2(t)$ is not equal to $s_1(t)$ plus $s_2(t)$. However, the departure from linearity in AM is of a mild sort, such that many of the mathematical procedures applicable to linear modulation may be retained. For example, the band-pass representation is still applicable to an AM wave, with the in-phase and quadrature components defined by, respectively,

$$s_I(t) = 1 + k_a m(t)$$

and

$$s_Q(t) = 0$$

where k_a is the amplitude sensitivity of the modulator.

6. In both SSB and VSB modulation schemes, the role of the quadrature component is merely to interfere with the in-phase component, so as to eliminate power in one of the sidebands. Herein lies the reason for the fact that SSB- and VSB-modulated waves have favorable spectral properties. Note, however, that regardless of the nature of the quadrature component, the message signal $m(t)$ may be recovered from the modulated signal $s(t)$ with the use of coherent detection.
7. The band-pass representation may also be used to describe quadrature amplitude modulation. In this case, we have (assuming a carrier of unit

amplitude)

$$s_I(t) = m_1(t)$$

and

$$s_Q(t) = -m_2(t)$$

where $m_1(t)$ and $m_2(t)$ are the two independent message signals at the quadrature-modulator input (see Eq. 7.38).

8. The *complex envelope* of the linearly modulated wave $s(t)$ equals

$$\tilde{s}(t) = s_I(t) + js_Q(t)$$

This compact notation retains complete information about the modulation process.

7.7 FREQUENCY TRANSLATION

In the processing of signals in communication systems, it is often convenient or necessary to translate the modulated wave upward or downward in frequency, so that it occupies a new frequency band. This frequency translation is accomplished by multiplication of the signal by a locally generated sine wave, and subsequent filtering. For example, consider the DSBSC wave

$$s(t) = m(t)\cos(2\pi f_c t) \tag{7.76}$$

The modulating wave $m(t)$ is limited to the frequency band $-W \leqslant f \leqslant W$. The spectrum of $s(t)$ therefore occupies the bands $f_c - W \leqslant f \leqslant f_c + W$ and $-f_c - W \leqslant f \leqslant -f_c + W$, as in Fig. 7.26*a*. Suppose that it is required to translate this modulated wave downward in frequency, so that its carrier frequency is changed from f_c to a new value f_o, where $f_o < f_c$. To accomplish this requirement, we first multiply the incoming modulated wave $s(t)$ by a sinusoidal wave of frequency f_l supplied by a local oscillator to obtain

$$\begin{aligned} v_1(t) &= s(t)\cos(2\pi f_l t) \\ &= m(t)\cos(2\pi f_c t)\cos(2\pi f_l t) \\ &= \tfrac{1}{2}m(t)\cos[2\pi(f_c - f_l)t] + \tfrac{1}{2}m(t)\cos[2\pi(f_c + f_l)t] \end{aligned} \tag{7.77}$$

The multiplier output $v_1(t)$ consists of two DSBSC waves, one with a carrier frequency of $f_c - f_l$ and the other with a carrier frequency of $f_c + f_l$. The spectrum of $v_1(t)$ is therefore as shown in Fig. 7.26*b*. Let the frequency f_l of the local oscillator be chosen so that

$$f_c - f_l = f_o \tag{7.78}$$

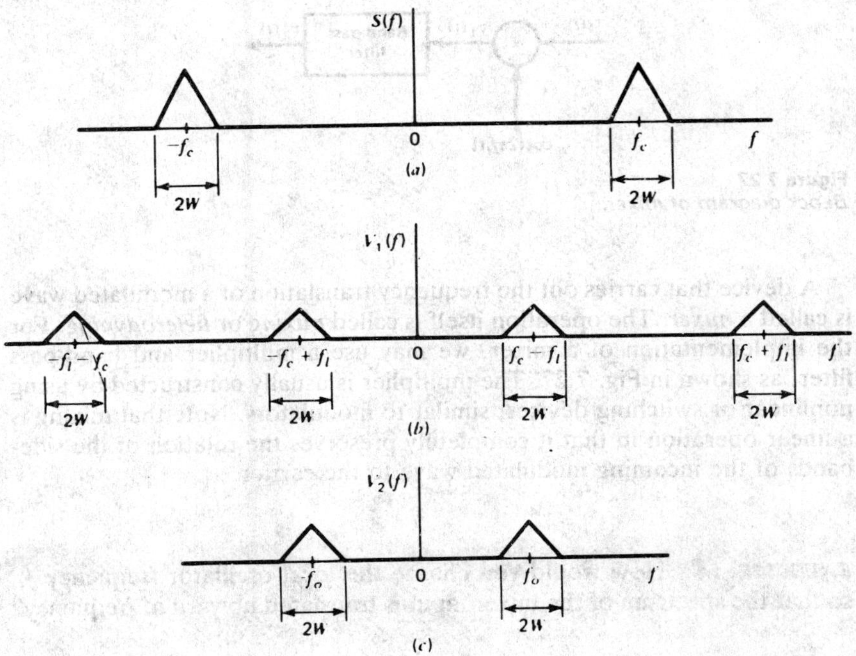

Figure 7.26
The frequency translation process. (a) *Spectrum of DSBSC wave.* (b) *Spectrum of signal obtained by multiplying DSBSC wave with a local carrier.* (c) *Spectrum of desired DSBSC wave, translated downward in frequency.*

Then from Fig. 7.26*b* we see that the modulated wave with the desired carrier frequency f_o may be extracted by passing the multiplier output $v_1(t)$ through a band-pass filter of midband frequency f_o and bandwidth $2W$, provided

$$f_c + f_l - W > f_c - f_l + W$$

or

$$f_l > W \tag{7.79}$$

The filter output is therefore

$$\begin{aligned} v_2(t) &= \tfrac{1}{2}m(t)\cos[2\pi(f_c - f_l)t] \\ &= \tfrac{1}{2}m(t)\cos(2\pi f_o t) \end{aligned} \tag{7.80}$$

This output is the desired modulated wave, translated downward in frequency, as shown in Fig. 7.26*c*.

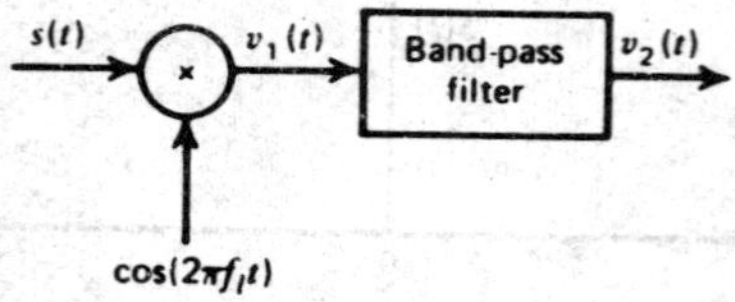

Figure 7.27
BLock diagram of mixer.

A device that carries out the frequency translation of a modulated wave is called a *mixer*. The operation itself is called *mixing* or *heterodyning*. For the implementation of a mixer, we may use a multiplier and band-pass filter, as shown in Fig. 7.27. The multiplier is usually constructed by using nonlinear or switching devices, similar to modulators. Note that mixing is a linear operation in that it completely preserves the relation of the sidebands of the incoming modulated wave to the carrier.

EXERCISE 14 How would you choose the local oscillator frequency f_l, so that the spectrum of the mixer input is translated upward in frequency?

EXAMPLE 4

Consider an incoming narrow-band signal of bandwidth 10 kHz, and mid-band frequency that may lie in the range 0.535–1.605 MHz. It is required to translate this signal to a fixed frequency band centered at 0.455 MHz. The problem is to determine the range of tuning that must be provided in the local oscillator. (The frequencies used in this example pertain to the AM broadcast band of frequencies, on which more will be said in Section 7.9.)

Let f_c denote the midband frequency of the incoming signal, and f_l denote the local oscillator frequency. Then we may write

$$0.535 < f_c < 1.605$$

and

$$f_c - f_l = 0.455$$

where both f_c and f_l are expressed in MHz. That is,

$$f_l = f_c - 0.455$$

When $f_c = 0.535$ MHz, we get $f_l = 0.08$ MHz; and when $f_c = 1.605$ MHz, we get $f_l = 1.15$ MHz. Thus the required range of tuning of the local oscillator is 0.08–1.15 MHz.

7.8 FREQUENCY-DIVISION MULTIPLEXING

Multiplexing is a technique whereby a number of independent signals can be combined into a composite signal suitable for transmission over a common channel. This operation requires that the signals be kept apart so that they do not interfere with each other, and thus they can be separated at the receiving end. This is accomplished by separating the signals either in frequency or in time. The technique of separating the signals in frequency is referred to as *frequency-division multiplexing* (FDM), whereas the technique of separating the signals in time is called *time-division multiplexing* (TDM). In this section, we discuss FDM systems, whereas TDM systems were discussed in Section 5.10.

A block diagram of an FDM system is shown in Fig. 7.28. The incoming message signals are assumed to be of the low-pass type, but their spectra do not necessarily have nonzero values all the way down to zero frequency. Following each signal input, we have shown a low-pass filter, which is designed to remove high-frequency components that do not contribute significantly to signal representation but are capable of disturbing other message signals that share the common channel. These low-pass filters may be omitted only if the input signals are sufficiently band-limited initially. The filtered signals are applied to modulators that shift the frequency ranges of the signals so as to occupy mutually exclusive frequency intervals. The necessary carrier frequencies, to perform these frequency translations, are obtained from a carrier supply. For the modulation, we may use any one of the processes described in previous sections of this chapter. However,

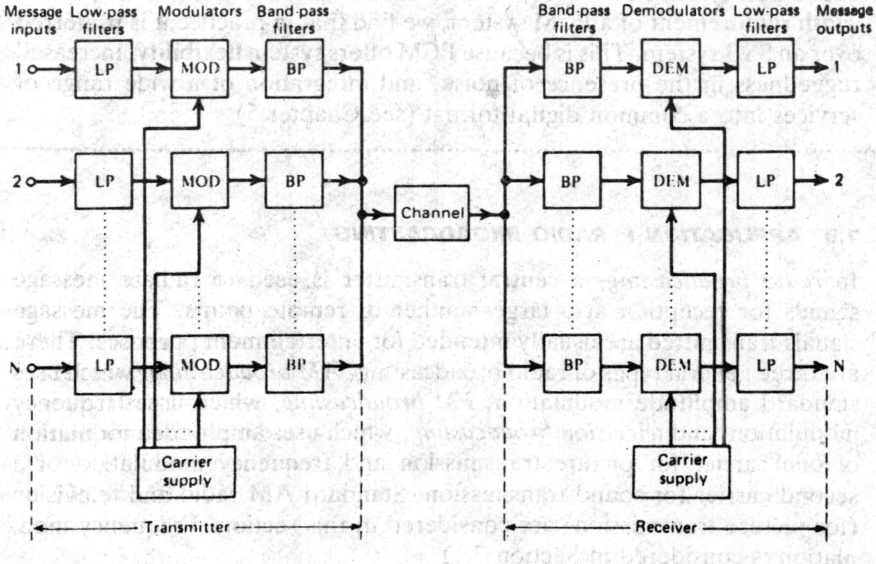

Figure 7.28
Block diagram of FDM system.

the most widely used method of modulation in frequency-division multiplexing is single-sideband modulation, which requires a bandwidth that is approximately equal to that of the original message signal. The band-pass filters following the modulators are used to restrict the band of each modulated wave to its prescribed range. The resulting band-pass filter outputs are next combined in parallel to form the input to the common channel. At the receiving terminal, a bank of band-pass filters, with their inputs connected in parallel, is used to separate the message signals on a frequency-occupancy basis. Finally, the original message signals are recovered by individual demodulators.

EXAMPLE 5 COMPARISON OF SSB/FDM WITH PCM/TDM

Consider an FDM system using SSB modulation to transmit 24 independent voice inputs. Assume a bandwidth of 4 kHz for each voice input. Thus, in order to accommodate an FDM system using SSB modulation to transmit the 24 voice inputs, the communication channel must provide the transmission bandwidth:

$$B = 24 \times 4 = 96 \text{ kHz}$$

In Example 1, Chapter 6, we showed that for the T1 system (based on the combined use of PCM and TDM), the minimum channel bandwidth required to transmit 24 voice inputs is equal to 772 kHz. This is an order of magnitude larger than the bandwidth requirement of the corresponding SSB/FDM system. However, in spite of the excessive transmission bandwidth requirement of a PCM system, we find that in practice it is preferred over an SSB system. This is because PCM offers system flexibility, increased ruggedness in the presence of noise, and integration of a wide range of services into a common digital format (see Chapter 5).

7.9 APPLICATION I: RADIO BROADCASTING

In *radio broadcasting,* a central transmitter is used to radiate message signals for reception at a large number of remote points. The message signals transmitted are usually intended for entertainment purposes. There are three general types of radio broadcasting, *AM broadcasting,* which uses standard amplitude modulation; *FM broadcasting,* which uses frequency modulation; and *television broadcasting,* which uses amplitude modulation of one carrier for picture transmission and frequency modulation of a second carrier for sound transmission. Standard AM radio and television (for picture transmission) are considered in this section. Frequency modulation is considered in Section 7.11.

AM RADIO

The usual AM radio receiver is of the *superheterodyne* type, which is represented schematically in Fig. 7.29. Basically, the receiver consists of a radio frequency (RF) section, a mixer and local oscillator, an intermediate frequency (IF) section, and a demodulator. Typical frequency parameters of commercial AM radio are:

RF carrier range = 0.535–1.605 MHz
Midband frequency of IF section = 455 kHz
IF bandwidth = 10 kHz

The incoming amplitude modulated wave is picked up by the receiving antenna and amplified in the RF section, which is tuned to the carrier frequency of the incoming wave. The combination of mixer and local oscillator (of adjustable frequency) provides a *frequency conversion* or *heterodyning* function, whereby the incoming signal is converted to a predetermined fixed *intermediate frequency,* usually lower than the signal frequency. This frequency conversion is achieved without disturbing the relation of the sidebands to the carrier. The result of this conversion is to produce an intermediate-frequency carrier defined by

$$f_{IF} = f_{RF} - f_{LO}$$

where f_{LO} is the frequency of the local oscillator and f_{RF} is the carrier frequency of the incoming RF signal. We refer to f_{IF} as the intermediate frequency (IF), because the signal is neither at the original input frequency nor at the final baseband frequency. The mixer–local oscillator combination is sometimes referred to as the *first detector,* in which case the demodulator is called the *second detector.*

The IF section consists of one or more stages of tuned amplification, with a bandwidth corresponding to that required for the particular type of signal that the receiver is intended to handle. This section provides most

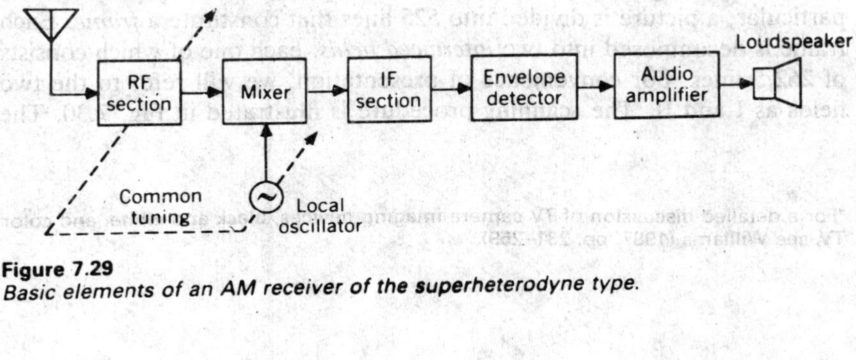

Figure 7.29
Basic elements of an AM receiver of the superheterodyne type.

of the amplification and selectivity in the receiver. The output of the IF section is applied to an envelope detector, the purpose of which is to recover the baseband signal. The final operation in the receiver is the power amplification of the recovered message. The loudspeaker constitutes the load of the power amplifier.

The superheterodyne operation refers to the frequency conversion from the *variable* carrier frequency of the incoming RF signal to the *fixed* IF signal.

In a superheterodyne receiver the mixer will develop an intermediate frequency output when the input signal frequency is greater or less than the local oscillator frequency by an amount equal to the intermediate frequency. That is, there are two input frequencies, namely, $|f_{LO} \pm f_{IF}|$, which will result in f_{IF} at the mixer output. This introduces the possibility of simultaneous reception of two signals differing in frequency by twice the intermediate frequency. Accordingly, it is necessary to employ selective stages in the RF section (i.e., between the antenna and the mixer) in order to favor the desired signal and discriminate against the undesired or *image signal.* The effectiveness of suppressing unwanted image signals increases as the number of selective stages in the radio frequency section increases, and as the ratio of intermediate-to-signal frequency increases.

TELEVISION

Television (TV) refers to the transmission of pictures in motion by means of electrical signals. To accomplish this transmission, each complete picture has to be *sequentially scanned.* The scanning process is carried out in a TV *camera.*[5] In a *black-and-white* TV, the camera contains optics designed to focus an image on a *photocathode* that consists of a large number of photosensitive elements. The charge pattern so generated on the photosensitive surface is scanned by an *electron beam,* thereby producing an output current that varies *temporally* in accordance with the way in which the brightness of the original picture varies *spatially* from one point to another. The resulting output current is called the *video signal.*

The type of scanning used in television is called a *raster scan;* it is somewhat analogous to the manner in which we read a printed paper in that the scanning is performed from left to right on a line-by-line basis. In particular, a picture is divided into 525 lines that constitute a *frame.* Each frame is decomposed into two *interlaced fields,* each one of which consists of 262.5 lines. For convenience of presentation, we will refer to the two fields as I and II. The scanning procedure is illustrated in Fig. 7.30. The

[5]For a detailed discussion of TV camera imaging devices, black and white, and color TV, see Williams (1987, pp. 231–259).

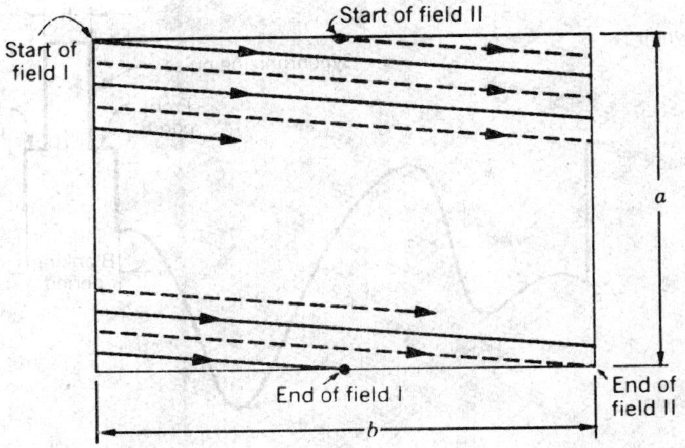

Figure 7.30
Interlaced raster scan.

lines of field I are depicted as solid lines, and those of field II are depicted as dashed lines. The *start* and *end* of each field are also included in the figure. Field I is scanned first. The scanning spot of the TV camera moves with constant velocity across each line of the field from left to right. When the end of a particular line is reached, the scanning spot quickly flies back (in a horizontal direction) to the start of the next line down in the field. This flyback is called the *horizontal retrace*. The scanning process described here is continued until the whole field has been accounted for. When this condition is reached, the scanning spot moves quickly (in a vertical direction) from the end of field I to the start of field II. This second flyback is called the *vertical retrace*. Field II is treated in the same fashion as field I. The time taken for each field to be scanned is 1/60 second. Correspondingly, the time taken for a frame or a complete picture to be scanned is 1/30 second. With 525 lines in a frame, the *line scanning frequency* equals 15.75 kHz.

Thus, by flashing 30 still pictures per second on the display tube of the TV receiver, the human eye perceives them to be moving pictures. This effect is due to a phenomenon known as the *persistence of vision*.

During the horizontal- and vertical-retrace intervals, the picture tube is made inoperative by means of *blanking pulses* that are generated at the transmitter. Moreover, synchronization between the various scanning operations at both the transmitter and receiver is accomplished by means of special pulses that are transmitted during the blanking periods; as such, the synchronizing pulses do not show on the reproduced picture. Figure 7.31 illustrates the use of blanking periods and synchronizing pulses for one full line of a video waveform.

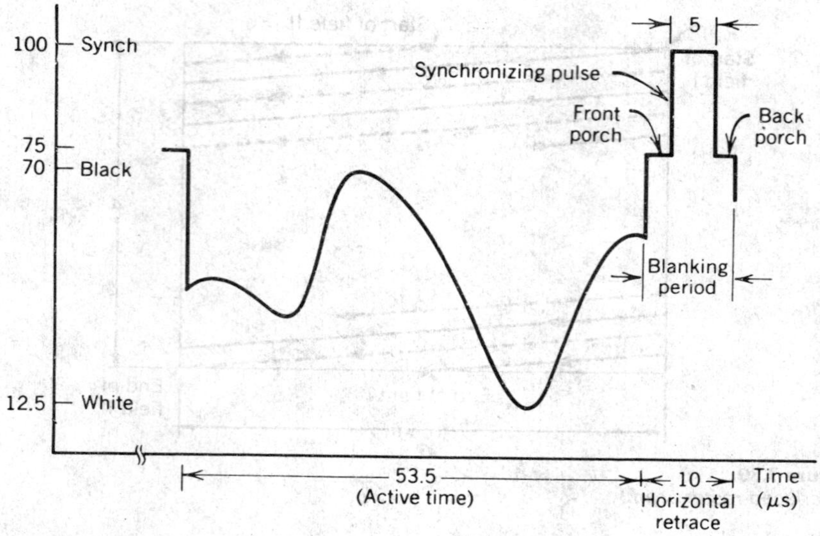

Figure 7.31
Video waveform for one full line of TV picture.

Video Bandwidth The reproduction quality of a TV picture is limited by two basic factors:

1. The number of lines available in a raster scan, which limits resolution of the picture in the vertical direction.
2. The channel bandwidth available for transmitting the video signal, which limits resolution of the picture in the horizontal direction.

For each direction, *resolution* is expressed in terms of the maximum number of lines alternating between black and white that can be resolved in the TV image along the pertinent direction by a human observer.

Consider first the image resolution in the vertical direction, denoted by R_v. It is tempting to equate the vertical resolution R_v to the total number of scan lines per frame minus those lines in the vertical interval that are not used for display. In practice, however, this is not so, because the scanning process that changes the image into a video signal in the camera (at the transmitter) and then reconstructs the image on the display (at the receiver) is in reality a *sampling process*. From our discussion of the sampling process in Section 5.3, we know that a message signal must be strictly band-limited or else distortion due to aliasing will occur. Consequently, we find that the vertical resolution in a TV picture is reduced not only by the vertical retrace, but also by aliasing, as shown by

$$R_v = k(N - 2N_{vr}) \tag{7.81}$$

where N is the *total* number of raster scan lines, and N_{vr} is the number of lines per field that are lost during the vertical retrace. The fact that the vertical resolution R_v in Eq. 7.81 is a fraction of $(N - 2N_{vr})$ is called the *Kell effect;* correspondingly, k is called the *Kell factor.* Normally, the Kell factor ranges between 0.6 and 0.7.

Let a denote the raster height, as in Fig. 7.30. Then, we may express the vertical resolution in a TV picture in terms of *vertical lines per unit distance* as

$$\frac{R_v}{a} = \frac{k}{a}(N - 2N_{vr}) \text{ lines/unit distance} \tag{7.82}$$

Consider next the horizontal resolution, denoted as R_h; this resolution is expressed in terms of the maximum number of lines that can be resolved in a TV picture along the horizontal direction. To determine R_h, we assume that the picture elements or *pixels* are arranged as alternate black and white squares along the scanning line. The corresponding video signal is a square wave with a fundamental frequency equal to the video bandwidth. Since there are two pixels per cycle of the square wave, we may express the horizontal resolution of a TV picture as

$$R_h = 2B(T - T_{hr}) \tag{7.83}$$

where B is the *video bandwidth,* T is the total duration of one scanning line, and T_{hr} is the duration of a horizontal retrace.

Let b denote the raster width, as in Fig. 7.30. We may then express the horizontal resolution of a TV picture in terms of *horizontal lines per unit distance* as

$$\frac{R_h}{b} = \frac{2B}{b}(T - T_{hr}) \text{ lines/unit distance} \tag{7.84}$$

A natural choice for the video bandwidth B is to make the vertical resolution equal the horizontal resolution, as shown by

$$\frac{R_v}{a} = \frac{R_h}{b} \tag{7.85}$$

Hence, using Eqs. 7.82, 7.84 and 7.85 to solve for the bandwidth B, we get the desired result

$$B = \frac{k}{2}\left(\frac{b}{a}\right)\left(\frac{N - 2N_{vr}}{T - T_{hr}}\right) \tag{7.86}$$

The ratio of raster width b to raster height a is called the *aspect ratio.*

In the NTSC[6] system, we have the following parameter values:

Aspect ratio $= \frac{b}{a} = \frac{4}{3}$
Total lines per frame $= N = 525$
Vertical retrace $= N_{vr} = 21$ lines/field
Kell factor $= k = 0.7$
Total line time $= T = 63.5\ \mu s$
Horizontal retrace time $= T_{hr} = 10\ \mu s$

Substituting these values in Eq. 7.86, we get the video bandwidth:

$$B = 4.21 \text{ MHz}$$

This result is very close to the actual maximum frequency in the standard video signal, which is 4.2 MHz.

Choice of Modulation The type of modulation chosen to transmit the video signal is influenced by two factors:

1. The video signal exhibits a large bandwidth and significant low-frequency content. This suggests the use of vestigial sideband modulation.
2. The circuitry used for demodulation in the receiver should be simple and therefore cheap. This suggests the use of envelope detection, which requires the addition of a carrier to the VSB modulated wave.

With regard to point 1, although there is a basic desire to conserve bandwidth, nevertheless in commercial TV broadcasting the transmitted signal is not quite VSB-modulated. The reason is that at the transmitter the power levels are high, with the result that it would be expensive to rigidly control the transition region. Instead, a VSB filter is inserted in each receiver where the power levels are low. The overall performance is the same as conventional vestigial-sideband modulation except for some wasted power and bandwidth. These remarks are illustrated in Fig. 7.32. In particular, part *a* of the figure shows the idealized specturm of a transmitted TV signal. The upper sideband, 25% of the lower sideband, and the picture carrier are transmitted. The frequency response of the VSB filter used to do the required spectrum shaping in the receiver is shown in part *b* of the figure.

With regard to point 2, the use of envelope detection (applied to a VSB-modulated wave plus carrier) produces *waveform distortion* in the message signal recovered at the detector output. The distortion is contributed by

[6]NTSC is the abbreviation for National Television System Committee.

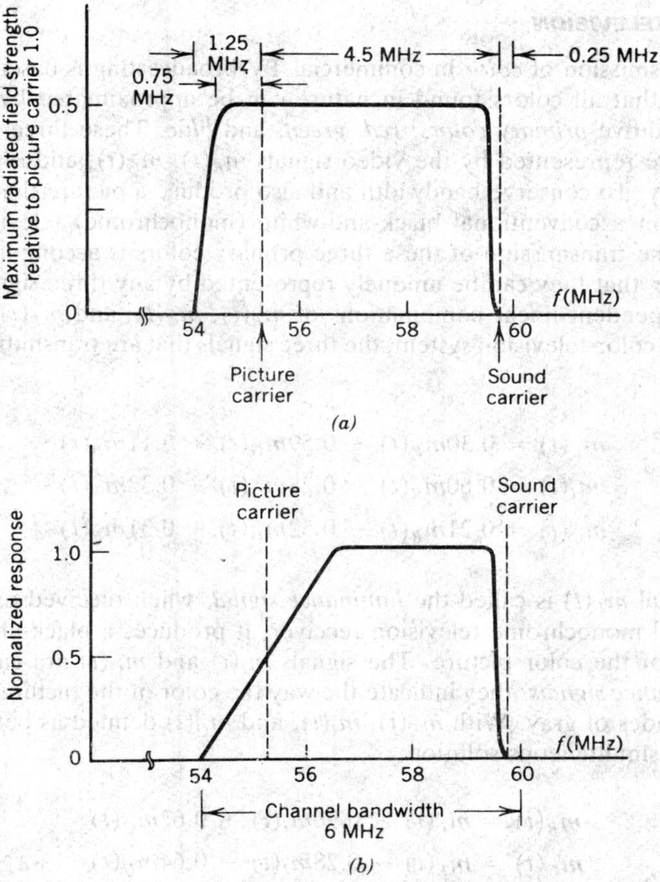

Figure 7.32
(a) *Idealized amplitude spectrum of transmitted TV signal.* (b) *Amplitude response of VSB shaping filter in the receiver.*

the quadrature component of the VSB wave. This issue was discussed in Section 7.5.

The channel bandwidth used for NTSC TV broadcast is 6 MHz; see Fig. 7.32*b*. This channel bandwidth not only accommodates the bandwidth requirement of the VSB-modulated video signal but also provides for the accompanying sound signal that modulates a carrier of its own.

The values presented on the frequency axis in parts (*a*) and (*b*) of Fig. 7.32 pertain to a specific TV channel. According to this figure, the picture carrier frequency is at 55.75 MHz, and the sound carrier frequency is at 59.75 MHz. Note, however, that the information content of the TV signal lies in a *baseband spectrum* extending from 1.25 MHz below the picture carrier to 4.5 MHz above it.

COLOR TELEVISION

The transmission of *color* in commercial TV broadcasting is based on the premise that all colors found in nature can be approximated by mixing three additive *primary colors: red, green,* and *blue.* These three primary colors are represented by the video signals $m_R(t)$, $m_G(t)$, and $m_B(t)$, respectively. To conserve bandwidth and also produce a picture that can be viewed on a conventional black-and-white (monochrome) television receiver, the transmission of these three primary colors is accomplished by observing that they can be uniquely represented by any three signals that are independent linear combinations of $m_R(t)$, $m_G(t)$, and $m_B(t)$. In the standard color-television system, the three signals that are transmitted have the form

$$\begin{aligned} m_L(t) &= 0.30m_R(t) + 0.59m_G(t) + 0.11m_B(t) \\ m_I(t) &= 0.60m_R(t) - 0.28m_G(t) - 0.32m_B(t) \\ m_Q(t) &= 0.21m_R(t) - 0.52m_G(t) + 0.31m_B(t) \end{aligned} \tag{7.87}$$

The signal $m_L(t)$ is called the *luminance signal;* when received on a conventional monochrome television receiver, it produces a black-and-white version of the color picture. The signals $m_I(t)$ and $m_Q(t)$ are called the *chrominance signals;* they indicate the way the color of the picture departs from shades of gray. With $m_L(t)$, $m_I(t)$, and $m_Q(t)$ defined as before, we have by simultaneous solution:

$$\begin{aligned} m_R(t) &= m_L(t) - 0.96m_I(t) + 0.62m_Q(t) \\ m_G(t) &= m_L(t) - 0.28m_I(t) - 0.64m_Q(t) \\ m_B(t) &= m_L(t) - 1.10m_I(t) + 1.70m_Q(t) \end{aligned} \tag{7.88}$$

The luminance signal $m_L(t)$ is assigned the entire 4.2 MHz bandwidth. Owing to certain properties of human vision, tests show that if the nominal bandwidths of the chrominance signals $m_I(t)$ and $m_Q(t)$ are 1.6 MHz and 0.6 MHz, respectively, then satisfactory color reproduction is possible.

Figure 7.33*a* shows a simplified block diagram of the color-television transmitter. The chrominance signals $m_I(t)$ and $m_Q(t)$ are combined using a variation of quadrature-multiplexing with a subcarrier having a frequency denoted by f_{cc}. The output resulting from the quadrature-multiplexing operation is next superimposed on the luminance signal $m_L(t)$ to give a combined video signal $m(t)$. The composite video signal $m(t)$ is thus described by

$$\begin{aligned} m(t) = m_L(t) &+ m_I(t)\cos(2\pi f_{cc}t) + m_Q(t)\sin(2\pi f_{cc}t) \\ &+ \hat{m}_{IH}(t)\sin(2\pi f_{cc}t) \end{aligned} \tag{7.89}$$

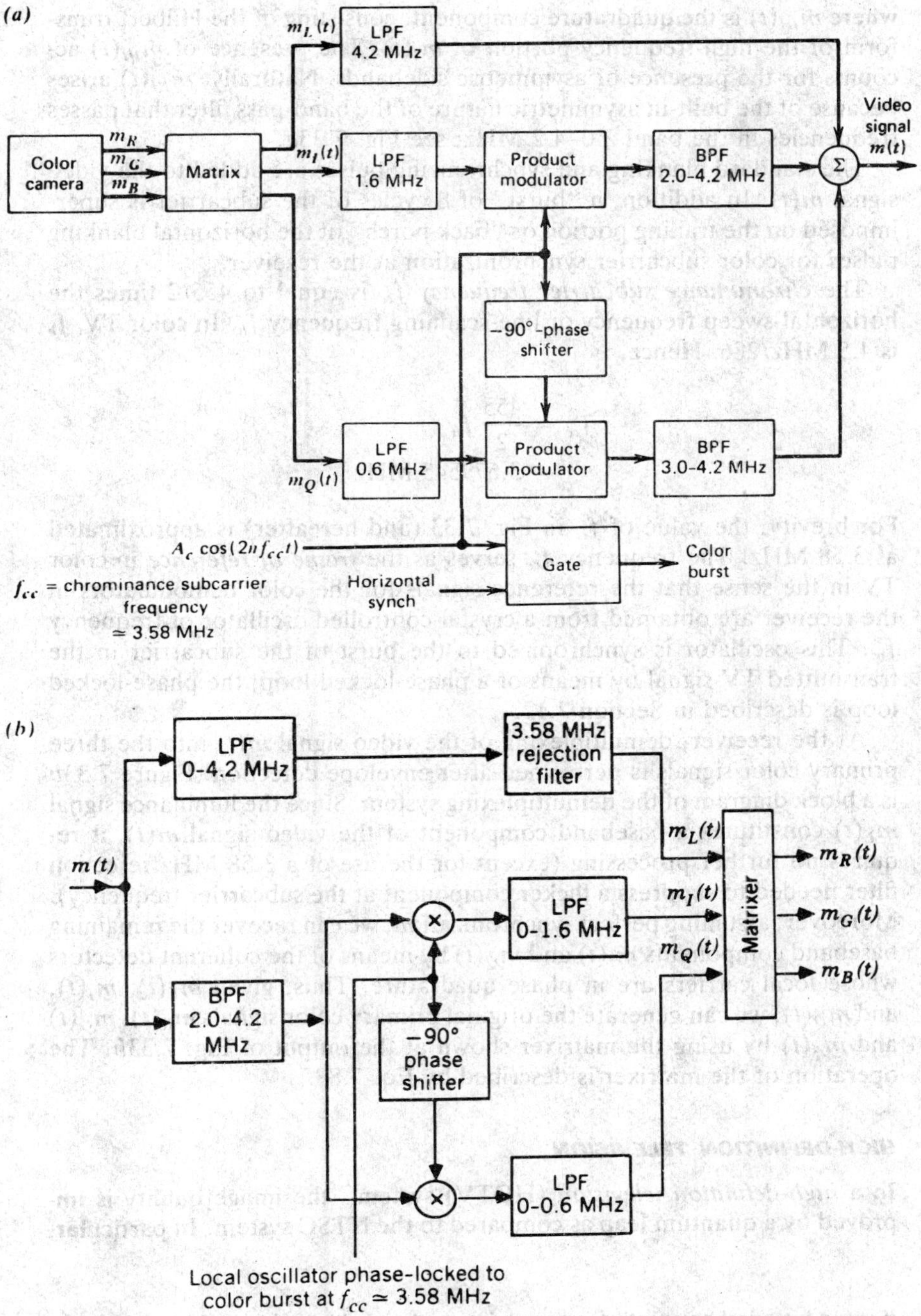

Figure 7.33
(a) *Block diagram of multiplexer in TV transmitter.* (b) *Block diagram of demultiplexer in TV receiver.*

where $\hat{m}_{IH}(t)$ is the quadrature component, consisting of the Hilbert transform of the high-frequency portion of $m_I(t)$. The presence of $\hat{m}_{IH}(t)$ accounts for the presence of asymmetric sidebands. Naturally, $\hat{m}_{IH}(t)$ arises because of the built-in asymmetric nature of the band-pass filter that passes frequencies in the band 2.0–4.2 MHz; see Fig. 7.33*a*.

The standard blanking and synchronizing pulses are added to the video signal $m(t)$. In addition, a "burst" of 8 cycles of the subcarrier is superimposed on the trailing portion or "back porch" of the horizontal blanking pulses for color subcarrier synchronization at the receiver.

The *chrominance subcarrier frequency* f_{cc} is equal to 455/2 times the horizontal-sweep frequency or line-scanning frequency f_h. In color TV, f_h is 4.5 MHz/286. Hence,

$$\begin{aligned} f_{cc} &= \frac{455}{2} f_h \\ &= 3.579545 \text{ MHz} \end{aligned}$$

For brevity, the value of f_{cc} in Fig. 7.33 (and hereafter) is approximated as 3.58 MHz. The frequency f_{cc} serves as the *frame of reference* in color TV in the sense that the reference signals for the color demodulators in the receiver are obtained from a crystal-controlled oscillator of frequency f_{cc}. This oscillator is synchronized to the burst of the subcarrier in the transmitted TV signal by means of a phase-locked loop; the phase-locked loop is described in Section 7.12.

At the receiver, demultiplexing of the video signal $m(t)$ into the three primary color signals is performed after envelope detection. Figure 7.33*b* is a block diagram of the demultiplexing system. Since the luminance signal $m_L(t)$ constitutes a baseband component of the video signal $m(t)$, it requires no further processing (except for the use of a 3.58 MHz rejection filter needed to suppress a flicker component at the subcarrier frequency). Moreover, assuming perfect synchronization, we can recover the remaining baseband components $m_I(t)$ and $m_Q(t)$ by means of the coherent detectors whose local carriers are in phase quadrature. Thus, given $m_L(t)$, $m_I(t)$, and $m_Q(t)$, we can generate the original primary color signals $m_R(t)$, $m_G(t)$ and $m_B(t)$ by using the matrixer shown at the output of Fig. 7.33*b*. The operation of the matrixer is described by Eq. 7.88.

HIGH-DEFINITION TELEVISION

In a *high-definition television* (HDTV) system,[7] the image quality is improved by a quantum leap as compared to the NTSC system. In particular,

[7]From a historical perspective, research into high-definition wide-screen television started in Japan in 1968; the outstanding contributor here is Takashi Fugio. The material presented herein is based on Rzeszewski (1983). This paper and several others on HDTV are reproduced in Rzeszewski (1985).

HDTV offers the following improvements:

1. Improved vertical resolution.
2. Improved horizontal resolution.
3. Less crosstalk between the components of the signal.

The improved image quality together with a large screen size provides the viewer with a feeling of realism and involvement that is unattainable otherwise.

However, for HDTV to be widely acceptable, two requirements are critical. First, there should be *receiver compatibility,* which means that the signal must be able to feed an HDTV and NTSC TV simultaneously and be received on the NTSC receiver with substantially the same picture quality as that achievable by conventional means. Meanwhile, the HDTV receiver realizes the full benefits, including increased resolution. Second,

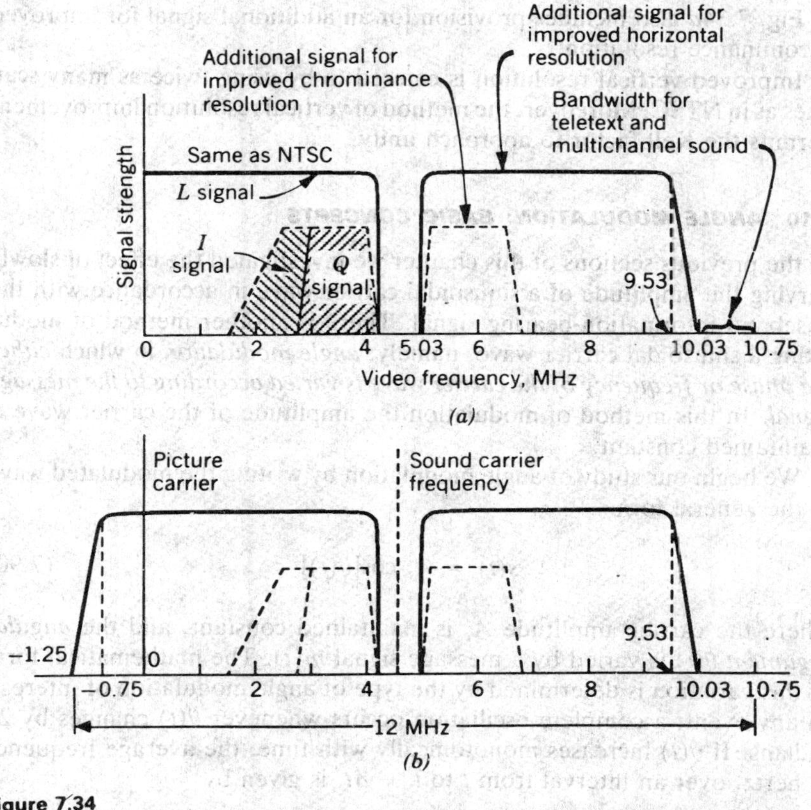

Figure 7.34
Split-luminance and split-chrominance high-definition television (a) *Baseband spectrum.* (b) *Idealized amplitude spectrum of broadcast picture transmission.*

a bandwidth of no more than twice the 6 MHz per channel for NTSC TV broadcast should be required.

Figure 7.34*a* shows the baseband format of a *split-luminance and split-chrominance* (SLSC) type of transmission system that satisfies both of these requirements. It uses a 10-MHz baseband composite signal that can be transmitted as a vestigial sideband modulated wave in a channel bandwidth of 12 MHz. Also, an NTSC receiver (tuned to the lower 6 MHz portion of the 12 MHz spectrum) will operate with the same quality achieved in a conventional system. Figure 7.34*b* shows the baseband version of the amplitude response of an idealized broadcast picture transmission system, measured with respect to the picture carrier frequency.

The composite signal of Fig. 7.34*a* is obtained by starting with a 1050-line scan source of high-bandwidth red, green, blue (R, G, B) signals. These signals are filtered and converted to a 525-line signal by a scan conversion technique that deletes every second line to obtain a 525-line signal suitable for transmission. Improved horizontal resolution is provided for by the use of the second 525-line signal that occupies a frequency range of approximately 5 to 10 MHz in the baseband. The baseband spectrum of Fig. 7.34*a* also includes provision for an additional signal for improved chrominance resolution.

Improved vertical resolution is catered to by using twice as many scan lines as in NTSC. Moreover, the method of vertical resolution improvement permits the Kell factor to approach unity.

7.10 ANGLE MODULATION: BASIC CONCEPTS

In the previous sections of this chapter we investigated the effect of slowly varying the amplitude of a sinusoidal carrier wave in accordance with the baseband information-bearing signal. There is another method of modulating a sinusoidal carrier wave, namely, *angle modulation* in which *either the phase or frequency of the carrier wave is varied according to the message signal.* In this method of modulation the amplitude of the carrier wave is maintained constant.

We begin our study of angle modulation by writing the modulated wave in the general form

$$s(t) = A_c \cos[\theta(t)] \tag{7.90}$$

where the carrier amplitude A_c is maintained constant, and the *angular argument* $\theta(t)$ is varied by a message signal $m(t)$. The mathematical form of this variation is determined by the type of angle modulation of interest. In any event, a complete oscillation occurs whenever $\theta(t)$ changes by 2π radians. If $\theta(t)$ increases monotonically with time, the average frequency in hertz, over an interval from t to $t + \Delta t$, is given by

$$f_{\Delta t}(t) = \frac{\theta(t + \Delta t) - \theta(t)}{2\pi\, \Delta t} \tag{7.91}$$

We define the *instantaneous frequency* of the angle-modulated wave $s(t)$ by

$$
\begin{aligned}
f_i(t) &= \lim_{\Delta t \to 0} f_{\Delta t}(t) \\
&= \lim_{\Delta t \to 0} \left[\frac{\theta(t + \Delta t) - \theta(t)}{2\pi \, \Delta t} \right] \\
&= \frac{1}{2\pi} \frac{d\theta(t)}{dt}
\end{aligned}
\tag{7.92}
$$

Thus, according to Eq. 7.90, we may interpret the angle-modulated wave $s(t)$ as a rotating phasor of length A_c and angle $\theta(t)$. The angular velocity of such a phasor is $d\theta(t)/dt$, in accordance with Eq. 7.92. In the simple case of an unmodulated carrier, the angle $\theta(t)$ is

$$\theta(t) = 2\pi f_c t + \phi_c$$

and the corresponding phasor rotates with a constant angular velocity equal to $2\pi f_c$. The constant ϕ_c is the value of $\theta(t)$ at $t = 0$.

There are an infinite number of ways in which the angle $\theta(t)$ may be varied in some manner with the message signal. However, we will consider only two commonly used methods, phase modulation and frequency modulation, as next defined:

1. *Phase modulation (PM) is that form of angle modulation in which the angular argument $\theta(t)$ is varied linearly with the message signal $m(t)$, as shown by*

$$\theta(t) = 2\pi f_c t + k_p m(t) \tag{7.93}$$

The term $2\pi f_c t$ represents the angular argument of the *unmodulated* carrier, and the constant k_p represents the *phase sensitivity* of the modulator, expressed in radians per volt. This assumes that $m(t)$ is a voltage waveform. For convenience, we have assumed in Eq. 7.93 that the angular argument of the unmodulated carrier is zero at $t = 0$. The phase-modulated wave $s(t)$ is thus described in the time domain by

$$s(t) = A_c \cos[2\pi f_c t + k_p m(t)] \tag{7.94}$$

2. *Frequency modulation (FM) is that form of angle modulation in which the instantaneous frequency $f_i(t)$ is varied linearly with the message signal $m(t)$, as shown by*

$$f_i(t) = f_c + k_f m(t) \tag{7.95}$$

The term f_c represents the frequency of the unmodulated carrier, and the constant k_f represents the *frequency sensitivity* of the modulator, expressed in hertz per volt. This assumes that $m(t)$ is a voltage waveform. Integrating Eq. 7.95 with respect to time and multiplying the result by 2π, we get

$$\theta(t) = 2\pi f_c t + 2\pi k_f \int_0^t m(t)\, dt \tag{7.96}$$

where, for convenience, we have assumed that the angular argument of the unmodulated carrier wave is zero at $t = 0$. The frequency-modulated wave is therefore described in the time domain by

$$s(t) = A_c \cos\left[2\pi f_c t + 2\pi k_f \int_0^t m(t)\, dt\right] \tag{7.97}$$

A consequence of allowing the angular argument $\theta(t)$ to become dependent on the message signal $m(t)$ as in Eq. 7.93 or on its integral as in Eq. 7.96 is that the *zero crossings* of a PM wave or FM wave no longer have a perfect regularity in their spacing; zero crossings refer to the instants of time at which a waveform changes from a negative to a positive value or vice versa. This is one important feature that distinguishes both PM and FM waves from an AM wave. Another important difference is that the envelope of a PM or FM wave is constant (equal to the carrier amplitude), whereas the envelope of an AM wave is dependent on the message signal.

Comparing Eq. 7.94 with 7.97 reveals that an FM wave may be regarded as a PM wave in which the modulating wave is $\int_0^t m(t)\, dt$ in place of $m(t)$. This means that an FM wave can be generated by first integrating $m(t)$ and then using the result as the input to a phase modulator, as in Fig. 7.35*a*. Conversely, a PM wave can be generated by first differentiating $m(t)$ and then using the result as the input to a frequency modulator, as in Fig. 7.35*b*. We may thus deduce all the properties of PM waves from those of FM waves, and vice versa.

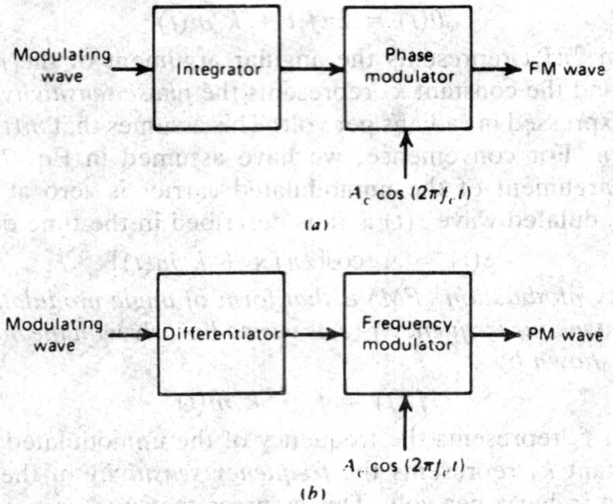

Figure 7.35
The relationship between frequency modulation and phase modulation. (a) *Scheme for generating an FM wave by using a phase modulator.* (b) *Scheme for generating a PM wave by using a frequency modulator.*

EXAMPLE 6 SINUSOIDAL MODULATION

Consider a *sinusoidal modulating wave* $m(t)$, two full cycles of which are plotted in Fig. 7.36*a*. The FM wave produced by this modulating wave is plotted in Fig. 7.36*b*.

To determine the PM wave for $m(t)$, we note that it is the same as the FM wave produced by $dm(t)/dt$, the derivative of $m(t)$ with respect to time (see Fig. 7.35*b*). In Fig. 7.36*c*, we plot the derivative $dm(t)/dt$, which consists of the original sinusoidal modulating wave shifted in phase by 90°. The desired PM wave is plotted in Fig. 7.36*d*.

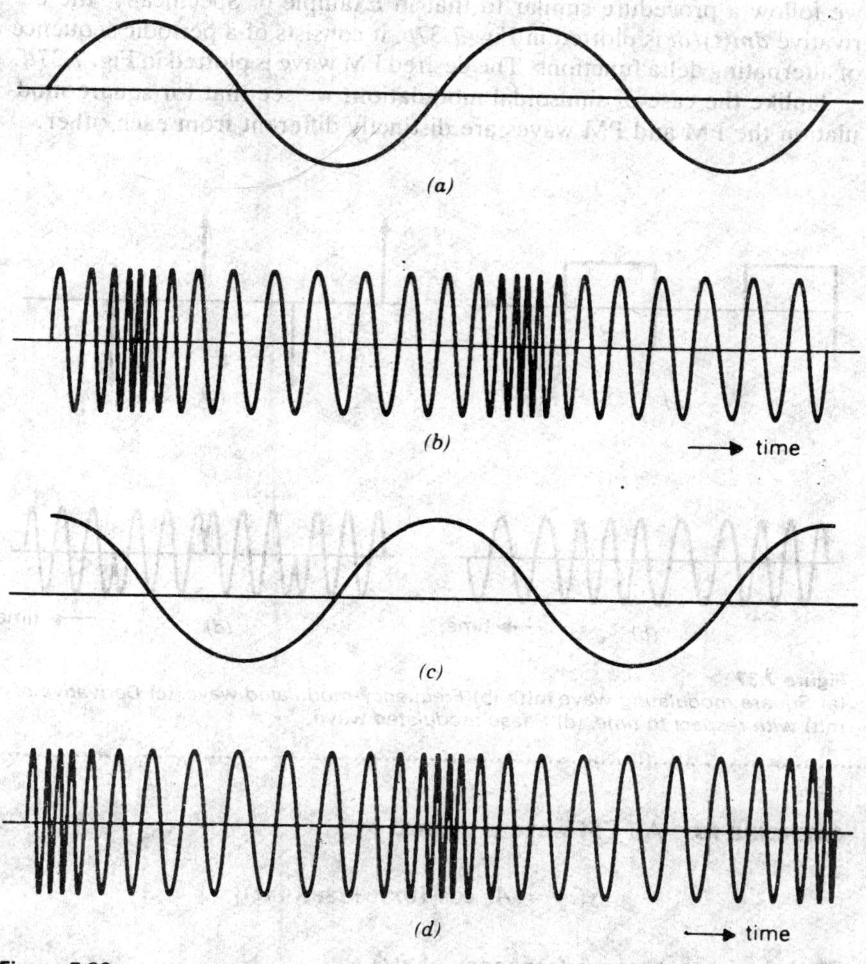

Figure 7.36
(a) *Sinusoidal modulating wave* m(t). (b) *Frequency-modulated wave.* (c) *Derivative of* m(t) *with respect to time.* (d) *Phase-modulated wave.*

From the waveforms of Fig. 7.36, we see that for sinusoidal modulation a distinction between FM and PM waves can be made only by comparing with the actual modulating waves.

EXAMPLE 7 SQUARE MODULATION

Consider next a *square modulating wave* $m(t)$, two full cycles of which are shown plotted in Fig. 7.37*a*. The FM wave produced by this modulating wave is plotted in Fig. 7.37*b*.

To plot the PM wave produced by the square modulating wave $m(t)$, we follow a procedure similar to that in Example 6. Specifically, the derivative $dm(t)/dt$ is plotted in Fig. 7.37*c*; it consists of a periodic sequence of alternating delta functions. The desired PM wave is plotted in Fig. 7.37*d*.

Unlike the case of sinusoidal modulation, we see that for square modulation the FM and PM waves are distinctly different from each other.

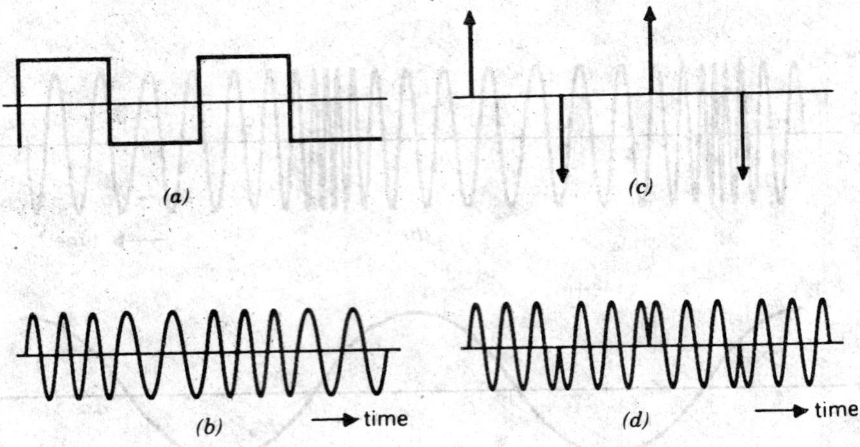

Figure 7.37
(a) *Square modulating wave* m(t). (b) *Frequency-modulated wave.* (c) *Derivative of* m(t) *with respect to time.* (d) *Phase-modulated wave.*

EXERCISE 15 An FM wave is defined by

$$s(t) = A_c \cos[10\pi t + \sin(4\pi t)]$$

Find the instantaneous frequency of $s(t)$.

EXERCISE 16 The square wave of Fig. 7.37*a* is applied to the scheme shown in Fig. 7.35*a*. Plot the waveforms at the input and output of the phase modulator in Fig. 7.35*a*.

7.11 FREQUENCY MODULATION

The FM wave $s(t)$ defined by Eq. 7.97 is a nonlinear function of the modulating wave $m(t)$. Hence, frequency modulation is a *nonlinear modulation process.* Consequently, unlike amplitude modulation, the spectrum of an FM wave is not related in a simple manner to that of the modulating wave. Thus, in order to study the spectral properties of an FM wave, the traditional approach is to start with single-tone modulation and build on the knowledge thus gained.

SINGLE-TONE FREQUENCY MODULATION

Consider then a sinusoidal modulating wave defined by

$$m(t) = A_m \cos(2\pi f_m t) \tag{7.98}$$

The instantaneous frequency of the resulting FM wave equals

$$\begin{aligned} f_i(t) &= f_c + k_f A_m \cos(2\pi f_m t) \\ &= f_c + \Delta f \cos(2\pi f_m t) \end{aligned} \tag{7.99}$$

where

$$\Delta f = k_f A_m \tag{7.100}$$

The quantity Δf is called the *frequency deviation,* representing the maximum departure of the instantaneous frequency of the FM wave from the carrier frequency f_c. A fundamental characteristic of an FM wave is that the frequency deviation Δf is proportional to the amplitude of the modulating wave and is independent of the modulation frequency.

Using Eq. 7.99, the regular argument $\theta(t)$ of the FM wave is obtained as

$$\begin{aligned} \theta(t) &= 2\pi \int_0^t f_i(t)\, dt \\ &= 2\pi f_c t + \frac{\Delta f}{f_m} \sin(2\pi f_m t) \end{aligned} \tag{7.101}$$

The ratio of the frequency deviation Δf to the modulation frequency f_m is commonly called the *modulation index* of the FM wave. We denote it by β, so that we may write

$$\beta = \frac{\Delta f}{f_m} \tag{7.102}$$

and

$$\theta(t) = 2\pi f_c t + \beta \sin(2\pi f_m t) \tag{7.103}$$

From Eq. 7.103 we see that, in a physical sense, the parameter β represents the phase deviation of the FM wave; that is, the maximum departure of the angular argument $\theta(t)$ from the angle $2\pi f_c t$ of the unmodulated carrier.

EXERCISE 17 A sinusoidal modulating wave of amplitude 5 V and frequency 1 kHz is applied to a frequency modulator. The frequency sensitivity of the modulator is 40 Hz/V. The carrier frequency is 100 kHz. Calculate (a) the frequency deviation, and (b) the modulation index.

SPECTRUM ANALYSIS OF SINUSOIDAL FM WAVE

The FM wave for sinusoidal modulation is given by

$$s(t) = A_c \cos[2\pi f_c t + \beta \sin(2\pi f_m t)] \tag{7.104}$$

Using a well-known trigonometric identity, we may expand this relation as

$$s(t) = A_c \cos(2\pi f_c t) \cos[\beta \sin(2\pi f_m t)] - A_c \sin(2\pi f_c t) \sin[\beta \sin(2\pi f_m t)] \tag{7.105}$$

From this expanded form, we see that the in-phase and quadrature components of the FM wave $s(t)$ for the case of sinusoidal modulation are as follows:

$$s_I(t) = A_c \cos[\beta \sin(2\pi f_m t)] \tag{7.106}$$

$$s_Q(t) = A_c \sin[\beta \sin(2\pi f_m t)] \tag{7.107}$$

Hence, the complex envelope of the FM wave equals

$$\begin{aligned} \tilde{s}(t) &= s_I(t) + j s_Q(t) \\ &= A_c \exp[j\beta \sin(2\pi f_m t)] \end{aligned} \tag{7.108}$$

The complex envelope $\tilde{s}(t)$ retains complete information about the modulation process. Indeed, we may readily express the FM wave $s(t)$ in terms of the complex envelope $\tilde{s}(t)$ by writing

$$\begin{aligned} s(t) &= \mathrm{Re}[A_c \exp(j2\pi f_c t + j\beta \sin(2\pi f_m t))] \\ &= \mathrm{Re}[\tilde{s}(t) \exp(j2\pi f_c t)] \end{aligned} \tag{7.109}$$

From Eq. 7.108 we see that the complex envelope is a periodic function of time, with a fundamental frequency equal to the modulation frequency f_m. We may therefore expand $\tilde{s}(t)$ in the form of a complex Fourier series as follows

$$\tilde{s}(t) = \sum_{n=-\infty}^{\infty} c_n \exp(j2\pi n f_m t) \tag{7.110}$$

where the complex Fourier coefficient c_n equals

$$\begin{aligned} c_n &= f_m \int_{-1/2f_m}^{1/2f_m} \tilde{s}(t) \exp(-j2\pi n f_m t)\, dt \\ &= f_m A_c \int_{-1/2f_m}^{1/2f_m} \exp[j\beta \sin(2\pi f_m t) - j2\pi n f_m t]\, dt \end{aligned} \tag{7.111}$$

For convenience, we define the variable

$$x = 2\pi f_m t \tag{7.112}$$

in terms of which we may rewrite Eq. 7.111 as

$$c_n = \frac{A_c}{2\pi} \int_{-\pi}^{\pi} \exp[j(\beta \sin x - nx)]\, dx \tag{7.113}$$

The integral on the right side of Eq. 7.113 is recognized as the nth order *Bessel function of the first kind* and argument β (see Appendix B). This function is commonly denoted by the symbol $J_n(\beta)$; that is,

$$J_n(\beta) = \frac{1}{2\pi} \int_{-\pi}^{\pi} \exp[j(\beta \sin x - nx)]\, dx \tag{7.114}$$

Hence, we may rewrite Eq. 7.113 as

$$c_n = A_c J_n(\beta) \tag{7.115}$$

Substituting Eq. 7.115 in 7.110, we get, in terms of the Bessel function $J_n(\beta)$, the following expansion for the complex envelope of the FM wave:

$$\tilde{s}(t) = A_c \sum_{n=-\infty}^{\infty} J_n(\beta) \exp(j2\pi n f_m t) \tag{7.116}$$

Next, substituting Eq. 7.116 in 7.109, we get

$$s(t) = A_c \operatorname{Re}\left[\sum_{n=-\infty}^{\infty} J_n(\beta) \exp[j2\pi(f_c + nf_m)t]\right] \tag{7.117}$$

Interchanging the order of summation and evaluating the real part of the right side of Eq. 7.117, we get

$$s(t) = A_c \sum_{n=-\infty}^{\infty} J_n(\beta) \cos[2\pi(f_c + nf_m)t] \tag{7.118}$$

This is the desired form for the Fourier series representation of the single-tone FM wave $s(t)$ for an arbitrary value of β. The discrete spectrum of $s(t)$ is obtained by taking the Fourier transforms of both sides of Eq. 7.118; thus

$$S(f) = \frac{A_c}{2} \sum_{n=-\infty}^{\infty} J_n(\beta)[\delta(f - f_c - nf_m) + \delta(f + f_c + nf_m)] \tag{7.119}$$

In Fig. 7.38 we have plotted the Bessel function $J_n(\beta)$ versus the modulation index β for $n = 0, 1, 2, 3, 4$. These plots show that for fixed n, $J_n(\beta)$ alternates between positive and negative values for increasing β and that $|J_n(\beta)|$ approaches zero as β approaches infinity. Note also that for fixed β, we have

$$J_{-n}(\beta) = \begin{cases} J_n(\beta), & n \text{ even} \\ -J_n(\beta), & n \text{ odd} \end{cases} \tag{7.120}$$

Accordingly, we need only plot or tabulate $J_n(\beta)$ for positive values of order n.

From Eqs. 7.97 and 7.118, we deduce the following properties of FM waves:

PROPERTY 1: NARROW-BAND FM

For small values of the modulation index β compared to one radian, the FM wave assumes a narrow-band form consisting essentially of a carrier, an upper side-frequency component, and a lower side-frequency component.

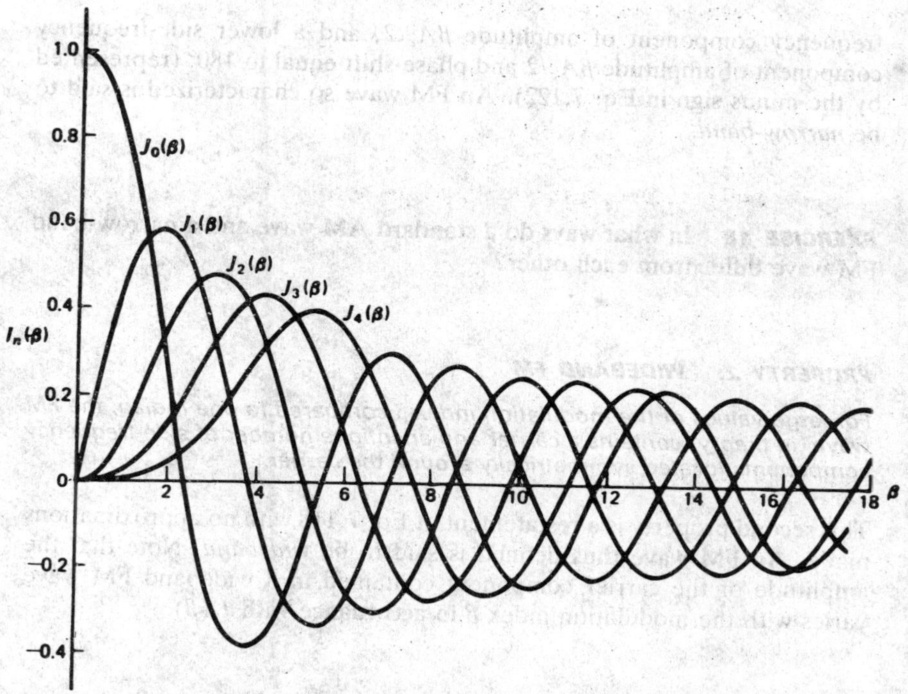

Figure 7.38
Plots of Bessel functions of the first kind.

This property follows from the fact that for small values of β, we have

$$\begin{aligned} J_0(\beta) &\simeq 1 \\ J_1(\beta) &\simeq \frac{\beta}{2} \\ J_n(\beta) &\simeq 0, \quad n > 1 \end{aligned} \tag{7.121}$$

The approximations indicated in Eqs. 7.121 are closely justified for values of the modulation index defined by $\beta \leqslant 0.3$ rad. Thus, substituting Eqs. 7.121 in 7.118, we get

$$\begin{aligned} s(t) \simeq A_c \cos(2\pi f_c t) &+ \frac{\beta A_c}{2} \cos[2\pi(f_c + f_m)t] \\ &- \frac{\beta A_c}{2} \cos[2\pi(f_c - f_m)t] \end{aligned} \tag{7.122}$$

This equation shows that for small β, the FM wave $s(t)$ may be closely approximated by the sum of a carrier of amplitude A_c, an upper side-frequency component of amplitude $\beta A_c/2$, and a lower side-frequency component of amplitude $\beta A_c/2$ and phase-shift equal to 180° (represented by the minus sign in Eq. 7.122). An FM wave so characterized is said to be *narrow-band*.

EXERCISE 18 In what ways do a standard AM wave and a narrow-band FM wave differ from each other?

PROPERTY 2: WIDEBAND FM

For large values of the modulation index β compared to one radian, the FM wave (in theory) contains a carrier and an infinite number of side-frequency components located symmetrically around the carrier.

This second property is a restatement of Eq. 7.118 with no approximations made. An FM wave thus defined is said to be *wideband*. Note that the amplitude of the carrier component contained in a wideband FM wave varies with the modulation index β in accordance with $J_0(\beta)$.

EXERCISE 19 In what ways do a standard AM wave and a wideband FM wave differ from each other?

PROPERTY 3: CONSTANT AVERAGE POWER

The envelope of an FM wave is constant, so that the average power of such a wave dissipated in a 1-ohm resistor is also constant.

This property follows directly from the definition given in Eq. 7.97 for an FM wave. Specifically, the FM wave $s(t)$ defined in Eq. 7.97 has a constant envelope equal to A_c. Accordingly, the average power dissipated by $s(t)$ in a 1-ohm resistor is given by

$$P = \frac{1}{2} A_c^2 \tag{7.123}$$

This result may also be derived from Eq. 7.118. In particular, we note from the series expansion of Eq. 7.118 that the average power of a single-tone FM wave $s(t)$ may be expressed in the form of a corresponding series as:

$$P = \frac{1}{2} A_c^2 \sum_{n=-\infty}^{\infty} J_n^2(\beta) \tag{7.124}$$

Next, we note that (see Appendix B)

$$\sum_{n=-\infty}^{\infty} J_n^2(\beta) = 1 \tag{7.125}$$

Thus, substituting Eq. 7.125 in 7.124, we get the result given in Eq. 7.123.

EXAMPLE 8

We wish to investigate the ways in which variations in the amplitude and frequency of a sinusoidal modulating wave affect the spectrum of the FM wave. Consider first the case when the frequency of the modulating wave is fixed, but its amplitude is varied, producing a corresponding variation in the frequency deviation Δf. Thus, keeping the modulation frequency f_m fixed, we find that the amplitude spectrum of the resulting FM wave is as plotted in Fig. 7.39 for $\beta = 1$, 2, and 5. In this diagram we have

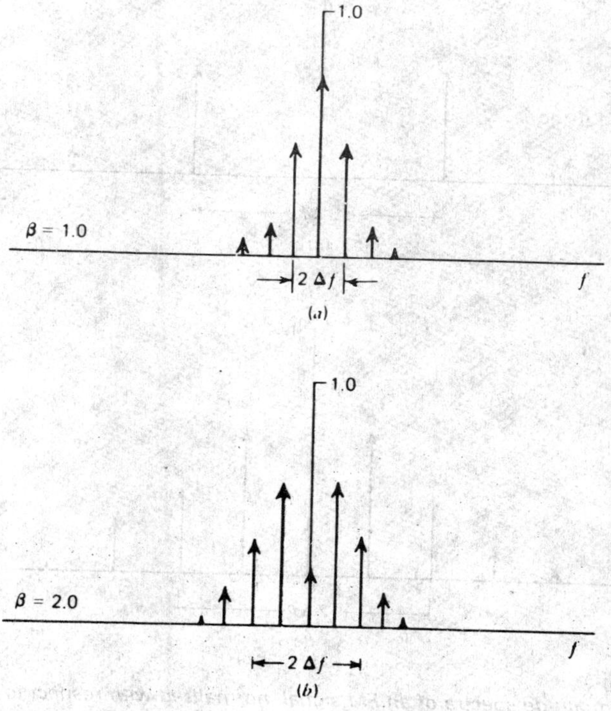

Figure 7.39
Discrete amplitude spectra of an FM signal, normalized with respect to the carrier amplitude, for the case of sinusoidal modulation of fixed frequency and varying amplitude. Only the spectra for positive frequencies are shown.

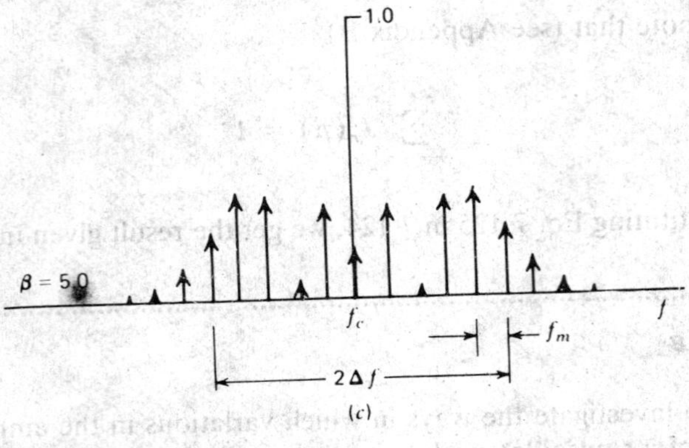

Figure 7.39 *(continued)*

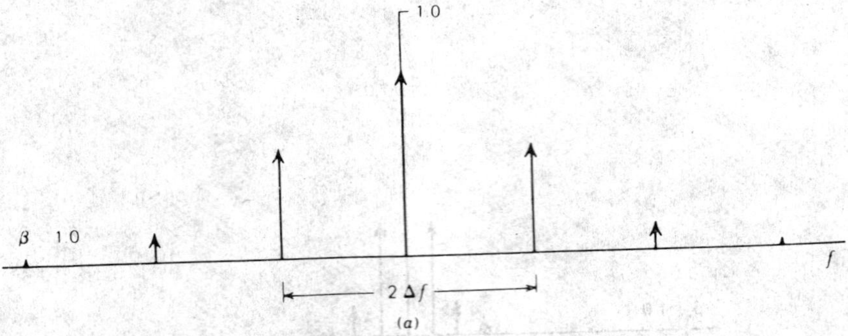

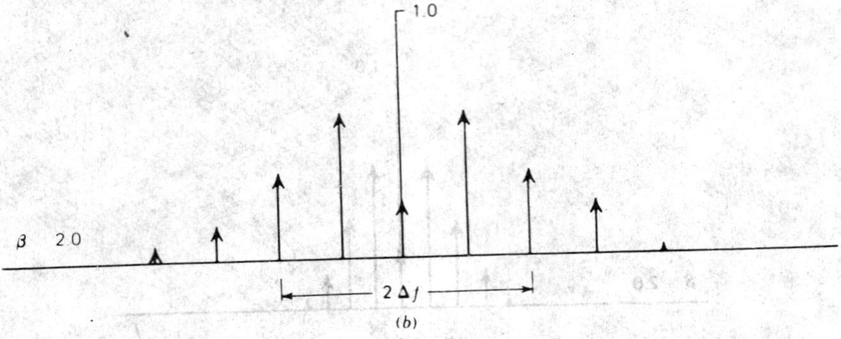

Figure 7.40
Discrete amplitude spectra of an FM signal, normalized with respect to the carrier amplitude, for the case of sinusoidal modulation of varying frequency and fixed amplitude. Only the spectra for positive frequencies are shown.

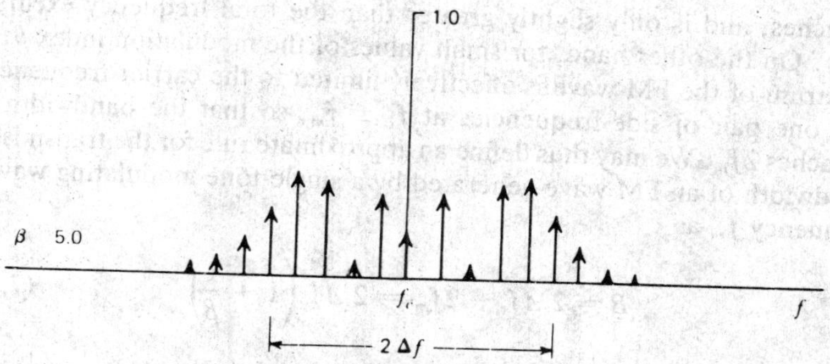

(c)

Figure 7.40 *(continued)*

normalized the spectrum with respect to the unmodulated carrier amplitude.

Consider next the case when the amplitude of the modulating wave is fixed; that is, the frequency deviation Δf is maintained constant and the modulation frequency f_m is varied. In this case we find that the amplitude spectrum of the resulting FM wave is as plotted in Fig. 7.40 for $\beta = 1, 2,$ and 5. We see that when Δf is fixed and β is increased, we have an increasing number of spectral lines crowding into a fixed frequency interval defined by $f_c - \Delta f < f < f_c + \Delta f$. That is, when β approaches infinity, the bandwidth of the FM wave approaches the limiting value of $2\Delta f$.

EXERCISE 20 Expand the discrete amplitude spectra shown in Figs. 7.39 and 7.40 by including the spectrum of an FM wave with $\beta = 0.2$.

TRANSMISSION BANDWIDTH OF FM WAVES

In theory, an FM wave contains an infinite number of side-frequencies so that the bandwidth required to transmit such a signal is similarly infinite in extent. In practice, however, we find that the FM wave is effectively limited to a finite number of significant side-frequencies compatible with a specified amount of distortion. We may therefore specify an effective bandwidth required for the transmission of an FM wave. Consider first the case of an FM wave generated by a single-tone modulating wave of frequency f_m. In such an FM wave, the side-frequencies that are separated from the carrier frequency f_c by an amount greater than the frequency deviation Δf decrease rapidly toward zero, so that the bandwidth always exceeds the total frequency excursion, but nevertheless is limited. Specifically, for large values of the modulation index β, the bandwidth approaches, and is only slightly greater than the total frequency excursion $2\,\Delta f$. On the other hand, for small values of the modulation index β, the spectrum of the FM wave is effectively limited to the carrier frequency f_c and one pair of side-frequencies at $f_c \pm f_m$, so that the bandwidth approaches $2f_m$. We may thus define an approximate rule for the transmission bandwidth of an FM wave generated by a single-tone modulating wave of frequency f_m as

$$B \simeq 2\,\Delta f + 2f_m = 2\,\Delta f\left(1 + \frac{1}{\beta}\right) \tag{7.126}$$

This relation is known as *Carson's rule.*

For a more accurate assessment of the bandwidth requirement of an FM wave, we may use a definition based on retaining the maximum number of significant side-frequencies with amplitudes all greater than some selected value. A convenient choice for this value is 1% of the unmodulated carrier amplitude. *We may thus define the 99 percent bandwidth of an FM wave as the separation between the two frequencies beyond which none of the side-frequencies is greater than 1% of the carrier amplitude obtained when the modulation is removed.* That is, we define the transmission bandwidth as $2n_{\max}f_m$, where f_m is the modulation frequency and $n_{\max}$ is the maximum value of the integer n that satisfies the requirement $|J_n(\beta)| > 0.01$. The value of $n_{\max}$ varies with the modulation index β and can be determined readily from tabulated values of the Bessel function $J_n(\beta)$. Table 7.2 shows the total number of significant side-frequencies (including both the upper and lower side-frequencies) for different values of β, calculated on the 1% basis just explained. The transmission bandwidth B calculated using this procedure can be presented in the form of a universal curve by normalizing it with respect to the frequency deviation Δf, and then plotting it versus

TABLE 7.2

Modulation index β	Number of significant side-frequencies $2n_{\max}$
0.1	2
0.3	4
0.5	4
1.0	6
2.0	8
5.0	16
10.0	28
20.0	50
30.0	70

β. This curve is shown in Fig. 7.41, which is drawn as a best fit through the set of points obtained by using Table 7.2. In Fig. 7.41 we note that as the modulation index β is increased, the bandwidth occupied by the significant side-frequencies drops toward that over which the carrier frequency actually deviates. This means that small values of the modulation index β are relatively more extravagant in transmission bandwidth than are the larger values of β.

Consider next an arbitrary modulating wave $m(t)$ with its highest frequency component denoted by W. The bandwidth required to transmit an FM wave generated by this modulating wave is estimated by using a worst-case tone-modulation analysis. Specifically, we first determine the so-called *deviation ratio D*, defined as the ratio of the frequency deviation Δf, which corresponds to the maximum possible amplitude of the modulating wave $m(t)$, to the highest modulation frequency W; these conditions represent the extreme cases possible. *The deviation ratio D plays the same role for nonsinusoidal modulation that the modulation index β plays for the case of sinusoidal modulation.* Then, replacing β by D and replacing f_m by W, we use Carson's rule given by Eq. 7.126 or the universal curve of Fig. 7.41 to obtain a value for the transmission bandwidth of the FM wave. From a practical viewpoint, Carson's rule somewhat underestimates the bandwidth requirement of an FM system, whereas using the universal curve of Fig. 7.41 yields a somewhat conservative result. Thus the choice of a trans-

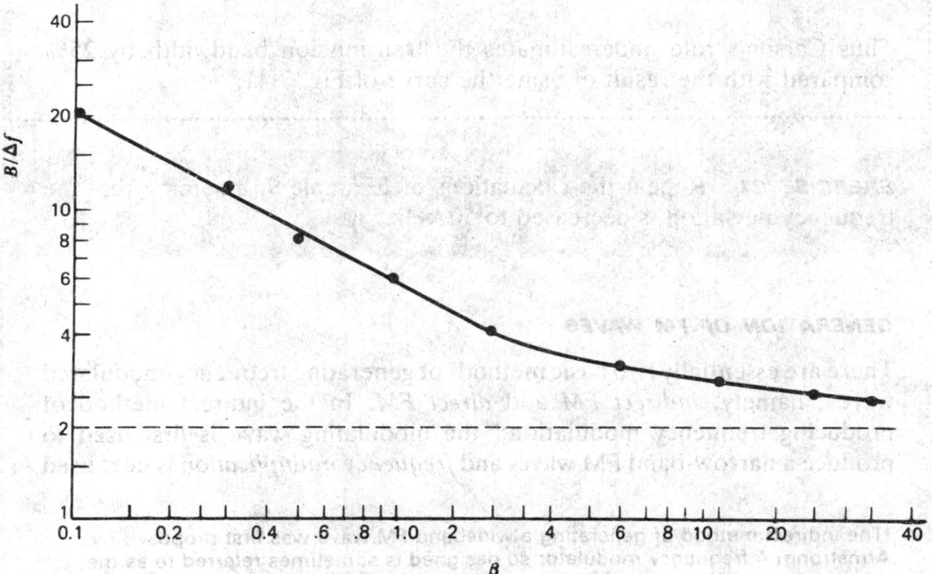

Figure 7.41
Universal curve for evaluating the 99% bandwidth of an FM wave.

mission bandwidth that lies between the bounds provided by these two rules of thumb is acceptable for most practical purposes.

EXAMPLE 9

In North America, the maximum value of frequency deviation Δf is fixed at 75 kHz for commercial FM broadcasting by radio. If we take the modulation frequency $W = 15$ kHz, which is typically the maximum audio frequency of interest in FM transmission, we find that the corresponding value of the deviation ratio is

$$D = \frac{75}{15} = 5$$

Using Carson's rule of Eq. 7.126, replacing β by D and replacing f_m by W, the approximate value of the transmission bandwidth of the FM wave is obtained as

$$B = 2(75 + 15) = 180 \text{ kHz}$$

On the other hand, use of the curve of Fig. 7.41 gives the transmission bandwidth of the FM wave to be

$$B = 3.2\,\Delta f = 3.2 \times 75 = 240 \text{ kHz}$$

Thus Carson's rule underestimates the transmission bandwidth by 25% compared with the result of using the curve of Fig. 7.41.

EXERCISE 21 Repeat the calculations of Example 9, assuming that the frequency deviation is decreased to 50 kHz.

GENERATION OF FM WAVES

There are essentially two basic methods of generating frequency-modulated waves, namely, *indirect FM* and *direct FM*. In the indirect method of producing frequency modulation,[8] the modulating wave is first used to produce a narrow-band FM wave, and *frequency multiplication* is next used

[8] The indirect method of generating a wideband FM wave was first proposed by Armstrong. A frequency modulator so designed is sometimes referred to as the *Armstrong modulator;* see Armstrong (1936). Armstrong was also the first to recognize the noise-cleaning properties of frequency modulation.

to increase the frequency deviation to the desired level. On the other hand, in the direct method of producing frequency modulation the carrier frequency is directly varied in accordance with the incoming message signal. In this subsection, we describe the important features of both methods.

Indirect FM Consider first the generation of a narrow-band FM wave. To do this, we begin with the expression for an FM wave $s_1(t)$ for the general case of a modulating wave $m(t)$, which is written in the form

$$s_1(t) = A_1 \cos[2\pi f_1 t + \phi_1(t)] \tag{7.127}$$

where f_1 is the carrier frequency and A_1 is the carrier amplitude. The angular argument $\phi_1(t)$ of $s_1(t)$ is related to $m(t)$ by

$$\phi_1(t) = 2\pi k_1 \int_0^t m(t)\, dt \tag{7.128}$$

where k_1 is the frequency sensitivity of the modulator. Provided that the angle $\phi_1(t)$ is small compared to one radian for all t, we may use the following *approximations:*

$$\cos[\phi(t)] \simeq 1 \tag{7.129}$$

$$\sin[\phi(t)] \simeq \phi(t) \tag{7.130}$$

Correspondingly, we may approximate Eq. 7.127 as follows

$$\begin{aligned} s_1(t) &\simeq A_1 \cos(2\pi f_1 t) - A_1 \sin(2\pi f_1 t)\phi_1(t) \\ &= A_1 \cos(2\pi f_1 t) - 2\pi k_1 A_1 \sin(2\pi f_1 t) \int_0^t m(t)\, dt \end{aligned} \tag{7.131}$$

Equation 7.131 defines a *narrow-band FM wave.* Indeed, we may use this equation to set up the scheme shown in Fig. 7.42*a* for the generation of a narrow-band FM wave; the scaling factor $2\pi k_1$ is taken care of by the product modulator. Moreover, bearing in mind the relationship that exists between frequency modulation and phase modulation (see Fig. 7.35), we see that the part of the frequency modulator that lies inside the dashed rectangle in Fig. 7.42*a* represents a *narrow-band phase modulator.*

The modulated wave produced by the narrow-band modulator of Fig. 7.42*a* differs from an *ideal* FM wave in two respects:

1. The envelope contains a *residual* amplitude modulation and, therefore, varies with time.

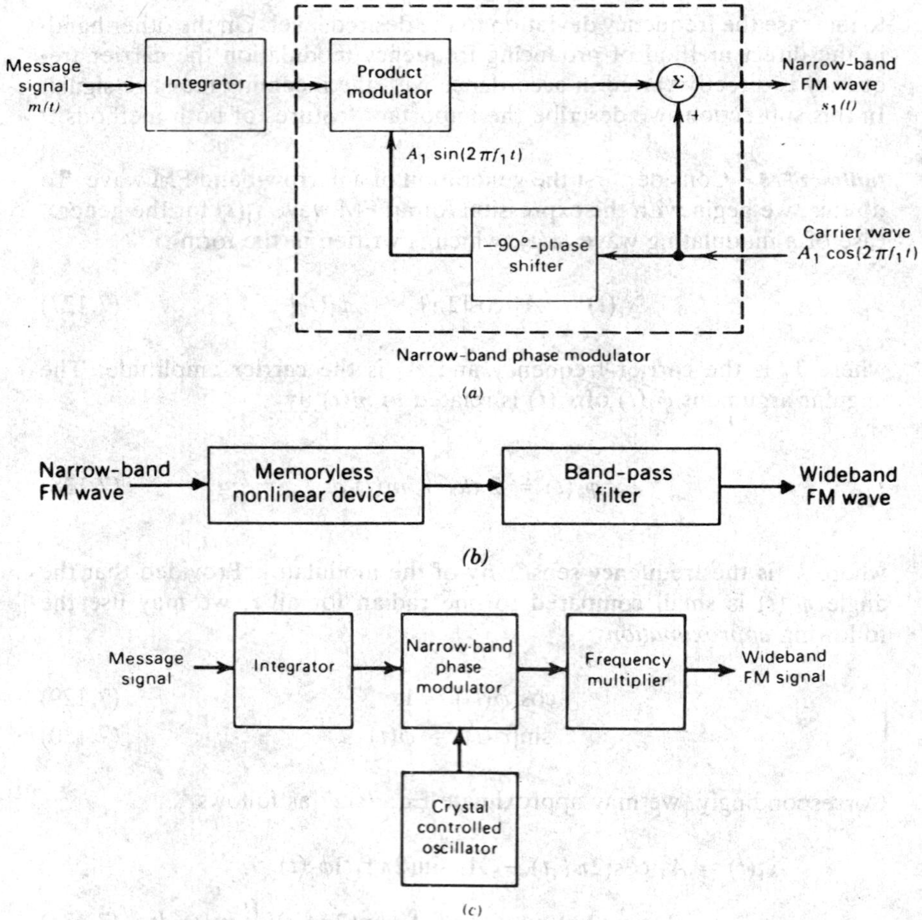

Figure 7.42
Block diagrams for (a) *narrow-band frequency modulator,* (b) *frequency multiplier, and* (c) *wideband frequency modulator.*

2. For a sinusoidal modulating wave, the phase of the FM wave contains *harmonic distortion* in the form of third- and higher-order harmonics of the modulation frequency f_m.

However, by restricting the modulation index to $\beta \leqslant 0.3$ rad, the effects of residual AM and harmonic PM are limited to negligible levels.

The next step in the indirect FM method is that of frequency multiplication. Basically, a *frequency multiplier* consists of a *nonlinear device* (e.g., diode or transistor) followed by a *band-pass filter,* as in Fig. 7.42*b*. The nonlinear device is assumed to be *memoryless,* which means that there is no energy storage. In general, a memoryless nonlinear device is represented by the input–output relation[9]

$$s_2(t) = a_1 s_1(t) + a_2 s_1^2(t) + \cdots + a_n s_1^n(t) \qquad (7.132)$$

where $a_1, a_2, \ldots, a_n$ are constant coefficients. Substituting Eq. 7.131 in 7.132, expanding and then collecting terms, we find that the output $s_2(t)$ has a dc component and n frequency-modulated waves with carrier frequencies $f_1, 2f_1, \ldots, nf_1$ and frequency deviations $\Delta f_1, 2\Delta f_1, \ldots, n\Delta f_1$, respectively. The value of Δf_1 is determined by the frequency sensitivity k_1 of the narrow-band frequency modulator and the maximum amplitude of the modulating wave $m(t)$. We now see the motivation for using the band-pass filter in Fig. 7.42*b*. Specifically, the filter is designed with two aims in mind:

1. To pass the FM wave centered at the carrier frequency nf_1 and with frequency deviation $n\,\Delta f_1$.
2. To suppress all other FM spectra.

Thus, connecting the narrow-band frequency modulator and the frequency multiplier as depicted in Fig. 7.42*c*, we may generate a wideband FM wave $s(t)$ with carrier frequency $f_c = nf_1$ and frequency deviation $\Delta f = n\,\Delta f_1$, as desired. Specifically, we may write

$$s(t) = A_c \cos\left[2\pi f_c t + 2\pi k_f \int_0^t m(t)\,dt\right] \qquad (7.133)$$

where

$$k_f = nk_1 \qquad (7.134)$$

In other words, the wideband frequency modulator of Fig. 7.42*c* has a frequency sensitivity n times that of the narrow-band frequency modulator of Fig. 7.42*a*, where n is the frequency multiplication ratio. In Fig. 7.42*c* we show a crystal-controlled oscillator as the source of carrier; this is done for frequency stability.

[9] *Nonlinearities,* in one form or another, are present in all electrical networks. There are two basic forms of nonlinearity to consider:

1. The nonlinearity is said to be *strong* when it is introduced intentionally and in a controlled manner for some specific application. Examples of strong nonlinearity include frequency multipliers, amplitude limiters, and square-law modulators.
2. The nonlinearity is said to be *weak* when a linear performance is desired, and any nonlinearities are viewed as parasitic in nature. The effect of such weak nonlinearities is to limit the useful signal levels in a system. Thus, weak nonlinearities become an important design consideration; see Problem 40.

EXERCISE 22 Consider a frequency multiplier that uses a square-law device defined by

$$s_2(t) = a_1 s_1(t) + a_2 s_1^2(t)$$

Specify the midband frequency and bandwidth of the band-pass filter used in the frequency multiplier for the resulting frequency deviation to be twice that at the input of the nonlinear device.

EXERCISE 23 An FM wave with a frequency deviation of 10 kHz at a modulation frequency of 5 kHz is applied to two frequency multipliers connected in cascade. The first multiplier doubles the frequency and the second multiplier triples the frequency. Determine the frequency deviation and the modulation index of the FM wave obtained at the second multiplier output. What is the frequency separation of the adjacent side-frequencies of this FM wave?

EXAMPLE 10

Figure 7.43 shows the simplified block diagram of a typical FM transmitter (based on the indirect method) used to transmit audio signals containing frequencies in the range 100 Hz to 15 kHz. The narrow-band phase modulator is supplied with a carrier wave of frequency $f_1 = 0.1$ MHz by a crystal-controlled oscillator. The desired FM wave at the transmitter output has a carrier frequency $f_c = 100$ MHz and frequency deviation $\Delta f = 75$ kHz.

In order to limit the harmonic distortion produced by the narrow-band phase modulator, we restrict the modulation index β_1 to a maximum value of 0.3 rad. Suppose then $\beta_1 = 0.2$ rad.

From Eq. 7.102, we see that for sinusoidal modulation, the frequency deviation equals the modulation index multiplied by the modulation frequency. Hence, for a fixed modulation index, the lowest modulation frequencies will limit the frequency deviation at the narrowband phase modulator output. Thus, with $\beta_1 = 0.2$, the 100-Hz modulation frequencies will limit the frequency deviation Δf_1 to 20 Hz.

To produce a frequency deviation of $\Delta f = 75$ kHz at the FM transmitter output, the use of frequency multiplication is required. Specifically, with $\Delta f_1 = 20$ Hz and $\Delta f = 75$ kHz, we require a total frequency multiplication ratio of 3750. However, using a straight frequency multiplication equal to this value would produce a much higher carrier frequency at the transmitter output than the desired value of 100 MHz. To generate an FM wave having

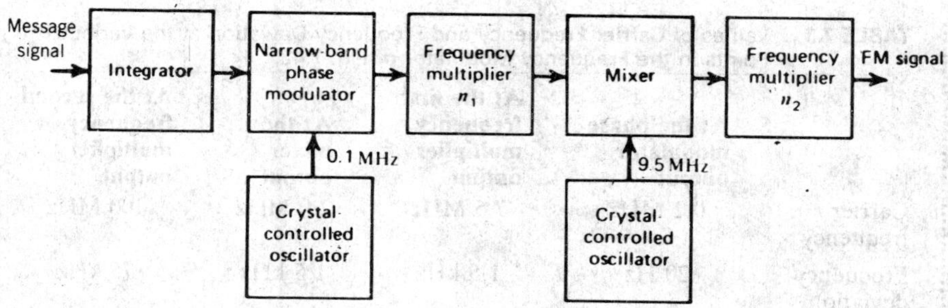

Figure 7.43
Block diagram of the wideband frequency modulator for Example 10.

both the desired frequency deviation and carrier frequency, we therefore need to use a *two-stage frequency multiplier* with an intermediate stage of frequency translation, as illustrated in Fig. 7.43.

Let n_1 and n_2 denote the respective frequency multiplication ratios, so that

$$n_1 n_2 = \frac{\Delta f}{\Delta f_1} = \frac{75{,}000}{20} = 3750 \qquad (7.135)$$

The carrier frequency at the first frequency multiplier output is translated downward in frequency to $(f_2 - n_1 f_1)$ by mixing it with a sinusoidal wave of frequency $f_2 = 9.5$ MHz, which is supplied by a second crystal-controlled oscillator. However, the carrier frequency at the input of the second frequency multiplier is equal to f_c/n_2. Equating these two frequencies, we get

$$f_2 - n_1 f_1 = \frac{f_c}{n_2}$$

Hence, with $f_1 = 0.1$ MHz, $f_2 = 9.5$ MHz, and $f_c = 100$ MHz, we have

$$9.5 - 0.1 n_1 = \frac{100}{n_2} \qquad (7.136)$$

Solving Eqs. 7.135 and 7.136 for n_1 and n_2, we obtain

$$n_1 = 75$$
$$n_2 = 50$$

Using these frequency multiplication ratios, we get the set of values indicated in Table 7.3.

TABLE 7.3 Values of Carrier Frequency and Frequency Deviation at the Various Points in the Frequency Modulator of Fig. 7.43.

	At the phase modulator output	At the first frequency multiplier output	At the mixer output	At the second frequency multiplier output
Carrier frequency	0.1 MHz	7.5 MHz	2.0 MHz	100 MHz
Frequency deviation	20 Hz	1.5 kHz	1.5 kHz	75 kHz

Direct FM In the *direct method* of FM generation, the instantaneous frequency of the carrier wave is varied directly in accordance with the message signal by means of a device known as a *voltage-controlled oscillator.* One way of implementing such a device is to use a sinusoidal oscillator having a relatively high-Q frequency-determining network and to control the oscillator by incremental variation of the reactive components. An example of this scheme is shown in Fig. 7.44, showing a *Hartley oscillator.* We assume that the capacitive component of the frequency-determining network consists of a fixed capacitor shunted by a voltage-variable capacitor. The resultant capacitance is represented by $C(t)$ in Fig. 7.44. A voltage-variable capacitor, commonly called a *varactor* or *varicap,* is one whose capacitance depends on the voltage applied across its electrodes. The variable-voltage capacitance may be obtained, for example, by using a *p-n* junction diode that is biased in the reverse direction; the larger the reverse voltage applied to such a diode, the smaller the transition capacitance of the diode. The frequency of oscillation of the Hartley oscillator of Fig. 7.44 is given by

$$f_i(t) = \frac{1}{2\pi\sqrt{(L_1 + L_2)C(t)}} \tag{7.137}$$

where $C(t)$ is the total capacitance of the fixed capacitor and the variable-voltage capacitor, and L_1 and L_2 are the two inductances in the frequency-determining network. Assume that for a modulating wave $m(t)$ the capacitance $C(t)$ is expressed as follows

$$C(t) = C_0 - k_c m(t) \tag{7.138}$$

where C_0 is the total capacitance in the absence of modulation, and k_c is the variable capacitor's sensitivity to voltage change. Substituting Eq. 7.138

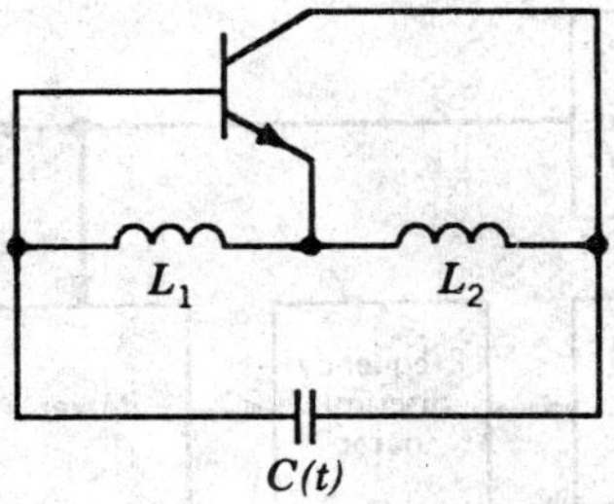

Figure 7.44
Hartley oscillator.

in 7.137, we get

$$f_i(t) = f_0\left[1 - \frac{k_c}{C_0}m(t)\right]^{-1/2} \tag{7.139}$$

where f_0 is the *unmodulated frequency of oscillation:*

$$f_0 = \frac{1}{2\pi\sqrt{C_0(L_1 + L_2)}} \tag{7.140}$$

Provided that the maximum change in capacitance produced by the modulating wave is small compared with the unmodulated capacitance C_0, we may approximate Eq. 7.139 as follows

$$f_i(t) \simeq f_0\left[1 + \frac{k_c}{2C_0}m(t)\right] \tag{7.141}$$

Define

$$k_f = \frac{f_0 k_c}{2C_0} \tag{7.142}$$

We then obtain the following relation for the instantaneous frequency of the oscillator:

$$f_i(t) \simeq f_0 + k_f m(t) \tag{7.143}$$

where k_f is the resultant frequency sensitivity of the modulator, defined by Eq. 7.142.

An FM transmitter using the direct method as described herein, however, has the disadvantage that the carrier frequency is not obtained from

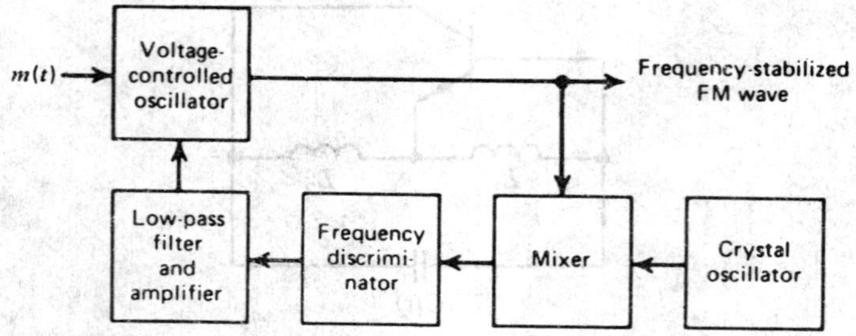

Figure 7.45
A feedback scheme for the frequency stabilization of a frequency modulator.

a highly stable oscillator. It is therefore necessary, in practice, to provide some auxiliary means by which a very stable frequency generated by a crystal will be able to control the carrier frequency. One method of effecting this control is illustrated in Fig. 7.45. The output of the FM generator is applied to a mixer together with the output of a crystal-controlled oscillator, and the difference frequency term is extracted. The mixer output is next applied to a frequency discriminator and then low-pass filtered. A frequency discriminator is a device whose output voltage has an instantaneous amplitude that is proportional to the instantaneous frequency of the FM wave applied to its input; this device is described later in the section. When the FM transmitter has exactly the correct carrier frequency, the low-pass filter output is zero. However, deviations of the transmitter carrier frequency from its assigned value will cause the frequency discriminator–filter combination to develop a dc output voltage with a polarity determined by the sense of the transmitter frequency drift. This dc voltage, after suitable amplification, is applied to the voltage-controlled oscillator of the FM transmitter in such a way as to modify the frequency of the oscillator in a direction that tends to restore the carrier frequency to its required value.

DEMODULATION OF FM WAVES

The process of *frequency demodulation* is the inverse of frequency modulation in the sense that it enables the original modulating wave to be recovered from a frequency-modulated wave. In particular, to perform frequency demodulation we require a two-port device that produces an *output signal with amplitude directly proportional to the instantaneous frequency of a frequency-modulated wave used as the input signal.* We refer to such a device as a *frequency demodulator.*

There are various methods of designing a frequency demodulator. They can be categorized into two broadly defined classes: (1) *direct* and (2) *indirect*. The direct methods distinguish themselves by the fact that their development is inspired by a direct application of the definition of instantaneous frequency. This class of frequency demodulators includes, as examples, *frequency-discriminators* and *zero crossing detectors*. On the other hand, indirect methods of frequency demodulation rely on the use of *feedback* to *track* variations in the instantaneous frequency of the input signal. The *phase-locked loop* is an example of this second class. In the remainder of this section, we describe the balanced frequency discriminator and zero-cross detector. The phase-locked loop is described in Section 7.12.

Balanced Frequency Discriminator To pave the way for the development of the balanced frequency discriminator, we begin by considering an idealized form of the circuit. In this context, we introduce the notion of an ideal *slope circuit* that is characterized by a purely imaginary transfer function, varying linearly with frequency inside a prescribed interval. Such a circuit includes the differentiator as a special case. To be specific, consider the transfer function depicted in Fig. 7.46*a*, which is defined by

$$H_1(f) = \begin{cases} j2\pi a\left(f - f_c + \dfrac{B}{2}\right), & f_c - \dfrac{B}{2} \leq f \leq f_c + \dfrac{B}{2} \\ j2\pi a\left(f + f_c - \dfrac{B}{2}\right), & -f_c - \dfrac{B}{2} \leq f \leq -f_c + \dfrac{B}{2} \\ 0, & \text{elsewhere} \end{cases} \tag{7.144}$$

where a is a constant. We wish to evaluate the response of this slope circuit, denoted by $s_1(t)$, for an input FM signal $s(t)$ of carrier frequency f_c and transmission bandwidth B. It is assumed that the spectrum of $s(t)$ is essentially zero outside the frequency band $f_c - B/2 \leq |f| \leq f_c + B/2$. For evaluation of the response $s_1(t)$, it is convenient to use the procedure described in Section 3.5, which involves replacing the slope circuit with an equivalent low-pass filter and driving this filter with the complex envelope of the input FM wave $s(t)$.

Let $\tilde{H}_1(f)$ denote the complex transfer function of the slope circuit defined by Fig. 7.46*a*. This complex transfer function is related to $H_1(f)$ by

$$\tilde{H}_1(f - f_c) = H_1(f), \qquad f > 0 \tag{7.145}$$

Hence, using Eqs. 7.144 and 7.145, we get

$$\tilde{H}_1(f) = \begin{cases} j2\pi a\left(f + \dfrac{B}{2}\right), & -\dfrac{B}{2} \leq f \leq \dfrac{B}{2} \\ 0, & \text{elsewhere} \end{cases} \tag{7.146}$$

which is shown in Fig. 7.46*b*.

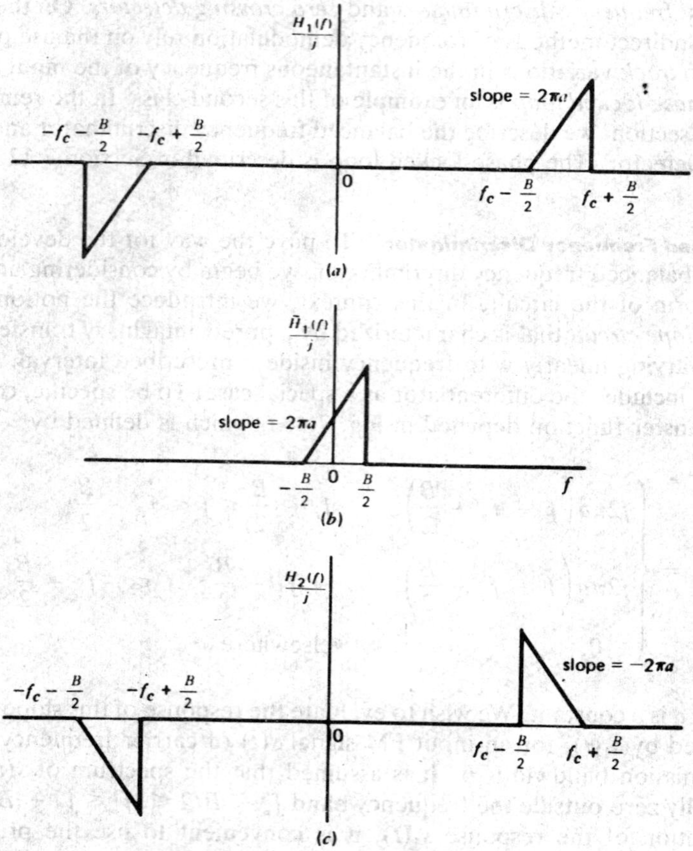

Figure 7.46
(a) *Frequency response of ideal slope circuit.* (b) *Frequency response of complex low-pass filter equivalent to the slope circuit response of part a.* (c) *Frequency response of ideal slope circuit complementary to that of part a.*

The incoming FM wave $s(t)$ is defined by Eq. 7.97, which is reproduced here for convenience:

$$s(t) = A_c \cos\left[2\pi f_c t + 2\pi k_f \int_0^t m(t)\, dt\right] \tag{7.147}$$

The complex envelope of this FM wave is

$$\tilde{s}(t) = A_c \exp\left[j2\pi k_f \int_0^t m(t)\, dt\right] \tag{7.148}$$

Let $\tilde{s}_1(t)$ denote the complex envelope of the response of the slope circuit defined by Fig. 7.46*a*. Then we may express the Fourier transform of $\tilde{s}_1(t)$ as

$$\tilde{S}_1(f) = \tilde{H}_1(f)\tilde{S}(f) = \begin{cases} j2\pi a\left(f + \dfrac{B}{2}\right)\tilde{S}(f), & -\dfrac{B}{2} \leq f \leq \dfrac{B}{2} \\ 0, & \text{elsewhere} \end{cases} \tag{7.149}$$

where $\tilde{S}(f)$ is the Fourier transform of $\tilde{s}(t)$. Now, from Section 2.3 we recall that the multiplication of the Fourier transform of a signal by the factor $j2\pi f$ is equivalent to differentiating the signal in the time domain. We thus deduce from Eq. 7.149 that

$$\tilde{s}_1(t) = a\left[\frac{d\tilde{s}(t)}{dt} + j\pi B\,\tilde{s}(t)\right] \tag{7.150}$$

Substituting Eq. 7.148 in 7.150, we get

$$\tilde{s}_1(t) = j\pi BaA_c\left[1 + \frac{2k_f}{B}m(t)\right]\exp\left(j2\pi k_f\int_0^t m(t)\,dt\right) \tag{7.151}$$

The response of the slope circuit is therefore

$$\begin{aligned} s_1(t) &= \text{Re}[\tilde{s}_1(t)\exp(j2\pi f_c t)] \\ &= \pi B\,aA_c\left[1 + \frac{2k_f}{B}m(t)\right]\cos\left(2\pi f_c t + 2\pi k_f\int_0^t m(t)\,dt + \frac{\pi}{2}\right) \end{aligned} \tag{7.152}$$

The signal $s_1(t)$ is a hybrid-modulated wave in which both the amplitude and frequency of the carrier wave vary with the message signal $m(t)$. However, provided that we choose

$$\left|\frac{2k_f}{B}m(t)\right| < 1$$

for all t, then we may use an envelope detector to recover the amplitude variations and thus, except for a bias term, obtain the original message signal. The resulting envelope detector output is therefore

$$|\tilde{s}_1(t)| = \pi BaA_c\left[1 + \frac{2k_f}{B}m(t)\right] \tag{7.153}$$

The bias term πBaA_c in the right side of Eq. 7.153 is proportional to the slope a of the transfer function of the slope circuit. This suggests that the bias may be removed by subtracting from the envelope detector output $|\tilde{s}_1(t)|$ the output of a second envelope detector preceded by the *complementary slope circuit* with a transfer function $H_2(f)$ as described in Fig. 7.46*c*. That is, the respective complex transfer functions of the two slope circuits are related by

$$\tilde{H}_2(f) = \tilde{H}_1(-f) \tag{7.154}$$

Let $s_2(t)$ denote the response of the complementary slope circuit produced by the incoming FM wave $s(t)$. Then, following a procedure similar to that described herein, we find that the envelope of $s_2(t)$ is

$$|\tilde{s}_2(t)| = \pi BaA_c\left[1 - \frac{2k_f}{B}m(t)\right] \tag{7.155}$$

where $\tilde{s}_2(t)$ is the complex envelope of the signal $s_2(t)$. The difference between the two envelopes in Eqs. 7.153 and 7.155 is

$$\begin{aligned} s_o(t) &= |\tilde{s}_1(t)| - |\tilde{s}_2(t)| \\ &= 4\pi k_f aA_c m(t) \end{aligned} \tag{7.156}$$

which is free from bias.

We may thus model the *ideal frequency discriminator* as a pair of slope circuits with their complex transfer functions related by Eq. 7.154, followed by envelope detectors and a summer, as in Fig. 7.47. This scheme is called a *balanced frequency discriminator* or *back-to-back frequency detector*.

The idealized scheme of Fig. 7.47 can be closely realized using the circuit shown in Fig. 7.48*a*. The upper and lower resonant filter sections of this circuit are tuned to frequencies above and below the unmodulated carrier frequency f_c, respectively. In Fig. 7.48*b* we have plotted the amplitude responses of these two tuned filters, together with their total response,

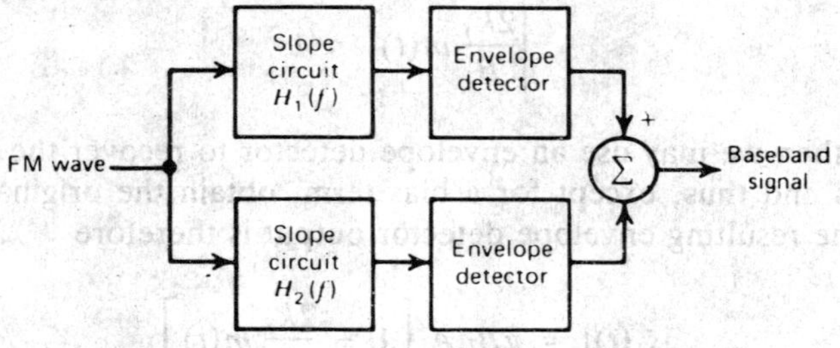

Figure 7.47
Idealized model of balanced frequency discriminator.

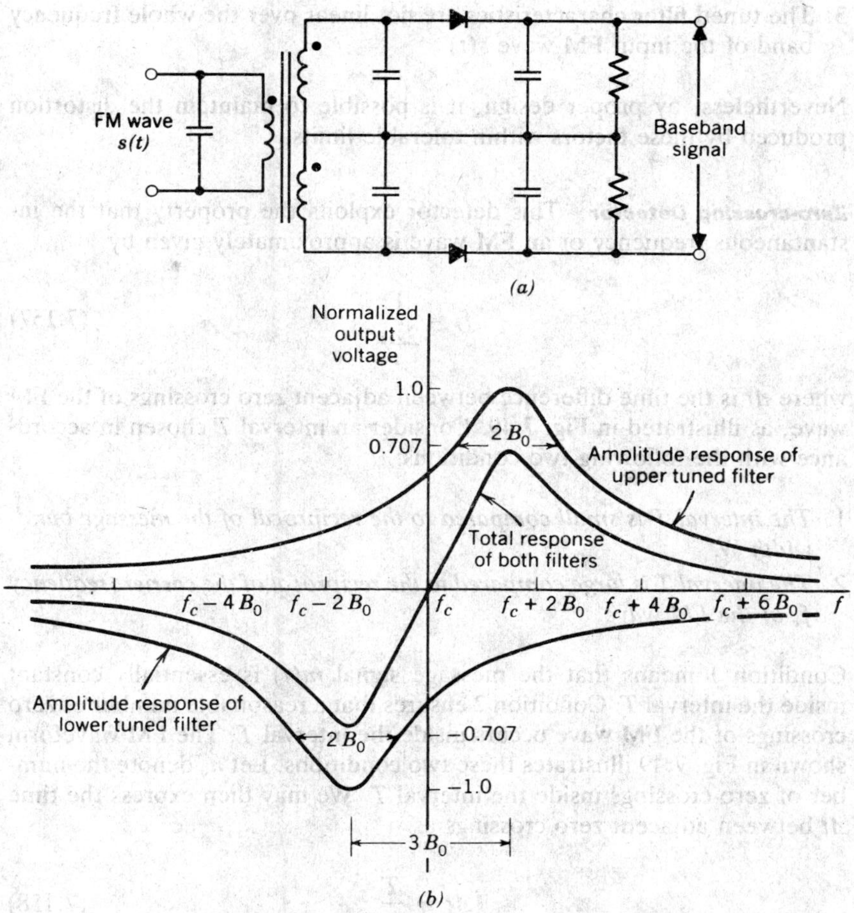

Figure 7.48
Balanced frequency discriminator. (a) *Circuit diagram.* (b) *Frequency response.*

assuming that both filters have a high-Q factor. The linearity of the useful portion of this total response, centered at f_c, is determined by the separation of the two resonant frequencies. As illustrated in Fig. 7.48*b*, a frequency separation of $3B_0$ gives satisfactory results, where $2B_0$ is the 3-dB bandwidth of either filter. However, there will be distortion in the output of this frequency discriminator due to the following factors:

1. The spectrum of the input FM wave $s(t)$ is not exactly zero for frequencies outside the range $f_c - B/2 \leq |f| \leq f_c + B/2$.
2. The tuned filter outputs are not strictly band-limited, and so some distortion is introduced by the low-pass RC filters following the diodes in the envelope detectors.

3. The tuned filter characteristics are not linear over the whole frequency band of the input FM wave $s(t)$.

Nevertheless, by proper design, it is possible to maintain the distortion produced by these factors within tolerable limits.

Zero-crossing Detector This detector exploits the property that the instantaneous frequency of an FM wave is approximately given by

$$f_i \simeq \frac{1}{2\Delta t} \tag{7.157}$$

where Δt is the time difference between adjacent zero crossings of the FM wave, as illustrated in Fig. 7.49. Consider an interval T chosen in accordance with the following two conditions:

1. *The interval T is small compared to the reciprocal of the message bandwidth W.*
2. *The interval T is large compared to the reciprocal of the carrier frequency f_c of the FM wave.*

Condition 1 means that the message signal $m(t)$ is essentially constant inside the interval T. Condition 2 ensures that a reasonable number of zero crossings of the FM wave occurs inside the interval T. The FM waveform shown in Fig. 7.49 illustrates these two conditions. Let n_0 denote the number of zero crossings inside the interval T. We may then express the time Δt between adjacent zero crossings as

$$\Delta t = \frac{T}{n_0} \tag{7.158}$$

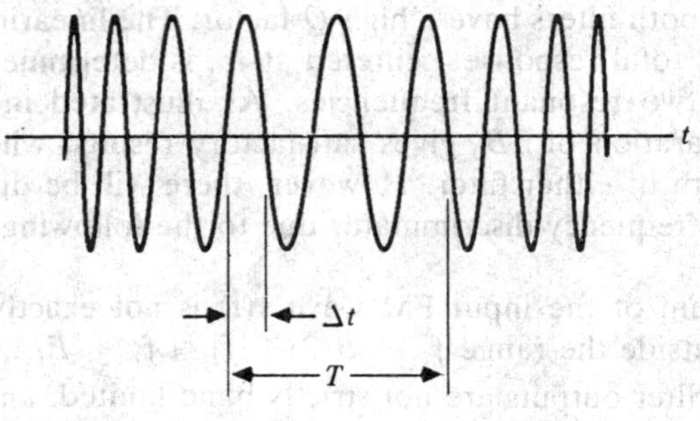

Figure 7.49
Illustrating Eq. 7.158.

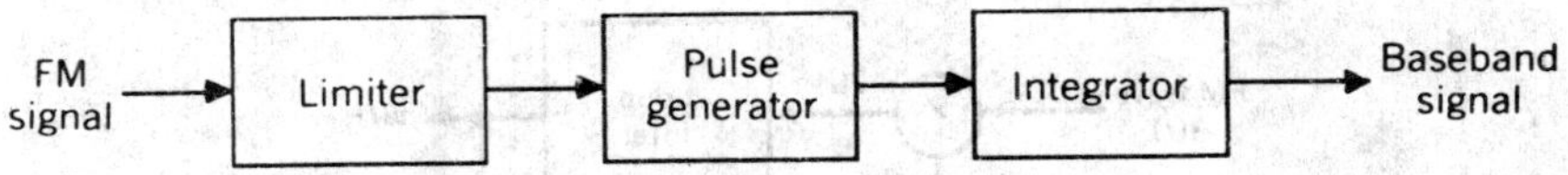

Figure 7.50
Block diagram of zero-crossing detector.

Hence, we may rewrite Eq. 7.157 as

$$f_i \simeq \frac{n_0}{2T} \tag{7.159}$$

Since, by definition, the instantaneous frequency is linearly related to the message signal $m(t)$, we see from Eq. 7.159 that $m(t)$ can be recovered from a knowledge of n_0. Figure 7.50 is the block diagram of a simplified form of the *zero-crossing detector* based on this principle. The limiter produces a square-wave version of the input FM wave; the limiting of FM waves is discussed later in Section 7.13. The pulse generator produces short pulses at the positive-going as well as negative-going edges of the limiter output. Finally, the integrator performs the averaging over the interval T as indicated in Eq. 7.159, thereby reproducing the original message signal $m(t)$ at its output.

EXERCISE 24 Consider an FM wave $s(t)$ that uses a linear modulating wave $m(t) = at$, where a is a constant. Show that the time difference between adjacent zero crossings of $s(t)$ varies inversely with time.

7.12 PHASE-LOCKED LOOP

The *phased-locked loop* (PLL) is a negative feedback system that consists of three major components: a multiplier, a loop filter, and a voltage-controlled oscillator (VCO) connected together in the form of a feedback loop, as in Fig. 7.51. The VCO is a sine-wave generator whose frequency is determined by a voltage applied to it from an external source. In effect, any frequency modulator may serve as a VCO.

We assume that initially we have adjusted the VCO so that when the control voltage is zero, two conditions are satisfied: (1) the frequency of the VCO is precisely set at the unmodulated carrier frequency f_c, and (2) the VCO output has a 90° phase-shift with respect to the unmodulated carrier wave. Suppose that the input signal applied to the phase-locked loop is an FM wave defined by

$$s(t) = A_c \sin[2\pi f_c t + \phi_1(t)] \tag{7.160}$$

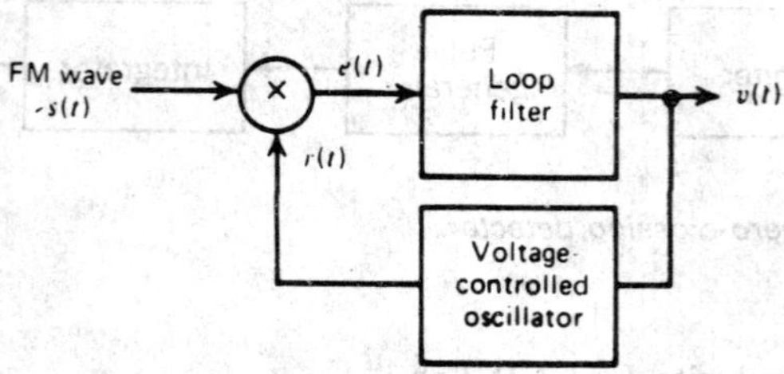

Figure 7.51
Phase-locked loop.

where A_c is the carrier amplitude. With a modulating wave $m(t)$, we have

$$\phi_1(t) = 2\pi k_f \int_0^t m(t)\, dt \tag{7.161}$$

where k_f is the frequency sensitivity of the frequency modulator. Let the VCO output be defined by

$$r(t) = A_v \cos[2\pi f_c t + \phi_2(t)] \tag{7.162}$$

where A_v is the amplitude. With a control voltage $v(t)$ applied to the VCO input, we have

$$\phi_2(t) = 2\pi k_v \int_0^t v(t)\, dt \tag{7.163}$$

where k_v is the frequency sensitivity of the VCO, measured in hertz per volt. The incoming FM wave $s(t)$ and the VCO output $r(t)$ are applied to the multiplier, producing two components:

1. A high-frequency component represented by
$$k_m A_c A_v \sin[4\pi f_c t + \phi_1(t) + \phi_2(t)]$$
2. A low-frequency component represented by $k_m A_c A_v \sin[\phi_1(t) - \phi_2(t)]$, where k_m is the *multiplier gain*, measured in volt^{-1}.

The high-frequency component is eliminated by the low-pass action of the filter and the VCO. Therefore, discarding the high-frequency component, the input to the loop filter is given by

$$e(t) = k_m A_c A_v \sin[\phi_e(t)] \tag{7.164}$$

where $\phi_e(t)$ is the *phase error* defined by

$$\begin{aligned}\phi_e(t) &= \phi_1(t) - \phi_2(t) \\ &= \phi_1(t) - 2\pi k_v \int_0^t v(t)\, dt\end{aligned} \tag{7.165}$$

The loop filter operates on its input $e(t)$ to produce the output

$$v(t) = \int_{-\infty}^{\infty} e(\tau)h(t - \tau)\, d\tau \tag{7.166}$$

where $h(t)$ is the impulse response of the filter.

Using Eqs. 7.164 through 7.166 to relate $\phi_e(t)$ and $\phi_1(t)$, and differentiating with respect to time, we obtain

$$\frac{d\phi_e(t)}{dt} = \frac{d\phi_1(t)}{dt} - 2\pi K_0 \int_{-\infty}^{\infty} \sin[\phi_e(\tau)]h(t - \tau)\, d\tau \tag{7.167}$$

where K_0 is a *loop parameter* defined by

$$K_0 = k_m k_v A_c A_v \tag{7.168}$$

Equation 7.167 suggests the representation or model of Fig. 7.52*a*. In this model we have also included the relationship between $v(t)$ and $e(t)$ as represented by Eqs. 7.164 and 7.166. We see that the block diagram of the model resembles Fig. 7.51. The multiplier is replaced by a subtractor and a sinusoidal nonlinearity, and the VCO by an integrator.

The loop parameter K_0 plays an important role in the operation of a phase-locked loop. It has the dimensions of frequency; this follows from Eq. 7.167, where we observe that the amplitudes A_c and A_m are both measured in volts and the multiplier gain k_m is measured in volt^{-1}.

LINEARIZED MODEL

When the phase error $\phi_e(t)$ is zero, the phase-locked loop is said to be in *phase-lock*. When $\phi_e(t)$ is at all times small compared with one radian, we may use the approximation

$$\sin[\phi_e(t)] \simeq \phi_e(t) \tag{7.169}$$

which is accurate to within 4% for $\phi_e(t)$ less than 0.5 rad. In this case the loop is said to be near phase-lock and the sinusoidal nonlinearity of Fig. 7.52*a* may be disregarded. Thus we may represent the phase-locked loop by the linearized model shown in Fig. 7.52*b*. According to this model, the

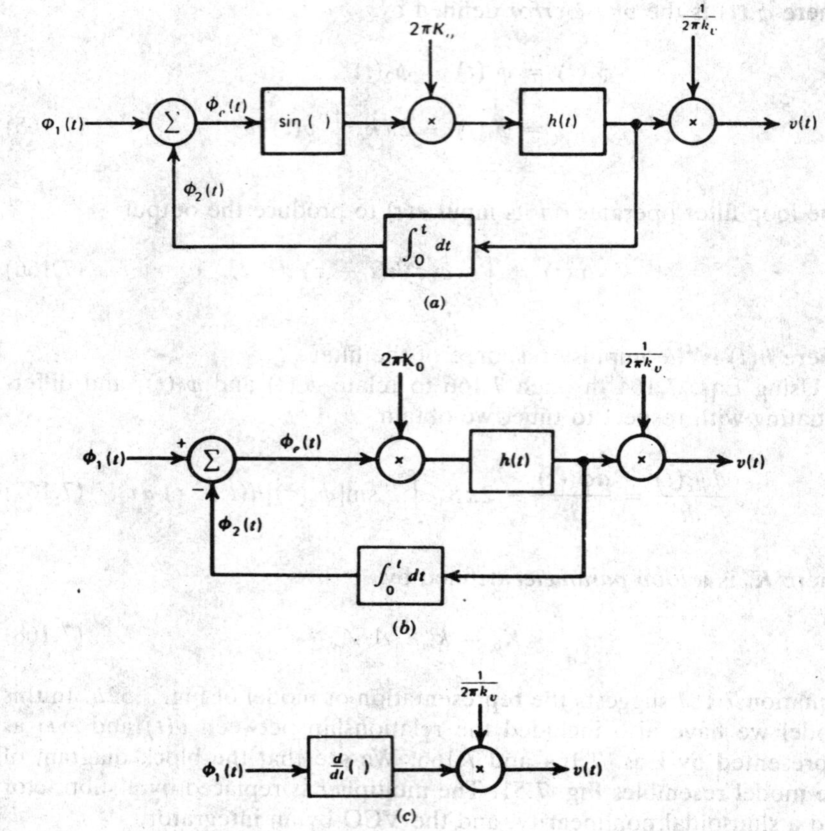

Figure 7.52
(a) *Nonlinear model of a phase-locked loop.* (b) *Linearized model.* (c) *Simplified model when the loop gain is very large compared to unity.*

phase error $\phi_e(t)$ is related to the input phase $\phi_1(t)$ by the integro-differential equation:

$$\frac{d\phi_e(t)}{dt} + 2\pi K_0 \int_{-\infty}^{\infty} \phi_e(\tau)h(t-\tau)\,d\tau = \frac{d\phi_1(t)}{dt} \qquad (7.170)$$

Transforming Eq. 7.170 into the frequency domain and solving for $\Phi_e(f)$, the Fourier transform of $\phi_e(t)$, in terms of $\Phi_1(f)$, the Fourier transform of $\phi_1(t)$, we get

$$\Phi_e(f) = \frac{1}{1 + L(f)}\Phi_1(f) \qquad (7.171)$$

The function $L(f)$ in Eq. 7.171 is defined by

$$L(f) = K_0 \frac{H(f)}{jf} \tag{7.172}$$

where $H(f)$ is the transfer function of the loop filter. The quantity $L(f)$ is called the *open-loop transfer function* of the phase-locked loop. Suppose that for all values of f inside the baseband we make the magnitude of $L(f)$ very large compared with unity. Then from Eq. 7.171 we find that $\Phi_e(f)$ approaches zero. That is, the phase of the VCO becomes asymptotically equal to the phase of the incoming wave, and phase-lock is thereby established.

From Fig. 7.52*b* we see that $V(f)$, the Fourier transform of the phase-locked loop output $v(t)$, is related to $\Phi_e(f)$ by

$$V(f) = \frac{K_0}{k_v} H(f)\Phi_e(f) \tag{7.173}$$

or, equivalently,

$$V(f) = \frac{jf}{k_v} L(f)\Phi_e(f) \tag{7.174}$$

Therefore, substituting Eq. 7.171 in 7.174, we may write

$$V(f) = \frac{(jf/k_v)L(f)}{1 + L(f)} \Phi_1(f) \tag{7.175}$$

Again, when we make $|L(f)| \gg 1$, we may approximate Eq. 7.175 as

$$V(f) \simeq \frac{jf}{k_v} \Phi_1(f) \tag{7.176}$$

The corresponding time-domain relation is

$$v(t) \simeq \frac{1}{2\pi k_v} \frac{d\phi_1(t)}{dt} \tag{7.177}$$

Thus, provided the magnitude of $L(f)$ is very large for all frequencies of interest, the phase-locked loop may be modeled as a differentiator with its output scaled by the factor $1/2\pi k_v$, as in Fig. 7.52*c*.

The simplified model of Fig. 7.52*c* provides the basis of using the phase-locked loop as a frequency demodulator. When the input signal is an FM wave as in Eq. 7.160, the phase $\phi_1(t)$ is related to the modulating wave

$m(t)$ as in Eq. 7.161. Therefore, substituting Eq. 7.161 in 7.177, we find that the resulting output signal of the phase-locked loop is

$$v(t) \simeq \frac{k_f}{k_v} m(t) \qquad (7.178)$$

That is, *the output $v(t)$ of the phase-locked loop is approximately the same, except for the scale factor k_f/k_v, as the original message signal $m(t)$, and the frequency demodulation is accomplished.*

A significant feature of the phase-locked loop demodulator is that the bandwidth of the incoming FM wave can be much wider than that of the loop filter characterized by $H(f)$. The transfer function $H(f)$ can and should be restricted to the baseband. Then the control signal of the VCO has the bandwidth of the message signal $m(t)$, whereas the VCO output is a wideband frequency modulated wave whose instantaneous frequency tracks that of the incoming FM wave.

The complexity of the phase-locked loop is determined by the transfer function $H(f)$ of the loop filter. The simplest form of a phase-locked loop is obtained when $H(f) = 1$; that is, there is no loop filter, and the resulting phase-locked loop is referred to as a *first-order phase-locked loop* (PLL). For higher-order loops, the transfer function $H(f)$ assumes a more complex form. The order of the PLL is determined by the order of the denominator polynomial of the *closed-loop transfer function,* which defines the output transform $V(f)$ in terms of the input transform $\Phi_1(f)$, as shown in Eq. 7.175. In the next sub-section we study the properties of a first-order phase-locked loop demodulator using the linear model of Fig. 7.52*a*.[10]

FIRST-ORDER PHASE-LOCKED LOOP

If the PLL has no loop filter, $H(f) = 1$, the linearized model of the loop simplifies as in Fig. 7.53, and Eq. 7.171 becomes

$$\Phi_e(f) = \frac{1}{1 + K_0/jf} \Phi_1(f) \qquad (7.179)$$

We wish to investigate the loop behavior in the presence of a frequency-modulated input. In particular, we assume a single-tone modulating wave

$$m(t) = A_m \cos(2\pi f_m t) \qquad (7.180)$$

[10]When a phase-locked loop is used to demodulate an FM wave, the loop must first lock onto the incoming FM wave and then follow the variations in its phase. During the lock-up operation, the phase error $\phi_e(t)$ between the incoming FM wave and the VCO output will be large, which therefore requires the use of the nonlinear model of Fig. 7.52*a*.

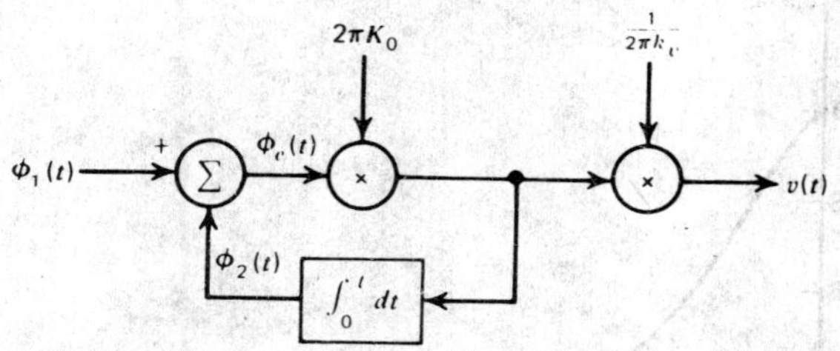

Figure 7.53
Linearized model of first-order phase-locked loop.

with the corresponding FM wave given by

$$s(t) = A_c \sin[2\pi f_c t + \beta \sin(2\pi f_m t)] \tag{7.181}$$

where β is the modulation index. Thus,

$$\phi_1(t) = \beta \sin(2\pi f_m t) \tag{7.182}$$

Therefore, using Eq. 7.182, we find that the phase error $\phi_e(t)$ of the loop produced by the phase input $\phi_1(t)$ of Eq. 7.179 varies sinusoidally with time, as shown by

$$\phi_e(t) = \phi_{e0} \cos(2\pi f_m t + \psi) \tag{7.183}$$

The amplitude ϕ_{e0} and phase ψ of the phase error $\phi_e(t)$ are defined by

$$\phi_{e0} = \frac{\Delta f / K_0}{[1 + (f_m/K_0)^2]^{1/2}} \tag{7.184}$$

and

$$\psi = -\tan^{-1}(f_m/K_0) \tag{7.185}$$

where Δf is the frequency deviation; that is, $\Delta f = \beta f_m$.

In Fig. 7.54 we have plotted the phase-error amplitude ϕ_{e0}, normalized with respect to $\Delta f/K_0$, versus the dimensionless parameter f_m/K_0. It is apparent that for a fixed frequency deviation Δf, the phase-error amplitude has its largest value of $\Delta f/K_0$ at $f_m = 0$, and it decreases with increasing modulation frequency f_m.

For the loop to track the frequency modulation sufficiently closely, the phase error $\phi_e(t)$ should remain within the linear region of operation of the loop for all t. This means that the largest phase-error amplitude should not exceed 0.5 rad, so that $\phi_e(t)$ satisfies the requirement of Eq. 7.169 for

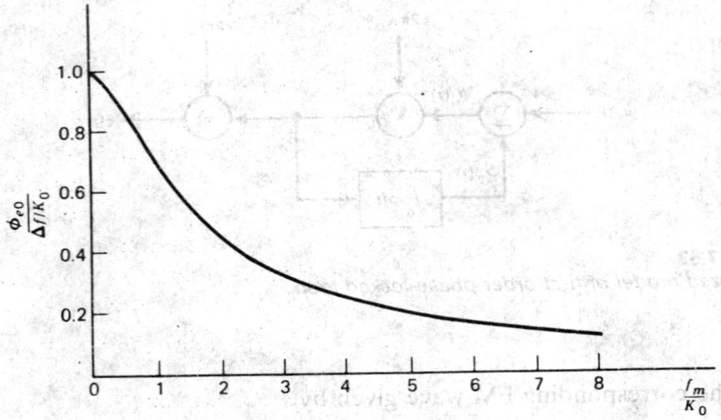

Figure 7.54
Phase-error amplitude characteristic of first-order phase-locked loop.

all t. That is, the frequency deviation of the incoming FM wave $s(t)$ should be bounded by

$$\Delta f \leqslant 0.5 K_0$$

The output signal $v(t)$ of the PLL is related to the phase error $\phi_e(t)$ by (see Fig. 7.53)

$$v(t) = \frac{K_0}{k_v} \phi_e(t) \tag{7.186}$$

Therefore, substituting Eq. 7.183 in 7.186, we get

$$v(t) = A_0 \cos(2\pi f_m t + \psi)$$

where the amplitude A_0 is defined by

$$A_0 = \frac{\Delta f / k_v}{[1 + (f_m/K_0)^2]^{1/2}} \tag{7.187}$$

and the phase ψ is given by Eq. 7.185. From Eq. 7.186 we see that at a modulation frequency $f_m = K_0$, the amplitude of the loop output $v(t)$ will have fallen by 3 dB below its value at $f_m = 0$. The loop bandwidth of a first-order PLL is therefore K_0. We also see from Eq. 7.187 that a first-order PLL demodulator introduces distortion between the original modulating wave $m(t)$ and the signal $v(t)$ obtained at the PLL output. This distortion is the same as the frequency distortion produced by passing the modulating wave $m(t)$ through a low-pass RC filter of time constant $1/2\pi K_0$.

We have thus far assumed that the phase error is sufficiently small to allow the loop to be considered linear in its operation. We next wish to evaluate the input frequency range over which the PLL will hold lock. Assume a constant input frequency, for which

$$\frac{d\phi_1(t)}{dt} = 2\pi\delta f$$

With this input applied to a first-order phase-locked loop, Eq. 7.167 becomes

$$\frac{d\phi_e(t)}{dt} + 2\pi K_0 \sin[\phi_e(t)] = 2\pi\delta f \tag{7.188}$$

The phase error $\phi_e(t)$ will have reached its steady-state value when the derivative $d\phi_e/dt$ is zero. Therefore, putting $d\phi_e/dt = 0$ in Eq. 7.188 we obtain

$$\sin\phi_e = \frac{\delta f}{K_0} \tag{7.189}$$

The sine of an angle cannot exceed unity in magnitude. Hence, Eq. 7.189 has no solution for $\delta f > K_0$. Instead, the loop falls out of lock and the phase error becomes a beat-note rather than a dc level. The *hold-in frequency range* of a first-order PLL is therefore equal to $\pm K_0$. In other words, a first-order PLL will lock to any constant input frequency, provided that it lies within the range $\pm K_0$ of the VCO's free-running frequency f_c.

EXERCISE 25 Let $\dot{\phi}_e = d\phi_e/dt$. Hence, we may rewrite Eq. 7.188 as

$$\dot{\phi}_e = 2\pi(\delta f - K_0 \sin\phi_e)$$

A plot of the derivative $\dot{\phi}_e$ versus the phase error ϕ_e for prescribed values of δf and K_0 is called a *phase-plane plot*.

(a) Sketch such a plot for $K_0 = 2\delta f$.

(b) Show that for initial values of ϕ_e inside the range 0 and 90°, the stable point of the PLL lies at $\phi_e = 30°$.

(c) Show that, in general, the stable points of the PLL lie at

$$\phi_e = 30° \pm n\,360°,$$

where n is an integer.

PRACTICAL CONSIDERATIONS

From the foregoing analysis of a first-order PLL, we conclude that the loop parameter K_0, defined by Eq. 7.168, uniquely determines the loop bandwidth as well as the hold-in frequency range of the PLL. This is a major limitation of first-order PLLs. In order to track variations in the instantaneous frequency of an FM wave, namely,

$$f_i(t) = f_c + k_f m(t)$$

the loop parameter K_0 must be large compared to the frequency deviation [i.e., the maximum departure of the instantaneous frequency $f_i(t)$ from the carrier frequency f_c]. In the case of a first-order PLL, such a choice for K_0 also results in a large loop bandwidth. This is undesirable because a large loop bandwidth lets in more noise power at the demodulator output than would normally be desired. Accordingly, we find that in practice a phase-locked loop used for frequency demodulation includes a *loop filter.*

Figure 7.55 shows a filter[11] often used in a *second-order PLL.* The filter consists of an integrator and a direct connection; its transfer function is given by

$$H(f) = 1 + \frac{f_0}{jf}$$

where f_0 is a constant. The inclusion of such a filter in the loop provides the designer with an additional degree of freedom, namely, f_0. It is now possible to exercise control over both the loop parameter K_0 and the loop bandwidth.[12] A second-order PLL is therefore capable of providing a good performance, and its use is adequate for most practical applications.

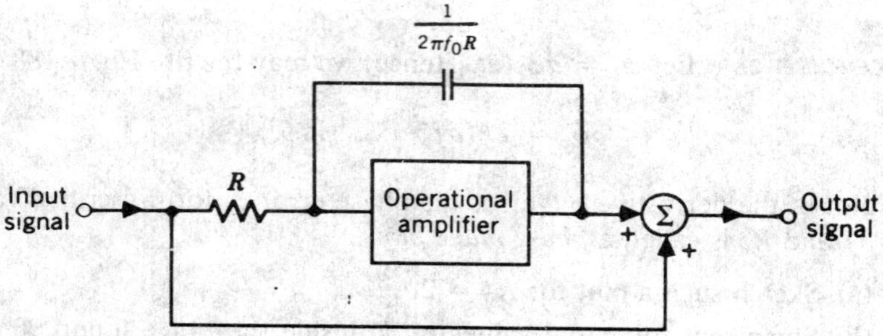

Figure 7.55
Loop filter for second-order phase-locked loop.

[11]In the theory of feedback systems, the filter of Fig. 7.55 is referred to as a *lead-lag filter.*

[12]For a detailed analysis of second-order phase-locked loops, see Gardner (1979).

7.13 LIMITING OF FM WAVES

When an FM wave is transmitted through a communication channel, in general, the output will not have a constant amplitude because of channel imperfections. At the receiver, it is essential to remove the amplitude fluctuations in the channel output prior to frequency demodulation. This is customarily done by means of an *amplitude limiter*. Figure 7.56 shows the input-output characteristic of an idealized form of amplitude limiter known as a *hard limiter*. The resulting output is essentially an *FM square wave*.

To analyze the FM output of a hard limiter, we assume that the limiter is in the form of a *memoryless device*. Accordingly, we may express the limiter output, in response to a frequency-modulated input $z(t)$, as

$$\begin{aligned} v(t) &= \operatorname{sgn}[z(t)] \\ &= \begin{cases} +1, & \text{if } z(t) > 0 \\ -1, & \text{if } z(t) < 0 \end{cases} \end{aligned} \tag{7.190}$$

We also assume that the amplitude fluctuations are slow compared to the zero-crossing rate of the frequency-modulated input $z(t)$. We may then take the sign changes of $z(t)$ as being proportional to the carrier phase shifts, as shown by

$$v(t) = \operatorname{sgn}\{\cos[\theta(t)]\} \tag{7.191}$$

where $\theta(t)$ is the angular argument of the FM wave. The function $\operatorname{sgn}\{\cos[\theta]\}$, viewed as a function of θ, is a periodic square wave when the modulation is zero. Hence, using the Fourier series representation of $\operatorname{sgn}\{\cos[\theta]\}$, we may write

$$\operatorname{sgn}\{\cos[\theta]\} = -\frac{4}{\pi}\sum_{k=1}^{\infty}(-1)^k \frac{\cos[(2k-1)\theta]}{(2k-1)} \tag{7.192}$$

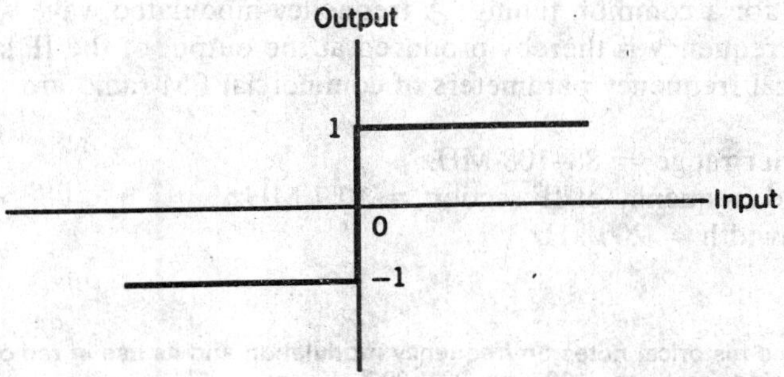

Figure 7.56
Input-output characteristic of a hard limiter.

This expansion holds for all θ. Thus, using $\theta(t)$ in place of θ in Eq. 7.192, we may express the hard limiter output as

$$v(t) = -\frac{4}{\pi}\sum_{k=1}^{\infty}(-1)^k\frac{\cos\{(2k-1)[2\pi f_c t + \theta(t)]\}}{(2k-1)} \tag{7.193}$$

where f_c is the carrier frequency, and the phase $\theta(t)$ is related to the message signal of interest.

Equation 7.193 shows that the hard limiting operation produces image FM sidebands at odd harmonics of the carrier frequency f_c. When the carrier frequency f_c is sufficiently large, we may use a band-pass filter (centered on f_c) to select the desired FM wave:

$$v(t) = \frac{4}{\pi}\cos[2\pi f_c t + \theta(t)]$$

In practice, the combination of hard limiter and band-pass filter is implemented as a single circuit commonly referred to as a *band-pass limiter.*

EXERCISE 26 Consider the periodic signum function sgn{cos[θ]} that is a real-valued, odd function of θ with period 2π. Show that this function may be expanded into a Fourier series as in Eq. 7.192.

7.14 APPLICATION II: FM RADIO

In Section 7.9 we described the standard AM radio format for audio signals and the television for video signals. In this section, we describe FM radio[13] that pertains to the remaining type of radio broadcasting.

As with standard AM radio, most FM radio receivers are of the *superhetrodyne* type. The block diagram of such an FM receiver is shown in Fig. 7.57. The RF section and the local oscillator are mechanically coupled to provide for a common tuning. A frequency-modulated wave with a fixed carrier frequency is thereby produced at the output of the IF section.

Typical frequency parameters of commercial FM radio are

RF carrier range = 88–108 MHz
Midband frequency of IF section = 10.7 MHz
IF bandwidth = 200 kHz

[13]For some historical notes on frequency modulation and its use in radio broadcasting, see Lathi (1983, pp. 301–302).

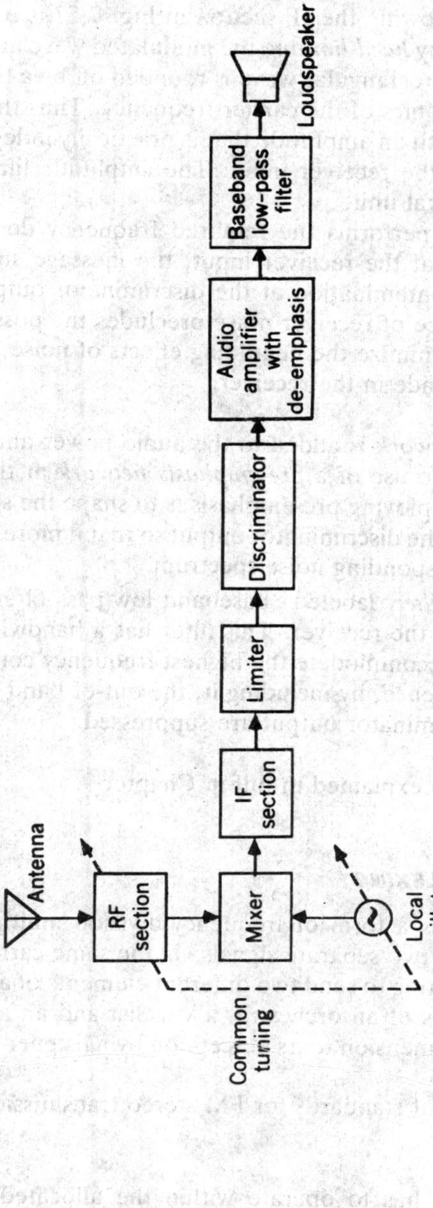

Figure 7.57
Basic elements of an FM receiver of the superheterodyne type.

In an FM radio, the message information is transmitted by variations of the instantaneous frequency of a sinusoidal carrier wave, and its amplitude is maintained constant. Therefore, any variations of the carrier amplitude at the receiver input must result from noise or interference. The *amplitude limiter,* following the IF section in Fig. 7.57 is used to remove amplitude variations by *hard-limiting* the modulated wave at the IF section output. The resulting rectangular wave is rounded off by a band-pass filter that suppresses harmonics of the carrier frequency. Thus the filter output is again sinusoidal, with an amplitude that is practically independent of the carrier amplitude at the receiver input. The amplitude limiter and filter usually form an integral unit.

The discriminator performs the required frequency demodulation. If there were no noise at the receiver input, the message signal would be recovered with no contamination at the discriminator output. However, the inevitable presence of receiver noise precludes the possibility of such an occurrence. To minimize the degrading effects of noise, two modifications are therefore made in the receiver:

1. A *de-emphasis network* is added to the audio power amplifier so as to compensate for the use of a *pre-emphasis network* at the transmitter. The reason for employing pre-emphasis is to shape the spectrum of the message signal at the discriminator output so that it more approximately matches the corresponding noise spectrum.
2. A *post-detection filter,* labeled "baseband low-pass filter," is added at the output end of the receiver. This filter has a bandwidth that is just large enough to accommodate the highest frequency component of the message signal. Hence, by including it, the out-of-band components of noise at the discriminator output are suppressed.

Both these issues are explained in full in Chapter 9.

FM STEREO MULTIPLEXING

Stereo multiplexing is a form of frequency-division multiplexing (FDM) designed to transmit two separate signals via the same carrier. It is widely used in FM broadcasting to send two different elements of a program (e.g., two different sections of an orchestra, a vocalist and an accompanist) so as to give a spatial dimension to its perception by a listener at the receiving end.

The specification of standards for FM stereo transmission is influenced by two factors:

1. The transmission has to operate within the allocated FM broadcast channels.
2. It has to be compatible with monophonic receivers.

The first requirement sets the permissible frequency parameters, including frequency deviation. The second requirement constrains the way in which the transmitted signal is configured.

Figure 7.58*a* shows the block diagram of the multiplexing system used in an FM stereo transmitter. Let $m_l(t)$ and $m_r(t)$ denote the signals picked up by left-hand and right-hand microphones at the transmitting end of the system. They are applied to a simple *matrixer* that generates the *sum signal*, $m_l(t) + m_r(t)$, and the *difference signal*, $m_l(t) - m_r(t)$. The sum signal is left unprocessed in its baseband form; it is available for monophonic reception. The difference signal and a 38-kHz subcarrier (derived from a 19-kHz crystal oscillator by frequency doubling) are applied to a product

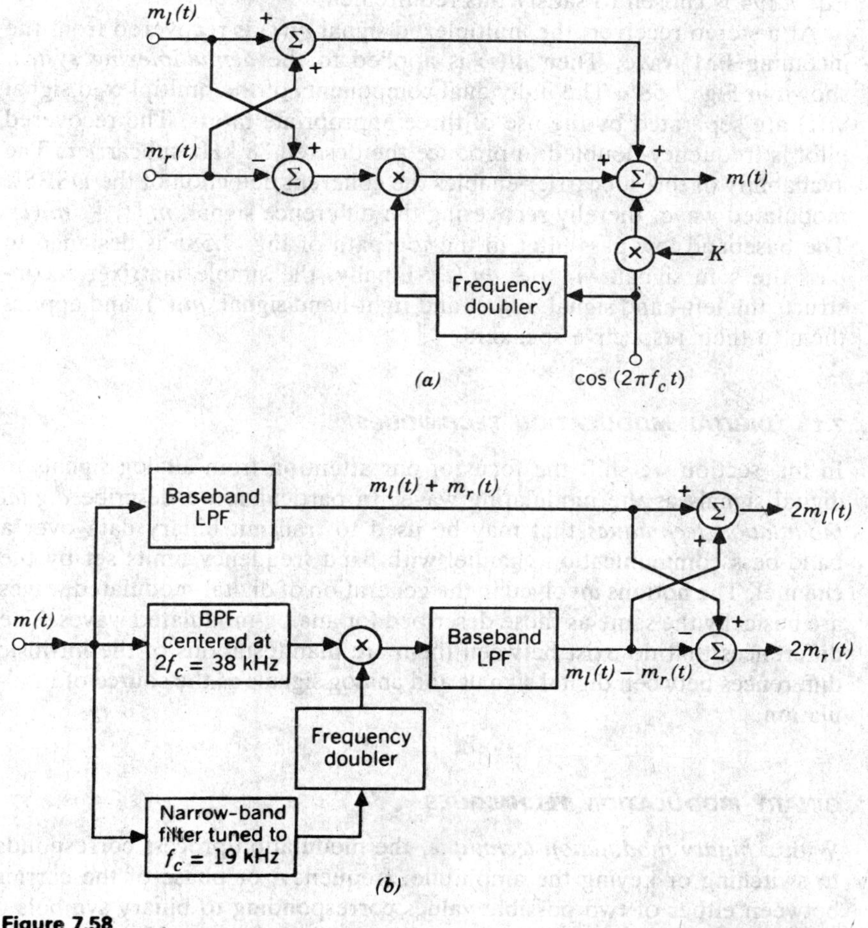

Figure 7.58
(a) *Multiplexer in transmitter of FM stereo.* (b) *Demultiplexer in receiver of FM stereo.*

modulator, thereby producing a DSBSC modulated wave. In addition to the sum signal and this DSBSC modulated wave, the multiplexed signal $m(t)$ also includes a 19-kHz pilot to provide a reference for the coherent detection of the difference signal at the stereo receiver. Thus the multiplexed signal is described by

$$m(t) = [m_l(t) + m_r(t)] + [m_l(t) - m_r(t)] \cos(4\pi f_c t) + K \cos(2\pi f_c t) \tag{7.194}$$

where $f_c = 19$ kHz. The multiplexed signal $m(t)$ then frequency modulates the main carrier to produce the transmitted signal. The pilot is allotted between 8 and 10% of the peak frequency deviation; the amplitude K in Eq. 7.194 is chosen to satisfy this requirement.

At a stereo receiver, the multiplexed signal $m(t)$ is recovered from the incoming FM wave. Then $m(t)$ is applied to the *demultiplexing system* shown in Fig. 7.58*b*. The individual components of the multiplexed signal $m(t)$ are separated by the use of three appropriate filters. The recovered pilot is frequency-doubled to produce the desired 38 kHz subcarrier. The availability of this subcarrier enables the coherent detection of the DSBSC modulated wave, thereby recovering the difference signal, $m_l(t) - m_r(t)$. The baseband low-pass filter in the top path of Fig. 7.58*b* is designed to pass the sum signal, $m_l(t) + m_r(t)$. Finally, the simple matrixer reconstructs the left-hand signal, $m_l(t)$, and right-hand signal, $m_r(t)$, and applies them to their respective speakers.

7.15 DIGITAL MODULATION TECHNIQUES

In this section we shift the focus of our attention from analog signals to digital signals as the modulating wave. In particular, we describe *digital modulation techniques* that may be used to transmit binary data over a band-pass communication channel with fixed frequency limits set by the channel. The notions involved in the generation of digital-modulated waves are basically the same as those described for analog-modulated waves. The differences that do exist between them are manifestations of the intrinsic differences between digital signals and analog signals as the source of modulation.

BINARY MODULATION TECHNIQUES

With a *binary modulation technique,* the modulation process corresponds to switching or keying the amplitude, frequency, or phase of the carrier between either of two possible values corresponding to binary symbols 0 and 1. This results in three basic signaling techniques, namely, *amplitude-*

shift keying (ASK), *frequency-shift keying* (FSK), and *phase-shift keying* (PSK), as described herein:

1. In an ASK system, binary symbol 1 is represented by transmitting a sinusoidal carrier wave of fixed amplitude A_c and fixed frequency f_c for the bit duration T_b seconds, whereas binary symbol 0 is represented by switching off the carrier for T_b seconds, as illustrated in Fig. 7.59*a*. In mathematical terms, we may express the binary ASK wave $s(t)$ as:

$$s(t) = \begin{cases} A_c \cos(2\pi f_c t), & \text{symbol 1} \\ 0, & \text{symbol 0} \end{cases} \tag{7.195}$$

2. In a PSK system, a sinusoidal carrier wave of fixed amplitude A_c and fixed frequency f_c is used to represent both symbols 1 and 0, except that the carrier phase for each symbol differs by 180°, as illustrated in Fig. 7.59*b*. In this case, we may express the binary PSK as:

$$s(t) = \begin{cases} A_c \cos(2\pi f_c t), & \text{symbol 1} \\ A_c \cos(2\pi f_c t + \pi), & \text{symbol 0} \end{cases} \tag{7.196}$$

3. In an FSK system, two sinusoidal waves of the same amplitude A_c but different frequencies f_1 and f_2 are used to represent binary symbols 1

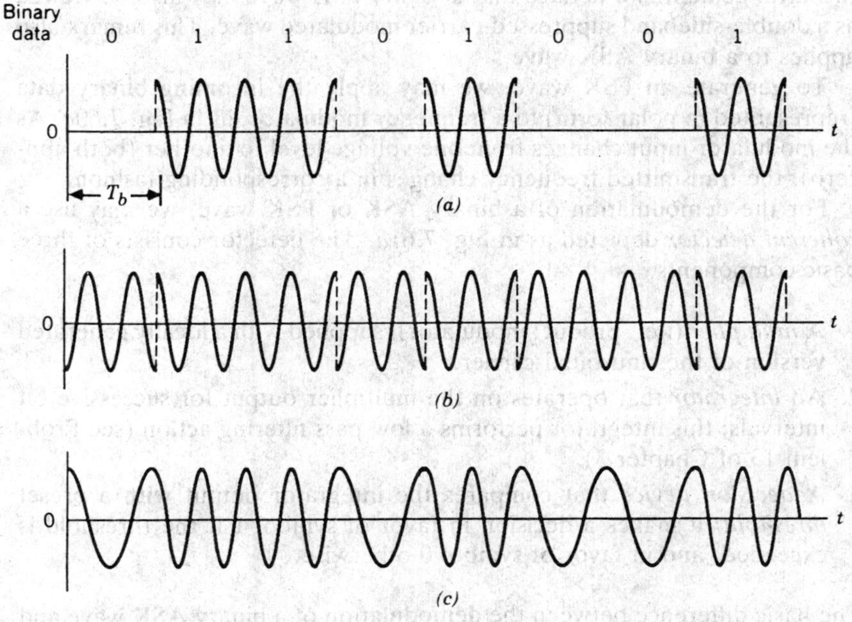

Figure 7.59
The three basic forms of signaling binary information. (a) *Amplitude-shift keying.* (b) *Phase-shift keying.* (c) *Frequency-shift keying with continuous phase.*

and 0, respectively, as in Fig. 7.59*c*. That is, we may express the binary FSK wave $s(t)$ as:

$$s(t) = \begin{cases} A_c \cos(2\pi f_1 t), & \text{symbol 1} \\ A_c \cos(2\pi f_2 t), & \text{symbol 0} \end{cases} \tag{7.197}$$

It is apparent, therefore, that ASK, PSK, and FSK signals are special cases of amplitude-modulated, phase-modulated, and frequency-modulated waves, respectively.

EXERCISE 27 Show that the binary FSK waveform of Fig. 7.59*c* may be viewed as the superposition of two binary ASK waveforms.

GENERATION AND DETECTION OF BINARY MODULATED WAVES

To generate an ASK wave, we may simply apply the incoming binary data (represented in unipolar form) and the sinusoidal carrier to a product modulator, as in Fig. 7.60*a*. The resulting output provides the desired ASK wave.

To generate a PSK wave, we may use the same scheme, except that the incoming binary data are represented in *polar* form, as in Fig. 7.60*b*. From this arrangement, we deduce that a binary PSK wave may also be viewed as a double-sideband suppressed-carrier modulated wave. This remark also applies to a binary ASK wave.

To generate an FSK wave, we may apply the incoming binary data (represented in polar form) to a frequency modulator, as in Fig. 7.60*c*. As the modulator input changes from one voltage level to another (both non-zero), the transmitted frequency changes in a corresponding fashion.

For the demodulation of a binary ASK or PSK wave, we may use a *coherent detector* depicted as in Fig. 7.61*a*. The detector consists of three basic components:

1. A *multiplier* (i.e., product modulator), supplied with a locally generated version of the sinusoidal carrier.
2. An *integrator* that operates on the multiplier output for successive bit intervals; this integrator performs a low-pass filtering action (see Problem 13 of Chapter 3).
3. A *decision device* that compares the integrator output with a preset *threshold;* it makes a decision in favor of symbol 1 if the threshold is exceeded, and in favor of symbol 0 otherwise.

The basic difference between the demodulation of a binary ASK wave and that of a binary PSK wave lies in the choice of the threshold level.

For the demodulation of a binary FSK wave, we may use a coherent detector as shown in Fig. 7.61*b*. This dectector consists of two correlators

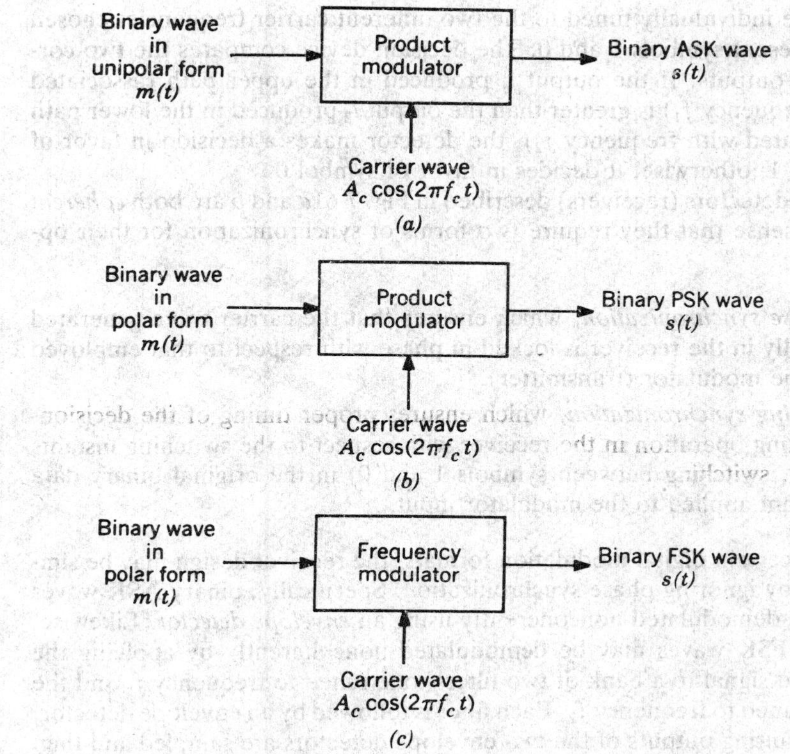

Figure 7.60
Generation schemes for (a) *binary ASK,* (b) *binary PSK, and* (c) *binary FSK.*

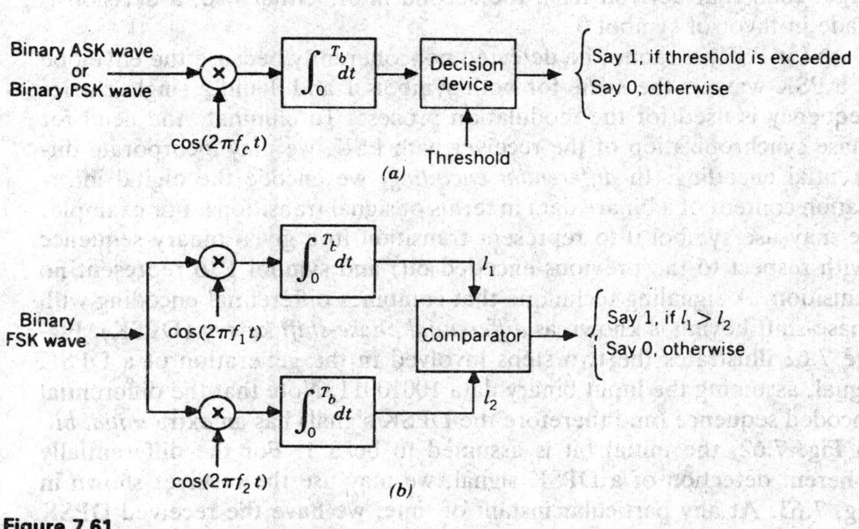

Figure 7.61
Coherent detectors for (a) *binary ASK or binary PSK, and* (b) *binary FSK.*

that are individually tuned to the two different carrier frequencies chosen to represent symbols 1 and 0. The decision device compares the two correlator outputs. If the output l_1 produced in the upper path (associated with frequency f_1) is greater than the output l_2 produced in the lower path (associated with frequency f_2), the detector makes a decision in favor of symbol 1; otherwise, it decides in favor of symbol 0.

The detectors (receivers) described in Fig. 7.61*a* and *b* are both *coherent* in the sense that they require two forms of synchronization for their operation:

1. *Phase synchronization,* which ensures that the carrier wave generated locally in the receiver is locked in phase with respect to that employed in the modulator (transmitter).
2. *Timing synchronization,* which ensures proper timing of the decision-making operation in the receiver with respect to the switching instants (i.e., switching between symbols 1 and 0) in the original binary data stream applied to the modulator input.

For certain digital modulation formats, the receiver design may be simplified by ignoring phase synchronization. Specifically, binary ASK waves may be demodulated noncoherently using an *envelope detector.* Likewise, binary FSK waves may be demodulated noncoherently by applying the received signal to a bank of two filters, one tuned to frequency f_1 and the other tuned to frequency f_2. Each filter is followed by an envelope detector. The resulting outputs of the two envelope detectors are sampled and then compared to each other. A decision is made in favor of symbol 1 if the envelope-detected output derived from the filter tuned to frequency f_1 is larger than that derived from the second filter. Otherwise, a decision is made in favor of symbol 0.

As for PSK, it cannot be detected noncoherently because the envelope of a PSK wave is the same for both symbols 1 and 0 and a single carrier frequency is used for the modulation process. To eliminate the need for phase synchronization of the receiver with PSK, we may incorporate differential encoding. In *differential encoding,* we encode the digital information content of a binary data in terms of signal transitions. For example, we may use symbol 0 to represent transition in a given binary sequence (with respect to the previous encoded bit) and symbol 1 to represent no transition. A signaling technique that combines differential encoding with phase-shift keying is known as *differential phase-shift keying* (DPSK). Figure 7.62 illustrates the two steps involved in the generation of a DPSK signal, assuming the input binary data 10010011. Note that the differential encoded sequence (and therefore the DPSK signal) has an extra *initial bit.* In Fig. 7.62, the initial bit is assumed to be a 1. For the differentially coherent detection of a DPSK signal, we may use the receiver shown in Fig. 7.63. At any particular instant of time, we have the received DPSK signal as one input into the multiplier in Fig. 7.63 and a delayed version

Binary data		1	0	0	1	0	0	1	1
Differentially encoded binary data	1 (Initial bit)	1	0	1	1	0	1	1	1
Phase of DPSK signal (radians)	0	0	π	0	0	π	0	0	0

Figure 7.62
The relationship betrween a binary sequence and its differentially encoded and DPSK versions.

of this signal, delayed by the bit duration T_b, as the other input. The integrator output is proportional to $\cos\phi$, where ϕ is the difference between the carrier phase angles in the received DPSK signal and its delayed version, measured in the same bit interval. Therefore, when $\phi = 0$ (corresponding to symbol 1), the integrator output is positive; on the other hand, when $\phi = \pi$ (corresponding to symbol 0), the integrator output is negative. Thus, by comparing the integrator output with a decision level of zero volts, the receiver of Fig. 7.63 can reconstruct the binary sequence, which, in the absence of noise, is exactly the same as the original binary data at the transmitter input.

DISCUSSION

The detectors shown in Fig. 7.61 are based on the use of a *correlator* that consists of a multiplier followed by an integrator. Digital communication receivers designed in this way are called *correlation receivers*. The correlator may be replaced by the combination of a multiplier, low-pass filter, and sampler; except for the sampler, such a combination parallels the scheme used for the coherent detection of amplitude-modulated waves.

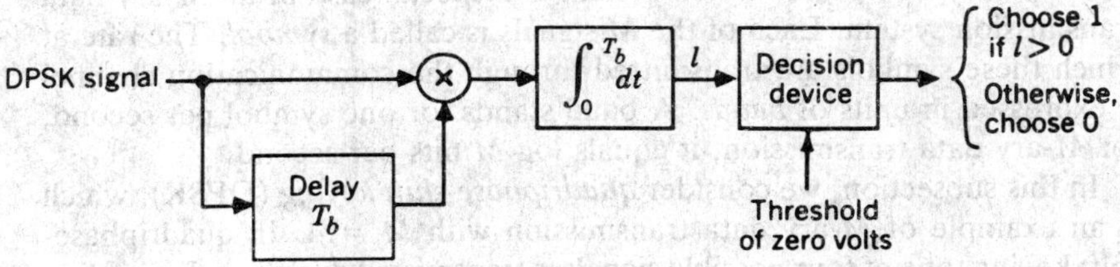

Figure 7.63
Receiver for the detection of DPSK signals.

However, in Chapter 10 it is shown that the correlation receiver is *optimum* for the detection of a pulse in a common type of channel noise called additive white Gaussian noise. Moreover, the combination of a multiplier and low-pass filter is suboptimum in comparison with the correlation receiver; hence, the preference for the use of a correlator in the detectors of Fig. 7.61.

The coherent detection of ASK, PSK, and FSK signals involves the use of *linear* operations and assumes the availability of local carriers (reference signals) that are in *perfect synchronism* with the carriers in the transmitter. On the other hand, the noncoherent detection of ASK and FSK signals involves *nonlinear* operations; the detection of DPSK signals involves the use of linear operations but the supply of a *noisy* reference signal. Accordingly, we find that the mathematical analysis of noise in the class of noncoherent receivers is much more complicated than the class of coherent receivers; more will be said on this issue in Chapter 10.

Another point that will emerge from the discussion presented in Chapter 10 is that receiver design simplification resulting from the use of noncoherent detection is achieved at the cost of some degradation in receiver performance in the presence of noise, compared to a coherent receiver.

It is also noteworthy that none of the digital modulation techniques described thus far is spectrally efficient, meaning that the available channel bandwidth is not fully used. To provide for *spectral efficiency* we may use baseband signal shaping combined with a bandwidth-conserving linear modulation scheme such as vestigial sideband modulation; we studied baseband shaping in Chapter 6 and vestigial sideband modulation in Section 7.5. In the next two sections we describe two other spectrally efficient modulation techniques known as quadriphase-shift keying and minimum shift keying, which are well suited for the transmission of digital data.

QUADRIPHASE-SHIFT KEYING

In binary data transmission, we send only one of two possible signals during each bit interval T_b. On the other hand, in an *M-ary data transmission* system we send any one of M possible signals, during each signaling interval T. For almost all applications, the number of possible signals $M = 2^n$, where n is an integer, and the signaling interval $T = nT_b$. It is apparent that a binary data transmission system is a special case of an M-ary data transmission system. Each of the M signals is called a *symbol.* The rate at which these symbols are transmitted through the communication channel is expressed in units of *bauds*. A baud stands for one symbol per second; for M-ary data transmission, it equals $\log_2 M$ bits per second.

In this subsection, we consider *quadriphase-shift keying* (QPSK), which is an example of M-ary data transmission with $M = 4$. In quadriphase-shift keying, one of four possible signals is transmitted during each signaling interval, with each signal uniquely related to a *dibit* (pairs of bits are termed dibits). For example, we may represent the four possible dibits 00, 10, 11, and 01 (in Gray-encoded form) by transmitting a sinusoidal carrier with one of four possible values, as follows:

$$s(t) = \begin{cases} A_c \cos\left(2\pi f_c t - \dfrac{3\pi}{4}\right), & \text{dibit 00} \\ A_c \cos\left(2\pi f_c t - \dfrac{\pi}{4}\right), & \text{dibit 10} \\ A_c \cos\left(2\pi f_c t + \dfrac{\pi}{4}\right), & \text{dibit 11} \\ A_c \cos\left(2\pi f_c t + \dfrac{3\pi}{4}\right), & \text{dibit 01} \end{cases} \tag{7.198}$$

where $0 \leq t \leq T$; we refer to T as the *symbol duration*. Figure 7.64 depicts the QPSK waveform (based on Eq. 7.198) for the binary sequence 01101000.

Clearly, QPSK represents a special form of phase modulation. This is done by expressing $s(t)$ succinctly as

$$s(t) = A_c \cos[2\pi f_c t + \phi(t)] \tag{7.199}$$

where the phase $\phi(t)$ assumes a constant value for each dibit of the incoming data stream. Specifically, we have (see Fig. 7.65)

$$\phi(t) = \begin{cases} -\dfrac{3\pi}{4}, & \text{dibit 00} \\ -\dfrac{\pi}{4}, & \text{dibit 10} \\ \dfrac{\pi}{4}, & \text{dibit 11} \\ \dfrac{3\pi}{4}, & \text{dibit 01} \end{cases} \tag{7.200}$$

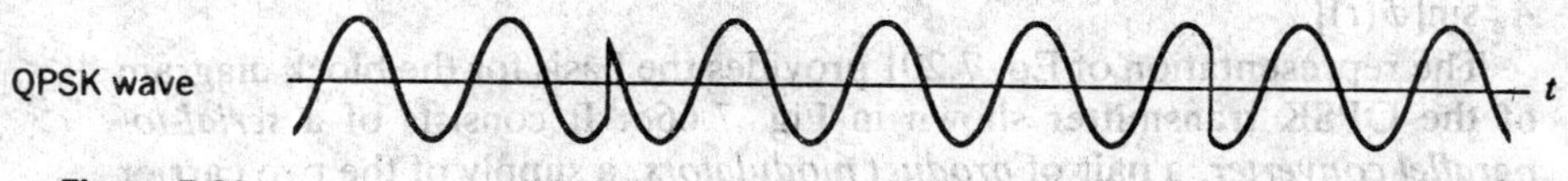

Figure 7.64
QPSK wave for the binary sequence 01101000, assuming the coding arrangement of Eq. 7.198.

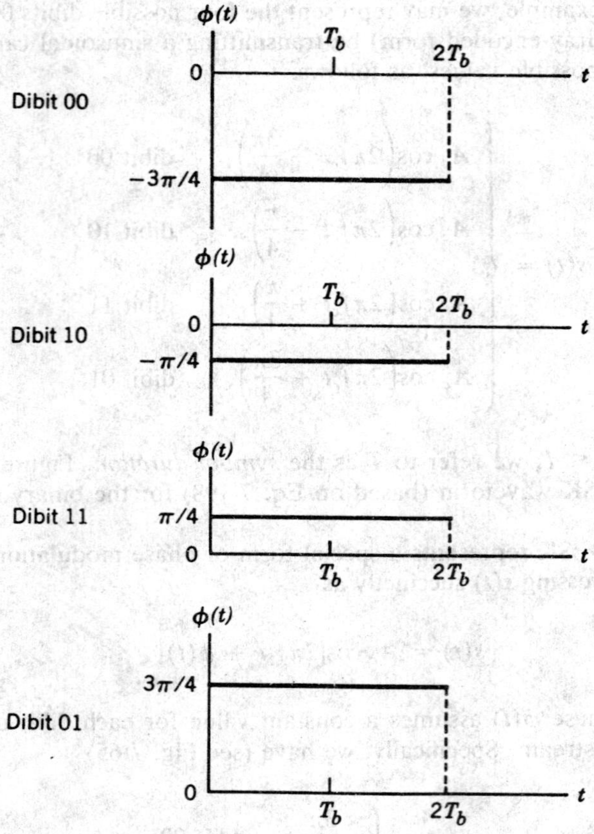

Figure 7.65
The coding of carrier phase of QPSK; the dibits are shown in Gray-coded form.

We may develop further insight into the representation of QPSK by expanding the cosine term in Eq. 7.199 and rewriting the expression for $s(t)$ as

$$s(t) = A_c \cos[\phi(t)] \cos(2\pi f_c t) - A_c \sin[\phi(t)] \sin(2\pi f_c t) \quad (7.201)$$

According to this representation, the QPSK wave $s(t)$ has an *in-phase component* equal to $A_c \cos[\phi(t)]$ and a *quadrature component* equal to $A_c \sin[\phi(t)]$.

The representation of Eq. 7.201 provides the basis for the block diagram of the QPSK transmitter shown in Fig. 7.66*a*. It consists of a *serial-to-parallel converter,* a pair of *product modulators,* a supply of the two carrier waves in phase quadrature, and a *summer*. The function of the serial-to-parallel converter is to represent each successive pair of bits of the incoming binary data stream $m(t)$ as two separate bits, with one bit applied to the

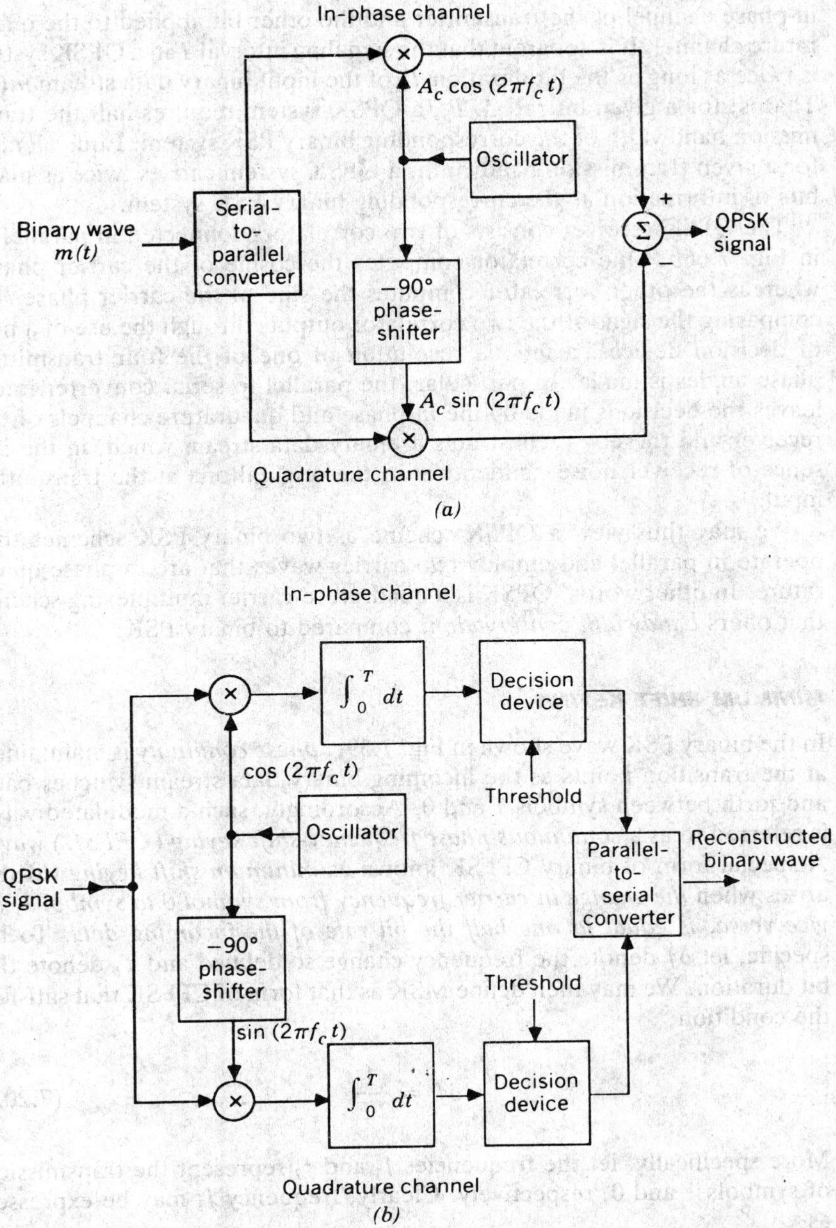

Figure 7.66
Block diagrams of (a) *QPSK transmitter, and* (b) *coherent QPSK receiver.*

in-phase channel of the transmitter and the other bit applied to the quadrature channel. It is apparent that the signaling interval T in a QPSK system is twice as long as the bit duration T_b of the input binary data stream $m(t)$. That is, for a given bit rate $1/T_b$, a QPSK system requires half the transmission bandwidth of the corresponding binary PSK system. Equivalently, for a given transmission bandwidth, a QPSK system carries twice as many bits of information as the corresponding binary PSK system.

The QPSK receiver consists of two correlators connected in parallel as in Fig. 7.66*b*. One correlator computes the cosine of the carrier phase, whereas the other correlator computes the sine of the carrier phase. By comparing the signs of the two correlator outputs through the use of a pair of decision devices, a unique resolution of one of the four transmitted phase angles is made. In particular, the parallel-to-serial converter interleaves the decisions made by the in-phase and quadrature channels of the receiver and thereby reconstructs a binary data stream which, in the absence of receiver noise, is identical to the original one at the transmitter input.

We may thus view a QPSK scheme as two binary PSK schemes that operate in parallel and employ two carrier waves that are in phase quadrature. In other words, QPSK is a quadrature-carrier multiplexing scheme that offers *bandwidth conservation*, compared to binary PSK.

MINIMUM SHIFT KEYING

In the binary FSK wave shown in Fig. 7.59*c*, *phase continuity* is maintained at the transition points as the incoming binary data stream switches back and forth between symbols 1 and 0. Accordingly, such a modulated wave is referred to as a *continuous-phase frequency-shift keying* (*CPFSK*) *wave.* A special form of binary CPFSK known as *minimum shift keying* (MSK) arises when *the change in carrier frequency from symbol 0 to symbol 1, or vice versa, is equal to one half the bit rate of the incoming data.* To be specific, let δf denote the frequency change so defined and T_b denote the bit duration. We may then define MSK as that form of CPFSK that satisfies the condition:

$$\delta f = \frac{1}{2T_b} \tag{7.202}$$

More specifically, let the frequencies f_1 and f_2 represent the transmission of symbols 1 and 0, respectively. Clearly, frequency f_1 may be expressed as

$$\begin{aligned} f_1 &= \frac{f_1 + f_2}{2} + \frac{f_1 - f_2}{2} \\ &= f_c + \frac{\delta f}{2} \end{aligned} \tag{7.203}$$

where

$$f_c = \frac{f_1 + f_2}{2} \tag{7.204}$$

and

$$\delta f = f_1 - f_2 \tag{7.205}$$

Similarly, we may express the second carrier frequency f_2 as

$$\begin{aligned} f_2 &= \frac{f_1 + f_2}{2} - \frac{f_1 - f_2}{2} \\ &= f_c - \frac{\delta f}{2} \end{aligned} \tag{7.206}$$

The "unmodulated" carrier frequency f_c represents the arithmetic mean of the two transmitted frequencies f_1 and f_2 as in Eq. 7.204.

Define the MSK signal as

$$s(t) = A_c \cos[2\pi f_c t + \phi(t)]$$

where

$$\phi(t) = \pm \pi \delta f t$$

Hence, under the condition specified by Eq. 7.202, the transmission of symbol 1 (i.e., frequency f_1) changes the phase of the MSK signal $s(t)$ by an amount defined by

$$\begin{aligned} \phi(t) &= \pi \delta f t \\ &= \frac{\pi t}{2T_b}, \qquad \text{symbol 1} \end{aligned} \tag{7.207}$$

From this relation we see that at the termination of the interval representing the transmission of symbol 1 at time $t = T_b$ the phase of an MSK wave increases by an amount equal to $\pi/2$ radians. On the other hand, the transmission of symbol 0 (i.e., frequency f_2) changes the phase of the MSK wave $s(t)$ by an amount defined by

$$\begin{aligned} \phi(t) &= -\pi \delta f t \\ &= -\frac{\pi t}{2T_b}, \qquad \text{symbol 0} \end{aligned} \tag{7.208}$$

This means that at the termination of the interval representing the transmission of symbol 0 at time $t = T_b$ the phase of an MSK wave decreases by an amount equal to $\pi/2$ radians.

We are now ready to demonstrate that MSK may be viewed as another example of quadrature multiplexing. First, we express the MSK wave $s(t)$ as a frequency-modulated wave as follows:

$$\begin{aligned} s(t) &= A_c \cos[2\pi f_c t + \phi(t)] \\ &= A_c \cos(2\pi f_c t) \cos[\phi(t)] - A_c \sin(2\pi f_c t) \sin[\phi(t)] \end{aligned} \quad (7.209)$$

This shows that $s(t)$ has an in-phase component equal to $A_c \cos[\phi(t)]$ and a quadrature component equal to $A_c \sin[\phi(t)]$. As with QPSK, there are four distinct dibits to be considered; they are 00, 10, 11, and 01. Consider first the transmission of dibit 00. In this case, the phase of the MSK wave

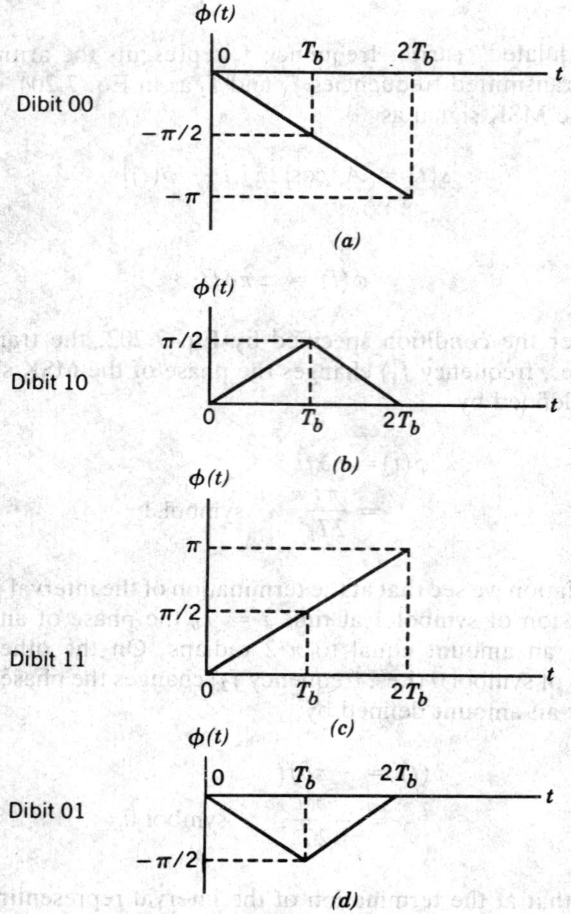

Figure 7.67
Coding of the carrier phase $\phi(t)$ for MSK; the dibits are shown in Gray-coded form.

TABLE 7.4

Dibit (Gray coded)	$\sin[\phi(T_b)]$	$\cos[\phi(2T_b)]$
00	-1	-1
10	$+1$	$+1$
11	$+1$	-1
01	-1	$+1$

experiencing a decrease (representing the first symbol 0) is followed by another decrease (representing the second symbol 0). Hence, the phase history of the MSK wave traces the path shown in Fig. 7.67*a*. Similarly, we find that the transmission of dibits 10, 11, and 01 traces the respective paths shown in parts *b*, *c*, and *d* of Fig. 7.67 for the phase history of the MSK wave. In Fig. 7.67 it is assumed that the *initial condition* is defined by $\phi(0) = 0$. Note that at time $t = T_b$ the phase of the MSK wave equals $+\pi/2$ or $-\pi/2$ radians, whereas at time $t = 2T_b$ it equals 0 or π radians, modulo 2π.

In Table 7.4 we show the pair of values, $\sin[\phi(T_b)]$ and $\cos[\phi(2T_b)]$, corresponding to each of the four possible dibits. This table shows that the identity of each dibit in MSK is uniquely defined by specifying the doublet $\{\sin[\phi(T_b)], \cos[\phi(2T_b)]\}$.

We thus see that QPSK and MSK are examples of quadrature multiplexing. They differ from each other in the sense that QPSK is a phase-modulated wave whereas MSK is a frequency-modulated wave. This basic difference manifests itself in the way in which the phase shift $\phi(t)$ of the sinusoidal carrier varies with time. In QPSK, the phase shift $\phi(t)$ assumes a distinct value that is constant for the entire duration of a symbol, depending on the dibit being transmitted, as in Fig. 7.65. In MSK, on the other hand, for each dibit the phase shift $\phi(t)$ varies with time along a distinct path made up of straight lines, depending on the dibit being transmitted, as in Fig. 7.67.

To generate an MSK wave, we may use a frequency modulator that fulfills the condition of Eq. 7.202. The coherent detection of MSK, however, involves a mathematical treatment that is beyond the scope of this introductory book.[14] Nevertheless, it suffices to say that the coherent detector consists of a pair of correlators with built-in *memory* and decisions made over successive pairs of bit intervals. The detector is designed in such a way that it can track the past history of the phase $\phi(t)$ as it evolves in time on a bit-by-bit basis, and thereby reconstruct a binary wave that (in the absence of receiver noise) is the same as that at the transmitter input.

[14]For a detailed treatment of minimum shift keying, see Haykin (1988, pp. 291–300).

7.16 APPLICATION III: DIGITAL COMMUNICATIONS BY SATELLITE

In this section we briefly describe the application of digital modulation for the transmission of binary data over a *satellite channel*. The satellite channel consists of an *uplink*, a *transponder*, and a *downlink*, as in Fig. 7.68. The uplink connects a transmitting station on the ground to the transponder on board a satellite positioned in geostationary orbit around the earth. The downlink connects the transponder to a receiving ground station (usually placed at a remote distance away from the transmitting ground station). The transponder is designed to provide adequate amplification to overcome the effects of channel noise. We may therefore view the satellite transponder as a repeater in the sky.

A satellite channel has a built-in *broadcast* capability. To exploit it, however, we require the use of a technique known as *multiple access*. A particular type of this technique, known as *time-division multiple access* (TDMA), is well suited for digital communications.[15] In TDMA, a number of ground stations are able to access a satellite by having their individual transmissions reach the satellite in *nonoverlapping time slots*. Hence, the radio frequency (RF) power amplifier at the output of the satellite transponder may be permitted to operate at or near saturation without having to introduce crosstalk between individual transmissions. Such a feature, which is essentially unique to TDMA, helps to optimize the noise performance of the receiver. Moreover, since only one modulated carrier is present in the nonlinear transponder at any one time, the generation of intermodulation products is avoided.

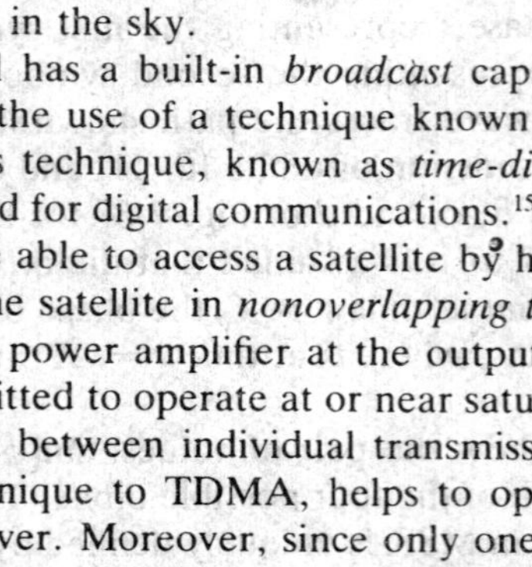

Figure 7.68
Satellite link.

[15]There are two other types of multiple access, namely, *frequency-division multiple access* (FDMA) and *code-division multiple access* (CDMA). The former is used for analog communications and the latter is used for secure communications. For discussions of the TDMA network, see Pratt and Bostian (1986, pp. 235–251).

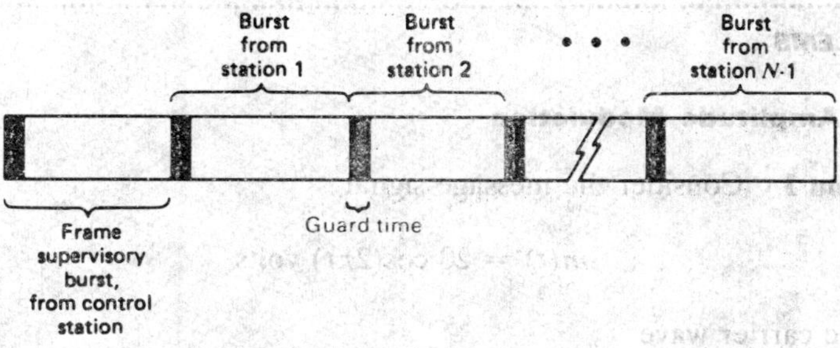

Figure 7.69
Structure of a TDMA frame.

Figure 7.69 illustrates the idea of a TDMA network, in which transmissions are organized into *frames*. A frame contains N *bursts*. To compensate for variations in satellite range, a *guard time* is inserted between successive bursts as in Fig. 7.69 to protect the system against overlap. One burst per frame is used as a *reference*. The remaining $N - 1$ bursts are allocated to ground stations on the basis of one burst per station. Thus, each station transmits once per frame. Typically, a burst consists of an initial portion called the *preamble*, which is followed by a *message* portion; in some systems a *postamble* is also included. The preamble consists of a part for carrier recovery, a part for symbol-timing recovery, a unique word for burst synchronization, a station identification code, and some housekeeping symbols. Two functionally different components may therefore be identified in each frame: a revenue-producing component represented by message portions of the bursts, and system overhead represented by guard times, the reference burst, preambles, and postambles (if included).

Two important points emerge from this brief discussion of the TDMA network:

1. Power efficiency in a satellite transponder is maximized by permitting the traveling-wave tube (responsible for power amplification) to operate at or near saturation.
2. The transmissions contain independent provisions for carrier synchronization and bit timing synchronization to occur simultaneously, thereby keeping overhead due to recovery time in the receiver to a minimum.

Therefore, only a limited set of digital modulation techniques is suitable for satellite communications. In particular, point 1 constrains the modulation format to have a constant envelope, thereby excluding ASK. Point 2 makes it feasible to employ coherent detection. We therefore find that in digital communications by satellite, primary interest is in the use of coherent binary PSK, coherent QPSK, and coherent MSK.

PROBLEMS

P7.1 Amplitude Modulation

Problem 1 Consider the message signal

$$m(t) = 20 \cos(2\pi t) \text{ volts}$$

and the carrier wave

$$c(t) = 50 \cos(100\pi t) \text{ volts}$$

(a) Sketch (to scale) the resulting AM wave for 75% modulation.
(b) Find the power developed across a load of 100 ohms due to this AM wave.

Problem 2 A carrier wave of frequency 1 MHz is modulated 50% by a sinusoidal wave of frequency 5 kHz. The resulting AM wave is transmitted through the resonant circuit of Fig. P7.1, which is tuned to the carrier frequency and has a Q factor of 175. Determine the modulated wave after transmission through this circuit. What is the percentage modulation of this modulated wave?

Problem 3 Using the message signal

$$m(t) = \frac{t}{1 + t^2}$$

determine and sketch the modulated wave for amplitude modulation whose percentage modulation equals the following values:

(a) 50%
(b) 100%
(c) 125%

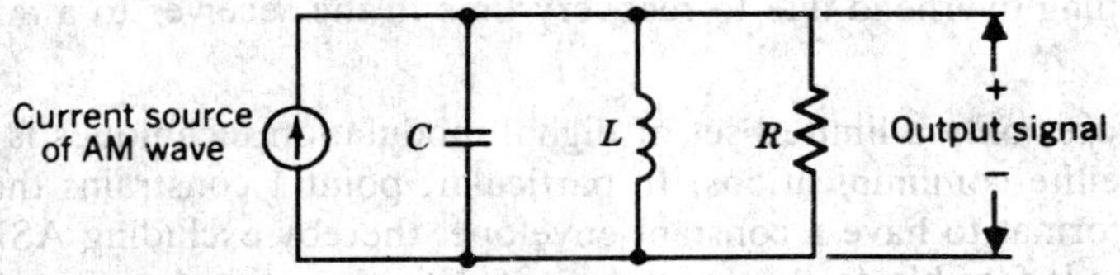

Figure P7.1

Problem 4 For a p-n junction diode, the current i through the diode and the voltage v across it are related by

$$i = I_0\left[\exp\left(-\frac{v}{V_T}\right) - 1\right]$$

where I_0 is the reverse saturation current and V_T is the thermal voltage defined by

$$V_T = \frac{kT}{e}$$

where k is Boltzmann's constant in joules per degree Kelvin, T is the absolute temperature in degrees Kelvin, and e is the charge of an electron. At room temperature $V_T = 0.026$ V.

(a) Expand i as a power series in v, retaining terms up to v^3.
(b) Let

$$v = 0.01\cos(2\pi f_m t) + 0.01\cos(2\pi f_c t) \text{ volts}$$

where $f_m = 1$ kHz and $f_c = 100$ kHz. Determine the spectrum of the resulting diode current i.
(c) Specify the band-pass filter required to extract from the diode current an AM wave with carrier frequency f_c.
(d) What is the percentage modulation of this AM wave?

Problem 5 Suppose nonlinear devices are available for which the output i_o and input voltage v_i are related by

$$i_o = a_1 v_i + a_3 v_i^3$$

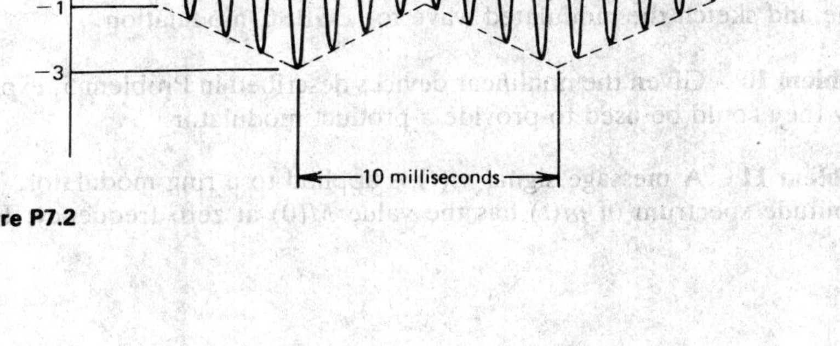

Figure P7.2

where a_1 and a_3 are constants. Explain how these devices could be used to provide an amplitude modulator.

Problem 6 Consider the amplitude-modulated wave of Fig. P7.2 with a periodic triangular envelope. This modulated wave is applied to an envelope detector with zero source resistance and a load resistance of 250 ohms. The carrier frequency $f_c = 40$ kHz. Suggest a suitable value for the capacitor C so that the distortion (at the envelope detector output) is negligible for frequencies up to and including the eleventh harmonic of the modulating wave.

P7.2 Double-Sideband Suppressed-Carrier Modulation

Problem 7 Consider the DSBSC modulated wave obtained by using the sinusoidal modulating wave

$$m(t) = A_m \cos(2\pi f_m t)$$

and the carrier wave

$$c(t) = A_c \cos(2\pi f_c t + \phi)$$

The phase angle ϕ, denoting the phase difference between $c(t)$ and $m(t)$ at time $t = 0$, is variable. Sketch this modulated wave for the following values of ϕ:

(a) $\phi = 0$
(b) $\phi = 45°$
(c) $\phi = 90°$
(d) $\phi = 135°$

Comment on your results.

Problem 8 A sinusoidal wave of frequency 5 kHz is applied to a product modulator, together with a carrier wave of frequency 1 MHz. The modulator output is next applied to the resonant circuit of Fig. P7.1. Determine the modulated wave after transmission through this circuit.

Problem 9 Using the message signal $m(t)$ described in Problem 3 determine and sketch the modulated wave for DSBSC modulation.

Problem 10 Given the nonlinear devices described in Problem 5, explain how they could be used to provide a product modulator.

Problem 11 A message signal $m(t)$ is applied to a ring modulator. The amplitude spectrum of $m(t)$ has the value $M(0)$ at zero frequency. Find the ring modulator output at $f = \pm f_c, \pm 3f_c, \pm 5f_c, \ldots$, where f_c is the fundamental frequency of the square carrier wave $c(t)$.

Problem 12 Consider a message signal $m(t)$ with the spectrum shown in Fig. P7.3. The message bandwidth $W = 1$ kHz. This signal is applied to a product modulator, together with a carrier wave $A_c \cos(2\pi f_c t)$, producing the DSBSC modulated wave $s(t)$. This modulated wave is next applied to a coherent detector. Assuming perfect synchronism between the carrier waves in the modulator and detector, determine the spectrum of the detector output when: (*a*) the carrier frequency $f_c = 1.25$ kHz and (*b*) the carrier frequency $f_c = 0.75$ kHz. What is the lowest carrier frequency for which each component of the modulated wave $s(t)$ is uniquely determined by $m(t)$?

Problem 13 A DSBSC wave is demodulated by applying it to a coherent detector.

(a) Evaluate the effect of a frequency error Δf in the local carrier frequency of the detector, measured with respect to the carrier frequency of the incoming DSBSC wave.

(b) For the case of a sinusoidal modulating wave, show that because of this frequency error, the demodulated wave exhibits *beats* at the error frequency. Illustrate your answer with a sketch of this demodulated wave.

Problem 14 Consider a composite wave obtained by adding a noncoherent carrier $A_c \cos(2\pi f_c t + \phi)$ to a DSBSC wave $\cos(2\pi f_c t)m(t)$. This composite wave is applied to an ideal envelope detector. Find the resulting detector output. Evaluate this output for

(a) $\phi = 0$.

(b) $\phi \neq 0$ and $|m(t)| \ll A_c/2$.

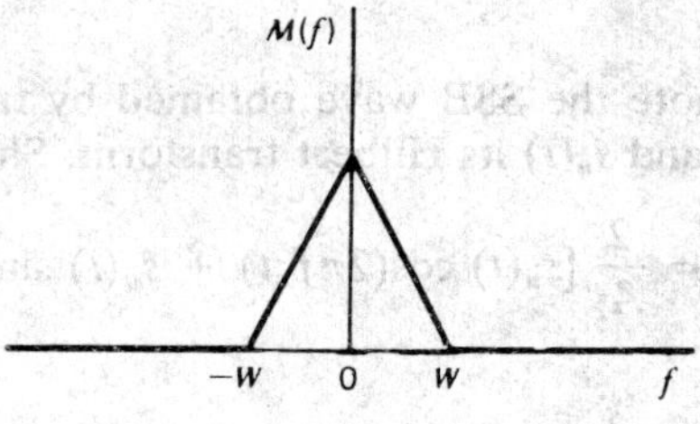

Figure P7.3

P7.3 Quadrature-Carrier Multiplexing

Problem 15 Consider the quadrature-carrier multiplex system of Fig. 7.16. The multiplexed signal $s(t)$ produced at the transmitter output in part *a* of this figure is applied to a communication channel of transfer function $H(f)$. The output of this channel is in turn applied to the receiver input in part *b* of Fig. 7.16. Prove that the condition

$$H(f_c + f) = H^*(f_c - f), \qquad 0 \leq f \leq W$$

is necessary for recovery of the message signals $m_1(t)$ and $m_2(t)$ at the receiver outputs; f_c is the carrier frequency, and W is the message bandwidth.

Hint: Evaluate the spectra of the two receiver outputs.

P7.4 Single-Sideband Modulation

Problem 16 Using the message signal $m(t)$ described in Problem 1, determine and sketch the modulated waves for single-sideband modulation with (a) only the upper sideband transmitted, and (b) only the lower sideband transmitted.

Problem 17 Consider a pulse of amplitude A and duration T. This pulse is applied to an SSB modulator, producing the modulated wave $s(t)$. Determine the envelope of $s(t)$, and show that this envelope exhibits peaks at the beginning and end of the pulse.

Problem 18 Consider the two-stage SSB modulator of Fig. 7.18*b*. The input signal consists of a voice signal occupying the frequency band 0.3 – 3.4 kHz. The two oscillator frequencies have the values f_1 = 100 kHz and f_2 = 10 MHz. Specify the following:

(a) The sidebands of the DSBSC modulated waves appearing at the two product modulator outputs.
(b) The sidebands of the SSB modulated waves appearing at the two band-pass filter outputs.
(c) The passbands and guardbands of the two band-pass filters.

Problem 19

(a) Let $s_u(t)$ denote the SSB wave obtained by transmitting only the upper sideband, and $\hat{s}_u(t)$ its Hilbert transform. Show that

$$m(t) = \frac{2}{A_c}[s_u(t)\cos(2\pi f_c t) + \hat{s}_u(t)\sin(2\pi f_c t)]$$

and

$$\hat{m}(t) = \frac{2}{A_c}[\hat{s}_u(t)\cos(2\pi f_c t) - s_u(t)\sin(2\pi f_c t)]$$

where $m(t)$ is the message signal, $\hat{m}(t)$ is its Hilbert transform, f_c the carrier frequency, and A_c is the carrier amplitude.

(b) Show that the corresponding equations in terms of the SSB wave $s_l(t)$ obtained by transmitting only the lower sideband are

$$m(t) = \frac{2}{A_c}[s_l(t)\cos(2\pi f_c t) + \hat{s}_l(t)\sin(2\pi f_c t)]$$

and

$$\hat{m}(t) = \frac{2}{A_c}[s_l(t)\sin(2\pi f_c t) - \hat{s}_l(t)\cos(2\pi f_c t)]$$

(c) Using the results of (a) and (b), set up the block diagram of a receiver for demodulating an SSB wave.

Problem 20

(a) Consider a message signal $m(t)$ containing frequency components at 100, 200, and 400 Hz. This signal is applied to an SSB modulator together with a carrier at 100 kHz, with only the upper sideband retained. In the coherent detector used to recover $m(t)$, the local oscillator supplies a sine wave of frequency 100.02 kHz. Determine the frequency components of the detector output.

(b) Repeat your analysis, assuming that only the lower sideband is transmitted.

P7.5 Vestigial Sideband Modulation

Problem 21 The single-tone modulating wave $m(t) = A_m \cos(2\pi f_m t)$ is used to generate the VSB modulated wave

$$s(t) = aA_m A_c \cos[2\pi(f_c + f_m)t] + A_m A_c(1 - a)\cos[2\pi(f_c - f_m)t]$$

where a is a constant, less than unity.

(a) Find the in-phase and quadrature components of the VSB modulated wave $s(t)$.

(b) What is the value of constant a for which $s(t)$ reduces to a DSBSC modulated wave?

(c) What are the values of constant a for which it reduces to an SSB modulated wave?

(d) The VSB wave $s(t)$, plus the carrier $A_c \cos(2\pi f_c t)$, is passed through an envelope detector. Determine the distortion produced by the quadrature component.

(e) What is the value of constant a for which this distortion reaches its worst possible value?

P7.7 Frequency Translation

Problem 22 Figure P7.4 shows the amplitude spectrum of an SSB-modulated signal $s(t)$. The signal $s(t)$ is applied to a mixer. Specify the parameters of the filter and local oscillator components of the mixer to do the following:

(a) Upconversion from 10 to 100 MHz.
(b) Downconversion from 10 to 1 MHz.

Problem 23 The spectrum of a voice signal $m(t)$ is zero outside the interval $f_a \leq |f| \leq f_b$. To ensure communication privacy, this signal is applied to a *scrambler* that consists of the following cascade of components: a product modulator, a high-pass filter, a second product modulator, and a low-pass filter. The carrier wave applied to the first product modulator has a frequency equal to f_c, whereas that applied to the second product modulator has a frequency equal to $f_b + f_c$; both of them have unity amplitude. The high-pass and low-pass filters have the same cutoff frequency at f_c. Assume that $f_c > f_b$.

(a) Derive an expression for the scrambler output $s(t)$, and sketch its spectrum.
(b) Show that the original voice signal $m(t)$ may be recovered from $s(t)$ by using a *descrambler* that is identical to the scrambler.

P7.8 Frequency-Division Multiplexing

Problem 24 The practical implementation of an FDM system usually involves many steps of modulation and demodulation. The first multiplexing step combines 12 voice inputs into a *basic group,* which is formed by having the nth input modulate a carrier at frequency $f_c = 112\text{ kHz} - 4n$, where $n = 1, 2, \ldots, 12$. The lower sidebands are then selected by band-pass filtering and are combined to form a group of 12 lower sidebands (one for each voice input). The next step in the FDM hierarchy involves the combination of 5 basic groups into a *supergroup*. This is

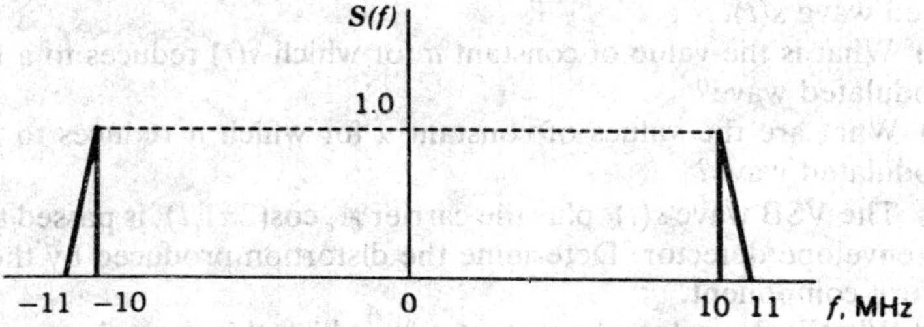

Figure P7.4

accomplished by using the nth group to modulate a carrier at frequency $f_c = 372 + 48n$ kHz, where $n = 1, 2, \ldots, 5$. Here again the lower sidebands are selected by filtering and are then combined to form a supergroup. In a similar manner, supergroups are combined into *mastergroups,* and mastergroups are combined into *very large groups.*

(a) Find the frequency band occupied by a basic group.
(b) Find the frequency band occupied by a supergroup.
(c) How many independent voice inputs does a supergroup accommodate?

P7.9 Application I

Problem 25 Figure P7.5 shows the block diagram of a *heterodyne spectrum analyzer.* It consists of a variable-frequency oscillator, multiplier, band-pass filter, and root mean-square (rms) meter. The oscillator has an amplitude A and operates over the range f_0 to $f_0 \pm W$, where f_0 is the midband frequency of the filter and W is the signal bandwidth. Assume that $f_0 = 2W$, the filter bandwidth Δf is small compared with f_0, and the passband amplitude response of the filter is one. Determine the value of the rms meter output for a low-pass input signal $g(t)$.

Problem 26 Figure P7.6 shows the block diagram of a *frequency synthesizer,* which enables the generation of many frequencies, each with the same high accuracy as the *master oscillator.* The master oscillator of frequency 1 MHz feeds two *spectrum generators,* one directly and the other through a *frequency divider.* Spectrum generator 1 produces a signal rich in the following harmonics: 1, 2, 3, 4, 5, 6, 7, 8, and 9 MHz. The frequency divider provides a 100-kHz output, in response to which spectrum generator 2 produces a second signal rich in the following harmonics: 100, 200, 300, 400, 500, 600, 700, 800, and 900 kHz. The harmonic selectors are designed to feed two signals into the mixer, one from spectrum generator 1 and the other from spectrum generator 2. Find the range of possible frequency outputs of this synthesizer and its resolution.

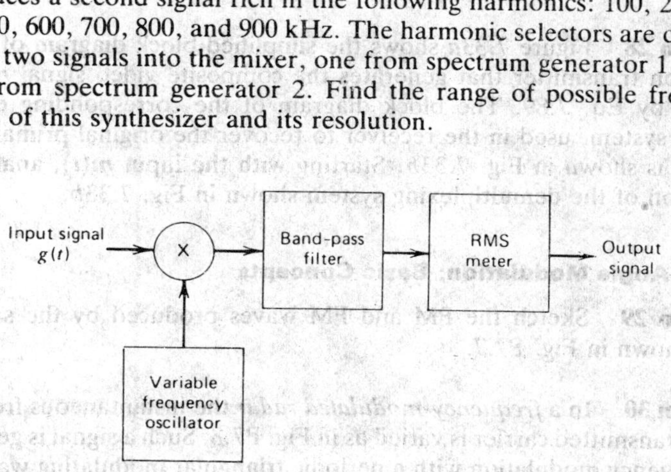

Figure P7.5

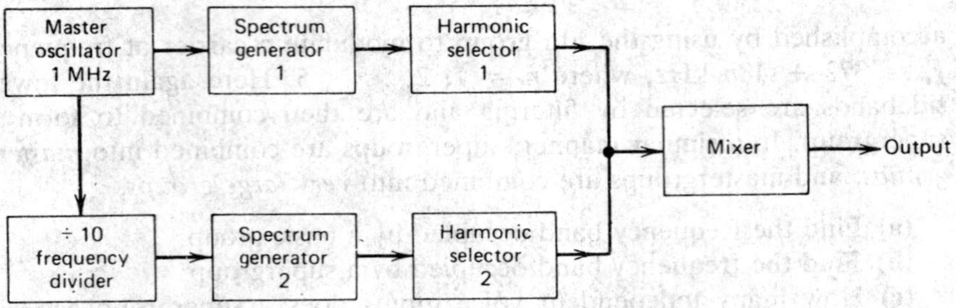

Figure P7.6

Problem 27 The use of quadrature-carrier multiplexing provides the basis for the generation of *AM stereo signals*. One particular form of such a signal is described by

$$s(t) = A_c[\cos(2\pi f_c t) + m_l(t)\cos(2\pi f_c t - \phi_0) + m_r(t)\cos(2\pi f_c t + \phi_0)]$$

where $A_c \cos(2\pi f_c t)$ is the unmodulated carrier, the phase difference $\phi_0 = 15°$, and $m_l(t)$ and $m_r(t)$ are the outputs of the left- and right-hand loudspeakers respectively. With $m_l(t)$ and $m_r(t)$ as inputs, do the following:

(a) Set up the block diagram of a system for generating the multiplexed signal $s(t)$.
(b) With $s(t)$ as input, set up the block diagram of a system for recovering $m_l(t)$ and $m_r(t)$.
(c) Suppose $s(t)$ is applied to an envelope detector. What is the resulting output?

Problem 28 Figure 7.33*a* shows the simplified block diagram of a color television transmitter that generates the composite video signal $m(t)$ described by Eq. 7.89. The block diagram of the corresponding demultiplexing system, used in the receiver to recover the original primary color signals, is shown in Fig. 7.33*b*. Starting with the input $m(t)$, analyze the operation of the demultiplexing system shown in Fig. 7.33*b*.

P7.10 Angle Modulation: Basic Concepts

Problem 29 Sketch the PM and FM waves produced by the sawtooth wave shown in Fig. P7.7.

Problem 30 In a *frequency-modulated radar* the instantaneous frequency of the transmitted carrier is varied as in Fig. P7.8. Such a signal is generated by frequency modulation with a periodic triangular modulating wave. The

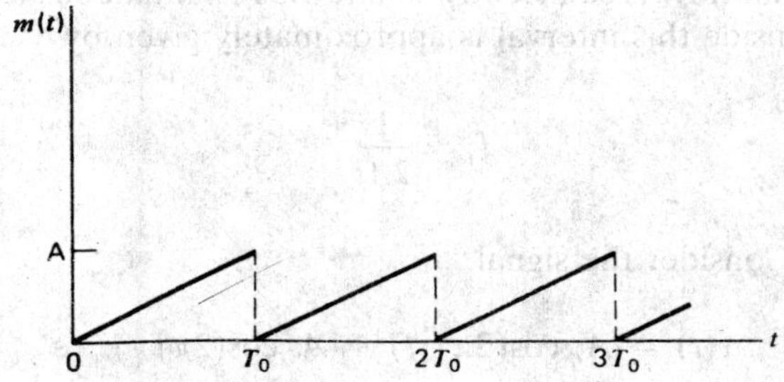

Figure P7.7

instantaneous frequency of the received echo signal is shown dashed in Fig. P7.8 where τ is the round-trip delay time. The transmitted and received echo signals are applied to a mixer, and the difference frequency component is retained. Assuming that $f_0\tau \ll 1$, determine the number of beat cycles at the mixer output, averaged over 1 s, in terms of the peak deviation Δf of the carrier frequency, the delay τ, and the repetition frequency f_0 of the transmitted signal.

Problem 31 The instantaneous frequency of a sine wave is equal to $f_c + \Delta f$ for $|t| \leq T/2$, and f_c for $|t| > T/2$. Determine the spectrum of this frequency-modulated wave.

Hint: Divide up the time interval of interest into three nonoverlapping regions: $-\infty < t < -T/2$, $-T/2 \leq t \leq T/2$, and $T/2 < t < \infty$.

Problem 32 Consider an interval Δt of an FM wave $s(t) = A_c \cos[\theta(t)]$ such that $\theta(t)$ satisfies the condition

$$\theta(t + \Delta t) - \theta(t) = \pi$$

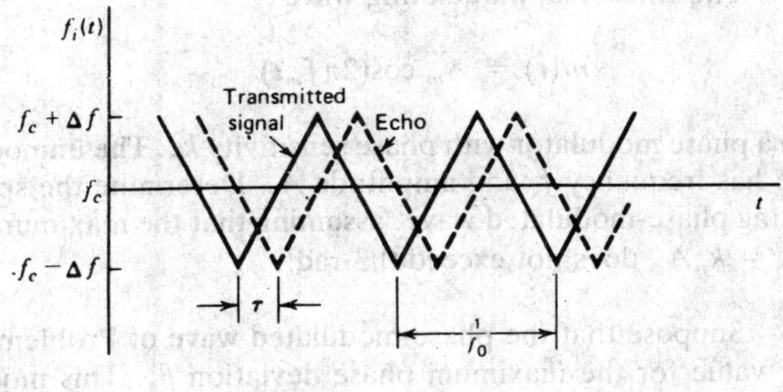

Figure P7.8

Hence, show that if Δt is sufficiently small, the instantaneous frequency of the FM wave inside this interval is approximately given by

$$f_i \simeq \frac{1}{2\Delta t}$$

Problem 33 Consider the signal

$$x(t) = A_c \cos(2\pi f_c t) + A_i \cos(2\pi f_i t)$$

where $A_c \cos(2\pi f_c t)$ represents an unmodulated carrier, and $A_i \cos(2\pi f_i t)$ represents an interfering signal. Assume that the amplitude ratio A_i/A_c is small compared to unity. Calculate the instantaneous frequency of $x(t)$ under this assumption.

Problem 34 Consider a narrow-band FM wave approximately defined by

$$s(t) \simeq A_c \cos(2\pi f_c t) - \beta A_c \sin(2\pi f_c t) \sin(2\pi f_m t)$$

(a) Determine the envelope of this modulated wave. What is the ratio of the maximum to the minimum value of this envelope? Plot this ratio versus β, assuming that β is restricted to the interval $0 \leqslant \beta \leqslant 0.3$.
(b) Determine the average power of the narrow-band FM wave, expressed as a percentage of the average power of the unmodulated carrier wave. Plot this result versus β, assuming that β is restricted to the interval $0 \leqslant \beta \leqslant 0.3$.
(c) By expanding the angular argument $\theta(t)$ of the narrow-band FM wave $s(t)$ in the form of a power series, and restricting the modulation index β to a maximum value of 0.3 rad, show that

$$\theta(t) \simeq 2\pi f_c t + \beta \sin(2\pi f_m t) - \frac{\beta^3}{3} \sin^3(2\pi f_m t)$$

What is the value of the harmonic distortion for $\beta = 0.3$?

Problem 35 The sinusoidal modulating wave

$$m(t) = A_m \cos(2\pi f_m t)$$

is applied to a phase modulator with phase sensitivity k_p. The unmodulated carrier wave has frequency f_c and amplitude A_c. Determine the spectrum of the resulting phase-modulated wave, assuming that the maximum phase deviation $\beta_p = k_p A_m$ does not exceed 0.3 rad.

Problem 36 Suppose that the phase-modulated wave of Problem 35 has an arbitrary value for the maximum phase deviation β_p. This modulated wave is applied to an ideal band-pass filter with midband frequency f_c and a passband extending from $f_c - 1.5f_m$ to $f_c + 1.5f_m$. Determine the envelope, phase, and instantaneous frequency of the modulated wave at the filter output as functions of time.

Problem 37 A carrier wave is frequency-modulated using a sinusoidal signal of frequency f_m and amplitude A_m.

(a) Determine the values of the modulation index β for which the carrier component of the FM wave is reduced to zero. For this calculation you may use the values of $J_0(\beta)$ given in Appendix B.
(b) In a certain experiment conducted with $f_m = 1$ kHz and increasing A_m (starting from 0 V), it is found that the carrier component of the FM wave is reduced to zero for the first time when $A_m = 2$ V. What is the frequency sensitivity of the modulator? What is the value of A_m for which the carrier component is reduced to zero for the second time?

Problem 38 A carrier wave of frequency 100 MHz is frequency-modulated by a sine wave of amplitude 20 V and frequency 100 kHz. The frequency sensitivity of the modulator is 25 kHz/V.

(a) Determine the approximate bandwidth of the FM wave, using Carson's rule.
(b) Determine the bandwidth by transmitting only those side-frequencies with amplitudes that exceed 1% of the unmodulated carrier amplitude. Use the universal curve of Fig. 7.41 for this calculation.
(c) Repeat your calculations, assuming that the amplitude of the modulating wave is doubled.
(d) Repeat your calculations, assuming that the modulation frequency is doubled.

Problem 39 Consider a wideband PM wave produced by a sinusoidal modulating wave $A_m \cos(2\pi f_m t)$, using a modulator with a phase sensitivity equal to k_p radians per volt.

(a) Show that if the maximum phase deviation of the PM wave is large compared with 1 rad, the bandwidth of the PM wave varies linearly with the modulation frequency f_m.
(b) Compare this characteristic of a wideband PM wave with that of a wideband FM wave.

Problem 40 In this problem we investigate the effect of a *weak nonlinearity* on frequency modulation. Specifically, consider a memoryless channel the transfer characteristic of which is described by the nonlinear relation:

$$v_o(t) = a_1 v_i(t) + a_2 v_i^2(t) + a_3 v_i^3(t)$$

where $v_i(t)$ and $v_o(t)$ are the input and output signals, respectively, and a_1, a_2, and a_3 are constant coefficients. Let

$$v_i(t) = A_c \cos[2\pi f_c t + \phi(t)]$$

where $\phi(t)$ is related to the message signal $m(t)$ by

$$\phi(t) = 2\pi k_f \int_0^t m(t)\, dt$$

(a) Show that the channel output $v_o(t)$ contains a dc component and three frequency-modulated waves with carrier frequencies f_c, $2f_c$ and $3f_c$.
(b) To extract an FM wave the same as that at the channel input, except for a change in carrier amplitude, show that by using Carson's rule the carrier frequency f_c must satisfy the following condition:

$$f_c > 3\Delta f + 2W$$

where W is the highest frequency component of the message signal $m(t)$ and Δf is the frequency deviation of the FM wave $v_i(t)$.
(c) Specify the band-pass filter required to do the extraction of the FM wave as specified in part (b).

Problem 41 Figure P7.9 shows the frequency-determining network of a voltage-controlled oscillator. Frequency modulation is produced by applying the modulating wave $A_m \sin(2\pi f_m t)$ plus a bias V_b to a pair of varactor diodes connected across the parallel combination of a 200 μH inductor and 100 pF capacitor. The capacitance of each varactor diode is related to the voltage V (in volts) applied across its electrodes by

$$C = 100V^{-1/2} \text{ pF}$$

The unmodulated frequency of oscillation is 1 MHz. The VCO output is applied to a frequency multiplier to produce an FM wave with a carrier frequency of 64 MHz and a modulation index of 5.

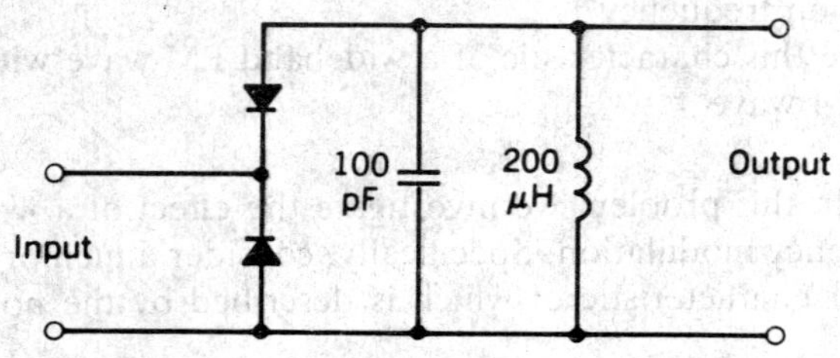

Figure P7.9

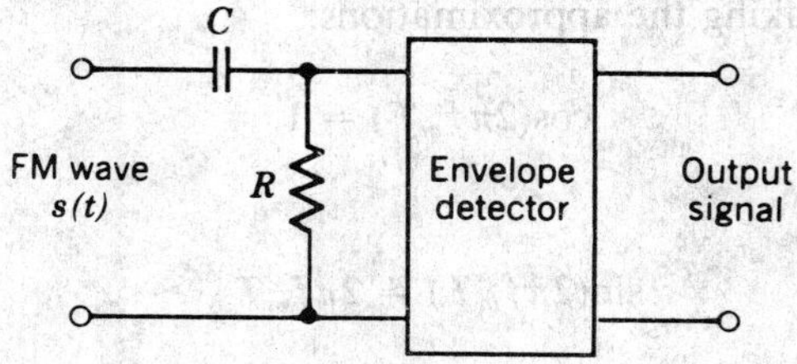

Figure P7.10

Determine: (a) the magnitude of the bias voltage V_b, and (b) the amplitude A_m of the modulating wave, given that $f_m = 10$ kHz.

Problem 42 The FM wave

$$s(t) = A_c \cos\left[2\pi f_c t + 2\pi k_f \int_0^t m(t)\, dt\right]$$

is applied to the system shown in Fig. P7.10 consisting of a high-pass RC filter and an envelope detector. Assume that: (a) the resistance R is small compared with the reactance of the capacitor C for all significant frequency components of $s(t)$, and (b) the envelope detector does not load the filter. Determine the resulting signal at the envelope detector output, assuming that $k_f|m(t)| < f_c$ for all t.

Problem 43 Consider the frequency demodulation scheme shown in Fig. P7.11 in which the incoming FM wave $s(t)$ is passed through a delay line that produces a phase shift of $-\pi/2$ radians at the carrier frequency f_c. The delay-line output is subtracted from the incoming FM wave, and the resulting composite wave is then envelope-detected. This demodulator finds wide application in demodulating FM waves at microwave frequencies. Assuming that

$$s(t) = A_c \cos[2\pi f_c t + \beta \sin(2\pi f_m t)]$$

analyze the operation of this demodulator when the modulation index β is less than unity and the delay T produced by the delay line is sufficiently

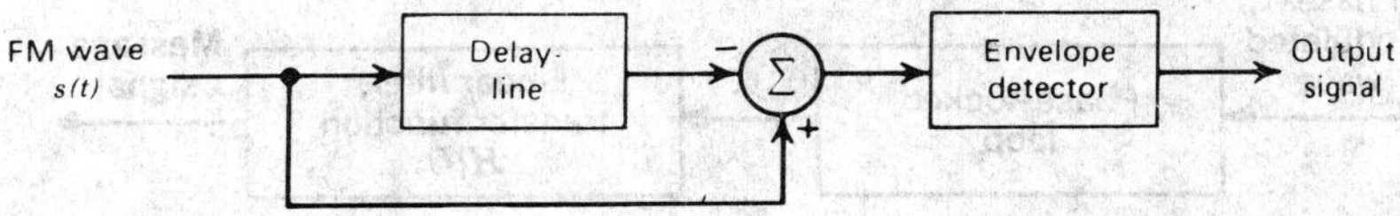

Figure P7.11

small to justify making the approximations:

$$\cos(2\pi f_m T) \simeq 1$$

and

$$\sin(2\pi f_m T) \simeq 2\pi f_m T$$

P7.12 Phase-Locked Loop

Problem 44 A first-order PLL is used to demodulate a single-tone FM wave that has the following characteristics:

$$\text{Modulation index } \beta = 5$$
$$\text{Modulation frequency } f_m = 15 \text{ kHz}$$

(a) Suggest a suitable value for the loop parameter K_0 of the PLL.
(b) For the value chosen in part (a), what is the corresponding value of the loop bandwidth?
(c) Suggest a method for reducing the loop bandwidth.

Problem 45 Show that a second-order PLL using the loop filter shown in Fig. 7.55 has the following closed-loop transfer function:

$$\frac{\Phi_e(f)}{\Phi_1(f)} = \frac{(jf/f_n)^2}{1 + 2\zeta(jf/f_n) + (jf/f_n)^2}$$

where f_n is the *natural frequency* of the loop and ζ is the *damping factor;* they are defined by

$$f_n = \sqrt{f_0 K_0}$$
$$\zeta = \sqrt{\frac{K_0}{4f_0}}$$

How does this PLL differ from a first-order PLL?

Problem 46 Figure P7.12 shows the cascade connection of a phase-locked loop and a linear filter. A phase-modulated wave is applied to the input

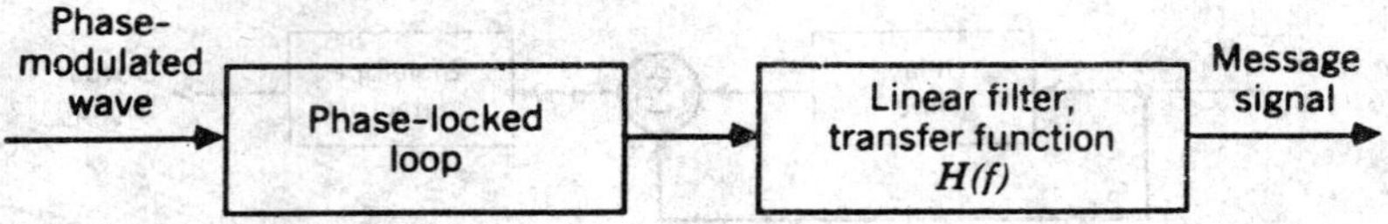

Figure P7.12

of the phase-locked loop. The requirement is to reproduce the message signal at the output of the filter. Find the transfer function $H(f)$ of the filter that satisfies this requirement, assuming that the phase-locked loop has a large loop gain.

P7.13 Limiting of FM Waves

Problem 47 Consider the modulated signal

$$s_1(t) = a(t)\cos[2\pi f_c t + 2\pi k_f \int_0^t m(t)\,dt]$$

where $a(t)$ is a slowly varying envelope function, f_c is the carrier frequency, k_f is a frequency sensitivity, and $m(t)$ is a message signal. The modulated signal $s(t)$ is processed by a band-pass limiter (consisting of a hard limiter followed by a band-pass filter) to remove amplitude fluctuations due to $a(t)$. Specify the parameters of the band-pass filter component so as to produce the FM wave

$$s_2(t) = A\cos[2\pi f_c t + 2\pi k_f \int_0^t m(t)\,dt]$$

where A is a constant amplitude.

P7.14 Application II

Problem 48 Consider the analysis of FM stereo transmission, assuming that the left-hand and right-hand signals consist of two tones of different frequencies but the same amplitude, as shown by

$$m_l(t) = A_m \cos(2\pi f_l t)$$

and

$$m_r(t) = A_m \cos(2\pi f_r t)$$

(a) Show that the amplitude of a composite signal consisting of the sum signal and the DSBSC modulated version of the difference signal is bounded by $2A_m$; that is:

$$|m_l(t) + m_r(t) + [m_l(t) - m_r(t)]\cos(4\pi f_{cc} t)| \leq 2A_m$$

where f_{cc} is the subcarrier frequency.

(b) Let $A_m = 0.45$, and let the pilot (of frequency f_{cc}) injected into the multiplexed FM stereo signal have amplitude $A_{cc} = 0.1$. Let the FM wave produced by this multiplexed signal have frequency deviation

Δf = 75 kHz. Find the effective frequency deviation that results from the reception of the FM wave by a monophonic receiver that responds only to the sum signal.

Problem 49 Figure P7.13 shows the block diagram of a real-time *spectrum analyzer* working on the principle of frequency modulation. The given signal $g(t)$ and a frequency-modulated signal $s(t)$ are applied to a multiplier and the output $g(t)s(t)$ is fed into a filter of impulse response $h(t)$. The $s(t)$ and $h(t)$ are *linear FM signals* whose instantaneous frequencies vary at opposite rates, as shown by

$$s(t) = \cos(2\pi f_c t - \pi k t^2)$$

and

$$h(t) = \cos(2\pi f_c t + \pi k t^2)$$

where k is a constant. Show that the envelope of the filter output is proportional to the amplitude spectrum of the input signal $g(t)$ with kt playing the role of frequency f.

Hint: Use the complex notations described in Section 3.5 for band-pass transmission.

P7.15 Digital Modulation Techniques

Problem 50 Sketch the binary ASK waveform for the sequence 1011010011. Assume that the carrier frequency f_c equals the bit rate $1/T_b$.

Problem 51 Repeat Problem 50 using binary PSK.

Problem 52 Sketch the binary FSK waveform for the sequence 1011010011. Assume that the two frequencies used to represent symbols 1 and 0 are given by, respectively,

$$f_1 = \frac{2}{T_b}$$

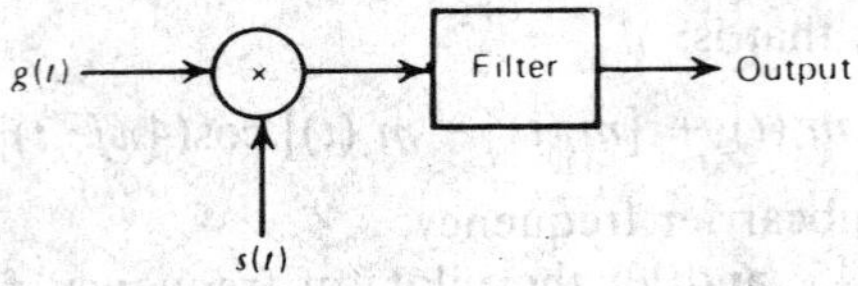

Figure P7.13

and

$$f_2 = \frac{1}{T_b}.$$

where T_b is the bit duration.

Problem 53 Both binary FSK and binary PSK signals have a constant envelope. Yet binary FSK signals can be noncoherently detected, whereas binary PSK signals cannot be. What are the reasons for this difference?

Problem 54 The binary sequence 1011010011 is transmitted over a communication channel using DPSK. The channel introduces a 180°-phase reversal.

(a) Sketch the transmitted DPSK waveform, assuming an initial bit of 1. What is the effect of changing the initial bit to a 0?
(b) Assuming that the channel is noise-free, show that the DPSK detector in the receiver reproduces the original binary sequence, despite the 180°-phase reversal in the channel.

Problem 55 Set up a circuit for generating a differentially encoded sequence (that includes the initial bit) in response to an incoming binary sequence. Is the structure of this circuit affected by the identity of the initial bit?

Problem 56 Sketch the QPSK waveform for the sequence 1011010011. You may assume the following:

(a) The carrier frequency equals the bit rate.
(b) The dibits 00, 10, 11, and 01 are represented by phase shifts equal to 0, $\pi/2$, π, $3\pi/2$ radians.

Problem 57 Sketch the waveform of the MSK signal for the sequence 1011010011. Assume that the carrier frequency equals the bit rate.

CHAPTER EIGHT

PROBABILITY THEORY AND RANDOM PROCESSES

The term "random" is used to describe erratic and apparently unpredictable variations of an observed signal. Indeed, random signals (in one form or another) are encountered in every practical communication system. Consider, for example, a radio communication system. The received signal in such a system is random in nature. Ordinarily, the received signal consists of an information-bearing signal component, a random-interference component, and receiver noise. The *information-bearing signal* component may represent, for example, a voice signal that, typically, consists of randomly spaced bursts of energy of random duration. The *interference* component represents the extraneous electromagnetic waves produced by other communication systems and atmospheric electricity. A major type of noise is *thermal noise,* which is caused by the random motion of the electrons in conductors and devices at the front end of the receiver.

The important point is that, regardless of the underlying causes of randomness, we cannot predict the exact value of the received signal. Nevertheless, the received signal can be described in terms of its statistical properties such as the average power, or the spectral distribution of the average power. The mathematical discipline that deals with the statistical characterization of random signals is probability theory.[1] We begin our discussion of random signals with a review of probability theory in the next section.

8.1 PROBABILITY THEORY

Probability theory is rooted in situations that involve performing an experiment with an outcome that is subject to *chance.* Moreover, if the experiment is repeated, the outcome can differ because of the influence of an underlying random phenomenon or chance mechanism. Such an experiment is referred to as a *random experiment.* For example, the experiment may be the observation of the result of the tossing of a fair coin. In this experiment, the possible outcomes of a trial are "heads" or "tails."

To be more precise in the description of a random experiment, we ask for three features:

1. The experiment is repeatable under identical conditions.
2. On any trial of the experiment, the outcome is unpredictable.
3. For a large number of trials of the experiment, the outcomes exhibit *statistical regularity.* That is, a definite *average* pattern of outcomes is observed if the experiment is repeated a large number of times.

RELATIVE-FREQUENCY APPROACH

Let *event* A denote one of the possible outcomes of a random experiment. For example, in the coin-tossing experiment, event A may represent "heads." Suppose that in n trials of the experiment, event A occurs n_A times. We may then assign the ratio n_A/n to the event A. This ratio is called the *relative frequency* of the event A. Clearly, the relative frequency

[1]For a detailed treatment of probability theory and the related subject of random processes, see Davenport and Root (1958), Fry (1965), Thomas (1986), Wozencraft and Jacobs (1965), Feller (1968), Fines (1973), Blake (1979), and Papoulis (1984).

is a *nonnegative real number less than or equal to one*. That is to say,

$$0 \leqslant \frac{n_A}{n} \leqslant 1 \tag{8.1}$$

If event A occurs in none of the trials, $(n_A/n) = 0$. If, on the other hand, event A occurs in all the n trials, $(n_A/n) = 1$.

We say that the experiment exhibits *statistical regularity* if for *any* sequence of n trials the relative frequency n_A/n *converges to the same limit* as n becomes large. Accordingly, it seems natural for us to define the *probability of event A* as

$$P(A) = \lim_{n\to\infty} \left(\frac{n_A}{n}\right) \tag{8.2}$$

Thus, in the coin-tossing experiment, we may expect that out of a million tosses of a fair coin, about one half of them will show up heads.

The probability of an event is intended to represent the *likelihood* that a trial of the experiment will result in the occurrence of that event. For many engineering applications and games of chance, the use of Eq. 8.2 to define the probability of an event is acceptable. However, for many other applications this definition is inadequate. Consider, for example, the statistical analysis of the stock market: How are we to achieve repeatability of such an experiment? A more satisfying approach is to state the properties that any measure of probability is expected to have, postulating them as *axioms,* and then use relative-frequency interpretations to justify them.

AXIOMS OF PROBABILITY

When we perform a random experiment, it is natural for us to be aware of the various outcomes that are likely to arise. In this context, it is convenient to think of an experiment and its possible outcomes as defining a space and its points. With each possible outcome of the experiment, we associate a point called the *sample point,* which we denote by s_k. The totality of sample points corresponding to the aggregate of all possible outcomes of the experiment, is called the *sample space,* which we denote by $\mathcal{S}$. An event corresponds to either a single sample point or a set of sample points. In particular, the entire sample space $\mathcal{S}$ is called the *sure event;* the null set $\emptyset$ is called the *null* or *impossible event;* and a single sample point is called an *elementary event.*

Consider, for example, an experiment that involves the throw of a die. In this experiment there are six possible outcomes: the showing of one, two, three, four, five and six dots on the upper face of the die. By assigning a sample point to each of these possible outcomes, we have a one-dimensional sample space that consists of six sample points, as shown in Fig. 8.1.

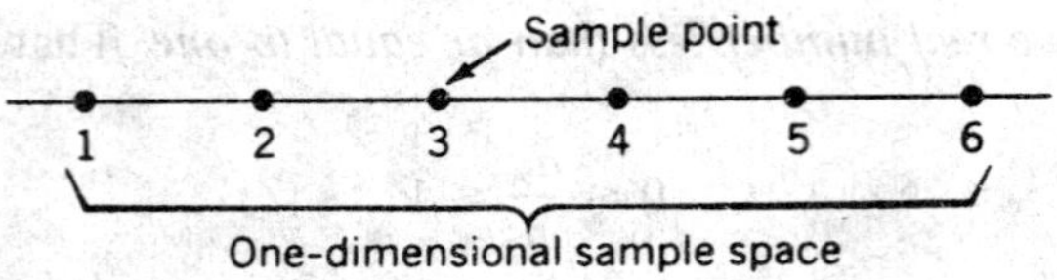

Figure 8.1
Sample space for the experiment of throwing a die.

The elementary event describing the statement "a six shows" corresponds to the sample point {6}. On the other hand, the event describing the statement "an even number of dots shows" corresponds to the subset {2,4,6} of the sample space. Note that the term "event" is used interchangeably to describe the subset or the statement.

We are now ready to make a formal definition of probability. A *probability system* consists of the triple:

1. *A sample space* $\mathcal{S}$ *of elementary events* (*outcomes*).
2. *A class* $\mathcal{E}$ *of events* that are subsets of $\mathcal{S}$.
3. *A probability measure* $P(\cdot)$ assigned to each event A in the class $\mathcal{E}$, which has the following properties:
   (i) $P(\mathcal{S}) = 1$ (8.3)
   (ii) $0 \leq P(A) \leq 1$ (8.4)
   (iii) If $A + B$ is the *union of two mutually exclusive events* in the class $\mathcal{E}$, then

$$P(A + B) = P(A) + P(B) \tag{8.5}$$

Properties (i), (ii), and (iii) are known as the *axioms of probability*. Axiom (i) states that the probability of the sure event is unity. Axiom (ii) states that the probability of an event is a nonnegative real number that is less than or equal to unity. Axiom (iii) states that the probability of the union of two mutually exclusive events is the sum of the probabilities of the individual events.

Although the axiomatic approach to probability theory is abstract in nature, all three axioms have relative-frequency interpretations of their own. Axiom (ii) corresponds to Eq. 8.1. Axiom (i) corresponds to the limiting case of Eq. 8.1 when the event A occurs in all the n trials. To interpret axiom (iii), we note that if event A occurs n_A times in n trials and event B occurs n_B times, then the union event "A or B" occurs in $n_A + n_B$ trials (since A and B can never occur on the same trial). Hence, $n_{A+B} = n_A + n_B$, and so we have

$$\frac{n_{A+B}}{n} = \frac{n_A}{n} + \frac{n_B}{n}$$

which has a mathematical form similar to that of axiom (iii).

ELEMENTARY PROPERTIES OF PROBABILITY

Axioms (i), (ii), and (iii) constitute an implicit definition of probability. We may use these axioms to develop some other basic properties of probability.

PROPERTY 1: $P(\bar{A}) = 1 - P(A)$ (8.6)

where $\bar{A}$ (denoting "not A") is the complement of event A.

The use of this property helps us investigate the *nonoccurrence of an event.* To prove it, we express the sample space $\mathcal{S}$ as the union of two mutually exclusive events A and $\bar{A}$:

$$S = A + \bar{A}$$

Then, the use of axioms (i) and (iii) yields

$$1 = P(A) + P(\bar{A})$$

from which Eq. 8.6 follows directly.

PROPERTY 2

If M mutually exclusive events $A_1, A_2, \ldots, A_M$ have the exhaustive property

$$A_1 + A_2 + \cdots + A_M = \mathcal{S} \tag{8.7}$$

then

$$P(A_1) + P(A_2) + \cdots + P(A_M) = 1 \tag{8.8}$$

To prove this property, we generalize axiom (iii) by writing

$$P(A_1 + A_2 + \cdots + A_M) = P(A_1) + P(A_2) + \cdots + P(A_M)$$

The use of axiom (i) in Eq. 8.7 yields

$$P(A_1 + A_2 + \cdots + A_M) = 1$$

Hence, the result of Eq. 8.8 follows.

When the M events are *equally likely* (i.e., they have equal probabilities), then Eq. 8.8 simplifies as

$$P(A_i) = \frac{1}{M}, \quad i = 1, 2, \ldots, M \tag{8.9}$$

PROPERTY 3

When events A and B are not mutually exclusive, then the probability of the union event "A or B" equals

$$P(A + B) = P(A) + P(B) - P(AB) \tag{8.10}$$

where P(AB) is the probability of the joint event "A and B".

The probability $P(AB)$ is called the *joint probability*. It has the following relative-frequency interpretation

$$P(AB) = \lim_{n\to\infty} \left(\frac{n_{AB}}{n}\right)$$

where n_{AB} denotes the number of times the events A and B occur simultaneously in n trials of the experiment. Axiom (iii) is a special case of Eq. 8.10; when A and B are mutually exclusive, $P(AB)$ is zero, and Eq. 8.10 reduces to the same form as Eq. 8.5.

EXERCISE 1 Consider an experiment in which two coins are thrown. What is the probability of getting one head and one tail?

EXERCISE 2 Consider an experiment in which two dice are thrown. What is the probability that the number of dots showing on the upper faces of the two dice add up to 6?

CONDITIONAL PROBABILITY

Suppose we perform an experiment that involves a pair of events A and B. Let $P(B|A)$ denote the probability of event B, given that event A has occurred. The probability $P(B|A)$ is called the *conditional probability of B given A*. Assuming that A has nonzero probability, the conditional probability $P(B|A)$ is defined by

$$P(B|A) = \frac{P(AB)}{P(A)} \tag{8.11}$$

where $P(AB)$ is the joint probability of A and B.

We justify the definition of conditional probability given in Eq. 8.11 by presenting a relative-frequency interpretation of it. Suppose that we perform an experiment and examine the occurrence of a pair of events A and B. Let n_{AB} denote the number of times the joint event AB occurs in n

trials. Suppose that in the same n trials the event A occurs n_A times. Since the joint event AB corresponds to both A and B occurring, it follows that n_A must include n_{AB}. In other words, we have

$$\frac{n_{AB}}{n_A} \leq 1$$

The ratio n_{AB}/n_A represents the relative frequency of B given that A has occurred. For large n, the ratio n_{AB}/n_A equals the conditional probability $P(B|A)$. That is,

$$P(B|A) = \lim_{n\to\infty}\left(\frac{n_{AB}}{n_A}\right)$$

or equivalently,

$$P(B|A) = \lim_{n\to\infty}\left(\frac{n_{AB}/n}{n_A/n}\right)$$

Recognizing that

$$P(AB) = \lim_{n\to\infty}\left(\frac{n_{AB}}{n}\right)$$

and

$$P(A) = \lim_{n\to\infty}\left(\frac{n_A}{n}\right)$$

the result of Eq. 8.11 follows.

We may rewrite Eq. 8.11 as

$$P(AB) = P(B|A)P(A) \tag{8.12}$$

It is apparent that we may also write

$$P(AB) = P(A|B)P(B) \tag{8.13}$$

Equations 8.12 and 8.13 state that the joint probability of two events may be expressed as the product of the conditional probability of one event, given the other, and the elementary probability of the other. Note that the conditional probabilities $P(B|A)$ and $P(A|B)$ have essentially the same properties as the various probabilities previously defined.

Situations may exist where the conditional probability $P(A|B)$ and the probabilities $P(A)$ and $P(B)$ are easily determined directly, but the conditional probability $P(B|A)$ is desired. From Eqs. 8.12 and 8.13, it follows that, provided $P(A) \neq 0$, we may determine $P(B|A)$ by using the relation

$$P(B|A) = \frac{P(A|B)P(B)}{P(A)} \tag{8.14}$$

This relation is a special form of *Bayes' rule.*

Suppose that the conditional probability $P(B|A)$ is simply equal to the elementary probability of occurrence of event B, that is,

$$P(B|A) = P(B) \tag{8.15}$$

Under this condition, the probability of occurrence of the joint event AB is equal to the product of the elementary probabilities of the events A and B:

$$P(AB) = P(A)P(B)$$

so that

$$P(A|B) = P(A)$$

That is, the conditional probability of the event A, assuming the occurrence of the event B, is simply equal to the elementary probability of the event A. We thus see that in this case a knowledge of the occurrence of one event tells us no more about the probability of occurrence of the other event than we knew without that knowledge. Events A and B that satisfy this condition are said to be *statistically independent.*

EXAMPLE 1 BINARY SYMMETRIC CHANNEL

Consider a *discrete memoryless channel* used to transmit binary data. The channel is said to be *discrete* in that it is designed to handle discrete messages. It is *memoryless* in the sense that the channel output at any time depends only on the channel input at that time. Owing to the unavoidable presence of *noise* in the channel, *errors* are made in the received binary data stream. Specifically, when symbol 1 is sent, *occasionally* an error is made and symbol 0 is received, and vice versa. The channel is assumed to be symmetric, which means that the probability of receiving symbol 1 when symbol 0 is sent is the same as the probability of receiving symbol 0 when symbol 1 is sent.

To describe the probabilistic nature of this channel fully, we need two sets of probabilities:

1. The *a priori probabilities* of sending binary symbols 0 and 1: They are

$$P(A_0) = p_0 \tag{8.16}$$

and

$$P(A_1) = p_1 \tag{8.17}$$

where A_0 and A_1 denote the events of transmitting symbols 0 and 1, respectively. Note that $p_0 + p_1 = 1$.

2. The *conditional probabilities of error:* They are

$$P(B_1|A_0) = P(B_0|A_1) = p \tag{8.18}$$

where B_0 and B_1 denote the events of receiving symbols 0 and 1, respectively. The conditional probability $P(B_1|A_0)$ is the probability of receiving symbol 1, given that symbol 0 is sent. The second conditional probability $P(B_0|A_1)$ is the probability of receiving symbol 0, given that symbol 1 is sent.

The requirement is to determine the *a posteriori probabilities* $P(A_0|B_0)$ and $P(A_1|B_1)$. The conditional probability $P(A_0|B_0)$ is the probability that symbol 0 was sent, given that symbol 0 is received. The second conditional probability $P(A_1|B_1)$ is the probability that symbol 1 was sent, given that symbol 1 is received. Both these conditional probabilities refer to events that are observed "after the fact"; hence, the name "a posteriori" probabilities.

Since the events B_0 and B_1 are mutually exclusive, and the probability of receiving symbol 0 or symbol 1 is unity, we have from axiom (iii):

$$P(B_0|A_0) + P(B_1|A_0) = 1$$

That is to say,

$$P(B_0|A_0) = 1 - p \tag{8.19}$$

Similarly, we may write

$$P(B_1|A_1) = 1 - p \tag{8.20}$$

Accordingly, we may use the *transition probability diagram* shown in Fig. 8.2 to represent the binary communication channel specified in this ex-

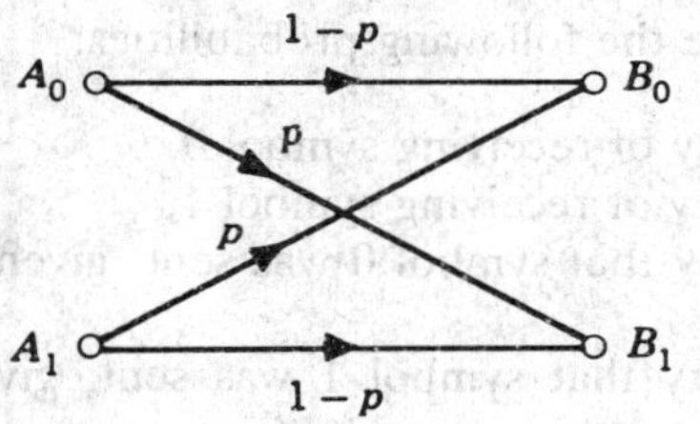

Figure 8.2
Transition probability diagram of binary symmetric channel.

ample; the term "transition probability" refers to the conditional probability of error. Figure 8.2 clearly depicts the (assumed) symmetric nature of the channel; hence, the name "binary symmetric channel."

From Fig. 8.2, we deduce the following results:

1. The probability of receiving symbol 0 is given by

$$\begin{aligned} P(B_0) &= P(B_0|A_0)P(A_0) + P(B_0|A_1)P(A_1) \\ &= (1 - p)p_0 + pp_1 \end{aligned} \tag{8.21}$$

2. The probability of receiving symbol 1 is given by

$$\begin{aligned} P(B_1) &= P(B_1|A_0)P(A_0) + P(B_1|A_1)P(A_1) \\ &= pp_0 + (1 - p)p_1 \end{aligned} \tag{8.22}$$

Therefore, applying Bayes' rule, we obtain

$$\begin{aligned} P(A_0|B_0) &= \frac{P(B_0|A_0)P(A_0)}{P(B_0)} \\ &= \frac{(1 - p)p_0}{(1 - p)p_0 + pp_1} \end{aligned} \tag{8.23}$$

$$\begin{aligned} P(A_1|B_1) &= \frac{P(B_1|A_1)P(A_1)}{P(B_1)} \\ &= \frac{(1 - p)p_1}{pp_0 + (1 - p)p_1} \end{aligned} \tag{8.24}$$

These are the desired results.

EXERCISE 3 Continuing with Example 1, find the following conditional probabilities: $P(A_0|B_1)$ and $P(A_1|B_0)$.

EXERCISE 4 Consider a binary symmetric channel for which the conditional probability of error $p = 10^{-4}$, and symbols 0 and 1 occur with equal probability. Calculate the following probabilities:

(a) The probability of receiving symbol 0.
(b) The probability of receiving symbol 1.
(c) The probability that symbol 0 was sent, given that symbol 0 is received.
(d) The probability that symbol 1 was sent, given that symbol 0 is received.

EXAMPLE 2 CHAIN OF PCM REGENERATIVE REPEATERS

In Section 5.6 we described the use of *regenerative repeaters* in a pulse-code modulation (PCM) system as a means of combatting the effects of channel noise. Specifically, the function of a regenerative repeater is two-fold: (1) to detect the presence of symbol 0 or 1 before the pulses representing these symbols become too weak and therefore lost in channel noise, and (2) to retransmit new clean pulses (representing the symbols detected) on to the next regenerative repeater. Consider a binary PCM system that uses a chain of $(k - 1)$ regenerative repeaters, followed by one last regeneration at the receiver input, as illustrated in Fig. 8.3. Given that the average probability of error incurred in each regeneration process is P_e, we wish to calculate the average probability of error P_E for the entire system.

The system may be viewed as the cascade connection of k identical links, with each link responsible for an average probability of error P_e. A binary symbol 1 or 0 sent over such a system is detected correctly at the receiver if either the symbol in question is detected correctly over each link in the system or it experiences errors over an even number of links. We may thus express the probability of *correct reception* P_C at the receiver output as

$$\begin{aligned} P_C &= 1 - P_E \\ &= P(\text{correct detection over all links in the system}) \\ &\quad + P(\text{error over any two links in the system}) \\ &\quad + P(\text{error over any four links in the system}) \\ &\quad\ \vdots \\ &\quad + P(\text{error over } l \text{ links in the system}) \end{aligned} \tag{8.25}$$

where, in the last term, we have

$$l = \begin{cases} k, & \text{if } k \text{ is even} \\ k - 1, & \text{if } k \text{ is odd} \end{cases}$$

Given that the probability of error over each link is P_e, we may write

$$P(\text{correct detection over all links in the system}) = (1 - P_e)^k$$

$$P(\text{error over any } j \text{ links in the system}) = \frac{k!}{j!\,(k-j)!} P_e^j (1 - P_e)^{k-j}$$

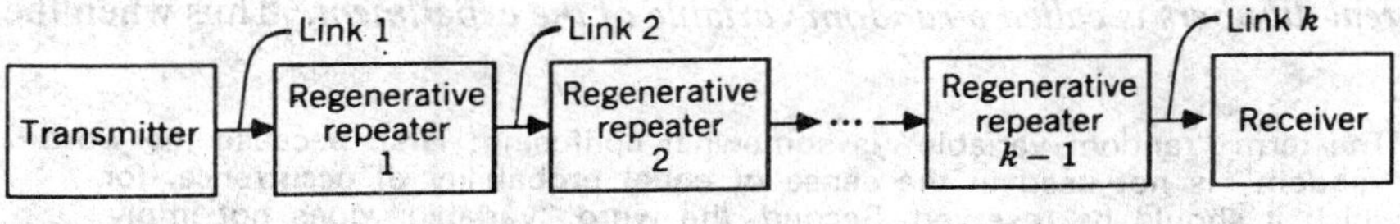

Figure 8.3
Chain of PCM regenerative repeaters.

Using these results in Eq. 8.25, we get

$$1 - P_E = (1 - P_e)^k + \sum_{j=2,4,\ldots}^{l} \frac{k!}{j!\,(k-j)!} P_e^j (1 - P_e)^{k-j} \qquad (8.26)$$

In practice, we usually find that P_e is very small compared to unity, so that we may make the following two approximations:

1. We may approximate the first term in Eq. 8.26 as

$$(1 - P_e)^k \simeq 1 - kP_e + \frac{k(k-1)}{2} P_e^2 \qquad (8.27)$$

2. We may approximate the second term in Eq. 8.26 by retaining only that term in the summation that corresponds to $j = 2$, and also writing $(1 - P_e)^{k-j} \simeq 1$.

Accordingly, we may approximate Eq. 8.26 as:

$$1 - P_E \simeq 1 - kP_e + k(k-1) P_e^2$$

or equivalently

$$P_E \simeq kP_e - k(k-1)P_e^2 \qquad (8.28)$$

If the number of links k in the system is such that kP_e is small compared to unity, we may further approximate Eq. 8.28 as

$$P_E \simeq kP_e \qquad (8.29)$$

That is, the average probability of error in the entire PCM system of Fig. 8.3 is equal to the average probability of error in a single link of the system times the total number of links in the system.

8.2 RANDOM VARIABLES

In conducting an experiment it is convenient to assign a variable to the experiment whose outcome determines the value of the variable. We do so because we may have no a priori knowledge of the outcome of the experiment other than it may take on a value within a certain range. *A function whose domain is a sample space and whose range is some set of real numbers is called a random variable of the experiment.*[2] Thus when the

[2]The term "random variable" is somewhat confusing: First, because the word "random" is not used in the sense of equal probability of occurrence, for which it should be reserved. Second, the word "variable" does not imply dependence on the experimental outcome, which is an essential part of the meaning. Nevertheless, the term is so deeply imbedded in the literature of probability that its usage has persisted.

outcome of the experiment is s, the random variable is denoted as $X(s)$ or simply X. For example, the sample space representing the outcomes of the throw of a die is a set of six sample points that may be taken to be the integers 1, 2, . . . , 6. Then if we identify the sample point k with the event that k dots show when the die is thrown, the function $X(k) = k$ is a random variable such that $X(k)$ equals the number of dots that show when the die is thrown. In this example, the random variable takes on only a discrete set of values. In such a case we say that we are dealing with a *discrete random variable.* More precisely, *the random variable X is a discrete random variable if X can take on only a finite number of values in any finite observation interval.* If, however, *the random variable X can take on any value in a finite observation interval, X is called a continuous random variable.* For example, the random variable that represents the amplitude of a noise voltage at a particular instant of time is a continuous random variable because, in theory, it may take on any value between plus and minus infinity.

To proceed further, we need a probabilistic description of random variables that works equally well for both discrete and continuous random variables. Let us consider the random variable X and the probability of the event $X \leq x$, when x is given. We denote the probability of this event by $P(X \leq x)$. It is apparent that this probability is a function of the *dummy variable* x. To simplify our notation, we write.

$$F_X(x) = P(X \leq x) \tag{8.30}$$

The function $F_X(x)$ is called the *cumulative distribution function* or simply the *distribution function* of the random variable X. Note that $F_X(x)$ is a function of x, not of the random variable X. However, it depends on the assignment of the random variable X, which accounts for the use of X as subscript. For any point x, the distribution function $F_X(x)$ expresses a probability.

The distribution function $F_X(x)$ has the following properties, which follow directly from Eq. 8.30:

1. The distribution function $F_X(x)$ is bounded between zero and one.
2. The distribution function $F_X(x)$ is a monotone nondecreasing function of x; that is,

$$F_X(x_1) \leq F_X(x_2), \qquad \text{if } x_1 < x_2 \tag{8.31}$$

An alternative description of the probability distribution of the random variable X is often useful. This is the derivative of the distribution function, as shown by

$$f_X(x) = \frac{d}{dx} F_X(x) \tag{8.32}$$

which is called the *probability density function.* Note that the differentiation in Eq. 8.32 is with respect to the dummy variable x. The name, density function, arises from the fact that the probability of the event $x_1 < X \leq x_2$ equals

$$\begin{aligned} P(x_1 < X \leq x_2) &= P(X \leq x_2) - P(X \leq x_1) \\ &= F_X(x_2) - F_X(x_1) \\ &= \int_{x_1}^{x_2} f_X(x)\, dx \end{aligned} \tag{8.33}$$

Since $F_X(\infty) = 1$, corresponding to the probability of the certain event, and $F_X(-\infty) = 0$, corresponding to the probability of the impossible event, it follows immediately from Eq. 8.33 that

$$\int_{-\infty}^{\infty} f_X(x)\, dx = 1 \tag{8.34}$$

Also, as mentioned earlier, a distribution function must always be monotone nondecreasing. Hence, its derivative, the probability density function, must always be nonnegative. *A probability density function must always be a nonnegative function with the total area under its curve equal to one.*

EXAMPLE 3 UNIFORM DISTRIBUTION

Consider a random variable X defined by (assuming $b > a$)

$$f_X(x) = \begin{cases} \dfrac{1}{b-a}, & a \leq x \leq b \\ 0, & \text{elsewhere} \end{cases} \tag{8.35}$$

This function, shown in Fig. 8.4*a*, satisfies the requirements of a probability density because $f_X(x) \geq 0$, and the area under the curve is unity. A random variable having the probability density function of Eq. 8.35 is said to be *uniformly distributed.*

The corresponding distribution function of the uniformly distributed random variable X is continuous everywhere, as shown by

$$F_X(x) = \begin{cases} 0, & x < a \\ \dfrac{x-a}{b-a}, & a \leq x \leq b \\ 1, & x > b \end{cases} \tag{8.36}$$

This distribution function is plotted in Fig. 8.4*b*.

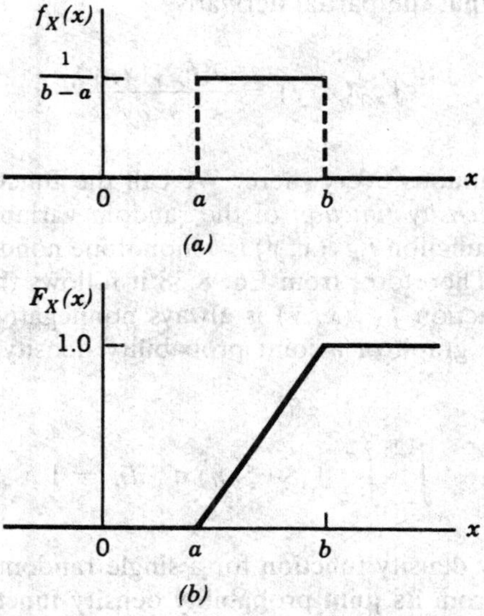

Figure 8.4
The uniform distribution. (a) Probability density function. (b) Distribution function.

SEVERAL RANDOM VARIABLES

Thus far we have focused attention on situations involving a single random variable. However, we find frequently that the outcome of an experiment requires several random variables to describe the experiment. In the sequel we consider situations involving two random variables. The probabilistic description developed in this way may be readily extended to any number of random variables.

Consider two random variables X and Y. We define *the joint distribution function $F_{X,Y}(x, y)$ as the probability that the random variable X is less than or equal to a specified value x and that the random variable Y is less than or equal to a specified value y*. The variables X and Y may be two distinct one-dimensional random variables or the components of a single two-dimensional random variable. The joint distribution function $F_{X,Y}(x, y)$ is the probability that the outcome of an experiment will result in a sample point lying inside the quadrant $(-\infty < X \leq x, -\infty < Y \leq y)$ of the joint-sample space. That is,

$$F_{X,Y}(x, y) = P(X \leq x, Y \leq y) \tag{8.37}$$

Suppose that the joint distribution function $F_{X,Y}(x, y)$ is continuous everywhere, and that the partial derivative

$$f_{X,Y}(x, y) = \frac{\partial^2 F_{X,Y}(x, y)}{\partial x \, \partial y} \tag{8.38}$$

exists and is continuous everywhere. We call the function $f_{X,Y}(x, y)$ the *joint probability density function* of the random variables X and Y. The joint distribution function $F_{X,Y}(x, y)$ is a monotone nondecreasing function of both x and y. Therefore, from Eq. 8.38 it follows that the joint probability density function $f_{X,Y}(x, y)$ is always nonnegative. Also, the total volume under the graph of a joint probability density function must be unity, as shown by

$$\int_{-\infty}^{\infty} \int_{-\infty}^{\infty} f_{X,Y}(\xi, \eta) \, d\xi \, d\eta = 1 \tag{8.39}$$

The probability density function for a single random variable (X, say) can be obtained from its joint probability density function with a second random variable (Y, say) in the following way. We first note that

$$F_X(x) = \int_{-\infty}^{\infty} \int_{-\infty}^{x} f_{X,Y}(\xi, \eta) \, d\xi \, d\eta \tag{8.40}$$

Therefore, differentiating both sides of Eq. 8.40 with respect to x, we get the desired relation:

$$f_X(x) = \int_{-\infty}^{\infty} f_{X,Y}(x, \eta) \, d\eta \tag{8.41}$$

Thus the probability density function $f_X(x)$ may be obtained from the joint probability density function $f_{X,Y}(x, y)$ by simply integrating over all possible values of the undesired random variable, Y. The use of similar arguments in the context of the other random variable Y yields $f_Y(y)$. The probability density functions $f_X(x)$ and $f_Y(y)$ are called *marginal densities*. Hence, the joint probability density function $f_{X,Y}(x, y)$ contains all the possible information about the joint random variables X and Y.

Suppose that X and Y are two continuous random variables with joint probability density function $f_{X,Y}(x, y)$. The *conditional probability density function* of Y given that $X = x$ is defined by

$$f_Y(y \mid X = x) = \frac{f_{X,Y}(x, y)}{f_X(x)} \tag{8.42}$$

provided that $f_X(x) > 0$, where $f_X(x)$ is the marginal density of X. The function $f_Y(y|X = x)$ may be thought of as a function of the variable y, with the variable x arbitrary, but fixed. Accordingly, it satisfies all the requirements of an ordinary probability density function, as shown by

$$f_Y(y|X = x) \geqslant 0 \tag{8.43}$$

and

$$\int_{-\infty}^{\infty} f_Y(y|X = x)\, dy = 1 \tag{8.44}$$

If the random variables X and Y are *statistically independent,* then knowledge of the outcome of X can in no way affect the distribution of Y. The result is that the condition probability density function $f_Y(y\ |X = x)$ reduces to the marginal density $f_Y(y)$, as shown by

$$f_Y(y|X = x) = f_Y(y)$$

In such a case, we may express the joint probability density function of the random variables X and Y as the product of their respective marginal densities, as shown by

$$f_{X,Y}(x, y) = f_X(x) f_Y(y) \tag{8.45}$$

This relation holds only when the random variables X and Y are statistically independent.

STATISTICAL AVERAGES

Having discussed probability and some of its ramifications, we now seek ways for determining the *average* behavior of the outcomes arising in random experiments.

The *mean* or *expected value* of a random variable X is commonly defined by

$$m_X = E[X] = \int_{-\infty}^{\infty} x f_X(x)\, dx \tag{8.46}$$

where E denotes the *expectation operator*. That is, the mean m_X locates the center of gravity of the area under the probability density curve of the random variable X. Similarly, the mean of a function of X, denoted by $g(X)$, is defined by

$$E[g(X)] = \int_{-\infty}^{\infty} g(x) f_X(x)\, dx \tag{8.47}$$

For the special case of $g(X) = X^n$ we obtain the nth *moment* of the probability distribution of the random variable X; that is,

$$E[X^n] = \int_{-\infty}^{\infty} x^n f_X(x)\, dx \tag{8.48}$$

By far the most important moments of X are the first two moments. Thus putting $n = 1$ in Eq. 8.48 gives the mean of the random variable as discussed herein, whereas putting $n = 2$ gives the *mean-square value* of X:

$$E[X^2] = \int_{-\infty}^{\infty} x^2 f_X(x)\, dx \tag{8.49}$$

We may also define *central moments,* which are simply the moments of the difference between a random variable X and its mean m_X. Thus the nth central moment is

$$E[(X - m_X)^n] = \int_{-\infty}^{\infty} (x - m_X)^n f_X(x)\, dx \tag{8.50}$$

For $n = 1$, the central moment is, of course, zero, whereas for $n = 2$ the second central moment is referred to as the *variance* of the random variable:

$$\text{Var}[X] = E[(X - m_X)^2] = \int_{-\infty}^{\infty} (x - m_X)^2 f_X(x)\, dx \tag{8.51}$$

The variance of a random variable X is commonly denoted as σ_X^2. The square root of the variance, namely, σ_X, is called the *standard deviation* of the random variable X.

The variance σ_X^2 of a random variable X is in some sense a measure of the variable's "dispersion." By specifying the variance σ_X^2, we essentially constrain the effective width of the probability density function $f_X(x)$ of the random variable X about the mean m_X. A precise statement of this constraint was developed by Chebyshev. The *Chebyshev inequality* states that for any positive number ε, we have

$$P(|X - m_X| \geq \varepsilon) \leq \frac{\sigma_X^2}{\varepsilon^2} \tag{8.52}$$

From this inequality we see that the mean and variance of a random variable give a partial description of its probability distribution.

The expectation operator is *linear* in that the expectation of *the sum of two random variables is equal to the sum of their individual expectations.*

Hence, expanding $E[(X - m_X)^2]$ and using the linearity of the expectation operator, we find that the variance σ_X^2 and the mean-square value $E[X^2]$ are related by

$$\begin{aligned}\sigma_X^2 &= E[X^2 - 2m_X X + m_X^2] \\ &= E[X^2] - 2m_X E[X] + m_X^2 \\ &= E[X^2] - m_X^2 \end{aligned} \tag{8.53}$$

Therefore, if the mean m_X is zero, then the variance σ_X^2 and the mean-square value $E[X^2]$ of the random variable X are equal.

Another important statistical average is the *characteristic function* $\phi_X(v)$ of the probability distribution of the random variable X, which is defined as the expectation of $\exp(jvX)$, as shown by

$$\begin{aligned}\phi_X(v) &= E[\exp(jvX)] \\ &= \int_{-\infty}^{\infty} f_X(x)\exp(jvx)\,dx \end{aligned} \tag{8.54}$$

where v is real. In other words, the characteristic function $\phi_X(v)$ is (except for a sign change in the exponent) the Fourier transform of the probability density function $f_X(x)$. In this relation we have used $\exp(jvx)$ rather than $\exp(-jvx)$, so as to conform with the convention adopted in probability theory. Recognizing that v and x play analogous roles to the variables $2\pi f$ and t of Fourier transforms, respectively, we deduce the following inverse relation from analogy with the inverse Fourier transform:

$$f_X(x) = \frac{1}{2\pi}\int_{-\infty}^{\infty} \phi_X(v)\exp(-jvx)\,dv \tag{8.55}$$

This relation may be used to evaluate the probability density function $f_X(x)$ of the random variable X from its characteristic function $\phi_X(v)$.

EXERCISE 5 Given the Chebyshev inequality of Eq. 8.52, what is the probability $P(|X - m_X| < \varepsilon)$?

EXAMPLE 4 UNIFORM DISTRIBUTION (CONTINUED)

Consider again the uniformly distributed random variable X, described in Example 3. We wish to evaluate the mean and variance of X.

The probability density function of the random variable X is given in Eq. 8.35. Therefore, substituting Eq. 8.35 in Eq. 8.46, we get the mean of X as

$$m_X = \int_a^b \frac{x}{b-a}\,dx$$

$$= \frac{b^2 - a^2}{2(b-a)}$$

$$= \frac{1}{2}(b + a) \tag{8.56}$$

Thus the mean of a uniformly distributed random variable is the arithmetic mean of its limits a and b, which is intuitively satisfying. The mean-square value of the random variable X is obtained by substituting Eq. 8.35 in 8.49; we thus get

$$E[X^2] = \int_a^b \frac{x^2}{b-a}\,dx$$

$$= \frac{b^3 - a^3}{3(b-a)}$$

$$= \frac{1}{3}(b^2 + ab + a^2) \tag{8.57}$$

Hence, the use of Eq. 8.53 yields the variance of the random variable X as

$$\sigma_X^2 = \frac{1}{3}(b^2 + ab + a^2) - \frac{1}{4}(b + a)^2$$

$$= \frac{1}{12}(b - a)^2 \tag{8.58}$$

As an application of these results, we may consider the quantizing error in pulse-code modulation. Assuming that the quantizing error is uniformly distributed inside the interval $(-\frac{1}{2}\Delta, \frac{1}{2}\Delta)$, we find from Eq. 8.56 that it has zero mean. Moreover, from Eqs. 8.57 and 8.58, we find that the mean-square value and the variance of the quantizing error are both equal to $\Delta^2/12$. These results are the same as those we used in discussing quantizing error in Section 5.4.

EXAMPLE 5 SUM OF INDEPENDENT RANDOM VARIABLES

As an application of the characteristic function, consider the problem of evaluating the probability density function of a random variable Z defined as the sum of two statistically independent random variables X and Y, that is, $Z = X + Y$. The characteristic function of Z is

$$\begin{aligned}\phi_Z(v) &= E[\exp(jv(X + Y))] \\ &= E[\exp(jvX) \cdot \exp(jvY)] \end{aligned} \tag{8.59}$$

Since X and Y are statistically independent, we may express $\phi_Z(v)$ as

$$\begin{aligned}\phi_Z(v) &= E[\exp(jvX)] \cdot E[\exp(jvY)] \\ &= \phi_X(v)\phi_Y(v) \end{aligned} \tag{8.60}$$

By analogy with the result in Fourier analysis, that the convolution of two functions of time corresponds to the multiplication of their Fourier transforms, we deduce that the probability density function of the random variable $Z = X + Y$ is given by the convolution of the probability density functions of X and Y, as shown by

$$f_Z(z) = \int_{-\infty}^{\infty} f_X(z - \eta) f_Y(\eta)\, d\eta \tag{8.61}$$

JOINT MOMENTS

Consider next a pair of random variables X and Y. A set of statistical averages of importance in this case are the *joint moments,* namely, the expected value of $X^j Y^k$, where j and k may assume any positive integer values. We may thus write

$$E[X^j Y^k] = \int_{-\infty}^{\infty} \int_{-\infty}^{\infty} x^j y^k f_{X,Y}(x, y)\, dx\, dy \tag{8.62}$$

A joint moment of particular importance is the *correlation* defined by $E[XY]$, which corresponds to $j = k = 1$ in Eq. 8.62.

The correlation of the two centered random variables $X - E[X]$ and $Y - E[Y]$, that is, the joint moment

$$\text{Cov}[XY] = E[(X - E[X])(Y - E[Y])] \tag{8.63}$$

is called the *covariance* of X and Y. Letting $m_X = E[X]$ and $m_Y = E[Y]$, we may expand Eq. 8.63 to obtain

$$\text{Cov}[XY] = E[XY] - m_X m_Y \tag{8.64}$$

Let σ_X^2 and σ_Y^2 denote the variances of X and Y, respectively. Then the covariance of X and Y normalized with respect to $\sigma_X \sigma_Y$ is called the *correlation coefficient* of X and Y:

$$\rho_{XY} = \frac{\text{Cov}[XY]}{\sigma_X \sigma_Y} \tag{8.65}$$

We say that *the two random variables X and Y are uncorrelated if and only if their covariance is zero, that is, if and only if*

$$\text{Cov}[XY] = 0$$

We say that *they are orthogonal if and only if their correlation is zero, that is, if and only if*

$$E[XY] = 0$$

From Eq. 8.64 we observe that if one or both of the random variables X and Y have zero means, and if they are orthogonal random variables, then they are uncorrelated, and vice versa. Note also that if X and Y are statistically independent, then they are uncorrelated. However, the converse of this statement is not necessarily true, as illustrated by the following example.

EXAMPLE 6

Let Z be a uniformly distributed random variable, defined by

$$f_Z(z) = \begin{cases} \frac{1}{2}, & -1 \leq z \leq 1 \\ 0, & \text{otherwise} \end{cases}$$

Let the random variable $X = Z$ and the random variable $Y = Z^2$. It is apparent that X and Y are not statistically independent because $Y = X^2$. We wish to show, however, that X and Y are uncorrelated.

Since $X = Z$, the mean of X is

$$E[X] = E[Z] = \int_{-1}^{1} \frac{1}{2} z \, dz = 0$$

Also, since $Y = Z^2$, the mean of Y is

$$E[Y] = E[Z^2] = \int_{-1}^{1} \frac{1}{2} z^2 \, dz = \frac{1}{3}$$

The covariance of X and Y is therefore

$$\begin{aligned}
\text{Cov}[XY] &= E[X(Y - \tfrac{1}{3})] \\
&= E[XY] - \tfrac{1}{3}E[X] \\
&= E[XY] \\
&= E[Z^3] \\
&= \int_{-1}^{1} \frac{1}{2} z^3 \, dz \\
&= 0
\end{aligned}$$

Hence, the random variables X and Y are uncorrelated despite the fact that they are statistically dependent.

8.3 GAUSSIAN DISTRIBUTION

The *Gaussian random variable*[3] is by far the most widely encountered random variable in the statistical analysis of communication systems. A Gaussian random variable X of mean m_X and variance σ_X^2 has the probability density function:

$$f_X(x) = \frac{1}{\sqrt{2\pi}\sigma_X} \exp\left[-\frac{1}{2\sigma_X^2}(x - m_X)^2 \right] \tag{8.66}$$

The fact that Eq. 8.66 is a probability density function is easily shown. First, note that $f_X(x) \geqslant 0$. Second, form the integral

$$\int_{-\infty}^{\infty} f_X(x)\, dx = \frac{1}{\sqrt{2\pi}\sigma_X} \int_{-\infty}^{\infty} \exp\left[-\frac{1}{2\sigma_X^2}(x - m_X)^2 \right] dx \tag{8.67}$$

[3]The Gaussian distribution is named after the great mathematician C. G. Gauss. At age 18, Gauss invented the *method of least squares* for finding the best estimate of a quantity based on a sequence of measurements. Gauss later used the method of least squares in estimating orbits of planets with noisy measurements, a procedure that was published in 1809 in his book *Theory of Motion of the Heavenly Bodies.* In connection with the error of observation, he developed the *Gaussian distribution.* This distribution is also known as the *normal distribution.* Partly for historical reasons, mathematicians commonly use normal, whereas engineers and physicists commonly use Gaussian.

We now make the change of variable $t = (x - m_X)/\sqrt{2\pi}\sigma_X$, so Eq. 8.67 becomes

$$\int_{-\infty}^{\infty} f_X(x)\,dx = \int_{-\infty}^{\infty} \exp(-\pi t^2)\,dt = 1 \tag{8.68}$$

For the last step in Eq. 8.68, see Exercise 6 of Chapter 2.

The distribution function of a Gaussian random variable X of mean m_X and variance σ_X^2 is defined by

$$F_X(x) = \frac{1}{\sqrt{2\pi}\sigma_X}\int_{-\infty}^{x} \exp\left[-\frac{1}{2\sigma_X^2}(\xi - m_X)^2\right] d\xi \tag{8.69}$$

Unfortunately, this distribution function is not expressible in terms of elementary functions. Nevertheless, it may be evaluated for a specified value of x by making use of tables of the *error function,*[4] which is defined as

$$\text{erf}(u) = \frac{2}{\sqrt{\pi}}\int_{0}^{u} \exp(-z^2)\,dz \tag{8.70}$$

Note that erf(0) = 0 and erf(∞) = 1. In Table 6 of Appendix D, we present a short set of values for the error function erf(u) for u in the range 0 to 3.3.

By using the symmetry of $f_X(x)$ and by a simple change of variables, we may express the distribution function of Eq. 8.69 in terms of the error function as follows:

$$F_X(x) = \frac{1}{2}\left[1 + \text{erf}\left(\frac{x - m_X}{\sqrt{2}\sigma_X}\right)\right] \tag{8.71}$$

The functions $f_X(x)$ and $F_X(x)$ are plotted in Fig. 8.5 for the *standardized* case when the mean m_X is 0 and the variance σ_X^2 is 1. Note that (1) the probability density function is symmetric about the mean, (2) values of x near the mean are most frequently encountered, and (3) the width of the probability density curve is proportional to the standard deviation σ_X.

[4]The error function is tabulated extensively in several references; see, for example, Abramowitz and Stegun (1965).

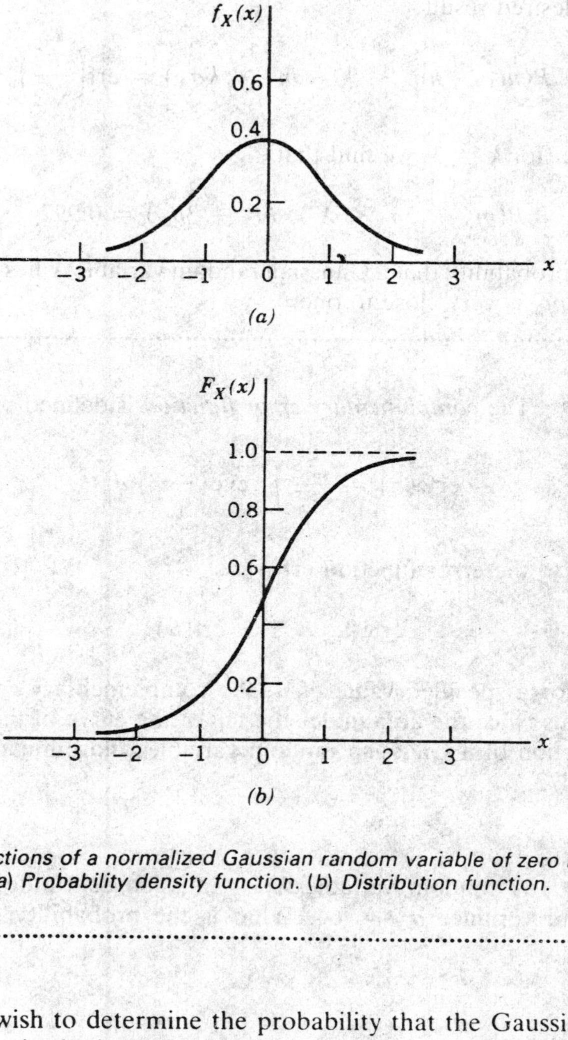

Figure 8.5
Probability functions of a normalized Gaussian random variable of zero mean and unit variance. (a) Probability density function. (b) Distribution function.

EXAMPLE 7

Suppose we wish to determine the probability that the Gaussian random variable X lies in the interval $m_X - k\sigma_X < X \leq m_X + k\sigma_X$, where k is a constant. In terms of the probability density function of X, we may use the second line of Eq. 8.33 and Eq. 8.71 to express this probability as

$$P(m_X - k\sigma_X < X \leq m_X + k\sigma_X) = F_X(m_X + k\sigma_X) - F_X(m_X - k\sigma_X)$$

$$= \frac{1}{2}\left[\operatorname{erf}\left(\frac{k}{\sqrt{2}}\right) - \operatorname{erf}\left(-\frac{k}{\sqrt{2}}\right)\right]$$

Noting that the error function erf(u) has the property that

$$\operatorname{erf}(-u) = -\operatorname{erf}(u),$$

we get the desired result

$$P(m_X - k\sigma_X < X \leq m_X + k\sigma_X) = \text{erf}\left(\frac{k}{\sqrt{2}}\right) \tag{8.72}$$

For example, for $k = 3$, we find that

$$P(m_X - 3\sigma_X < X \leq m_X + 3\sigma_X) = 0.997$$

That is, the probability that a Gaussian random variable X lies within $\pm 3\sigma_X$ of its mean m_X is very close to one.

EXERCISE 6 The *complimentary error function* is defined by

$$\text{erfc}(u) = \frac{2}{\sqrt{\pi}} \int_u^{\infty} \exp(-z^2)\, dz$$

It is related to the error function erf(u) as

$$\text{erfc}(u) = 1 - \text{erf}(u)$$

Show that for a specified value of u, the complementary error function erfc(u) equals twice the area under the tail of the curve of the probability density function of a Gaussian random variable whose mean is zero and variance is 1/2.

EXERCISE 7 A random variable X is Gaussian distributed with mean $m_X = 5$ and variance $\sigma_X^2 = 64$. What is the probability of the event $-3 < X \leq 13$?

CENTRAL LIMIT THEOREM

An important result in probability theory that is closely related to the Gaussian distribution is the *central limit theorem.*[5] Let $X_1, X_2, \ldots, X_n$ be a set of random variables that satisfies the following requirements:

1. The X_k, with $k = 1, 2, \ldots, n$, are statistically independent.
2. The X_k all have the same probability density function.
3. Both the mean and variance exist for each X_k.

[5] For a proof of the central limit theorem, see the references listed in footnote 1.

Define a new random variable Y as

$$Y = \sum_{k=1}^{n} X_k \tag{8.73}$$

Then, according to the central limit theorem, the standardized random variable:

$$Z = \frac{Y - E[Y]}{\sigma_Y} \tag{8.74}$$

approaches a Gaussian random variable with zero mean and unit variance as the number of the random variables $X_1, X_2, \ldots, X_n$ increases without limit. Note that from the definitions of expectation and variance of a random variable, we may relate the mean and variance of Y to the corresponding moments of the X_k as follows:

$$E[Y] = \sum_{k=1}^{n} E[X_k] \tag{8.75}$$

and

$$\text{Var}[Y] = \sum_{k=1}^{n} \text{Var}[X_k] \tag{8.76}$$

It is important to realize that the central limit theorem gives only the "limiting" form of the distribution function of the standardized sum Z as n tends to infinity. When n is finite, it is sometimes found that the Gaussian limit gives a relatively poor approximation for the actual distribution function of Z, even though n may be large. The accuracy of this approximation depends on the nature of the distribution of the X_k.

EXAMPLE 8 SUM OF n UNIFORMLY DISTRIBUTED RANDOM VARIABLES

Consider the random variable

$$Y = \sum_{k=1}^{n} X_k$$

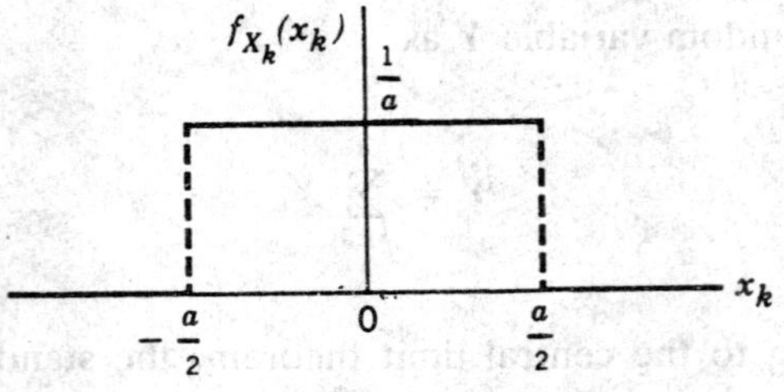

Figure 8.6
Uniform distribution.

where the X_k are uniformly distributed random variables defined by (see Fig. 8.6)

$$f_{X_k}(x_k) = \begin{cases} \frac{1}{a}, & -\frac{a}{2} \leqslant x_k \leqslant \frac{a}{2} \\ 0, & \text{elsewhere} \end{cases} \tag{8.77}$$

From Example 4, we find that the mean and variance of the X_k are given by

$$m_{X_k} = 0$$

$$\sigma^2_{X_k} = \frac{a^2}{12} \tag{8.78}$$

Therefore, according to the central limit theorem, we may use a Gaussian random variable of zero mean and variance $na^2/12$ to approximate the sum of n *independent and identically distributed* (iid) random variables, assuming a uniform distribution and large n.

8.4 TRANSFORMATION OF RANDOM VARIABLES

Consider the problem of determining the probability density function of a random variable Y, which is obtained by a one-to-one transformation of a given random variable X. The simplest possible case is when the new random variable Y is a monotone increasing differentiable function g of the random variable X (see Fig. 8.7):

$$Y = g(X)$$

In this case we have

$$\begin{aligned} F_Y(y) &= P(Y \leqslant y) \\ &= P(X \leqslant h(y)) \\ &= F_X(h(y)) \end{aligned} \tag{8.79}$$

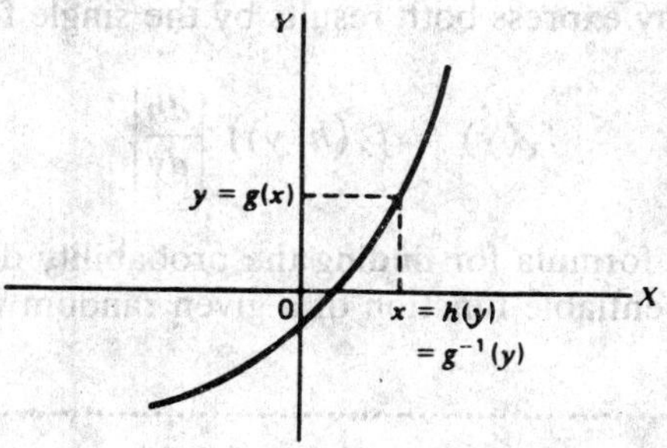

Figure 8.7
A one-to-one transformation of a random variable X.

where h is the inverse transformation

$$h(y) = g^{-1}(y) \tag{8.80}$$

This inverse transformation exists for all y, because x and y are related one-to-one. Assuming that the given random variable X has a probability density function $f_X(x)$, we may write

$$F_Y(y) = \int_{-\infty}^{h(y)} f_X(x)\, dx$$

Differentiating both sides of this relation with respect to the variable y, we get

$$f_Y(y) = f_X(h(y)) \frac{dh}{dy} \tag{8.81}$$

Consider next the case when g is a differentiable monotone decreasing function with an inverse h. We may then write

$$F_Y(y) = \int_{h(y)}^{\infty} f_X(x)\, dx$$

which, on differentiation, yields

$$f_Y(y) = -f_X(h(y)) \frac{dh}{dy} \tag{8.82}$$

Since the derivative dh/dy is negative in Eq. 8.82, whereas it is positive in Eq. 8.81, we may express both results by the single formula

$$f_Y(y) = f_X(h(y)) \left| \frac{dh}{dy} \right| \tag{8.83}$$

This is the desired formula for finding the probability density function of a one-to-one differentiable function of a given random variable.

EXAMPLE 9 SQUARE-LAW TRANSFORMATION

Consider a Gaussian random variable X of zero mean and variance σ_X^2, which is transformed by a square-law device defined by

$$Y = X^2 \tag{8.84}$$

as illustrated in Fig. 8.8. We wish to find the probability density function of the new random variable Y.

First, we see from Fig. 8.8 that Y can never be negative. Therefore,

$$P(Y \leq y) = 0, \qquad y < 0$$

and so

$$F_Y(y) = 0, \qquad y < 0$$

Furthermore, we note that the inverse transformation is not single-valued, as shown by

$$x = h(y) = \pm\sqrt{y} \tag{8.85}$$

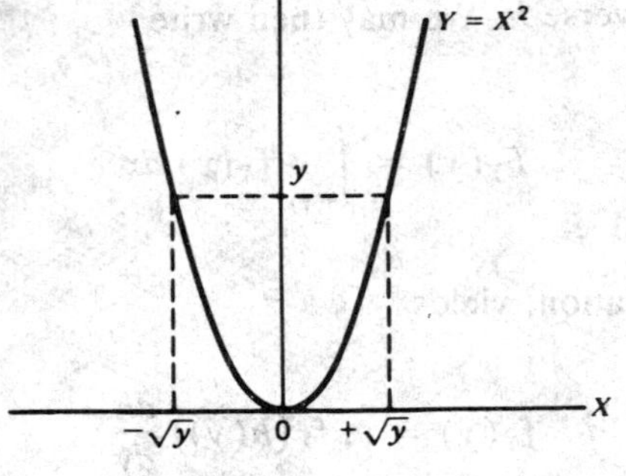

Figure 8.8
Square-law transformation.

Consequently, both positive and negative values of x contribute to y. Suppose that we are interested in the probability that $Y \leq y$, where $y \geq 0$. We may then write

$$
\begin{aligned}
P(Y \leq y) &= P(-\sqrt{y} \leq X \leq \sqrt{y}) \\
&= P(X \leq \sqrt{y}) - P(X \leq -\sqrt{y}) \\
&= \int_{-\infty}^{\sqrt{y}} f_X(x)\,dx - \int_{-\infty}^{-\sqrt{y}} f_X(x)\,dx
\end{aligned}
$$

Differentiating both sides of this relation with respect to y, we obtain

$$
f_Y(y) = \frac{1}{2\sqrt{y}}\left[f_X(\sqrt{y}) + f_X(-\sqrt{y})\right]
$$

Noting that

$$
f_X(x) = \frac{1}{\sqrt{2\pi}\sigma_X} \exp\left(-\frac{x^2}{2\sigma_X^2}\right)
$$

we obtain

$$
f_Y(y) = \frac{1}{\sqrt{2\pi y}\,\sigma_X} \exp\left(-\frac{y}{2\sigma_X^2}\right), \qquad y \geq 0
$$

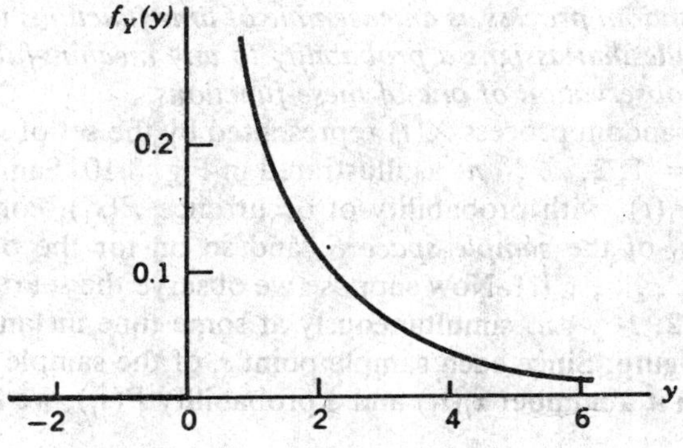

Figure 8.9
Probability density function of random variable Y at the output of a square-law device with a Gaussian random variable as input.

We thus find that the complete probability density function of the transformed random variable Y is given by

$$f_Y(y) = \begin{cases} \dfrac{1}{\sqrt{2\pi y}\,\sigma_X} \exp\left(-\dfrac{y}{2\sigma_X^2}\right), & y \geq 0 \\ 0, & y < 0 \end{cases} \tag{8.86}$$

which is plotted in Fig. 8.9. The probability density function of Eq. 8.86 is called a *chi-squared density function* when it is written as a function of the variable $\chi^2 = y$.

8.5 RANDOM PROCESSES

A basic concern in the statistical analysis of communication systems is the characterization of random signals such as voice signals, television signals, digital computer data, and electrical noise. These random signals have two properties. First, the signals are functions of time, defined on some observation interval. Second, the signals are random in the sense that before conducting an experiment, it is not possible to describe exactly the waveforms that will be observed. Accordingly, in describing random signals we find that each sample point in our sample space is a function of time. For example, in studying the fluctuations in the output of a transistor, we may assume the simultaneous testing of an indefinitely large number of identical transistors as a conceptual model of our problem. The output (measured as a function of time) of a particular transistor in the collection is then one sample point in our sample space. The sample space ensemble comprised of functions of time is called a *random* or *stochastic*[6] *process*. As an integral part of this notion, we assume the existence of a probability distribution defined over an appropriate class of sets in the sample space, so that we may speak with confidence of the probability of various events. *We may thus define a random process as an ensemble of time functions together with a probability rule that assigns a probability to any meaningful event associated with an observation of one of these functions.*

Consider a random process $X(t)$ represented by the set of *sample functions* $\{x_j(t)\}$, $j = 1, 2, \ldots, n$, as illustrated in Fig. 8.10. Sample function or waveform $x_1(t)$, with probability of occurrence $P(s_1)$, corresponds to *sample point* s_1 of the *sample space* S, and so on for the other sample functions $x_2(t), \ldots, x_n(t)$. Now suppose we observe the set of waveforms $\{x_j(t)\}$, $j = 1, 2, \ldots, n$, simultaneously at some time instant, $t = t_1$, as shown in the figure. Since each sample point s_j of the sample space S has associated with it a number $x_j(t_1)$ and a probability $P(s_j)$, we find that the

[6]The word "stochastic" comes from Greek for "to aim (guess) at".

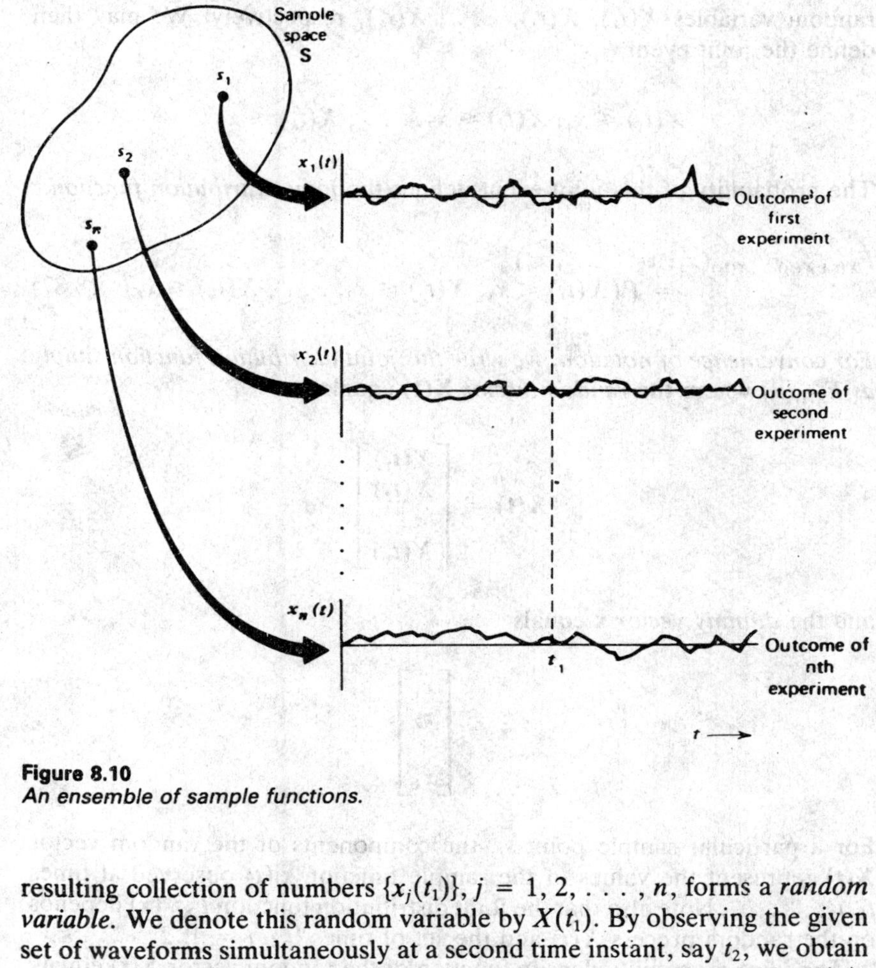

Figure 8.10
An ensemble of sample functions.

resulting collection of numbers $\{x_j(t_1)\}$, $j = 1, 2, \ldots, n$, forms a *random variable.* We denote this random variable by $X(t_1)$. By observing the given set of waveforms simultaneously at a second time instant, say t_2, we obtain a different collection of numbers, hence a different random variable $X(t_2)$. Indeed, the set of waveforms $\{x_j(t)\}$ defines a different random variable for each choice of observation instant. The difference between a random variable and a random process is that for a random variable the outcome of an experiment is mapped into a number, whereas for a random process the outcome is mapped into a waveform that is a function of time.

RANDOM VECTORS OBTAINED FROM RANDOM PROCESSES

By definition, a random process $X(t)$ implies the existence of an infinite number of random variables, one for each value of time t in the range $-\infty < t < \infty$. Thus we may speak of the distribution function $F_{X(t_1)}(x_1)$ of the random variable $X(t_1)$ obtained by observing the random process $X(t)$ at time t_1. In general, for k time instants $t_1, t_2, \ldots, t_k$ we define the k

random variables $X(t_1), X(t_2), \ldots, X(t_k)$, respectively. We may then define the joint event

$$X(t_1) \leq x_1, X(t_2) \leq x_2, \ldots, X(t_k) \leq x_k$$

The probability of this joint event defines the *joint distribution function:*

$$F_{X(t_1),X(t_2),\ldots,X(t_k)}(x_1, x_2, \ldots, x_k) = P(X(t_1) \leq x_1, X(t_2) \leq x_2, \ldots, X(t_k) \leq x_k) \quad (8.87)$$

For convenience of notation, we write this joint distribution function simply as $F_{\mathbf{X}(\mathbf{t})}(\mathbf{x})$ where the *random vector* $\mathbf{X}(\mathbf{t})$ equals

$$\mathbf{X}(\mathbf{t}) = \begin{bmatrix} X(t_1) \\ X(t_2) \\ \vdots \\ X(t_k) \end{bmatrix}$$

and the *dummy vector* $\mathbf{x}$ equals

$$\mathbf{x} = \begin{bmatrix} x_1 \\ x_2 \\ \vdots \\ x_k \end{bmatrix}$$

For a particular sample point s_j, the components of the random vector $\mathbf{X}(\mathbf{t})$ represent the values of the sample function $x_j(t)$ observed at times $t_1, t_2, \ldots, t_k$. Note also that the joint distribution function $F_{\mathbf{X}(\mathbf{t})}(\mathbf{x})$ depends on the random process $X(t)$ and the set of times $\{t_i\}$, $i = 1, 2, \ldots, k$.

The joint probability density function of the random vector $\mathbf{X}(\mathbf{t})$ equals

$$f_{\mathbf{X}(\mathbf{t})}(\mathbf{x}) = \frac{\partial^k}{\partial x_1 \partial x_2 \ldots \partial x_k} F_{\mathbf{X}(\mathbf{t})}(\mathbf{x}) \quad (8.88)$$

This function is always nonnegative, with a total volume underneath its curve in k-dimensional space that is equal to one.

EXAMPLE 10

Consider the probability of obtaining a sample function or waveform $x(t)$ of the random process $X(t)$ that passes through a set of k "windows," as

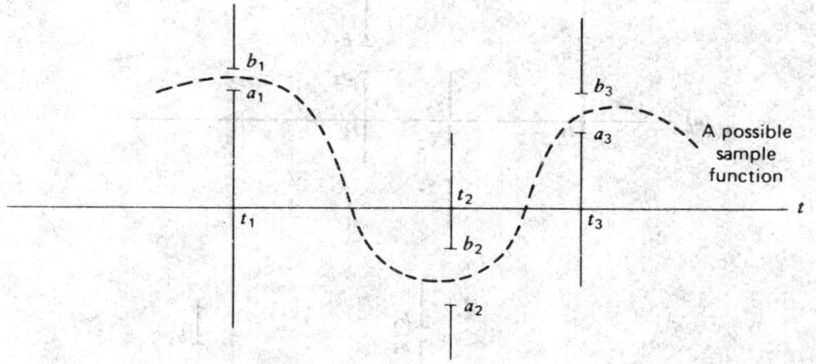

Figure 8.11
The probability of a joint event.

illustrated in Fig. 8.11 for the case of $k = 3$. That is, we wish to find the probability of the joint event

$$A = \{a_i < X(t_i) \leq b_i\}, \qquad i = 1, 2, \ldots, k$$

Given the joint probability density function $f_{\mathbf{X(t)}}(\mathbf{x})$, this probability equals

$$P(A) = \int_{a_1}^{b_1} \int_{a_2}^{b_2} \cdots \int_{a_k}^{b_k} f_{\mathbf{X(t)}}(\mathbf{x})\, dx_1\, dx_2 \ldots dx_k$$

8.6 STATIONARITY

Consider a set of times $t_1, t_2, \ldots, t_k$ in the interval in which a random process $X(t)$ is defined. A complete characterization of the random process $X(t)$ enables us to specify the joint probability density function $f_{\mathbf{X(t)}}(\mathbf{x})$. The random process $X(t)$ is said to be *strictly stationary* if the joint probability density function $f_{\mathbf{X(t)}}(\mathbf{x})$ is invariant under shifts of the time origin. In other words, the process $X(t)$ is strictly stationary if the equality

$$f_{\mathbf{X(t)}}(\mathbf{x}) = f_{\mathbf{X(t}+T)}(\mathbf{x}) \tag{8.89}$$

holds for every finite set of time instants $\{t_i\}$, $i = 1, 2, \ldots, k$, and for every time-shift T. The components of the random vector $\mathbf{X(t)}$ are obtained by observing the random process $X(t)$ at times $t_1, t_2, \ldots, t_k$. Correspondingly, the components of the random vector $\mathbf{X}(\mathbf{t} + T)$ are obtained by observing the random process $X(t)$ at times $t_1 + T, t_2 + T, \ldots, t_k + T$, where T is a time shift.

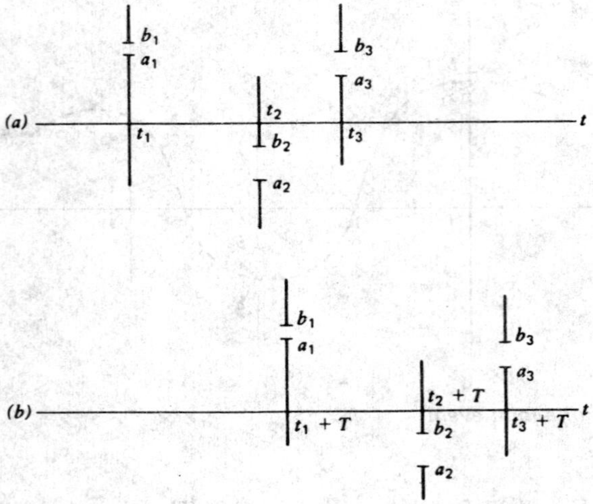

Figure 8.12
The concept of stationarity.

Stationary processes are of great importance for at least two reasons:

1. They are frequently encountered in practice or approximated to a high degree of accuracy. It is not necessary that a random process be stationary for all time, but only for some observation interval that is long enough for the particular situation of interest.
2. Many of the important properties of commonly encountered stationary processes are described by first and second moments. Consequently, it is relatively easy to develop a simple but useful theory to describe these processes.

Random processes that are not stationary are called *nonstationary*.

EXAMPLE 11

Suppose we have a random process $X(t)$ that is known to be strictly stationary. An implication of stationarity is that the probability that a set of sample functions of this process pass through the windows of Fig. 8.12*a* is equal to the probability that a set of the same number of sample functions pass through the corresponding time-shifted windows of Fig. 8.12*b*. Note, however, that it is not necessary that these two sets consist of the same sample functions.

8.7 MEAN, CORRELATION, AND COVARIANCE FUNCTIONS

In many practical situations we find that it is not possible to determine (by means of suitable measurements, say) the probability distribution of a random process. Then we must content outselves with a *partial description* of the distribution of the process. Ordinarily, the mean, autocorrelation function, and autocovariance function of the random process are taken to give a crude but, nevertheless, useful description of the distribution; these terms are defined in the following paragraphs.

Consider a random process $X(t)$ assumed to be strictly stationary. Let $X(t_k)$ denote the random variable obtained by observing the process $X(t)$ at time t_k. The *mean* of the process $X(t)$ is a constant, defined by

$$m_X = E[X(t_k)] \qquad \text{for any } t_k$$

where E denotes the expectation operator. We may simplify the notation by writing

$$m_X = E[X(t)] \tag{8.90}$$

where $X(t)$ is treated as a random variable for a fixed value of t.

The *autocorrelation function* of a stationary process $X(t)$ is defined as

$$R_X(t_k - t_j) = E[X(t_k)X(t_j)] \qquad \text{for any } t_k \text{ and } t_j$$

where $X(t_k)$ and $X(t_j)$ are the random variables obtained by observing the process $X(t)$ at times t_k and t_j, respectively. Note that the autocorrelation function depends only on the time difference $t_k - t_j$. We may simplify the notation by using the variable τ to denote the time difference $t_k - t_j$ and redefining the autocorrelation function of the process $X(t)$ as

$$R_X(\tau) = E[X(t)X(t - \tau)] \tag{8.91}$$

where insofar as the expectation is concerned, $X(t)$ and $X(t - \tau)$ are treated as random variables. The variable τ is commonly referred to as a *time lag* or *time delay;* the terms are used interchangeably. Equation 8.91 shows that for a stationary process, the autocorrelation function $R_X(\tau)$ is independent of a shift of the time origin. Note also that the argument of $R_X(\tau)$ is obtained by subtracting the argument of the second factor $X(t - \tau)$ from that of the first factor $X(t)$.

Yet another characteristic of a stationary process $X(t)$ is the *autocovariance function* defined by

$$K_X(t_k - t_i) = E[(X(t_k) - m_X)(X(t_j) - m_X)] \qquad \text{for any } t_k \text{ and } t_j$$

As with the autocorrelation function, we may simplify the notation by redefining the autocovariance function of the process $X(t)$ as

$$K_X(\tau) = E[(X(t) - m_X)(X(t - \tau) - m_X)] \tag{8.92}$$

It is a straightforward matter to show that the autocovariance function $K_X(\tau)$, the autocorrelation function $R_X(\tau)$, and the mean m_X of a stationary process $X(t)$ are related as follows

$$K_X(\tau) = R_X(\tau) - m_X^2 \tag{8.93}$$

Clearly, if the process $X(t)$ has zero mean (i.e., m_X is zero), then the autocovariance and autocorrelation functions of the process are the same.

From here on we will use the mean and autocorrelation function as a *partial description* of a random process. Moreover, we assume that

1. The mean of the process is constant.
2. The autocorrelation function of the process is independent of a shift of the time origin.
3. The autocorrelation function at a lag of zero is finite.

These three conditions, however, are not sufficient to guarantee that the random process in question is strictly stationary. A random process that is not strictly stationary but for which these conditions hold is said to be *wide-sense stationary* (WSS). Naturally, all strictly stationary processes are wide-sense stationary, but the converse is not necessarily true.

PROPERTIES OF THE AUTOCORRELATION FUNCTION

The autocorrelation function $R_X(\tau)$ of a wide-sense stationary process $X(t)$ has several important properties that follow from the definition given in Eq. 8.91. In particular, we may state:

PROPERTY 1

The autocorrelation function of a wide-sense stationary process is an even function of the time lag.

That is to say, the autocorrelation function $R_X(\tau)$ satisfies the *symmetry* condition:

$$R_X(\tau) = R_X(-\tau) \tag{8.94}$$

For $R_X(\tau)$, we write (see Eq. 8.91):

$$R_X(\tau) = E[X(t)X(t - \tau)]$$

Clearly, the product $X(t)X(t - \tau)$ is unaffected by an interchange of the two terms $X(t)$ and $X(t - \tau)$; hence,

$$\begin{aligned} R_X(\tau) &= E[X(t - \tau)X(t)] \\ &= R_X(-\tau) \end{aligned}$$

which is the desired result.

PROPERTY 2

The mean-square value of a wide-sense stationary process equals the autocorrelation function of the process for zero time lag.

In mathematical terms, we may write

$$R_X(0) = E[X^2(t)] \tag{8.95}$$

This result follows directly from Eq. 8.91 by putting the time lag $\tau = 0$.

PROPERTY 3

The autocorrelation function of a wide-sense stationary process has its maximum magnitude at zero time lag.

In mathematical terms, Property 3 states that

$$|R_X(\tau)| \leqslant R_X(0) \tag{8.96}$$

To prove this result, we first note that the mean-square value of the difference between $X(t)$ and $X(t - \tau)$ is always nonnegative, as shown by

$$E[(X(t) - X(t - \tau))^2] \geqslant 0$$

Since we have

$$(X(t) - X(t - \tau))^2 = X^2(t) - 2X(t)X(t - \tau) + X^2(t - \tau),$$

and the expectation is a linear operator, we may write

$$E[X^2(t)] - 2E[X(t)X(t - \tau)] + E[X^2(t - \tau)] \geqslant 0 \tag{8.97}$$

We next note that for a wide-sense stationary process $X(t)$:

$$\begin{aligned} E[X^2(t)] &= E[X^2(t - \tau)] = R_X(0) \\ E[X(t)X(t - \tau)] &= R_X(\tau) \end{aligned}$$

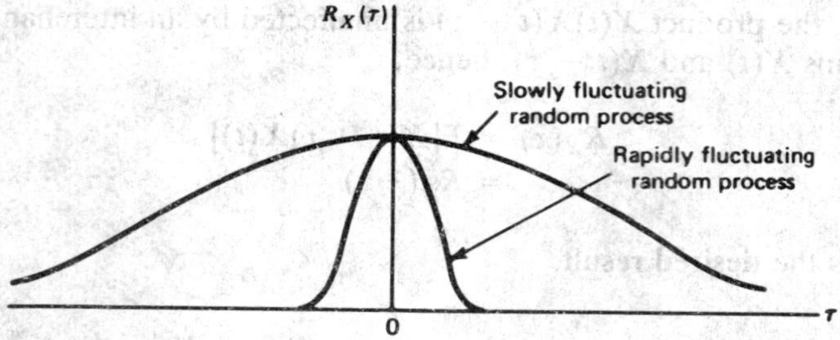

Figure 8.13
The autocorrelation functions of slowly and rapidly fluctuating random processes.

Substituting these values in Eq. 8.97 and simplifying, we get the result given in Eq. 8.96.

PHYSICAL SIGNIFICANCE OF THE AUTOCORRELATION FUNCTION

The physical significance of the autocorrelation function $R_X(\tau)$ is that it provides a means of describing the interdependence of two random variables obtained by observing a random process $X(t)$ at times τ seconds apart. It is therefore apparent that the more rapidly the random process $X(t)$ changes with time, the more rapidly will the autocorrelation function $R_X(\tau)$ decrease from its maximum $R_X(0)$ as τ increases, as illustrated in Fig. 8.13. This decrease may be characterized by a *decorrelation time* τ_0, such that for $\tau > \tau_0$, the magnitude of the autocorrelation function $R_X(\tau)$ remains below some prescribed value. We may thus define the decorrelation time τ_0 of a wide-sense stationary process $X(t)$ of zero mean as the time taken for the magnitude of the autocorrelation function $R_X(\tau)$ to decrease to 1% of its maximum value $R_X(0)$; the choice of 1% is arbitrary.

EXAMPLE 12 SINUSOIDAL WAVE WITH RANDOM PHASE

Consider a sinusoidal process with random phase. The process is denoted by

$$X(t) = A\cos(2\pi f_c t + \Theta) \tag{8.98}$$

where A and f_c are constants, and the random variable Θ denotes the phase. We assume that Θ is *uniformly distributed* over a range of 0 to 2π, that is,

$$f_\Theta(\theta) = \begin{cases} \dfrac{1}{2\pi}, & 0 \le \theta \le 2\pi \\ 0, & \text{elsewhere} \end{cases} \tag{8.99}$$

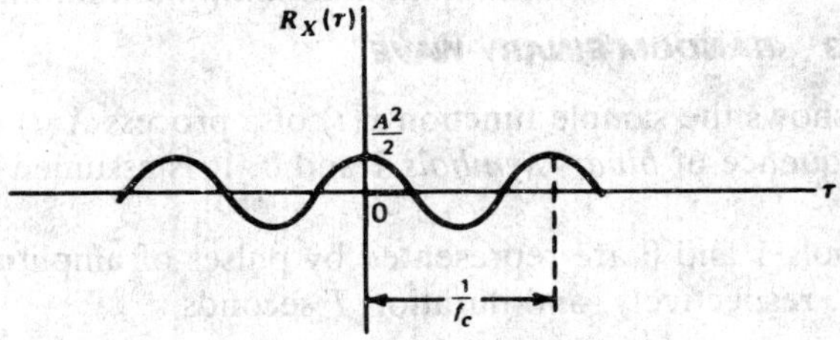

Figure 8.14
Autocorrelation function of a sinusoidal wave with random phase.

This means that the random variable Θ is equally likely to have any value in the range 0 to 2π. A sample function of the random process $X(t)$ is given by

$$x(t) = A \cos(2\pi f_c t + \theta)$$

where θ lies inside the interval $[0, 2\pi]$. Note that for each sample function, θ remains constant.

The autocorrelation function of $X(t)$ is

$$\begin{aligned} R_X(\tau) &= E[X(t + \tau)X(t)] \\ &= E[A^2 \cos(2\pi f_c t + 2\pi f_c \tau + \Theta) \cos(2\pi f_c t + \Theta)] \\ &= \frac{A^2}{2} E[\cos(4\pi f_c t + 2\pi f_c \tau + 2\Theta)] + \frac{A^2}{2} E[\cos(2\pi f_c \tau)] \end{aligned}$$

Since the expectation is with respect to the random variable Θ, we get

$$R_X(\tau) = \frac{A^2}{2} \int_0^{2\pi} \frac{1}{2\pi} \cos(4\pi f_c t + 2\pi f_c \tau + 2\theta)\, d\theta + \frac{A^2}{2} \cos(2\pi f_c \tau)$$

The first term integrates to 0, so we get

$$R_X(\tau) = \frac{A^2}{2} \cos(2\pi f_c \tau) \qquad (8.100)$$

which is plotted in Fig. 8.14. We see, therefore, that the autocorrelation function of a sinusoidal process with random phase is another sinusoid at the same frequency in the "time-lag domain" rather than the time domain.

EXAMPLE 13 RANDOM BINARY WAVE

Figure 8.15 shows the sample function $x(t)$ of a process $X(t)$ consisting of a random sequence of *binary symbols* 1 and 0. It is assumed that:

1. The symbols 1 and 0 are represented by pulses of amplitude $+A$ and $-A$ volts, respectively, and duration T seconds.
2. The pulse sequence is not synchronized so that the starting time of the first pulse, t_d, is equally likely to lie anywhere between 0 and T seconds. That is, t_d is the sample value of a uniformly distributed random variable T_d, with its probability density function defined by

$$f_{T_d}(t_d) = \begin{cases} \dfrac{1}{T}, & 0 \leqslant t_d \leqslant T \\ 0, & \text{elsewhere} \end{cases} \tag{8.101}$$

3. During any time interval $(n - 1)T < t - t_d < nT$, where n is an integer, we have $P(0) = P(1)$. That is, the two symbols 0 and 1 are equally likely, and the presence of a 1 or 0 in any one interval is independent of all other intervals.

Since the amplitude levels $-A$ and $+A$ occur with equal probability, it follows immediately that $E[X(t)] = 0$, for all t, and the mean of the process is therefore zero.

To find the autocorrelation function $R_X(t_k - t_i)$, we have to evaluate $E[X(t_k)X(t_i)]$, where $X(t_k)$ and $X(t_i)$ are random variables obtained by observing the random process $X(t)$ at times t_k and t_i, respectively.

Consider the first case when $|t_k - t_i| > T$. Then the random variables $X(t_k)$ and $X(t_i)$ occur in different pulse intervals and are therefore independent. We thus have

$$E[X(t_k)X(t_i)] = E[X(t_k)]E[X(t_i)] = 0, \qquad |t_k - t_i| > T$$

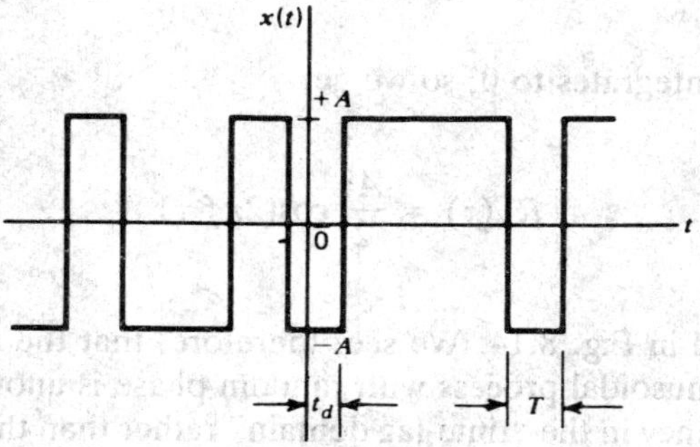

Figure 8.15
Sample function of random binary wave.

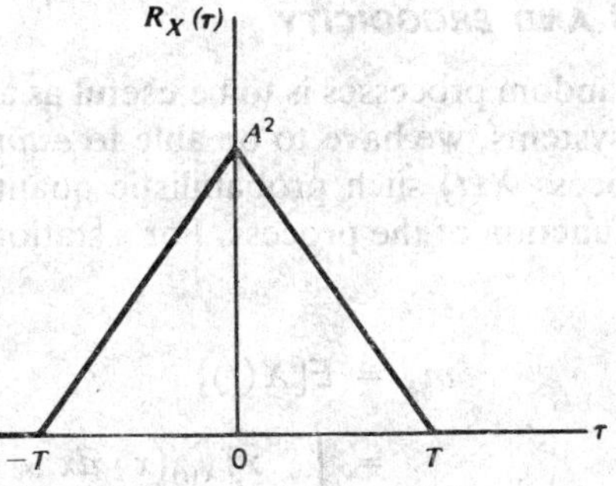

Figure 8.16
Autocorrelation function of random binary wave.

Consider next the case when $|t_k - t_i| < T$. In such a situation we observe from Fig. 8.15 that the random variables $X(t_k)$ and $X(t_i)$ occur in the same pulse interval if and only if the delay t_d is less than $T - |t_k - t_i|$. We thus obtain the *conditional expectation*:

$$E[X(t_k)X(t_i)|t_d] = \begin{cases} A^2, & t_d < T - |t_k - t_i| \\ 0, & \text{elsewhere} \end{cases}$$

Averaging this result over all possible values of t_d, we get

$$\begin{aligned} E[X(t_k)X(t_i)] &= \int_0^{T-|t_k-t_i|} A^2 f_{T_d}(t_d)\, dt_d \\ &= \int_0^{T-|t_k-t_i|} \frac{A^2}{T}\, dt_d \\ &= A^2\left(1 - \frac{|t_k - t_i|}{T}\right), \qquad |t_k - t_i| < T \end{aligned}$$

We therefore conclude that the autocorrelation function of a random binary wave, represented by the sample function shown in Fig. 8.15 is only a function of the time difference $\tau = t_k - t_i$, as shown by

$$R_X(\tau) = \begin{cases} A^2\left(1 - \dfrac{|\tau|}{T}\right), & |\tau| < T \\ 0, & |\tau| \geq T \end{cases} \tag{8.102}$$

This result is plotted in Fig. 8.16.

EXERCISE 8 **What is the mean-square value of the random binary wave described in Example 13? Use physical arguments to justify your answer.**

TIME AVERAGES AND ERGODICITY

If the theory of random processes is to be useful as a method for describing communication systems, we have to be able to *estimate* from observations of a random process $X(t)$ such probabilistic quantities as the mean and autocorrelation function of the process. For a stationary process, the mean is defined by

$$\begin{aligned} m_X &= E[X(t)] \\ &= \int_{-\infty}^{\infty} x f_{X(t)}(x)\, dx \end{aligned} \tag{8.103}$$

and the autocorrelation function is defined by

$$\begin{aligned} R_X(\tau) &= E[X(t)X(t-\tau)] \\ &= \int_{-\infty}^{\infty}\int_{-\infty}^{\infty} xy f_{X(t),X(t-\tau)}(x, y)\, dx\, dy \end{aligned} \tag{8.104}$$

To compute m_X and $R_X(\tau)$ by *ensemble averaging*, as defined in Eqs. 8.103 and 8.104, we have to average across all the sample functions of the process. In particular, this computation requires complete knowledge of the first-order and second-order joint probability density functions of the process. In many practical situations, however, these probability density functions are simply not available. Indeed, the only thing that we may usually find available is the recording of one sample function of the random process. It seems natural then to consider also *time averages* of individual sample functions of the process.

We define the *time-averaged mean* of the sample function $x(t)$ of a random process $X(t)$ as

$$\langle x(t) \rangle = \lim_{T\to\infty} \frac{1}{2T} \int_{-T}^{T} x(t)\, dt \tag{8.105}$$

where the symbol $\langle \cdot \rangle$ denotes *time-averaging* and $2T$ is the total observation interval. In a similar way, we may define the *time-averaged autocorrelation function* of the sample function $x(t)$ as

$$\langle x(t)x(t-\tau) \rangle = \lim_{T\to\infty} \frac{1}{2T} \int_{-T}^{T} x(t)x(t-\tau)\, dt \tag{8.106}$$

The two time averages $\langle x(t) \rangle$ and $\langle x(t)x(t-\tau) \rangle$ are random variables in that their values depend on which sample function of the random process $X(t)$ is used in the time-averaging evaluations. On the other hand, m_X is a constant, and $R_X(\tau)$ is an ordinary function of the variable τ.

In general, ensemble averages and time averages are not equal except for a very special class of random process known as *ergodic processes.*[7] *A random process $X(t)$ is said to be ergodic in the most general form if all of its statistical properties can be determined from a sample function representing one possible realization of the process.* We note here that it is necessary for a random process to be strictly stationary for it to be ergodic. However, the converse is not always true; that is, not all stationary processes are ergodic.

Usually, we are not interested in estimating all the ensemble averages of a random process but rather only certain averages such as the mean and the autocorrelation function of the process. Accordingly, we may define ergodicity in a more limited sense, as next described.

Ergodicity in the Mean The time average $(1/2T) \int_{-T}^{T} x(t)\, dt$ is a random variable with a mean and variance of its own. For a stationary process, we find that its mean is equal to

$$\begin{aligned} E\left[\frac{1}{2T}\int_{-T}^{T} x(t)\, dt\right] &= \frac{1}{2T}\int_{-T}^{T} E[x(t)]\, dt \\ &= \frac{1}{2T}\int_{-T}^{T} m_X\, dt \\ &= m_X \end{aligned} \tag{8.107}$$

Therefore, this time average provides an *unbiased estimate* of m_X. An estimator is said to be unbiased if the expected value of the estimate is exactly the same as the true value of the pertinent parameter. We say that the random process $X(t)$ is *ergodic in the mean* if

$$\lim_{T\to\infty} \frac{1}{2T}\int_{-T}^{T} x(t)\, dt = m_X \tag{8.108}$$

with probability one. That is, for a random process to be ergodic in the mean, its time-averaged and ensemble-averaged mean values must be equal with probability one. The necessary and sufficient condition for the ergodicity of the mean is that the variance of the estimator $(1/2T) \int_{-T}^{T} x(t)\, dt$ approach zero as T approaches infinity.

[7]The problem of determining conditions under which time averages computed from a sample function of a random process can be ultimately identified with corresponding ensemble averages first arose in statistical mechanics. Physical systems possessing properties of this kind were called *ergodic* by L. Boltzmann in 1887. The term "ergodic" is of Greek origin. It comes from the Greek for "work path," which relates to the path of an energetic particle in a gas in the context of statistical mechanics (Gardner, 1987).

Equation 8.108 suggests that we may estimate the mean of an ergodic process by passing a *finite record* of the sample function of the process through an integrator. An estimate of the mean of the process is produced at the integrator output.

Ergodicity in the Autocorrelation Function Consider next the time average $(1/2T)\int_{-T}^{T} x(t)x(t-\tau)\,dt$, which is also a random variable. Its mean is equal to

$$\begin{aligned} E\left[\frac{1}{2T}\int_{-T}^{T} x(t)x(t-\tau)\,dt\right] &= \frac{1}{2T}\int_{-T}^{T} E[x(t)x(t-\tau)]\,dt \\ &= \frac{1}{2T}\int_{-T}^{T} R_X(\tau)\,dt \\ &= R_X(\tau) \end{aligned} \tag{8.109}$$

Accordingly, this time average provides an unbiased estimate of the ensemble-averaged autocorrelation function $R_X(\tau)$ of the random process $X(t)$. We say that the random process $X(t)$ is *ergodic in the autocorrelation function* if

$$\lim_{T\to\infty} \frac{1}{2T}\int_{-T}^{T} x(t)x(t-\tau)\,dt = R_X(\tau) \tag{8.110}$$

with probability one. The necessary and sufficient condition for a stochastic process to be ergodic in the autocorrelation function is that the variance of the estimator $(1/2T)\int_{-T}^{T} x(t)x(t-\tau)\,dt$ approach zero as T approaches infinity.

To test a sample function of a stochastic process for ergodicity in the mean, it suffices to know the mean m_X and autocorrelation function $R_X(\tau)$ of the process. However, to test it for ergodicity in the autocorrelation function, we have to know fourth-order moments of the process. Therefore, except for certain simple cases, it is usually very difficult to establish if a random process meets the conditions for the ergodicity in both the mean and the autocorrelation function. Thus, in practice, we are usually forced to consider the physical origin of the random process, and thereby make a somewhat intuitive judgment as to whether it is reasonable to interchange time and ensemble averages.

EXAMPLE 14 SINUSOIDAL WAVE WITH RANDOM PHASE (CONTINUED)

Consider again the sinusoidal process $X(t)$ defined by

$$X(t) = A\cos(2\pi f_c t + \Theta)$$

where A and f_c are constants and Θ is a uniformly distributed random variable:

$$f_\Theta(\theta) = \begin{cases} \dfrac{1}{2\pi}, & 0 \leq \theta \leq 2\pi \\ 0, & \text{elsewhere} \end{cases}$$

The mean of this random process is

$$\begin{aligned} m_X &= \int_{-\infty}^{\infty} A \cos(2\pi f_c t + \theta) f_\Theta(\theta)\, d\theta \\ &= \int_0^{2\pi} \frac{A}{2\pi} \cos(2\pi f_c t + \theta)\, d\theta \\ &= 0 \end{aligned}$$

The autocorrelation function of the process was determined in Example 12; the result is reproduced here for convenience

$$R_X(\tau) = \frac{A^2}{2} \cos(2\pi f_c \tau)$$

Let $x(t)$ denote a sample function of the process; thus

$$x(t) = A \cos(2\pi f_c t + \theta)$$

The time-averaged mean of the process is

$$\begin{aligned} \langle x(t) \rangle &= \lim_{T\to\infty} \frac{1}{2T} \int_{-T}^{T} A \cos(2\pi f_c t + \theta)\, dt \\ &= 0 \end{aligned}$$

The time-averaged autocorrelation function of the process is

$$\langle x(t)x(t - \tau) \rangle = \lim_{T\to\infty} \frac{A^2}{2T} \int_{-T}^{T} \cos(2\pi f_c t + 2\pi f_c \tau + \theta) \cos(2\pi f_c t + \theta)\, dt$$

Using the trigonometric relation

$$\begin{aligned} \cos(2\pi f_c t + 2\pi f_c \tau + \theta)\cos(2\pi f_c t + \theta) &= \frac{1}{2} \cos(2\pi f_c \tau) \\ &\quad + \frac{1}{2} \cos(4\pi f_c t + 2\pi f_c \tau + \theta) \end{aligned}$$

and then integrating, the expression for the time-averaged autocorrelation function simplifies as

$$\langle x(t)x(t-\tau)\rangle = \frac{A^2}{2}\cos(2\pi f_c \tau)$$

Hence, the time-averaged mean and time-averaged autocorrelation function of the process are exactly the same as the corresponding ensemble averages. This random process is therefore ergodic in both the mean and the autocorrelation function.

8.8 RANDOM PROCESS TRANSMISSION THROUGH LINEAR FILTERS

Suppose that a random process $X(t)$ is applied as input to a linear time-invariant filter of impulse response $h(t)$, producing a random process $Y(t)$ at the filter output, as in Fig. 8.17. In general, it is difficult to describe the probability distribution of the output random process $Y(t)$, even when the probability distribution of the input random process $X(t)$ is completely specified for $-\infty < t < \infty$.

In this section, we determine the mean and autocorrelation functions of the output random process $Y(t)$ in terms of those of the input $X(t)$, assuming that $X(t)$ is a wide-sense stationary process.

Consider first the mean of the output random process $Y(t)$. By definition, we have

$$m_Y(t) = E[Y(t)] = E\left[\int_{-\infty}^{\infty} h(\tau)X(t-\tau)\,d\tau\right] \tag{8.111}$$

Provided that the expectation $E[X(t)]$ is finite for all t, and the system is stable, we may interchange the order of the expectation and the integration with respect to τ in Eq. 8.111, and so write

$$\begin{aligned} m_Y(t) &= \int_{-\infty}^{\infty} h(\tau)E[X(t-\tau)]\,d\tau \\ &= \int_{-\infty}^{\infty} h(\tau)m_X(t-\tau)\,d\tau \end{aligned} \tag{8.112}$$

When the input random process $X(t)$ is wide-sense stationary, the mean $m_X(t)$ is a constant m_X, so that we may simplify Eq. 8.112 as

$$\begin{aligned} m_Y &= m_X \int_{-\infty}^{\infty} h(\tau)\,d\tau \\ &= m_X H(0) \end{aligned} \tag{8.113}$$

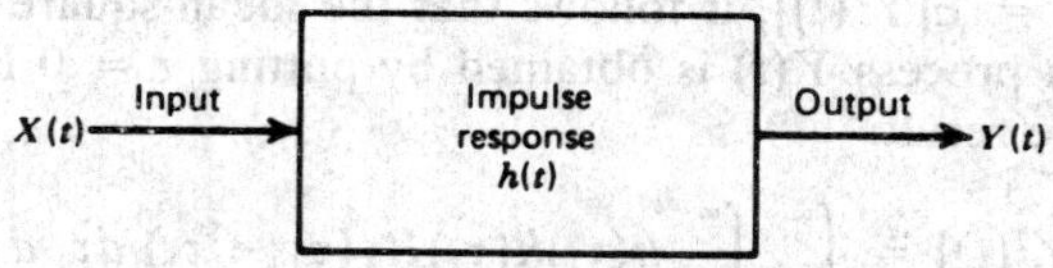

Figure 8.17
Transmission of a random process through a linear filter.

where $H(0)$ is the *zero-frequency response* of the system. Equation 8.113 states that the mean of the output process of a stable linear time-invariant system is equal to the mean of the input process multiplied by the zero-frequency response of the system.

Consider next the autocorrelation function of the output random process $Y(t)$. By definition, we have

$$R_Y(t, u) = E[Y(t)Y(u)]$$

where t and u denote two values of time at which the output process is observed. We may therefore use the convolution integral to write

$$R_Y(t, u) = E\left[\int_{-\infty}^{\infty} h(\tau_1)X(t - \tau_1)\, d\tau_1 \int_{-\infty}^{\infty} h(\tau_2)X(u - \tau_2)\, d\tau_2\right] \tag{8.114}$$

Here again, provided that $E[X^2(t)]$ is finite for all t and the system is stable, we may interchange the order of the expectation and the integrations with respect to τ_1 and τ_2 in Eq. 8.114, obtaining

$$\begin{aligned} R_Y(t, u) &= \int_{-\infty}^{\infty} d\tau_1 h(\tau_1) \int_{-\infty}^{\infty} d\tau_2 h(\tau_2) E[X(t - \tau_1)X(u - \tau_2)] \\ &= \int_{-\infty}^{\infty} d\tau_1 h(\tau_1) \int_{-\infty}^{\infty} d\tau_2 h(\tau_2) R_X(t - \tau_1, u - \tau_2) \end{aligned} \tag{8.115}$$

When the input $X(t)$ is a wide-sense stationary process, the autocorrelation function of $X(t)$ is only a function of the difference between the observation times $t - \tau_1$ and $u - \tau_2$. Thus, putting $\tau = t - u$ in Eq. 8.115, we may write

$$R_Y(\tau) = \int_{-\infty}^{\infty}\int_{-\infty}^{\infty} h(\tau_1)h(\tau_2)R_X(\tau - \tau_1 + \tau_2)\, d\tau_1\, d\tau_2 \tag{8.116}$$

On combining this result with that involving the mean m_Y, we see that if the input to a stable linear time-invariant filter is a wide-sense stationary process, then the output of the filter is also a wide-sense stationary process.

Since $R_Y(0) = E[Y^2(t)]$, it follows that the mean-square value of the output random process $Y(t)$ is obtained by putting $\tau = 0$ in Eq. 8.116. We thus get the result:

$$E[Y^2(t)] = \int_{-\infty}^{\infty} \int_{-\infty}^{\infty} h(\tau_1)h(\tau_2)R_X(\tau_2 - \tau_1)\, d\tau_1\, d\tau_2 \qquad (8.117)$$

which is a constant.

8.9 POWER SPECTRAL DENSITY

Thus far we have considered the characterization of wide-sense stationary processes in linear systems in the time domain. We turn next to the characterization of random processes in linear systems by using frequency-domain ideas. In particular, we wish to derive the frequency-domain equivalent to the result of Eq. 8.117 defining the mean-square value of the filter output.

By definition, the impulse response of a linear time-invariant filter is equal to the inverse Fourier transform of the transfer function of the system. We may thus write

$$h(\tau_1) = \int_{-\infty}^{\infty} H(f) \exp(j2\pi f\tau_1)\, df \qquad (8.118)$$

Substituting this expression for $h(\tau_1)$ in Eq. 8.117, and rearranging the resultant triple integration, we get

$$E[Y^2(t)] = \int_{-\infty}^{\infty} df\, H(f) \int_{-\infty}^{\infty} d\tau_2 h(\tau_2) \int_{-\infty}^{\infty} R_X(\tau_2 - \tau_1) \exp(j2\pi f\tau_1)\, d\tau_1 \qquad (8.119)$$

Define a new variable

$$\tau = \tau_2 - \tau_1$$

Then we may rewrite Eq. 8.119 in the form

$$E[Y^2(t)] = \int_{-\infty}^{\infty} df\, H(f) \int_{-\infty}^{\infty} d\tau_2 h(\tau_2) \exp(j2\pi f\tau_2) \times \int_{-\infty}^{\infty} R_X(\tau) \exp(-j2\pi f\tau)\, d\tau \qquad (8.120)$$

The middle integral on the right side in Eq. 8.120 is simply $H^*(f)$, the complex conjugate of the transfer function $H(f)$ of the filter; hence, we may simplify this equation as

$$E[Y^2(t)] = \int_{-\infty}^{\infty} df |H(f)|^2 \int_{-\infty}^{\infty} R_X(\tau) \exp(-j2\pi f\tau)\, d\tau \qquad (8.121)$$

We may further simplify Eq. 8.121 by recognizing that the last integral is simply the Fourier transform of the autocorrelation function $R_X(\tau)$ of the input random process $X(t)$. Let this transform be denoted by $S_X(f)$, written in expanded form as

$$S_X(f) = \int_{-\infty}^{\infty} R_X(\tau) \exp(-j2\pi f\tau)\, d\tau \qquad (8.122)$$

The function $S_X(f)$ is called the *power spectral density* or *power spectrum* of the wide-sense stationary process $X(t)$. Thus substituting Eq. 8.122 in 8.121, we obtain the desired relation

$$E[Y^2(t)] = \int_{-\infty}^{\infty} |H(f)|^2 S_X(f)\, df \qquad (8.123)$$

Equation 8.123 states that *the mean-square value of the output of a stable linear time-invariant filter in response to a wide-sense stationary input process is equal to the integral over all frequencies of the power spectral density of the input random process multiplied by the squared magnitude of the transfer function of the filter.* This is the desired frequency-domain equivalent to the time-domain relation of Eq. 8.117.

PROPERTIES OF THE POWER SPECTRAL DENSITY

The power spectral density $S_X(f)$ and the autocorrelation function $R_X(\tau)$ of a wide-sense stationary process $X(t)$ form a Fourier transform pair, as shown by the pair of relations:

$$S_X(f) = \int_{-\infty}^{\infty} R_X(\tau) \exp(-j2\pi f\tau)\, d\tau \qquad (8.124)$$

$$R_X(\tau) = \int_{-\infty}^{\infty} S_X(f) \exp(j2\pi f\tau)\, df \qquad (8.125)$$

This pair of equations constitutes the *Einstein–Wiener–Khintchine* relations for wide-sense stationary processes

The power spectral density of a wide-sense stationary process has a number of important properties that follow directly from Eqs. 8.124 and 8.125, as next described.

PROPERTY 1

The zero-frequency value of the power spectral density of a wide-sense stationary process equals the total area under the graph of the autocorrelation function; that is,

$$S_X(0) = \int_{-\infty}^{\infty} R_X(\tau)\, d\tau \tag{8.126}$$

This property follows directly from Eq. 8.124 by putting $f = 0$.

PROPERTY 2

The mean-square value of a wide-sense stationary process equals the total area under the graph of the power spectral density; that is,

$$E[X^2(t)] = \int_{-\infty}^{\infty} S_X(f)\, df \tag{8.127}$$

This property follows directly from Eq. 8.125 by putting $\tau = 0$, and noting that $R_X(0) = E[X^2(t)]$.

PROPERTY 3

The power spectral density of a wide-sense stationary process is always nonnegative; that is,

$$S_X(f) \geq 0, \qquad \text{for all } f \tag{8.128}$$

This is a necessary and sufficient condition for the mean-square value of a random process (which equals the total area under the curve of the power spectral density of the process) to be nonnegative.

PROPERTY 4

The power spectral density of a wide-sense stationary process is an even function of the frequency; that is

$$S_X(-f) = S_X(f) \tag{8.129}$$

This property is readily obtained by substituting $-f$ for f in Eq. 8.124.

$$S_X(-f) = \int_{-\infty}^{\infty} R_X(\tau) \exp(j2\pi f\tau)\, d\tau$$

Next, substituting $-\tau$ for τ, and recognizing that $R_X(-\tau) = R_X(\tau)$, we get

$$S_X(-f) = \int_{-\infty}^{\infty} R_X(\tau) \exp(-j2\pi f\tau)\, d\tau = S_X(f)$$

which is the desired result.

These properties parallel those for periodic signals, which we described in Chapter 4. Indeed, we may use ideas similar to those described therein to measure the autocorrelation function and power spectral density of a wide-sense stationary process.

EXERCISE 9 Consider the function $\sigma(f)$ defined by

$$\sigma(f) = \frac{S(f)}{R(0)}$$

where $S(f)$ is the power spectral density of a random process and $R(0)$ is the value of its autocorrelation function for a lag of zero (i.e., $\tau = 0$). Explain why $\sigma(f)$ has the properties usually associated with a probability density function.

EXAMPLE 15 SINE WAVE WITH RANDOM PHASE (CONTINUED)

Consider the sinusoidal process $X(t) = A\cos(2\pi f_c t + \Theta)$, where the phase Θ is a uniformly distributed random variable over the range 0 to 2π. The autocorrelation function of this process is given by Eq. 8.100, which is reproduced here for convenience:

$$R_X(\tau) = \frac{A^2}{2}\cos(2\pi f_c \tau)$$

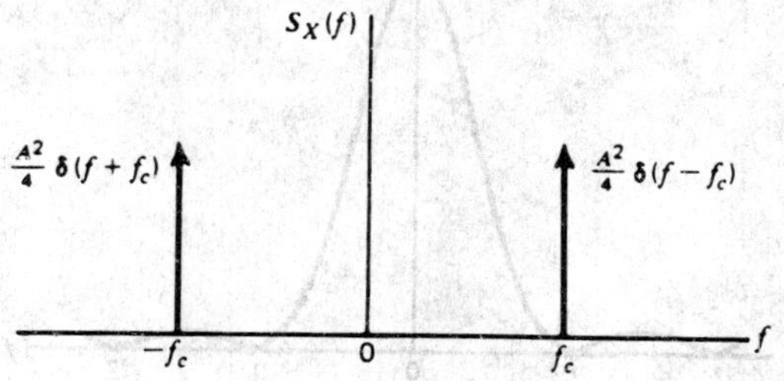

Figure 8.18
Power spectral density of a sinusoidal process.

Taking the Fourier transform of both sides of this relation, we find that the power spectral density of the sinusoidal process $X(t)$ is given by

$$S_X(f) = \frac{A^2}{4}[\delta(f - f_c) + \delta(f + f_c)] \tag{8.130}$$

The power spectral density $S_X(f)$ consists of a pair of delta functions weighted by the factor $A^2/4$ and located at $\pm f_c$ as in Fig. 8.18. We note that the total area under a delta function is 1. Hence, the total area under the $S_X(f)$ of Eq. 8.130 is equal to $A^2/2$, as expected.

EXAMPLE 16 RANDOM BINARY WAVE (CONTINUED)

Consider again a random binary wave consisting of a sequence of 1's and 0's represented by the values $+A$ and $-A$, respectively. In Example 13 we showed that the autocorrelation function of this random process (see Eq. 8.102) is

$$R_X(\tau) = \begin{cases} A^2\left(1 - \dfrac{|\tau|}{T}\right), & |\tau| < T \\ 0, & |\tau| \geqslant T \end{cases}$$

The power spectral density of the process is therefore

$$S_X(f) = \int_{-T}^{T} A^2\left(1 - \frac{|\tau|}{T}\right)\exp(-j2\pi f\tau)\,d\tau$$

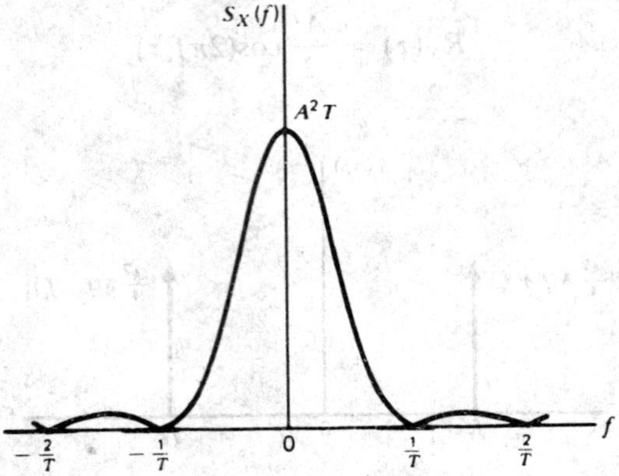

Figure 8.19
Power spectral density of random binary wave.

Using the Fourier transform of a triangular function evaluated in Example 10 of Chapter 2, we obtain

$$S_X(f) = A^2T \operatorname{sinc}^2(fT) \tag{8.131}$$

which is plotted in Fig. 8.19. Here again, we see that the power spectral density is nonnegative for all f and that it is an even function of f. We note from exercise 7 of Chapter 2, that

$$\int_{-\infty}^{\infty} \operatorname{sinc}^2(fT)\, df = \frac{1}{T} \tag{8.132}$$

Therefore, the total area under $S_X(f)$, or the average power of the random binary wave is A^2.

The result of Eq. 8.131 may be generalized as follows. We note that the energy spectral density of a rectangular pulse $g(t)$ of amplitude A and duration T is given by

$$\Psi_g(f) = A^2T^2 \operatorname{sinc}^2(fT) \tag{8.133}$$

We may therefore rewrite Eq. 8.131 in terms of $\Psi_g(f)$ as

$$S_X(f) = \frac{\Psi_g(f)}{T} \tag{8.134}$$

Equation 8.134 states that, for a random binary wave in which binary symbols 1 and 0 are represented by pulses $g(t)$ and $-g(t)$, respectively, the power spectral density $S_X(f)$ is equal to the energy spectral density $\Psi_g(f)$ of the *symbol shaping pulse* $g(t)$ divided by the *symbol duration* T.

EXERCISE 10 Sketch the autocorrelation function and power spectral density of a random binary wave alternating between -1 and $+1$ V for the following values of pulse duration T:

(a) $T = \frac{1}{2}s$
(b) $T = 1s$
(c) $T = 2s$

Comment on your results.

EXAMPLE 17 LINEAR MAXIMAL SEQUENCES

There exists a class of deterministic sequences known as *maximum length sequences* with many of the properties of a random binary sequence and

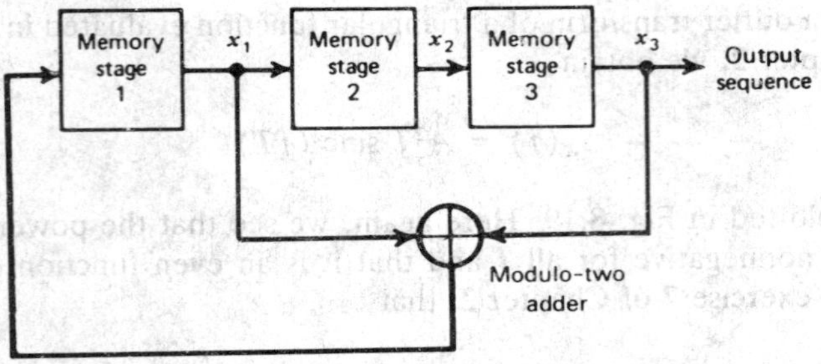

Figure 8.20
Linear-maximal-sequence generator.

yet requiring simple instrumentation. A *maximum-length* sequence is a periodic binary sequence generated by a *feedback shift register* that has the longest possible period for this particular method of generation. A shift register of length m is a device consisting of m consecutive 2-state memory stages (flip-flops) regulated by a single timing clock. At each clock pulse, the state (represented by binary symbol 1 or 0) of each memory stage is shifted to the next stage down the line. To prevent the shift register from emptying by the end of m clock pulses, we use a logical (i.e., Boolean) function of the states of the m memory stages to compute a *feedback term*, and apply it to the first memory stage of the shift register. The most important special form of this feedback shift register is the *linear* case in which the feedback function is obtained by using *modulo-two adders* to combine the outputs of the various memory stages. This operation is illustrated in Fig. 8.20 for $m = 3$. Representing the states of the three memory stages as x_1, x_2, and x_3, we see that in Fig. 8.20 the feedback function is equal to the modulo-two sum of x_1 and x_3.[8] A maximum length sequence generated by a feedback shift register using a linear feedback function is called a *linear maximal sequence*. This sequence is always periodic with a period defined by

$$N = 2^m - 1 \tag{8.135}$$

where m is the length of the shift register. Assuming, for example, that the three memory stages of the shift register shown in Fig. 8.20 are in the initial states 0, 0, and 1, respectively, we find that the resulting output sequence is 1001110, repeating with period 7.

Representing the symbols 1 and 0 by the values $+A$ and $-A$, respectively, we find that the autocorrelation function of a linear maximal sequence-

[8]In modulo-two addition, the sum of x_1 and x_3 takes the value 1 only when x_1 or x_3, but not both, takes the value 1. In other words, the carry is ignored. This operation is equivalent to the logical EXCLUSIVE OR.

quence is periodic with period NT, and that for values of time lag τ lying in the interval $-NT/2 \leq \tau \leq NT/2$, it is defined by

$$R_X(\tau) = \begin{cases} A^2\left(1 - \dfrac{N+1}{NT}|\tau|\right), & |\tau| \leq T \\ -\dfrac{A^2}{N} & \text{for the remainder of the period} \end{cases} \tag{8.136}$$

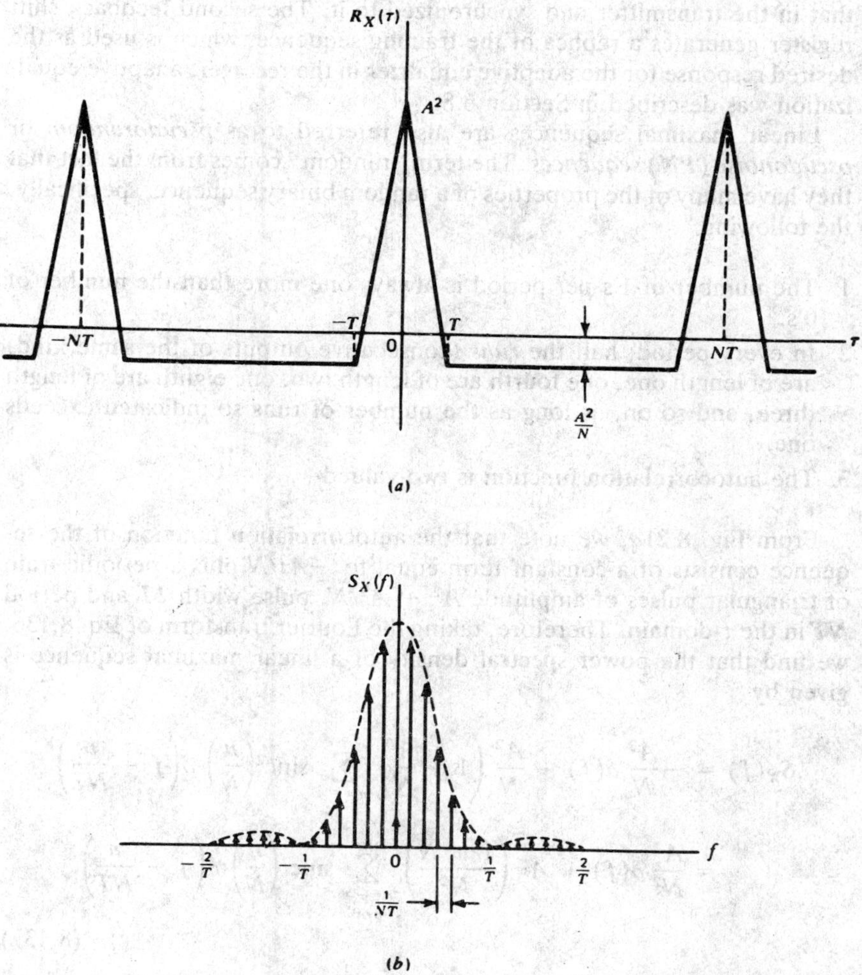

Figure 8.21
Characteristics of linear maximal sequence. (a) Autocorrelation function. (b) Power spectral density.

where T is the duration for which the symbol 1 or 0 is defined. This result is plotted in Fig. 8.21a for the case of $m = 3$ or $N = 7$.

The autocorrelation function depicted in Fig. 8.21a exhibits two characteristics: a distinct peak value and a periodic nature. These two characteristics make linear maximal sequences well-suited for use in synchronous digital communications. For example, we may use a linear maximal sequence as the training sequence for adaptive equalization in a data transmission system operating over an unknown channel. Specifically, we use a feedback shift register in the transmitter to generate a linear maximal sequence for probing the channel during the training mode of the system, and use a second feedback shift register in the receiver that is identical to that in the transmitter and synchronized to it. The second feedback shift register generates a replica of the training sequence, which is used as the desired response for the adaptive equalizer in the receiver; adaptive equalization was described in Section 6.8.

Linear maximal sequences are also referred to as *pseudorandom* or *pseudonoise (PN) sequences*. The term "random" comes from the fact that they have many of the properties of a random binary sequence, specifically, the following:[9]

1. The number of 1's per period is always one more than the number of 0's.
2. In every period, half the *runs* (consecutive outputs of the same kind) are of length one, one fourth are of length two, one eighth are of length three, and so on, as long as the number of runs so indicated exceeds one.
3. The autocorrelation function is two-valued.

From Fig. 8.21a, we note that the autocorrelation function of the sequence consists of a constant term equal to $-A^2/N$ plus a periodic train of triangular pulses of amplitude $A^2 + A^2/N$, pulse width $2T$ and period NT in the τ-domain. Therefore, taking the Fourier transform of Eq. 8.136, we find that the power spectral density of a linear maximal sequence is given by

$$S_X(f) = -\frac{A^2}{N}\,\delta(f) + \frac{A^2}{N}\left(1 + \frac{1}{N}\right) \sum_{n=-\infty}^{\infty} \operatorname{sinc}^2\left(\frac{n}{N}\right) \delta\left(f - \frac{n}{NT}\right)$$

$$= \frac{A^2}{N^2}\,\delta(f) + A^2\left(\frac{1+N}{N^2}\right) \sum_{\substack{n=-\infty \\ n\neq 0}}^{\infty} \operatorname{sinc}^2\left(\frac{n}{N}\right) \delta\left(f - \frac{n}{NT}\right) \tag{8.137}$$

[9] For further details of linear maximal sequences, see Golomb (1964), pp. 1–32. See also the review paper by Sarwate and Pursley (1980).

which is plotted in Fig. 8.21*b* for $m = 3$ or $N = 7$. Comparing this power spectral density characteristic with that of Fig. 8.19 for a random binary sequence, we see that they both have an envelope of the same form, namely, $\text{sinc}^2(fT)$, which depends only on the duration T. The fundamental difference, of course, is that whereas the random binary sequence has a continuous spectral density characteristic, the corresponding characteristic of a linear maximal sequence consists of delta functions spaced $1/NT$ hertz apart.

EXERCISE 11 Find the limiting value of the power spectral density of the linear maximal sequence considered in Example 17 as the period of the sequence becomes large. Compare your result with the power spectral density of a random binary wave of similar characteristics.

EXAMPLE 18 MODULATED RANDOM PROCESS

A situation that often arises in practice is that of *mixing* (i.e., multiplication) of a wide-sense stationary process $X(t)$ with a sinusoidal wave denoted by $\cos(2\pi f_c t + \Theta)$, where the phase Θ is a random variable that is uniformly distributed over the interval 0 to 2π. The addition of the random phase Θ in this manner merely recognizes the fact that the time origin is arbitrarily chosen when $X(t)$ and $\cos(2\pi f_c t + \Theta)$ come from physically independent sources, as is usually the case. We are interested in determining the power spectral density of the random process $Y(t)$ defined by

$$Y(t) = X(t)\cos(2\pi f_c t + \Theta) \tag{8.138}$$

We note that the autocorrelation of $Y(t)$ is given by

$$\begin{aligned} R_Y(\tau) &= E[Y(t+\tau)Y(t)] \\ &= E[X(t+\tau)\cos(2\pi f_c t + 2\pi f_c \tau + \Theta)X(t)\cos(2\pi f_c t + \Theta)] \\ &= E[X(t+\tau)X(t)] \\ &\quad \times E[\cos(2\pi f_c t + 2\pi f_c \tau + \Theta)\cos(2\pi f_c t + \Theta)] \\ &= \tfrac{1}{2}R_X(\tau)E[\cos(2\pi f_c \tau) + \cos(4\pi f_c t + 2\pi f_c \tau + 2\Theta)] \\ &= \tfrac{1}{2}R_X(\tau)\cos(2\pi f_c \tau) \end{aligned}$$

Because the power spectral density is the Fourier transform of the autocorrelation function, we find that the power spectral densities of the random process $X(t)$ and $Y(t)$ are related as follows:

$$S_Y(f) = \tfrac{1}{4}[S_X(f - f_c) + S_X(f + f_c)] \tag{8.139}$$

That is, to obtain the power spectral density of the random process $Y(t)$, we shift the given power spectral density $S_X(f)$ of random process $X(t)$ to the right by f_c, shift it to the left by f_c, add the two shifted power spectra, and divide the result by 4.

RELATION AMONG THE POWER SPECTRAL DENSITIES OF THE INPUT AND OUTPUT RANDOM PROCESSES

Let $S_Y(f)$ denote the power spectral density of the output random process $Y(t)$ obtained by passing the random process $X(t)$ through a linear filter of transfer function $H(f)$. Then, recognizing by definition that the power spectral density of a random process is equal to the Fourier transform of its autocorrelation function and substituting Eq. 8.116 for $R_Y(\tau)$, we obtain

$$\begin{aligned} S_Y(f) &= \int_{-\infty}^{\infty} R_Y(\tau) \exp(-j2\pi f\tau)\, d\tau \\ &= \int_{-\infty}^{\infty}\int_{-\infty}^{\infty}\int_{-\infty}^{\infty} h(\tau_1)h(\tau_2)R_X(\tau - \tau_1 + \tau_2) \exp(-j2\pi f\tau)\, d\tau_1\, d\tau_2\, d\tau \end{aligned} \tag{8.140}$$

Let $\tau - \tau_1 + \tau_2 = \tau_0$, or, equivalently, $\tau = \tau_0 + \tau_1 - \tau_2$. Then, by making this substitution in Eq. 8.140, we find that $S_Y(f)$ may be expressed as the product of three terms: the transfer function $H(f)$ of the filter, the complex conjugate of $H(f)$, and the power spectral density $S_X(f)$ of the input process $X(t)$, as shown by

$$S_Y(f) = H(f)H^*(f)S_X(f) \tag{8.141}$$

However, $|H(f)|^2 = H(f)H^*(f)$. We thus find that the relationship among the power spectral densities of the input and output random processes is simply expressed in the frequency domain by writing

$$S_Y(f) = |H(f)|^2 S_X(f) \tag{8.142}$$

That is, *the output power spectral density equals the input power spectral density multiplied by the squared magnitude of the transfer function of the filter*. By using this relation, we can determine the effect of passing a wide-sense stationary process through a linear time-invariant filter.

It is of interest to note that Eq. 8.142 may also be deduced from Eq. 8.123 simply by recognizing that the mean-square value of a wide-sense stationary process equals the total area under the curve of power spectral density of the process in accordance with Property 2 (i.e., Eq. 8.127).

EXERCISE 12 Consider the *comb filter*[10] of Fig. 8.22 consisting of a delay line and a summing device. Evaluate the power spectral density $S_Y(f)$ of the filter output $Y(t)$, given that the power spectral density of the filter input $X(t)$ is $S_X(f)$. What is the approximate value of $S_Y(f)$ for small values of frequency f?

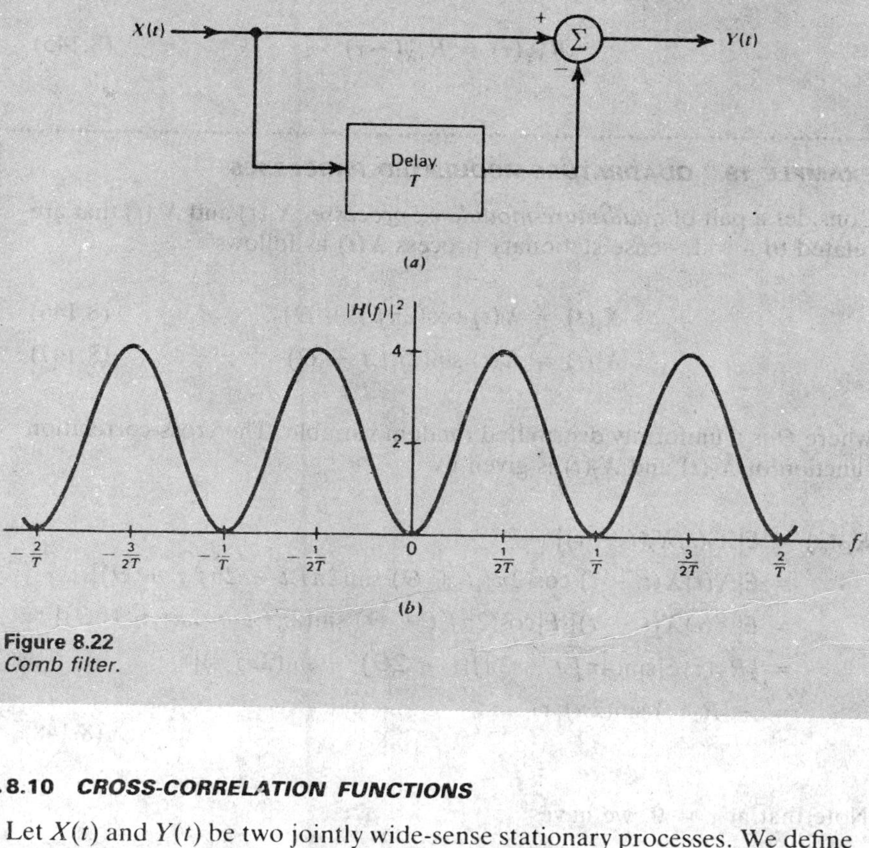

Figure 8.22
Comb filter.

8.10 CROSS-CORRELATION FUNCTIONS

Let $X(t)$ and $Y(t)$ be two jointly wide-sense stationary processes. We define the *cross-correlation function* $R_{XY}(\tau)$ of these two processes as:

$$R_{XY}(\tau) = E[X(t)Y(t - \tau)] \tag{8.143}$$

[10]The filter of Fig. 8.22 is referred to as a "comb" filter because a graph of its frequency response is somewhat comb-like in appearance.

Similarly, we define the *second* cross-correlation function $R_{YX}(\tau)$ of the processes $X(t)$ and $Y(t)$ as

$$R_{YX}(\tau) = E[Y(t)X(t-\tau)] \tag{8.144}$$

A cross-correlation function is not generally an even function of τ, as is true for an autocorrelation function, nor does it have a maximum at the origin. However, it does obey a certain *symmetry* relationship:

$$R_{XY}(\tau) = R_{YX}(-\tau) \tag{8.145}$$

EXAMPLE 19 QUADRATURE-MODULATED PROCESSES

Consider a pair of *quadrature-modulated processes* $X_1(t)$ and $X_2(t)$ that are related to a wide-sense stationary process $X(t)$ as follows

$$X_1(t) = X(t)\cos(2\pi f_c t + \Theta) \tag{8.146}$$

$$X_2(t) = X(t)\sin(2\pi f_c t + \Theta) \tag{8.147}$$

where Θ is a uniformly distributed random variable. The cross-correlation function of $X_1(t)$ and $X_2(t)$ is given by

$$\begin{aligned} R_{12}(\tau) &= E[X_1(t)X_2(t-\tau)] \\ &= E[X(t)X(t-\tau)\cos(2\pi f_c t + \Theta)\sin(2\pi f_c t - 2\pi f_c \tau + \Theta)] \\ &= E[X(t)X(t-\tau)]\,E[\cos(2\pi f_c t + \Theta)\sin(2\pi f_c t - 2\pi f_c \tau + \Theta)] \\ &= \tfrac{1}{2}R_X(\tau)E[\sin(4\pi f_c t - 2\pi f_c \tau + 2\Theta) - \sin(2\pi f_c \tau)] \\ &= -\tfrac{1}{2}R_X(\tau)\sin(2\pi f_c \tau) \end{aligned} \tag{8.148}$$

Note that at $\tau = 0$, we have

$$\begin{aligned} R_{12}(0) &= E[X_1(t)X_2(t)] \\ &= 0 \end{aligned} \tag{8.149}$$

This shows that the random variables $X_1(t)$ and $X_2(t)$ obtained by observing the quadrature-modulated processes $X_1(t)$ and $X_2(t)$ at some fixed value of time t are orthogonal to each other.

EXERCISE 13

(a) Prove the property of cross-correlation functions of a wide-sense stationary process described in Eq. 8.145.
(b) Demonstrate the validity of this property for the quadrature-modulated processes of Example 19.

8.11 CROSS-SPECTRAL DENSITIES

Just as the power spectral density provides a measure of the frequency distribution of a single random process, cross-spectral densities provide a measure of the frequency interrelationship between two random processes.

We define the *cross-spectral densities* $S_{XY}(f)$ and $S_{YX}(f)$ of the pair of random processes $X(t)$ and $Y(t)$ to be the Fourier transforms of the respective cross-correlation functions, as shown by

$$S_{XY}(f) = \int_{-\infty}^{\infty} R_{XY}(\tau) \exp(-j2\pi f\tau)\, d\tau \qquad (8.150)$$

and

$$S_{YX}(f) = \int_{-\infty}^{\infty} R_{YX}(\tau) \exp(-j2\pi f\tau)\, d\tau \qquad (8.151)$$

The cross-correlation functions and cross-spectral densities thus form Fourier transform pairs. Accordingly, we may write

$$R_{XY}(\tau) = \int_{-\infty}^{\infty} S_{XY}(f) \exp(j2\pi f\tau)\, df \qquad (8.152)$$

and

$$R_{YX}(\tau) = \int_{-\infty}^{\infty} S_{YX}(f) \exp(j2\pi f\tau)\, df \qquad (8.153)$$

The cross-spectral densities $S_{XY}(f)$ and $S_{YX}(f)$ are not necessarily real functions of the frequency f. However, substituting the relationship

$$R_{XY}(\tau) = R_{YX}(-\tau)$$

in Eq. 8.150, we find that $S_{XY}(f)$ and $S_{YX}(f)$ are related by

$$S_{XY}(f) = S_{YX}(-f) = S_{YX}^{*}(f) \qquad (8.154)$$

That is to say, the cross-spectral densities of a pair of jointly wide-sense stationary processes are the complex conjugate of each other. Because of this property, the sum of $S_{XY}(f)$ and $S_{YX}(f)$ is real.

EXAMPLE 20

Suppose that the random processes $X(t)$ and $Y(t)$ have zero mean, and they are individually stationary in the wide sense. Consider the sum random process

$$Z(t) = X(t) + Y(t) \tag{8.155}$$

The problem is to determine the power spectral density of $Z(t)$.

The autocorrelation function of $Z(t)$ is given by

$$\begin{aligned} R_Z(t, u) &= E[Z(t)Z(u)] \\ &= E[(X(t) + Y(t))(X(u) + Y(u))] \\ &= E[X(t)X(u)] + E[X(t)Y(u)] + E[Y(t)X(u)] + E[Y(t)Y(u)] \\ &= R_X(t, u) + R_{XY}(t, u) + R_{YX}(t, u) + R_Y(t, u) \end{aligned} \tag{8.156}$$

Defining $\tau = t - u$, we may therefore write

$$R_Z(\tau) = R_X(\tau) + R_{XY}(\tau) + R_{YX}(\tau) + R_Y(\tau) \tag{8.157}$$

when the random processes $X(t)$ and $Y(t)$ are also jointly stationary in the wide sense. Accordingly, taking the Fourier transform of both sides of Eq. 8.157, we get

$$S_Z(f) = S_X(f) + S_{XY}(f) + S_{YX}(f) + S_Y(f) \tag{8.158}$$

We thus see that the cross-spectral densities $S_{XY}(f)$ and $S_{YX}(f)$ represent the spectral components that must be added to the individual power spectral densities of a pair of correlated random processes in order to obtain the power spectral density of their sum.

When the wide-sense stationary processes $X(t)$ and $Y(t)$ are uncorrelated, the cross-spectral densities $S_{XY}(f)$ and $S_{YX}(f)$ are zero, so Eq. 8.158 reduces to

$$S_Z(f) = S_X(f) + S_Y(f) \tag{8.159}$$

We may generalize this result by stating that when there is a multiplicity of zero-mean wide-sense stationary processes that are uncorrelated with each other, the power spectral density of their sum is equal to the sum of their individual power spectral densities.

EXAMPLE 21

Consider next the problem of passing two jointly wide-sense stationary random processes through a pair of separate, stable, linear, time-invariant filters, as shown in Fig. 8.23. In particular, suppose that the random process $X(t)$ is the input to the filter of impulse response $h_1(t)$ and that the random process $Y(t)$ is the input to the filter of impulse response $h_2(t)$. Let $V(t)$ and $Z(t)$ denote the random processes at the respective filter outputs. The cross-correlation function of $V(t)$ and $Z(t)$ is therefore,

$$\begin{aligned} R_{VZ}(t, u) &= E[V(t)Z(u)] \\ &= E\left[\int_{-\infty}^{\infty} h_1(\tau_1)X(t-\tau_1)\,d\tau_1 \int_{-\infty}^{\infty} h_2(\tau_2)Y(u-\tau_2)\,d\tau_2\right] \\ &= \int_{-\infty}^{\infty}\int_{-\infty}^{\infty} h_1(\tau_1)h_2(\tau_2)E[X(t-\tau_1)Y(u-\tau_2)]\,d\tau_1\,d\tau_2 \\ &= \int_{-\infty}^{\infty}\int_{-\infty}^{\infty} h_1(\tau_1)h_2(\tau_2)R_{XY}(t-\tau_1, u-\tau_2)\,d\tau_1\,d\tau_2 \qquad (8.160) \end{aligned}$$

where $R_{XY}(t, u)$ is the cross-correlation function of $X(t)$ and $Y(t)$. Because the input random processes are jointly wide-sense stationary (by hypothesis), we may put $\tau = t - u$ and so rewrite Eq. 8.160 as

$$R_{VZ}(\tau) = \int_{-\infty}^{\infty}\int_{-\infty}^{\infty} h_1(\tau_1)h_2(\tau_2)R_{XY}(\tau - \tau_1 + \tau_2)\,d\tau_1\,d\tau_2 \qquad (8.161)$$

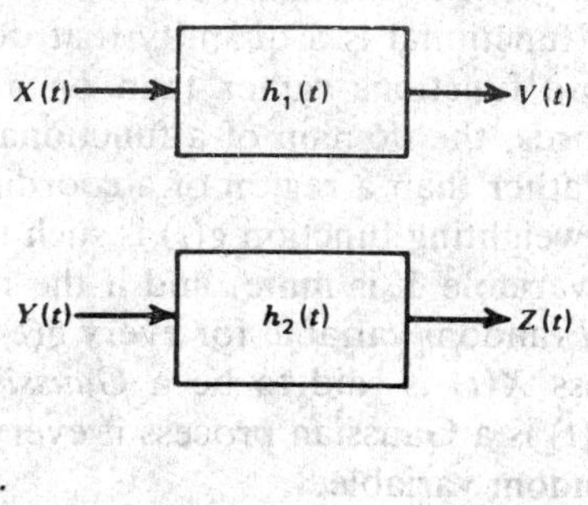

Figure 8.23
A pair of separate filters.

Taking the Fourier transform of both sides of Eq. 8.161 and using a procedure similar to that which led to the development of Eq. 8.141, we finally get

$$S_{VZ}(f) = H_1(f)H_2^*(f)S_{XY}(f) \tag{8.162}$$

where $H_1(f)$ and $H_2(f)$ are the transfer functions of the respective filters in Fig. 8.23 and $H_2^*(f)$ is the complex conjugate of $H_2(f)$. This is the desired relationship between the cross-spectral density of the output processes and that of the input processes. Equation 8.162 includes the relation of Eq. 8.142 as a special case.

8.12 GAUSSIAN PROCESS

Up to this point in our discussion, we have presented the theory of random processes in general terms. In the remainder of the chapter, we consider this theory in the context of some important random processes that are commonly encountered in the study of communication systems.

Let us suppose that we observe a random process $X(t)$ for an interval that starts at time $t = 0$ and lasts until $t = T$. Suppose also that we weight the random process $X(t)$ by some function $g(t)$ and then integrate the product $g(t)X(t)$ over this observation interval, thereby obtaining a random variable Y defined by

$$Y = \int_0^T g(t)X(t)\,dt \tag{8.163}$$

We refer to Y as a *linear functional* of $X(t)$. The distinction between a function and a functional should be carefully noted. For example, the sum $Y = \sum_{i=1}^{N} a_i X_i$, where the a_i are constants and the X_i are random variables, is a linear *function* of the X_i; for each observed set of values for the random variables X_i, we have a corresponding value for the random variable Y. On the other hand, in Eq. 8.163 the value of the random variable Y depends on the course of the *argument function* $g(t)X(t)$ over the observation interval 0 to T. Thus a functional is a quantity that depends on the entire course of one or more functions rather than on a number of discrete variables. In other words, the domain of a functional is a set or space of admissible functions rather than a region of a coordinate space.

If in Eq. 8.163 the weighting function $g(t)$ is such that the mean-square value of the random variable Y is finite, and if the random variable Y is a *Gaussian-distributed* random variable for every $g(t)$ in this class of functions, then the process $X(t)$ is said to be a *Gaussian process*. In other words, the process $X(t)$ is a Gaussian process if every linear functional of $X(t)$ is a Gaussian random variable.

Naturally, when a Guassian process $X(t)$ is *sampled* at time t_i for example, the result is a Gaussian random variable $X(t_i)$. Let $m(t_i)$ denote the mean of $X(t_i)$, and $\sigma^2(t_i)$ denote its variance. We may then express the probability density function of the sample $X(t_i)$ as

$$f_{X(t_i)}(x_i) = \frac{1}{\sqrt{2\pi}\sigma(t_i)} \exp\left[-\frac{(x_i - m(t_i))^2}{2\sigma^2(t_i)}\right] \tag{8.164}$$

A Gaussian process has two main virtues. First, the Gaussian process has many properties that make analytic results possible. Second, the random processes produced by physical phenomena are often such that a Gaussian model is appropriate. The central limit theorem provides the mathematical justification for using a Gaussian process as a model of a large number of different physical phenomena in which the observed random variable, at a particular instant of time, is the result of a large number of individual random events. Furthermore, the use of a Gaussian model to describe such physical phenomena is usually confirmed by experiments. Thus the widespread occurrence of physical phenomena for which a Gaussian model is appropriate, together with the ease with which a Gaussian process is handled mathematically, make the Gaussian process very important in the study of communication systems.

Some of the important properties of a Gaussian process are as follows:

PROPERTY 1

If a Gaussian process $X(t)$ is applied to a stable linear filter, then the random process $Y(t)$ developed at the output of the filter is also Gaussian.

This property is readily derived by using the definition of a Gaussian process based on Eq. 8.163. Consider the situation depicted in Fig. 8.17, where we have a linear time-invariant filter of impulse response $h(t)$, with the random process $X(t)$ as input and the random process $Y(t)$ as output. We assume that $X(t)$ is a Gaussian process. The random processes $Y(t)$ and $X(t)$ are related by the convolution integral

$$Y(t) = \int_0^T h(t - \tau)X(\tau)\, d\tau, \qquad 0 \le t < \infty \tag{8.165}$$

where $0 \le t \le T$ is the observation interval of the input $X(t)$. We assume that the impulse response $h(t)$ is such that the mean-square value of the output random process $Y(t)$ is finite for all t in the time interval $0 \le t < \infty$ for which $Y(t)$ is defined. To demonstrate that the output process $Y(t)$ is Gaussian, we must show that any linear functional of it is a Gaussian random variable. That is, if we define the random variable

$$Z = \int_0^\infty g_Y(t)Y(t)\, dt$$

or, equivalently,

$$Z = \int_0^\infty g_Y(t) \int_0^T h(t - \tau)X(\tau)\, d\tau\, dt \tag{8.166}$$

then Z must be a Gaussian random variable for every function $g_Y(t)$, such that the mean-square value of Z is finite. Interchanging the order of integration in Eq. 8.166, we get

$$Z = \int_0^T g(\tau)X(\tau)\, d\tau \tag{8.167}$$

where

$$g(\tau) = \int_0^\infty g_Y(t)h(t - \tau)\, dt \tag{8.168}$$

Since $X(t)$ is a Gaussian process by hypothesis, it follows from Eq. 8.167 that Z must be a Gaussian random variable. We have thus shown that if the input $X(t)$ to a linear filter is a Gaussian process, then the output $Y(t)$ is also a Gaussian process. Note, however, that although our proof was carried out assuming a time-invariant linear filter, this property is true for any arbitrary stable linear system.

PROPERTY 2

Consider the set of random variables or samples $X(t_1), X(t_2), \ldots, X(t_n)$, obtained by observing a random process $X(t)$ at times $t_1, t_2, \ldots, t_n$. If the process $X(t)$ is Gaussian, then this set of random variables are jointly Gaussian for any n, with their n-fold joint probability density function[11] being completely determined by specifying the set of means

$$m_X(t_i) = E[X(t_i)], \qquad i = 1, 2, \ldots, n$$

and the set of autocorrelation functions

$$R_X(t_k - t_i) = E[X(t_k)X(t_i)], \qquad k, i = 1, 2, \ldots, n$$

Property 2 is frequently used as the definition of a Gaussian process. However, this definition is more difficult to use than that based on Eq. 8.163 for evaluating the effects of filtering on a Gaussian process.

[11] For a detailed discussion of Property 2, see Davenport and Root (1958), pp. 147–154; Sakrison (1968) pp. 87–97.

PROPERTY 3

If a Gaussian process is wide-sense stationary, then the process is also stationary in the strict sense.

This follows directly from Property 2.

PROPERTY 4

If the set of random variables $X(t_1), X(t_2), \ldots, X(t_n)$, obtained by sampling a Gaussian process $X(t)$ at times $t_1, t_2, \ldots, t_n$ are uncorrelated, that is,

$$E[(X(t_k) - m_{X(t_k)})(X(t_i) - m_{X(t_i)})] = 0, \qquad i \neq k$$

then this set of random variables are statistically independent.

The implication of this property is that the joint probability density function of the set of random variables $X(t_1), X(t_2), \ldots, X(t_n)$ can be expressed as the product of the probability density functions of the individual random variables in the set.

8.13 NARROW-BAND RANDOM PROCESS

The receiver of a communication system usually includes some provision for *preprocessing* the received signal. The preprocessing may take the form of a narrow-band filter designed to restrict noise at the receiver input to a band of frequencies just wide enough to accommodate the detection of the modulated wave in the received signal. The signal appearing at the output of the narrow-band filter represents the sample function of a *narrow-band random process*. In this section, we present a canonical representation of such a process and its statistical characteristics.

By analogy with the canonical representation of a narrow-band signal (See Section 3.5), we may likewise represent a narrow-band random process $X(t)$, centered at some frequency f_c, in the *canonical form*:

$$X(t) = X_I(t)\cos(2\pi f_c t) - X_Q(t)\sin(2\pi f_c t) \qquad (8.169)$$

where $X_I(t)$ is the *in-phase component* of $X(t)$, and $X_Q(t)$ is its *quadrature component*. Given the random process $X(t)$, we may extract the in-phase component $X_I(t)$ and the quadrature components $X_Q(t)$, except for scaling factors, using the arrangement depicted in Fig. 8.24*a*.

Suppose the narrow-band random process $X(t)$ is known to have the following characteristics:

1. The power spectral density $S_X(f)$ of the process $X(t)$ satisfies the condition:

$$S_X(f) = 0 \quad \text{for} \quad |f| \leq f_c - W \quad \text{and} \quad |f| \geq f_c + W \qquad (8.170)$$

   This condition is illustrated in Fig. 8.25.

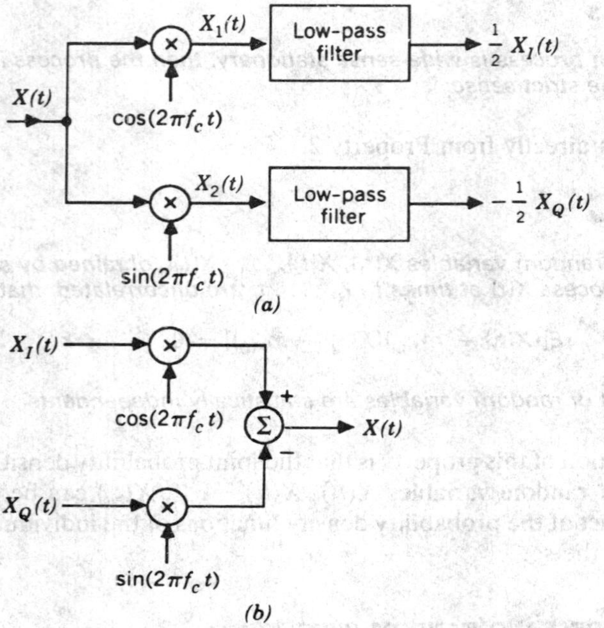

Figure 8.24
(a) Extraction of in-phase and quadrature components of a narrow-band process. (b) Generation of a narrow-band process from its in-phase and quadrature components.

2. The process $X(t)$ is Gaussian with zero mean and variance σ_X^2; the zero-mean characteristic is a direct consequence of the fact that $X(t)$ is narrow-band.

We then find that the random processes $X_I(t)$ and $X_Q(t)$ have the following properties:

PROPERTY 1

The in-phase component $X_I(t)$ and the quadrature component $X_Q(t)$ of a narrow-band random process $X(t)$ are both low-pass random processes.

This property follows directly from the scheme of Fig. 8.24*a*. Both the in-phase component $X_I(t)$ and the quadrature component $X_Q(t)$ appear in Fig. 8.24*a* as the outputs of low-pass filters.

PROPERTY 2

The in-phase component $X_I(t)$ and the quadrature component $X_Q(t)$ of a narrow-band random process $X(t)$ have identical power spectral densities related

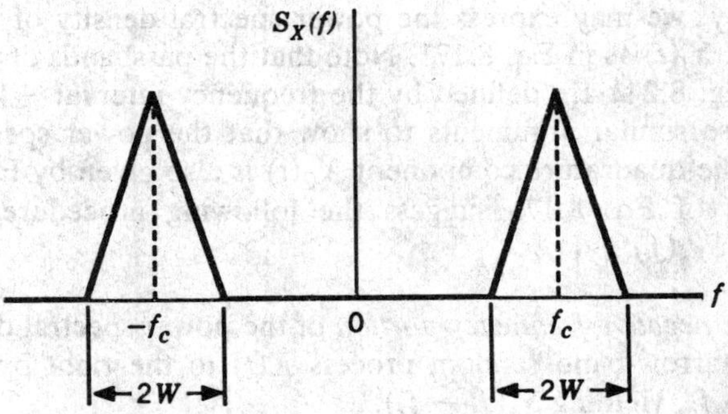

Figure 8.25
Power spectrum of a narrow-band random process.

to that of X(t) as follows:

$$S_{X_I}(f) = S_{X_Q}(f) = \begin{cases} S_X(f - f_c) + S_X(f + f_c), & -W < f < W \\ 0, & \text{otherwise} \end{cases} \tag{8.171}$$

The proof of this property also follows from Fig. 8.24*a*. We first recognize that $X_I(t)$ and $X_Q(t)$ may be extracted from $X(t)$ as follows:

1. The narrow-band random process $X(t)$ is multiplied alternately by the sinusoidal carriers $\cos(2\pi f_c t)$ and $\sin(2\pi f_c t)$ to generate the pair of quadrature-modulated processes:
$$X_1(t) = X(t)\cos(2\pi f_c t) \tag{8.172}$$
$$X_2(t) = X(t)\sin(2\pi f_c t) \tag{8.173}$$
where we have set the phase of the two sinusoidal carriers to be zero for convenience of presentation.
2. The modulated process $X_1(t)$ is passed through a low-pass filter of bandwidth W, yielding $\frac{1}{2}X_I(t)$.
3. The modulated process $X_2(t)$ is passed through a second low-pass filter of bandwidth W, yielding $-\frac{1}{2}X_Q(t)$.

Next we recognize that the power spectral density of the modulated process $X_1(t)$ is related to that of the narrow-band random process $X(t)$ as follows (see Example 18)

$$S_{X_1}(f) = \tfrac{1}{4}[S_X(f - f_c) + S_X(f + f_c)] \tag{8.174}$$

The part of $S_{X_1}(f)$ that lies inside the passband of the low-pass filter in the upper path of Fig. 8.24*a* defines the power spectral density of $\frac{1}{2}X(t)$.

Accordingly, we may express the power spectral density of the in-phase component $X_I(t)$ as in Eq. 8.171. Note that the passbands of the low-pass filters in Fig. 8.24*a* are defined by the frequency interval $-W < f < W$. We may use similar arguments to show that the power spectral density $S_{X_Q}(f)$ of the quadrature component $X_Q(t)$ is also given by Eq. 8.171.

The use of Eq. 8.171 suggest the following procedure for finding $S_{X_I}(f)$ and $S_{X_Q}(f)$:

1. Shift the *negative-frequency portion* of the power spectral density $S_X(f)$ of the narrow-band random process $X(t)$ to the right by an amount equal to f_c, yielding $S_X(f - f_c)$.
2. Shift the *positive-frequency portion* of the power spectral density $S_X(f)$ of the narrow-band random process $X(t)$ to the left by an amount equal to f_c, yielding $S_X(f + f_c)$.
3. Add the shifted power spectra found in (1) and (2), thereby obtaining the desired $S_{X_I}(f)$ or $S_{X_Q}(f)$.

PROPERTY 3

The in-phase component $X_I(t)$ and the quadrature component $X_Q(t)$ have the same mean and variance as the narrow-band random process $X(t)$.

Since the narrow-band random process $X(t)$ has zero mean, the modulated processes $X_1(t)$ and $X_2(t)$ (defined in Eqs. 8.172 and 8.173) must also have zero mean. Moreover, Fig. 8.24*a* reveals that $X_I(t)$ and $X_Q(t)$, low-pass filtered versions of $X_1(t)$ and $X_2(t)$, also have zero mean.

To prove the remaining part of Property 3, we first observe that when a random process has zero mean, its variance and mean-square value assume a common value. Since both $X_I(t)$ and $X_Q(t)$ have zero mean, their mean-square value and therefore variance equals the total area under the curves of their respective power spectra, as shown by

$$\begin{aligned}\sigma_{X_I}^2 = \sigma_{X_Q}^2 &= \int_{-W}^{W} [S_X(f - f_c) + S_X(f - f_c)]\, df \\ &= \int_{-f_c-W}^{-f_c+W} S_X(f)\, df + \int_{f_c-W}^{f_c+W} S_X(f)\, df \\ &= \sigma_X^2 \end{aligned} \tag{8.175}$$

where σ_X^2 is the variance of the zero-mean narrow-band process $X(t)$.

PROPERTY 4

The in-phase component $X_I(t)$ and the quadrature component $X_Q(t)$ of the narrow-band random process $X(t)$ are uncorrelated with each other.

To prove this property, we first observe from Eqs. 8.172 and 8.173 that the modulated processes $X_1(t)$ and $X_2(t)$ are obtained from $X(t)$ by the use of a pair of carriers, $\cos(2\pi f_c t)$ and $\sin(2\pi f_c t)$, that are in-phase-quadrature. Hence, $X_1(t)$ and $X_2(t)$ are orthogonal to each other (see Example 19). Since they both have zero mean, they are also uncorrelated with each other. Accordingly, the in-phase component $X_I(t)$ and the quadrature component $X_Q(t)$, low-pass filtered versions of $X_1(t)$ and $X_2(t)$, are also uncorrelated with each other.

PROPERTY 5

If a narrow-band random process $X(t)$ is Gaussian, then the in-phase component $X_I(t)$ and the quadrature component $X_Q(t)$ are also Gaussian.

This property follows directly from the definition of a Gaussian process. Specifically, we observe from Fig. 8.24*a* that both the in-phase component $X_I(t)$ and the quadrature component $X_Q(t)$ are derived by performing *linear* operations on the narrow-band random process $X(t)$. If therefore $X(t)$ is Gaussian, then so are $X_I(t)$ and $X_Q(t)$.

These properties have an important implication. Specifically, if the narrow-band random process $X(t)$ is Gaussian, then the in-phase component $X_I(t)$ and the quadrature component $X_Q(t)$ are uncorrelated with each other (Property 4) and they are both Gaussian (Property 5). Consequently, $X_I(t)$ and $X_Q(t)$ are *statistically independent* of each other. Let Y and Z denote the Gaussian random variables obtained by observing the Gaussian processes $X_I(t)$ and $X_Q(t)$ at some fixed value of time t. The probability density functions of these two random variables with zero mean and variance σ_X^2 (Property 3) are

$$f_Y(y) = \frac{1}{\sqrt{2\pi}\sigma_X} \exp\left(-\frac{y^2}{2\sigma_X^2}\right) \tag{8.176}$$

$$f_Z(z) = \frac{1}{\sqrt{2\pi}\sigma_X} \exp\left(-\frac{z^2}{2\sigma_X^2}\right) \tag{8.177}$$

With Y and Z representing statistically independent random variables, the joint probability density function of Y and Z is equal to the product of their individual probability density functions, as shown by

$$\begin{aligned} f_{Y,Z}(y, z) &= f_Y(y) f_Z(z) \\ &= \frac{1}{2\pi\sigma_X^2} \exp\left(-\frac{y^2 + z^2}{2\sigma_X^2}\right) \end{aligned} \tag{8.178}$$

Another important implication of these properties is that we may construct a narrow-band Gaussian process $X(t)$ of prescribed statistical characteristics by means of the scheme shown in Fig. 8.24*b*. Specifically, we start with low-pass Gaussian processes $X_I(t)$ and $X_Q(t)$ derived from two independent sources. These two processes have zero mean and the same variance as the process $X(t)$. The processes $X_I(t)$ and $X_Q(t)$ are modulated individually by a pair of sinusoidal carriers that are in phase quadrature. The resulting modulated processes are then added to produce the narrow-band Gaussian process $X(t)$.

EXAMPLE 22

Consider a noise process that is both *Gaussian and white*; the process is said to be white in the sense that it has a constant power spectral density (see Section 4.7). A white Gaussian noise process represents the ultimate in "randomness" in the sense that any two of its samples are statistically independent. Suppose then a white Gaussian noise process of zero mean and power spectral density $N_0/2$ is passed through an ideal narrow-band filter, resulting in a narrowband Gaussian process $X(t)$ with zero mean and power spectral density as shown in Fig. 8.26*a*. The requirement is to find

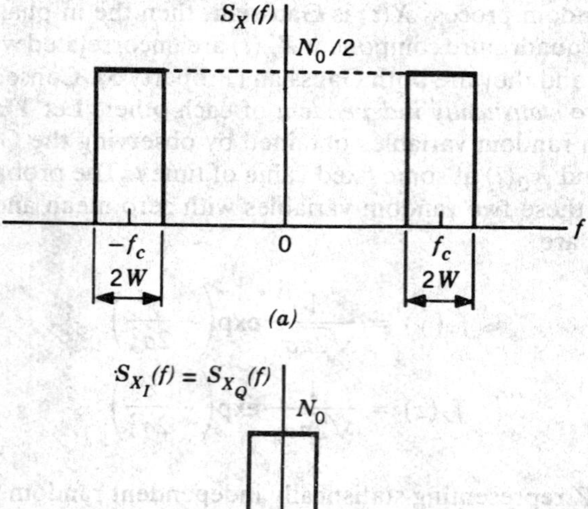

Figure 8.26
(a) Power spectral density of a narrow-band Gaussian process. (b) Power spectral density of in-phase and quadrature components.

the statistical characteristics of the in-phase and quadrature components of the process $X(t)$.

Following the procedure described previously (see Property 2), we find that the power spectra of the in-phase component $X_I(t)$ and the quadrature component $X_Q(t)$ are as shown in Fig. 8.26*b*.

From Fig. 8.26 we deduce that the processes $X(t)$, $X_I(t)$, and $X_Q(t)$ have a common variance:

$$\sigma_X^2 = 2N_0W$$

Moreover, they all have zero mean. Hence, the probability density functions of the random variables Y and Z, obtained by observing $X_I(t)$ and $X_Q(t)$ at some fixed time, are:

$$f_Y(y) = \frac{1}{2\sqrt{\pi N_0 W}} \exp\left(-\frac{y^2}{4N_0W}\right)$$

$$f_Z(z) = \frac{1}{2\sqrt{\pi N_0 W}} \exp\left(-\frac{z^2}{4N_0W}\right)$$

EXERCISE 14 Find the probability density function of a random variable obtained by observing the narrow-band random process $X(t)$ of Example 22 at some fixed time.

EXERCISE 15 Continuing with Example 22, do the following:

(a) Find the autocorrelation function of the narrow-band random process $X(t)$.
(b) Find the autocorrelation functions of the in-phase component $X_I(t)$ and quadrature component $X_Q(t)$.

EXERCISE 16 Consider a narrow-band random process $X(t)$ whose power spectral density $S_X(f)$ is symmetric with respect to the midband frequency f_c. Show that, for this special case, the power spectral densities of the in-phase component $X_I(t)$ and quadrature component $X_Q(t)$ are:

$$S_{X_I}(f) = S_{X_Q}(f) = \begin{cases} 2S_X(f - f_c) & -W < f < W \\ 0, & \text{otherwise} \end{cases} \tag{8.179}$$

where $2W$ is the bandwidth of $X(t)$.

8.14 ENVELOPE AND PHASE OF NARROW-BAND RANDOM PROCESS

As with narrow-band signals, we may also represent a narrow-band random process $X(t)$ in terms of its *envelope* and *phase* components. Specifically, we may write

$$X(t) = A(t)\cos[2\pi f_c t + \Phi(t)] \tag{8.180}$$

where $A(t)$ is the envelope and $\Phi(t)$ is the phase of $X(t)$. These two components are related to the in-phase component $X_I(t)$ and quadrature component $X_Q(t)$ of the process $X(t)$ as follows

$$A(t) = [X_I^2(t) + X_Q^2(t)]^{1/2} \tag{8.181}$$

$$\Phi(t) = \tan^{-1}\left(\frac{X_Q(t)}{X_I(t)}\right) \tag{8.182}$$

Let R and Ψ denote the random variables obtained by observing the random processes $A(t)$ and $\Phi(t)$, respectively, at some fixed time. Let Y and Z denote the random variables obtained by observing the related processes $X_I(t)$ and $X_Q(t)$, respectively, at the same time. The probability density functions of R and Ψ may be related to those of Y and Z as follows. The joint-probability density function of Y and Z is given by Eq. 8.178. Accordingly, the joint probability that the random variable Y lies between

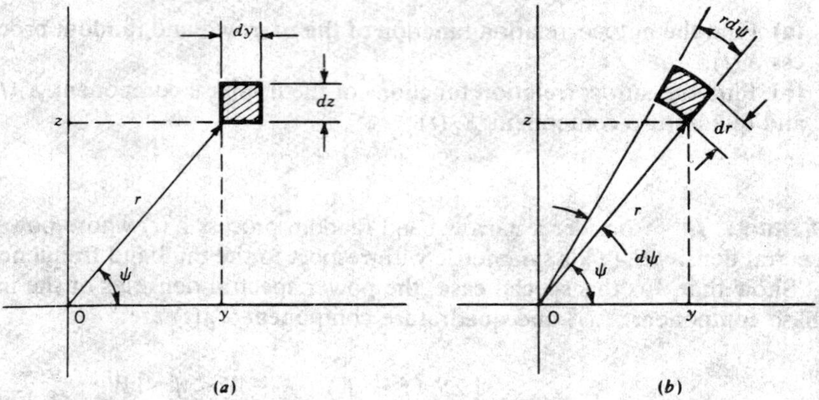

Figure 8.27
Illustrating the coordinate system for representation of a narrowband random process: (a) In terms of in-phase and quadrature components, and (b) in terms of envelope and phase.

y and $y + dy$ and that the random variable Z lies between z and $z + dz$ (i.e., the pair of random variables Y and Z lies inside the shaded area of Fig. 8.27*a*) is given by

$$f_{Y,Z}(y, z)\, dy\, dz = \frac{1}{2\pi\sigma_X^2} \exp\left(-\frac{y^2 + z^2}{2\sigma_X^2}\right) dy\, dz \tag{8.183}$$

However, from Fig. 8.27 we observe that

$$y = r \cos\psi \tag{8.184}$$

and

$$z = r \sin\psi \tag{8.185}$$

where r and ψ are sample values of the random variables R and Ψ, respectively. Also, in a limiting sense, we may equate the two areas shown shaded in parts *a* and *b* of Fig. 8.27, and so write

$$dy\, dz = r\, dr\, d\psi \tag{8.186}$$

Therefore, substituting Eqs. 8.184 through 8.186 in 8.183, we find that the probability that the random variables R and Ψ lie inside the shaded area of Fig. 8.27*b* is equal to

$$\frac{r}{2\pi\sigma_X^2} \exp\left(-\frac{r^2}{2\sigma_X^2}\right) dr\, d\psi$$

That is, the joint probability density function of R and Ψ is

$$f_{R,\Psi}(r, \psi) = \frac{r}{2\pi\sigma_X^2} \exp\left(-\frac{r^2}{2\sigma_X^2}\right) \tag{8.187}$$

This probability density function is independent of the angle ψ, which means that the random variables R and Ψ are statistically independent. We may thus express $f_{R,\Psi}(r, \psi)$ as the product of $f_R(r)$ and $f_\Psi(\psi)$. In particular, the random variable Ψ is uniformly distributed inside the range 0 to 2π, as shown by

$$f_\Psi(\psi) = \begin{cases} \dfrac{1}{2\pi}, & 0 \leq \psi \leq 2\pi \\ 0, & \text{elsewhere} \end{cases} \tag{8.188}$$

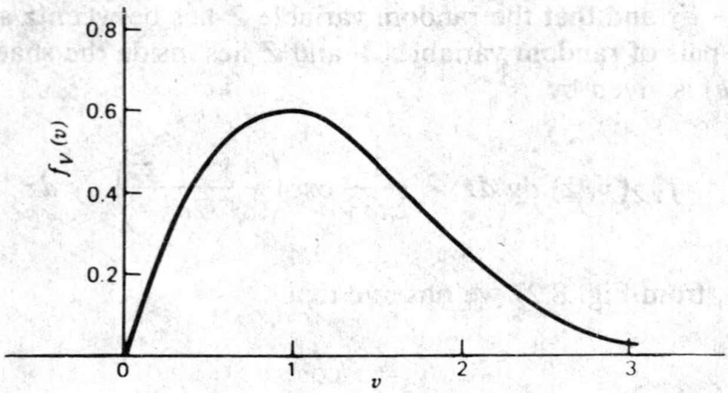

Figure 8.28
Rayleigh distribution.

This leaves the probability density function of R as

$$f_R(r) = \begin{cases} \dfrac{r}{\sigma_X^2} \exp\left(-\dfrac{r^2}{2\sigma_X^2}\right), & r > 0 \\ 0, & \text{elsewhere} \end{cases} \tag{8.189}$$

where σ_X^2 is the variance of the original narrow-band process $X(t)$. A random variable having the probability density function of Eq. 8.189 is said to be *Rayleigh-distributed.*[12]

For convenience of graphical presentation, let

$$v = \frac{r}{\sigma_X} \tag{8.190}$$

and

$$f_V(v) = \sigma_X f_R(r) \tag{8.191}$$

Then we may rewrite the Rayleigh distribution of Eq. 8.189 in the *standardized* form

$$f_V(v) = \begin{cases} v \exp\left(-\dfrac{v^2}{2}\right), & v > 0 \\ 0, & \text{elsewhere} \end{cases} \tag{8.192}$$

[12]The Rayleigh distribution is named after the English physicist J. W. Strutt, Lord Rayleigh.

Equation 8.192 is plotted in Fig. 8.28. The peak value of the distribution $f_V(v)$ occurs at $v = 1$ and is equal to 0.607. Note also that, unlike the Gaussian distribution, the Rayleigh distribution is zero for negative values of v. This is because an envelope function can only assume positive values.

PROBLEMS

P8.1 Probability Theory

Problem 1 Consider a deck of 52 cards, divided into 4 different suits, with 13 cards in each suit ranging from the two up through the ace. Assume that all cards are equally likely to be drawn.

(a) Suppose that a single card is drawn from a full deck. What is the probability that this card is the ace of diamonds? What is the probability that the single card drawn is an ace of any one of the four suits?
(b) Suppose next that two cards are drawn from a full deck. What is the probability that the cards drawn are an ace and a king, not necessarily of the same suit?

P8.2 Random Variables

Problem 2 Consider a random variable X that is uniformly distributed between the values 0 and 1 with probability 1/4, takes on the value 1 with probability 1/4, and is uniformly distributed between the values 1 and 2 with probability 1/2. Determine the distribution function of the random variable X.

Problem 3 Consider a random variable X defined by the double-exponential density:

$$f_X(x) = a \exp(-b|x|) \qquad -\infty < x < \infty$$

where a and b are positive constants.

(a) Determine the relationship between a and b so that $f_X(x)$ is a probability density function.
(b) Determine the corresponding distribution function $F_X(x)$.
(c) Find the probability that the random variable X lies between 1 and 2.

Problem 4 A random variable R is Rayleigh distributed with its probability density function given by

$$f_R(r) = \begin{cases} \dfrac{r}{b} \exp\left(-\dfrac{r^2}{2b}\right), & 0 \leq r < \infty \\ 0, & \text{otherwise} \end{cases}$$

(a) Determine the corresponding distribution function $F_R(r)$.
(b) Show that the mean of R is equal to $\sqrt{b\pi/2}$.
(c) What is the mean-square value of R?
(d) What is the variance of R?

Problem 5 Consider a uniformly distributed random variable Z defined by

$$f_Z(z) = \begin{cases} \dfrac{1}{2\pi}, & 0 \leqslant z \leqslant 2\pi \\ 0, & \text{otherwise} \end{cases}$$

The two random variables X and Y are related to Z by

$$X = \sin(Z)$$

and

$$Y = \cos(Z)$$

(a) Determine the probability density functions of X and Y.
(b) Show that X and Y are uncorrelated random variables.
(c) Are X and Y statistically independent? Why?

Problem 6 A random variable Z is defined by

$$Z = X + Y$$

where X and Y are statistically independent. Given that

$$f_X(x) = \begin{cases} \exp(-x), & 0 \leqslant x \leqslant \infty \\ 0, & \text{otherwise} \end{cases}$$

and

$$f_Y(y) = \begin{cases} 2\exp(-2y), & 0 \leqslant y \leqslant \infty \\ 0, & \text{otherwise} \end{cases}$$

determine the probability density function of Z.

P8.3 Gaussian Distribution

Problem 7

(a) The characteristic function of a random variable X is denoted by

$\phi_X(v)$. Show that the nth moment of X is related to $\phi_X(v)$ by

$$E[X^n] = (-j)^n \frac{d^n}{dv^n} \phi_X(v)\Bigg|_{v=0}$$

(b) Show that the characteristic function of a Gaussian random variable X of mean m_X and variance σ_X^2 is

$$\phi_X(v) = \exp(jvm_X - \tfrac{1}{4}v^2\sigma_X^2)$$

(c) Show that the nth central moment of this Gaussian random variable is

$$E[(X - m_X)^n] = \begin{cases} 1 \times 3 \times 5 \cdots (n-1)\sigma_X^n, & \text{for } n \text{ even} \\ 0, & \text{for } n \text{ odd} \end{cases}$$

Problem 8 A Gaussian random variable has zero mean and a standard deviation of 10 V. A constant voltage of 5 V is added to this random variable.

(a) Determine the probability that a measurement of this composite signal yields a positive value.
(b) Determine the probability that the arithmetic mean of two independent measurements of this signal is positive.

Problem 9 A random variable Z is defined by

$$Z = \sum_{i=1}^{4} X_i$$

where the X_i are identically distributed and statistically independent random variables. It is given that the probability density function of each X_i is

$$f_{X_i}(x_i) = \begin{cases} 1, & -\tfrac{1}{2} \leq x_i \leq \tfrac{1}{2} \\ 0, & \text{otherwise} \end{cases}$$

(a) Determine the probability density function $f_Z(z)$.
(b) Show that $f_Z(z)$ is closely approximated by a Gaussian probability density function with zero mean and variance 1/3, as predicted by the central limit theorem.

P8.4 Transformation of Random Variables

Problem 10 A Gaussian random variable X of zero mean and variance σ_X^2 is transformed by a piecewise-linear rectifier characterized by the input-

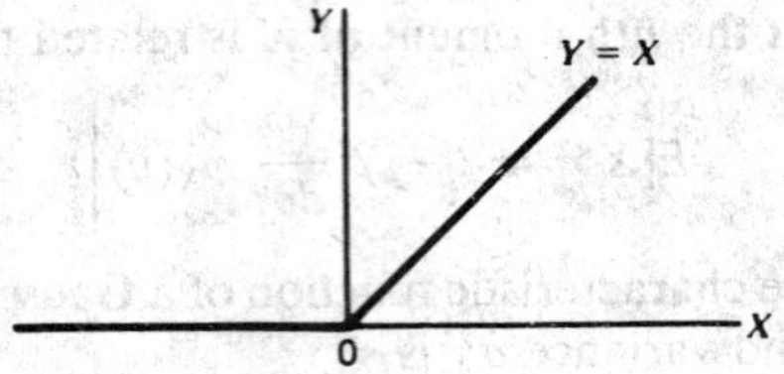

Figure P8.1

output relation (see Fig. P8.1):

$$Y = \begin{cases} X, & X > 0 \\ 0, & X \leqslant 0 \end{cases}$$

The probability density function of the new random variable Y is described by

$$f_Y(y) = \begin{cases} 0, & y < 0 \\ k\delta(f), & y = 0 \\ \dfrac{1}{\sqrt{2\pi}\sigma_X} \exp\left(-\dfrac{y^2}{2\sigma_X^2}\right), & y > 0 \end{cases}$$

(a) Explain the reasons for this result.
(b) Determine the value of the constant k by which the delta function $\delta(f)$ is weighted.

P8.5 Stationarity

Problem 11 Consider a random process $X(t)$ defined by

$$X(t) = \sin(2\pi F t)$$

in which the frequency F is a random variable with the probability density function

$$f_F(\nu) = \begin{cases} \dfrac{1}{W}, & 0 \leqslant \nu \leqslant W \\ 0, & \text{otherwise} \end{cases}$$

Show that $X(t)$ is nonstationary. (To avoid confusion, we have used ν to denote frequency in place of the standard symbol f.)

Hint: Examine specific sample functions of the random process $X(t)$ for the frequency $\nu = W/4$, $W/2$, and W, say.

Problem 12 Consider the sinusoidal process

$$X(t) = A\cos(2\pi f_c t)$$

where the frequency f_c is constant and the amplitude A is uniformly distributed:

$$f_A(a) = \begin{cases} 1, & 0 \leq a \leq 1 \\ 0, & \text{otherwise} \end{cases}$$

Determine whether or not this process is stationary in the strict sense.

Problem 13 A random process $X(t)$ is defined by

$$X(t) = A\cos(2\pi f_c t)$$

where A is a Gaussian random variable of zero mean and variance σ_A^2. This random process is applied to an ideal integrator, producing an output $Y(t)$ defined by

$$Y(t) = \int_0^t X(\tau)\, d\tau$$

(a) Determine the probability density function of the output $Y(t)$ at a particular time t_k.
(b) Determine whether or not $Y(t)$ is stationary.

P8.7 Mean, Correlation, and Covariance Functions

Problem 14 Prove the following two properties of the autocorrelation function $R_X(\tau)$ of a random process $X(t)$:

(a) If $X(t)$ contains a dc component equal to A, then $R_X(\tau)$ will contain a constant component equal to A^2.
(b) If $X(t)$ contains a sinusoidal component, then $R_X(\tau)$ will also contain a sinusoidal component of the same frequency.

Problem 15 The square wave $x(t)$ of Fig. P8.2 of constant amplitude A, period T_0, and delay t_d, represents the sample function of a random process $X(t)$. The delay is random, described by the probability density function

$$f_{T_d}(t_d) = \begin{cases} \dfrac{1}{T_0}, & -\frac{1}{2}T_0 \leq t_d \leq \frac{1}{2}T_0 \\ 0, & \text{otherwise} \end{cases}$$

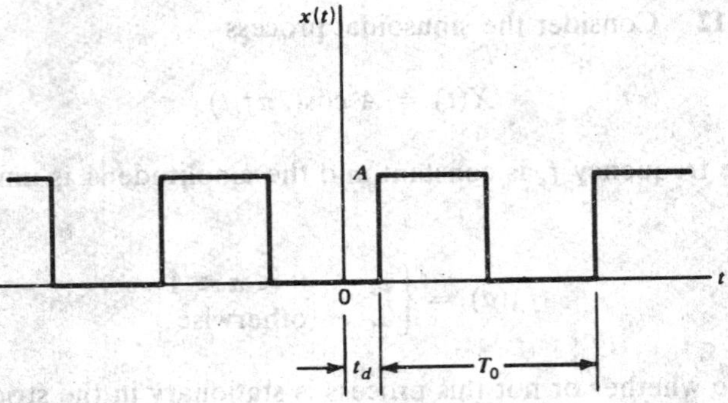

Figure P8.2

(a) Determine the probability density function of the random variable $X(t_k)$ obtained by observing the random process $X(t)$ at time t_k.
(b) Determine the mean and autocorrelation function of $X(t)$ using ensemble-averaging.
(c) Determine the mean and autocorrelation function of $X(t)$ using time-averaging.
(d) Establish whether or not $X(t)$ is wide-sense stationary. In what sense is it ergodic?

Problem 16 A binary wave consists of a random sequence of symbols 1 and 0, similar to that described in Example 13, with one basic difference: symbol 1 is now represented by a pulse of amplitude A volts and symbol 0 is represented by zero volt. All other parameters are the same as before. Show that for this new random binary wave $X(t)$, the autocorrelation

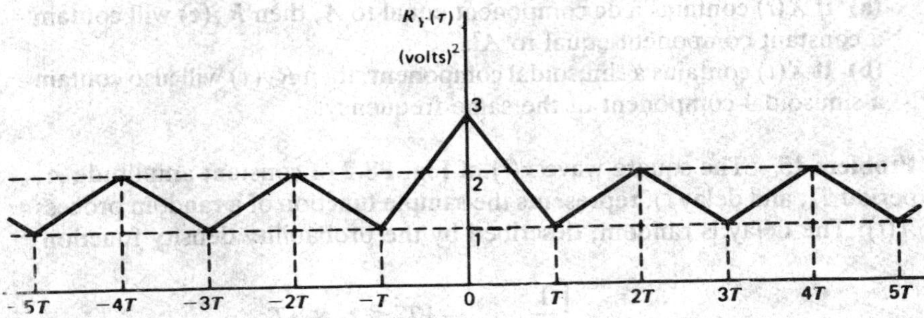

Figure P8.3

function is

$$R_X(\tau) = \begin{cases} \dfrac{A^2}{4} + \dfrac{A^2}{4}\left(1 - \dfrac{|\tau|}{T}\right), & |\tau| < T \\ \dfrac{A^2}{4}, & |\tau| \geq T \end{cases}$$

Problem 17 A random process $Y(t)$ consists of a dc component of $\sqrt{3/2}$ V, a periodic component $g_p(t)$, and a random component $X(t)$. The autocorrelation function of $Y(t)$ is shown in Fig. P8.3

(a) What is the average power of the periodic component $g_p(t)$?
(b) What is the average power of the random component $X(t)$?

P8.8 Random Process Transmission Through Linear Filters

Problem 18 A *random telegraph signal* $X(t)$, characterized by the autocorrelation function

$$R_X(\tau) = \exp(-2v|\tau|)$$

where v is a constant, is applied to the low-pass RC filter of Fig. P8.4. Determine the autocorrelation function of the random process at the filter output.

Problem 19 Let $X(t)$ be a stationary process with zero mean and autocorrelation function $R_X(\tau)$. We are required to find a linear filter with impulse response $h(t)$, such that the filter output is $X(t)$ when the input is white noise of zero mean and autocorrelation function $(N_0/2)\ \delta(\tau)$. Determine the condition that the impulse response $h(t)$ must satisfy in order to achieve this requirement.

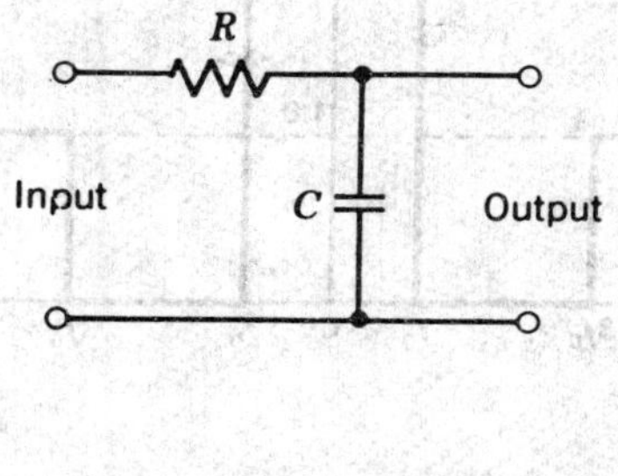

Figure P8.4

P8.9 Power Spectral Density

Problem 20 The output of an oscillator is described by

$$X(t) = A\cos(2\pi F t + \Theta),$$

where A is a constant, and F and Θ are independent random variables. The probability density function of F is denoted by $f_F(\nu)$, and that of Θ is defined by

$$f_\Theta(\theta) = \begin{cases} \dfrac{1}{2\pi}, & 0 \leq \theta \leq 2\pi \\ 0, & \text{otherwise} \end{cases}$$

Determine the power spectral density of $X(t)$. What happens to this power spectrum when the frequency ν assumes a constant value? (To avoid confusion, we have used ν to denote frequency in place of the standard symbol f.)

Problem 21 Continuing with the random binary wave considered in Problem 16, show that the power spectral density of the wave equals

$$S_X(f) = \frac{A^2}{4}\,\delta(f) + \frac{A^2 T}{4}\,\text{sinc}^2(fT)$$

What is the percentage power contained in the dc component of the binary wave?

Problem 22 Given that a stationary random process $X(t)$ has an autocorrelation function $R_X(\tau)$ and a power spectral density $S_X(f)$, show that:

(a) The autocorrelation function of $dX(t)/dt$, the first derivative of $X(t)$, is equal to minus the second derivative of $R_X(\tau)$.

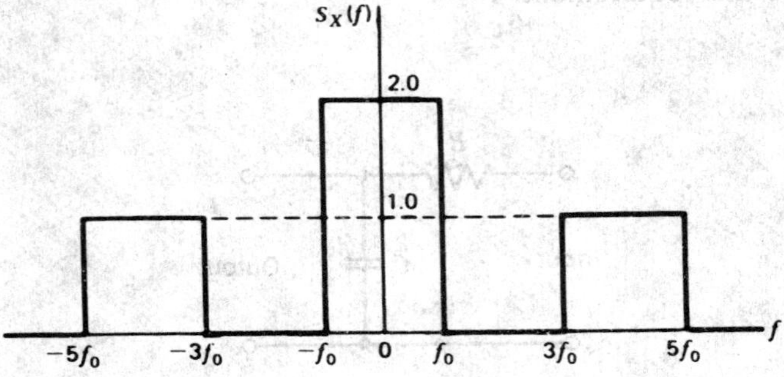

Figure P8.5

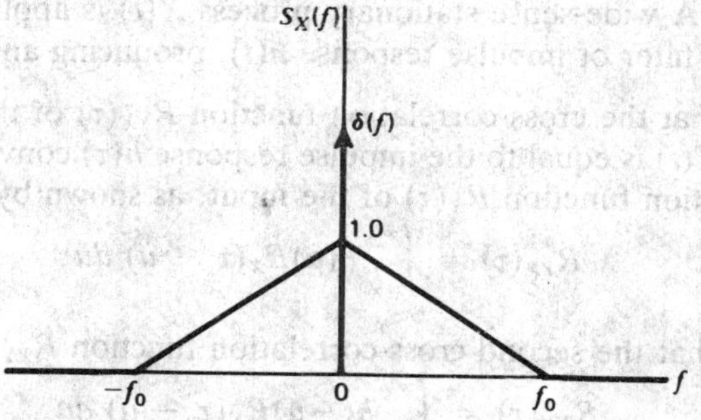

Figure P8.6

(b) The power spectral density of $dX(t)/dt$ is equal to $4\pi^2 f^2 S_X(f)$.

Problem 23 Consider a wide-sense stationary process $X(t)$ having the power spectral density $S_X(f)$ shown in Fig. P8.5. Find the autocorrelation function $R_X(\tau)$ of the process $X(t)$.

Problem 24 The power spectral density of a random process $X(t)$ is shown in Fig. P8.6.

(a) Determine and sketch the autocorrelation function $R_X(\tau)$ of $X(t)$.
(b) What is the dc power contained in $X(t)$?
(c) What is the ac power contained in $X(t)$?
(d) What sampling rates will give uncorrelated samples of $X(t)$? Are the samples statistically independent?

P8.10 Cross-Correlation Functions

Problem 25 Consider two linear filters connected in cascade as in Fig. P8.7. Let $X(t)$ be a wide-sense stationary process with autocorrelation function $R_X(\tau)$. The random process appearing at the first filter output is $V(t)$ and that at the second filter output is $Y(t)$.

(a) Find the autocorrelation function of $Y(t)$.
(b) Find the cross-correlation function $R_{VY}(\tau)$ of $V(t)$ and $Y(t)$.

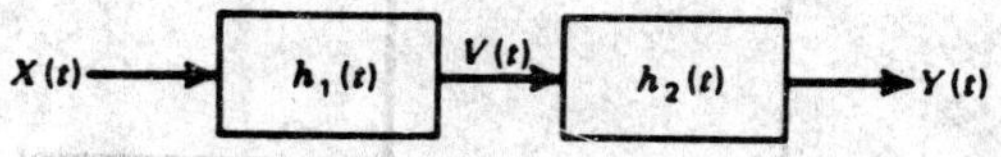

Figure P8.7

Problem 26 A wide-sense stationary process $X(t)$ is applied to a linear time-invariant filter of impulse response $h(t)$, producing an output $Y(t)$.

(a) Show that the cross-correlation function $R_{YX}(\tau)$ of the output $Y(t)$ and input $X(t)$ is equal to the impulse response $h(\tau)$ convolved with the autocorrelation function $R_X(\tau)$ of the input, as shown by

$$R_{YX}(\tau) = \int_{-\infty}^{\infty} h(u)R_X(\tau - u)\, du$$

(b) Show that the second cross-correlation function $R_{XY}(\tau)$ is

$$R_{XY}(\tau) = \int_{-\infty}^{\infty} h(-u)R_X(\tau - u)\, du$$

(c) Assuming that $X(t)$ is a white noise process with zero mean and power spectral density $N_0/2$, show that

$$R_{YX}(\tau) = \frac{N_0}{2}\, h(\tau)$$

Comment on the practical significance of this result.

P8.11 Cross-Spectral Densities

Problem 27 Let $S_{XY}(f)$ and $S_{YX}(f)$ denote the cross-spectral densities of two wide-sense stationary processes $X(t)$ and $Y(t)$. Show that $S_{XY}(f)$ and $S_{YX}(f)$ are related to each other as in Eq. 8.154.

P8.12 Gaussian Processes

Problem 28 A stationary, Gaussian process $X(t)$ with zero mean and power spectral density $S_X(f)$ is applied to a linear filter whose impulse response $h(t)$ is shown in Fig. P8.8. A sample Y is taken of the random process at the filter output at time T.

(a) Determine the mean and variance of Y.
(b) What is the probability density function of Y?

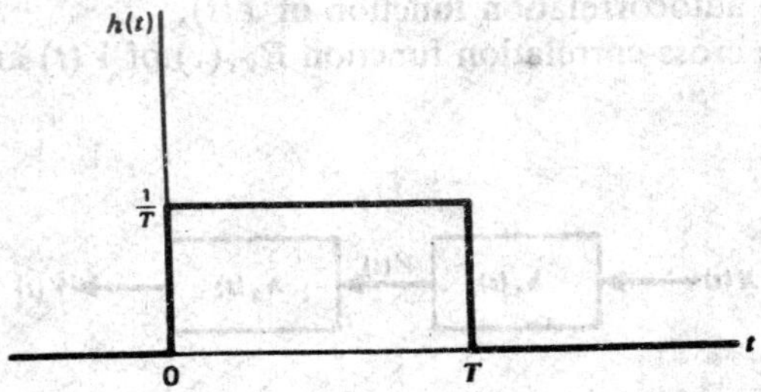

Figure P8.8

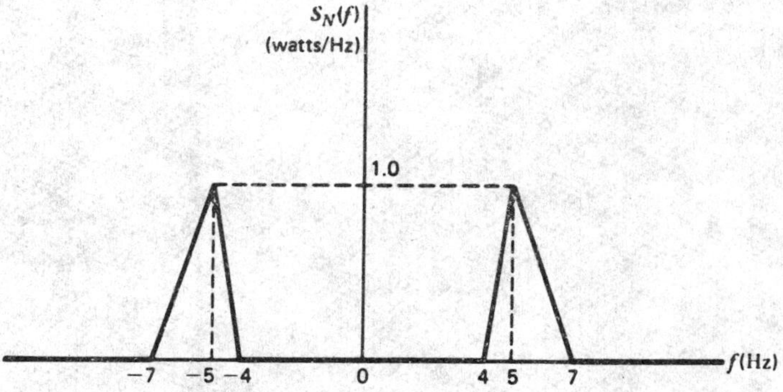

Figure P8.9

Problem 29 Continuing with the situation described in Problem 28, determine the autocorrelation function and power spectral density of the Gaussian process produced at the filter output.

P8.13 Narrow-band Random Process

Problem 30 The power spectral density of a narrow-band random process $X(t)$ is as shown in Fig. P8.9. Find the power spectral densities of the in-phase and quadrature components of $X(t)$, assuming that $f_c = 5$ kHz.

Problem 31 Assume that the narrow-band random process $X(t)$ described in Problem 30 is Gaussian with zero mean and variance σ_X^2.

(a) Calculate σ_X^2.
(b) Determine the joint probability density function of the random variables Y and Z obtained by observing the in-phase and quadrature components of $X(t)$ at some fixed time.

P8.14 Envelope and Phase of Narrow-band Random Process

Problem 32 Consider a narrow-band Gaussian process $X(t)$ with zero mean and power spectral density $S_X(f)$ as shown in Fig. 8.26*a*.

(a) Find the probability density function of the envelope of $X(t)$.
(b) What are the mean and variance of this envelope?

Problem 33 Continuing with Problem 32, find the probability of the event $R \geqslant A_c$, where R is the random variable obtained by observing the envelope of the narrow-band process $X(t)$ at some fixed time, and A_c is a prescribed positive constant. Plot this probability as a function of the ratio

$$\rho = \frac{A_c^2}{4WN_0}$$

where W and N_0 are defined in Fig. 8.26*a*.

CHAPTER 9

NOISE IN ANALOG MODULATION

The term "noise" is shorthand for random fluctuations of power in electrical systems. As such, noise is the *limiting factor* on the power required to transport information-bearing signals practically over all communication channels. To develop an understanding of this basic issue, we need to examine how noise affects the demodulation process intended to recover some message signal in a receiver. Another matter of related interest is the comparison of the noise performances of different modulation–demodulation schemes. In this chapter, we study the noise performance of analog (continuous-wave) modulation schemes. We defer discussion of the noise performance of digital modulation schemes until

Chapter 10, since its theoretical development follows a different approach.

To undertake an introductory treatment of the noise performance of analog communication receivers, we may assume that the *channel noise* or *front-end receiver noise* is *white*. This simplifying assumption not only is justified on physical grounds, but it also enables us to obtain a basic understanding of the way in which noise affects the performance of different receivers. We begin the study by describing signal-to-noise ratios that provide the basis for evaluating the noise performance of an analog communication receiver.

9.1 SIGNAL-TO-NOISE RATIOS

To carry out the noise analysis of analog modulation systems, we obviously need a criterion that describes in a meaningful way the noise performance of the system under study. In the case of analog modulation systems, the customary practice is to use the *output signal-to-noise ratio* as an intuitive measure for describing the fidelity with which the demodulation process in the receiver recovers the original message from the received modulated signal in the presence of noise. *Output signal-to-noise ratio is defined as the ratio of the average power of the message signal to the average power of the noise, both measured at the receiver output.* Let $(SNR)_O$ denote the output signal-to-noise ratio, expressed as

$$(SNR)_O = \frac{\text{average power of message signal at the receiver output}}{\text{average power of noise at the receiver output}} \tag{9.1}$$

The output signal-to-noise ratio is unambiguous as long as the recovered message and noise at the demodulator output are *additive*. This requirement is satisfied exactly in the case of linear receivers using coherent detection, and approximately in the case of nonlinear receivers (e.g., using envelope detection or frequency discrimination) provided that the average input noise power is small compared with the average carrier power.

The calculation of the output signal-to-noise ratio $(SNR)_O$ involves the use of an *idealized receiver model*, the details of which naturally depend on the channel noise and the type of demodulation used in the receiver. We will have more to say on these issues in subsequent sections of the Chapter. For the present, we wish to point out that knowledge of $(SNR)_O$ by itself may be insufficient, particularly when we have to compare the output signal-to-noise ratios of different analog modulation-demodulation systems. In order to make such a comparison meaningful, we introduce the idea of a *baseband transmission model*, as depicted in Fig. 9.1. In this model, two assumptions are made:

1. **The transmitted or modulated message signal power is fixed.**
2. **The baseband low-pass filter passes the message signal, and rejects out-of-band noise.**

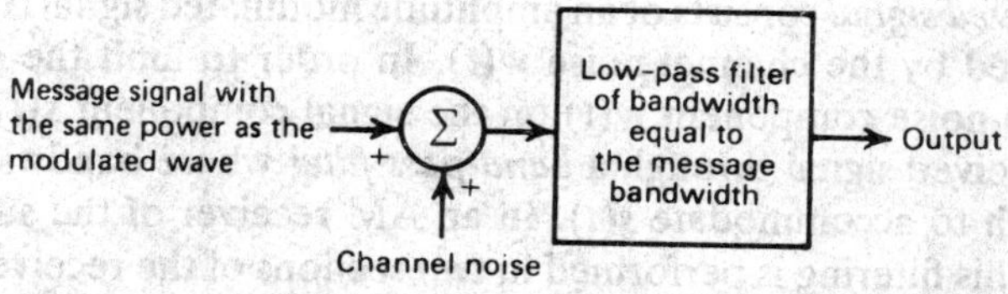

Figure 9.1
The baseband transmission of a message signal for calculating the channel signal-to-noise ratio.

Accordingly, we may define the *channel signal-to-noise ratio*, referred to the receiver input as

$$(SNR)_C = \frac{\text{average power of the modulated message signal}}{\text{average power of noise measured in the message bandwidth}} \tag{9.2}$$

This ratio is independent of the type of modulation or demodulation used.

The channel signal-to-noise ratio of Eq. 9.2 may be viewed as a *frame of reference* for comparing different modulation systems. Specifically, we may normalize the noise performance of a specific modulation-demodulation system by dividing the output signal-to-noise ratio of the system by the channel signal-to-noise ratio. We may thus define a *figure of merit* for the system as

$$\text{Figure of merit} = \frac{(SNR)_O}{(SNR)_C} \tag{9.3}$$

Clearly, the higher the value that the figure of merit has, the better the noise performance of the receiver.

9.2 AM RECEIVER MODEL

It is customary to model channel noise as a sample function of a *white noise process*[1] whose mean is zero and whose power spectral density is constant. We will denote the channel noise by $w(t)$, and denote its power spectral density by $N_0/2$ defined for both positive and negative frequencies. In other words, N_0 *is the average noise power per unit bandwidth measured at the front end of the receiver*.

[1]To be complete, the channel noise process is usually modeled as white and Gaussian. The Gaussian assumption relates to the probability distribution of a sample (random variable) drawn from the process. The Gaussian assumption does not enter the calculation of average noise power; hence, we do not need to involve it in this chapter, except for a situation described in Section 9.4 dealing with the so-called threshold phenomenon in amplitude modulation.

The *received signal* consists of an amplitude modulated signal component $s(t)$ corrupted by the channel noise $w(t)$. In order to limit the degrading effect of the noise component $w(t)$ on the signal component $s(t)$, we may pass the received signal through a *band-pass filter* whose bandwidth is just large enough to accommodate $s(t)$. In an AM receiver of the superhetrodyne type, this filtering is performed in two sections of the receiver: a radio frequency (RF) section and an intermediate freqμency (IF) section; for a description of an AM receiver, see Section 7.9. Figure 9.2a depicts an idealized receiver model for amplitude modulation. The *IF filter* shown in this model accounts for the combination of two effects: (1) the filtering effect of the actual IF section in the superhetrodyne AM receiver, and (2) the filtering effect of the actual RF section in the receiver translated down to the IF band. Typically, however, the IF section provides most of the amplification and selectivity in the receiver.

The IF filter has a bandwidth that is just wide enough to accommodate the bandwidth of the modulated signal $s(t)$. The IF filter is usually tuned so that its midband frequency is the same as the carrier frequency of the modulated signal $s(t)$. An exception to this is the single-sideband modulated wave, as will be explained later. For convenience in signal-to-noise analysis, we assume that the IF filter in the model of Fig. 9.2a has an ideal band-pass characteristic, as shown in Fig. 9.2*b*, where f_c is the midband frequency of the filter, and B refers to the transmission bandwidth of the modulated signal $s(t)$.

The composite signal $x(t)$, at the IF filter output, is defined by

$$x(t) = s(t) + n(t) \tag{9.4}$$

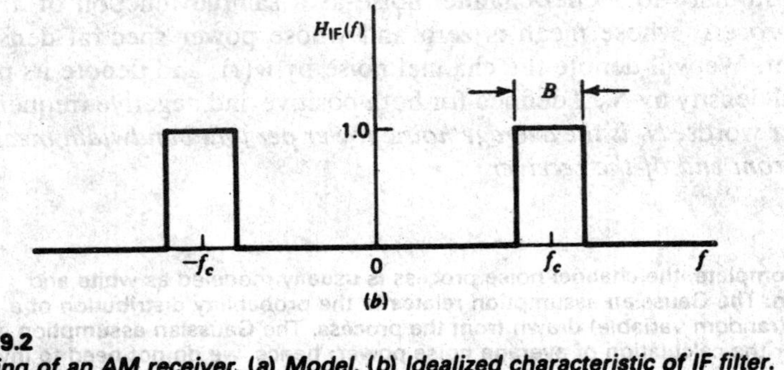

Figure 9.2
Modeling of an AM receiver. (a) Model. (b) Idealized characteristic of IF filter.

where $n(t)$ is a *band-limited version of the white noise* $w(t)$. In particular, $n(t)$ is the sample function of a noise process $N(t)$ with the following power spectral density:

$$S_N(f) = \begin{cases} \dfrac{N_0}{2}, & f_c - \dfrac{B}{2} < |f| < f_c + \dfrac{B}{2} \\ 0, & \text{otherwise} \end{cases} \tag{9.5}$$

The band-limited noise $n(t)$ may be regarded as being *narrow-band*, because the IF filter has a bandwidth that is usually small compared with its midband frequency.

The modulated wave $s(t)$ consists of a band-pass signal, the exact description of which depends on the type of modulation used. To perform a noise analysis of the receiver, we need a corresponding representation for the narrow-band noise $n(t)$. From the theory presented in Sections 8.13 and 8.14 on narrow-band random processes, we have two methods for the time representation of $n(t)$. In the first method, the narrow-band noise $n(t)$ is represented in terms of its *in-phase* and *quadrature components*. This method is well-suited for the noise analysis of AM receivers using coherent detection; it may also be used for AM receivers using envelope detection provided that the received signal-to-noise ratio is high enough. In the second method, the narrow-band noise $n(t)$ is represented in terms of its *envelope* and *phase*; this method is well-suited for the noise analysis of FM receivers.

9.3 SIGNAL-TO-NOISE RATIOS FOR COHERENT RECEPTION

We begin the noise analysis by evaluating the output and channel signal-to-noise ratios for an AM receiver using coherent detection, with an incoming DSBSC- or SSB-modulated wave. The use of coherent detection requires multiplication of the IF filter output $x(t)$ by a locally generated sinusoidal wave $\cos(2\pi f_c t)$ and then low-pass filtering the product, as in Fig. 9.3. For convenience, we assume that the amplitude of the locally generated sinusoidal wave is unity. For this demodulation scheme to operate satisfactorily, however, it is necessary that the local oscillator be synchronized both in phase and frequency with the oscillator generating the carrier wave in the transmitter. We assume that this synchronization has been achieved.

We show presently that coherent detection has the unique feature that for any input signal-to-noise ratio, an output strictly proportional to the original message signal is always present. It is this property of coherent detection, namely, that the output message component is unmutilated and the noise component always appears additively with the message irrespective of the input signal-to-noise ratio, that distinguishes coherent detection from all other demodulation techniques.

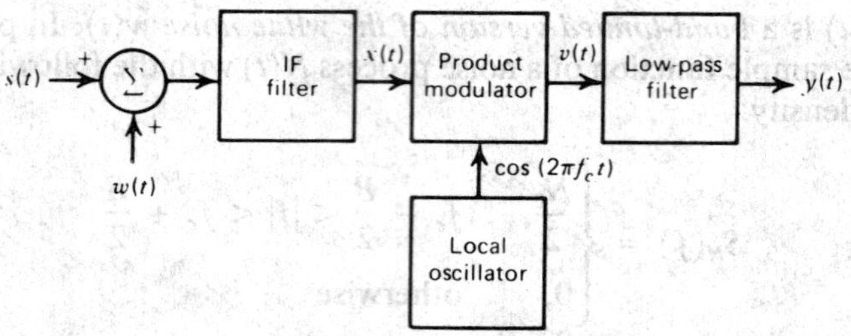

Figure 9.3
Model of DSBSC receiver using coherent detection.

DSBSC RECEIVER

Consider a DSBSC wave defined by

$$s(t) = A_c \cos(2\pi f_c t)m(t) \tag{9.6}$$

where $A_c \cos(2\pi f_c t)$ is the carrier wave and $m(t)$ is the message signal. Typically, the carrier frequency f_c is greater than the message bandwidth W. Accordingly, we find that the average power of the DSBSC modulated wave $s(t)$ equals $A_c^2 P/2$, where A_c is the carrier amplitude and P is the average power of the message signal $m(t)$. This result follows directly from the description of a modulated process as in Eq. 9.6. We also note that the transmission bandwidth B of the DSBSC modulated wave $s(t)$ equals twice the message bandwidth W.

With a noise power spectral density of $N_0/2$, defined for both positive and negative frequencies, the average noise power in the message bandwidth W is equal to WN_0. The channel signal-to-noise ratio of the system is therefore

$$(SNR)_{C,DSB} = \frac{A_c^2 P}{2WN_0} \tag{9.7}$$

Next, we determine the output signal-to-noise ratio of the system. Using the narrow-band representation of the filtered noise $n(t)$, the total signal at the coherent detector input may be expressed as:

$$\begin{aligned} x(t) &= s(t) + n(t) \\ &= A_c \cos(2\pi f_c t)m(t) + n_I(t)\cos(2\pi f_c t) - n_Q(t)\sin(2\pi f_c t) \end{aligned} \tag{9.8}$$

where $n_I(t)$ and $n_Q(t)$ are the in-phase and quadrature components of $n(t)$, with respect to the carrier $\cos(2\pi f_c t)$, respectively. The output of the product-modulator component of the coherent detector is therefore

$$
\begin{aligned}
v(t) &= x(t)\cos(2\pi f_c t) \\
&= \tfrac{1}{2}A_c m(t) + \tfrac{1}{2}n_I(t) + \tfrac{1}{2}[A_c m(t) + n_I(t)]\cos(4\pi f_c t) \\
&\quad - \tfrac{1}{2}A_c n_Q(t)\sin(4\pi f_c t)
\end{aligned} \tag{9.9}
$$

The low-pass filter in the coherent detector removes the high-frequency components of $v(t)$, yielding a receiver output

$$y(t) = \tfrac{1}{2}A_c m(t) + \tfrac{1}{2}n_I(t) \tag{9.10}$$

Equation 9.10 indicates that

1. The message $m(t)$ and in-phase noise component $n_I(t)$ of the narrow-band noise $n(t)$ appear additively at the receiver output.
2. The quadrature component $n_Q(t)$ of the noise $n(t)$ is completely rejected by the coherent detector.

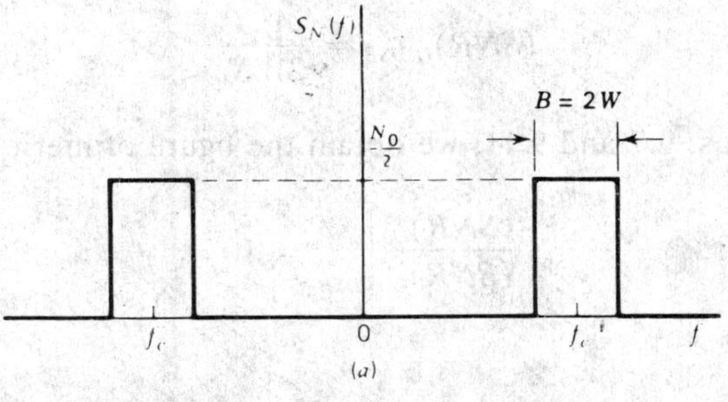

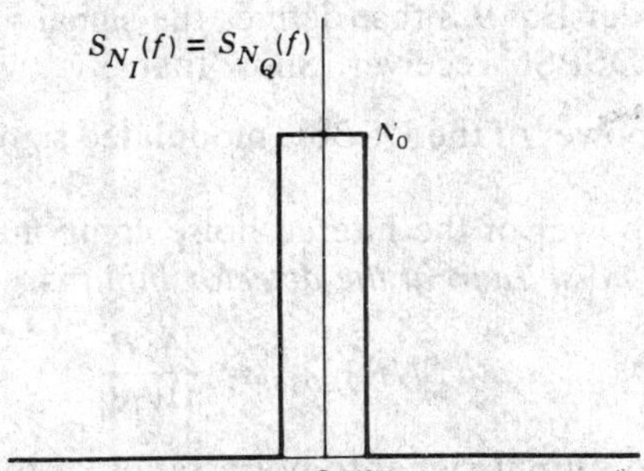

Figure 9.4.
Noise analysis of DSBSC modulation system using coherent detection. (a) Power spectral density of narrow-band noise n(t) *at IF filter output. (b) Power spectral density of in-phase components* $n_I(t)$ *and quadrature component* $n_Q(t)$ *of noise* n(t).

The message signal component at the receiver output equals $A_c m(t)/2$. Hence, the average power of message signal at the receiver output is equal to $A_c^2 P/4$, where P is the average power of the original message signal $m(t)$.

The noise component at the receiver output equals $n_I(t)/2$. Hence, the power spectral density of the output noise equals one quarter that of $n_I(t)$. To calculate the average power of the noise at the receiver output, we first determine the power spectral density of the in-phase noise component $n_I(t)$. In order to accommodate the upper and lower sidebands of the modulated wave $s(t)$, the IF filter has a bandwidth B equal to $2W$, twice the message bandwidth. The power spectral density $S_N(f)$ of the narrow-band noise $n(t)$ thus takes on the ideal form shown in Fig. 9.4*a*. Hence, the power spectral density of $n_I(t)$ is as shown in Fig. 9.4*b* (see Example 22 of Chapter 8). Evaluating the area under the curve of power spctral density of Fig. 9.4*b* and multiplying the result by $\frac{1}{4}$, we find that the average noise power at the receiver output equals $WN_0/2$.

Thus dividing the average power of the message signal by the average power of the noise at the receiver output, we find that the output signal-to-noise ratio for DSBSC modulation is given by

$$(SNR)_{O,DSB} = \frac{A_c^2 P}{2WN_0} \tag{9.11}$$

Next, using Eqs. 9.7 and 9.11, we obtain the figure of merit

$$\left.\frac{(SNR)_O}{(SNR)_C}\right|_{DSB} = 1 \tag{9.12}$$

EXERCISE 1 Consider Eq. 9.8 that defines the signal $x(t)$ at the detector input of a coherent DSBSC receiver. Show that:

(a) The average power of the DSBSC modulated signal component $s(t)$ is $A_c^2 P/2$.

(b) The average power of the filtered noise component $n(t)$ is $2WN_0$.

(c) The *signal-to-noise ratio at the detector input* is

$$(SNR)_{I,DSB} = \frac{A_c^2 P}{4WN_0}$$

(d) The input and output signal-to-noise ratios of the detector are related by

$$(SNR)_{I,DSB} = \frac{1}{2}(SNR)_{O,DSB}$$

Give physical reasons for this result.

SSB RECEIVER

Consider next the case of a coherent receiver with an incoming SSB wave. We assume that only the lower sideband is transmitted, so that we may express the modulated wave as

$$s(t) = \frac{A_c}{2}\cos(2\pi f_c t)m(t) + \frac{A_c}{2}\sin(2\pi f_c t)\hat{m}(t) \qquad (9.13)$$

where $\hat{m}(t)$ is the Hilbert transform of the message signal $m(t)$. We may make the following observations concerning the in-phase and quadrature components of $s(t)$ in Eq. 9.13:

1. The two components $m(t)$ and $\hat{m}(t)$ are uncorrelated with each other. Therefore, their power spectral densities are additive.
2. The Hilbert transform $\hat{m}(t)$ is obtained by passing $m(t)$ through a linear filter with transfer function $-j\ \mathrm{sgn}(f)$. The squared magnitude of this transfer function is equal to one for all f. Accordingly, $m(t)$ and $\hat{m}(t)$ have the same average power.

Thus, proceeding in a manner similar to that for the DSBSC receiver, we find that the in-phase and quadrature components of the SSB modulated wave $s(t)$ contribute an average power of $A_c^2P/8$ each. The average power of $s(t)$ is therefore $A_c^2P/4$. This result is half that in the DSBSC case, which is intuitively satisfying.

The average noise power in the message bandwidth W is WN_0. Thus the channel signal-to-noise ratio of a coherent-receiver with SSB modulation is

$$(SNR)_{C,SSB} = \frac{A_c^2P}{4WN_0} \qquad (9.14)$$

The transmission bandwidth $B = W$. The midband frequency of the power spectral density $S_N(f)$ of the narrow-band noise $n(t)$ differs from the carrier frequency f_c by $W/2$. Therefore, we may express $n(t)$ as

$$n(t) = n_I(t)\cos\left[2\pi\left(f_c - \frac{W}{2}\right)t\right] - n_Q(t)\sin\left[2\pi\left(f_c - \frac{W}{2}\right)t\right] \qquad (9.15)$$

The output of the coherent detector, due to the combined influence of the modulated signal $s(t)$ and noise $n(t)$, is thus given by

$$y(t) = \frac{A_c}{4}m(t) + \tfrac{1}{2}n_I(t)\cos(\pi Wt) + \tfrac{1}{2}n_Q(t)\sin(\pi Wt) \qquad (9.16)$$

As expected, we see that the quadrature component $\hat{m}(t)$ of the modulated message signal $s(t)$ has been eliminated from the detector output, but unlike the case of DSBSC modulation, the quadrature component of the narrow-band noise $n(t)$ now appears at the output.

The message component in the receiver output is $A_c m(t)/4$ so that the average power of the recovered message is $A_c^2 P/16$. The noise component in the receiver output is $[n_I(t)\cos(\pi Wt) + n_Q(t)\sin(\pi Wt)]/2$. Evaluating the average power of the output noise so defined, we find that it is equal to $WN_0/4$ (see Exercise 2). Accordingly, the output signal-to-noise ratio of a system using SSB modulation in the transmitter and coherent detection in the receiver is given by

$$(SNR)_{O,SSB} = \frac{A_c^2 P}{4WN_0} \tag{9.17}$$

Hence, from Eqs. 9.14 and 9.17, the figure of merit of such a system is

$$\left.\frac{(SNR)_O}{(SNR)_C}\right|_{SSB} = 1 \tag{9.18}$$

Comparing Eqs. 9.12 and 9.18, we conclude that insofar as noise performance is concerned, DSBSC and SSB modulation systems using coherent detection in the receiver have the same performance as baseband transmission. The only effect of the modulation process is to translate the message signal to a different frequency band.

EXERCISE 2 Consider the two elements of the noise component in the SSB receiver output of Eq. 9.16.

(a) Sketch the power spectral density of the in-phase noise component $n_I(t)$ and quadrature noise component $n_Q(t)$.
(b) Show that the average power of the modulated noise $n_I(t)\cos(\pi Wt)$ or $n_Q(t)\sin(\pi Wt)$ is $WN_0/2$.
(c) Hence, show that the average power of the output noise is $WN_0/4$.

EXERCISE 3 The signal $x(t)$ at the detector input of a coherent SSB receiver is defined by

$$x(t) = s(t) + n(t)$$

where the signal component $s(t)$ and noise component $n(t)$ are themselves defined by Eqs. 9.13 and 9.15, respectively. Show that:

(a) The average power of the signal component $s(t)$ is $A_c^2 P/4$.
(b) The average power of the noise component $n(t)$ is WN_0.

(c) The signal-to-noise ratio at the detector input is

$$(SNR)_{I,SSB} = \frac{A_c^2 P}{4WN_0}$$

(d) The input and output signal-to-noise ratios are related by

$$(SNR)_{I,SSB} = (SNR)_{O,SSB}$$

9.4 NOISE IN AM RECEIVERS USING ENVELOPE DETECTION

In a standard amplitude modulated (AM) wave both sidebands and the carrier are transmitted. The AM wave may be written as

$$s(t) = A_c[1 + k_a m(t)] \cos(2\pi f_c t) \tag{9.19}$$

where $A_c \cos(2\pi f_c t)$ is the carrier wave, $m(t)$ is the message signal, and k_a is a constant that determines the percentage modulation. In this section, we evaluate the noise performance of an AM receiver using an envelope detector. As explained in Section 7.1, an envelope detector consists simply of a nonlinear device (usually a diode) followed by a low-pass RC filter.

From Eq. 9.19, the average power in the modulated message signal $s(t)$ is equal to $A_c^2(1 + k_a^2 P)/2$, where P is the average power of the message signal. With an average noise power of WN_0 in the message bandwidth, W, the channel signal-to-noise ratio is therefore

$$(SNR)_{C,AM} = \frac{A_c^2(1 + k_a^2 P)}{2WN_0} \tag{9.20}$$

The received signal $x(t)$ at the envelope detector input consists of the modulated message signal $s(t)$ and narrow-band noise $n(t)$. Representing $n(t)$ in terms of its in-phase and quadrature components, namely, $n_I(t)$ and $n_Q(t)$, we may express $x(t)$ as

$$\begin{aligned} x(t) &= s(t) + n(t) \\ &= [A_c + A_c k_a m(t) + n_I(t)] \cos(2\pi f_c t) - n_Q(t) \sin(2\pi f_c t) \end{aligned} \tag{9.21}$$

It is informative to represent the components that comprise the signal $x(t)$ by means of phasors, as in Fig. 9.5. From this phasor diagram, the receiver output is obtained as

$$\begin{aligned} y(t) &= \text{envelope of } x(t) \\ &= \{[A_c + A_c k_a m(t) + n_I(t)]^2 + n_Q^2(t)\}^{1/2} \end{aligned} \tag{9.22}$$

The signal $y(t)$ defines the output of an ideal envelope detector. The phase of $x(t)$ is of no interest to us, because an ideal envelope detector is totally insensitive to variations in the phase of $x(t)$.

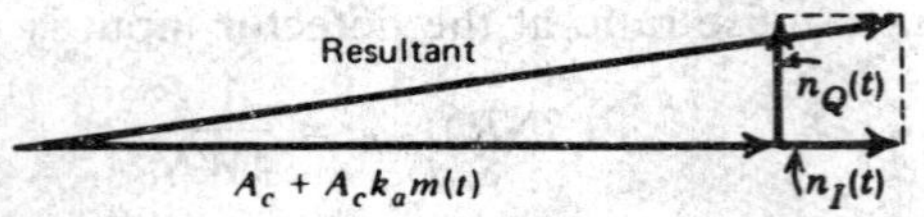

Figure 9.5
Phasor diagram for AM wave plus narrow-band noise for the case of high carrier-to-noise ratio.

The expression defining $y(t)$ is somewhat complex and needs to be simplified in some manner. Specifically, we would like to approximate the output $y(t)$ as the sum of a message term plus a term due to noise. In general, this is difficult to achieve. However, when the average carrier power is large compared with the average noise power, so that the receiver is operating satisfactorily, then the signal term $A_c[1 + k_a m(t)]$ will be large compared with the noise terms $n_I(t)$ and $n_Q(t)$, most of the time. Then we may approximate the output $y(t)$ as

$$y(t) \simeq A_c + A_c k_a m(t) + n_I(t) \tag{9.23}$$

The presence of the dc or constant term A_c in the envelope detector output $y(t)$ of Eq. 9.23 is due to demodulation of the transmitted carrier wave. We may ignore this term, however, because it bears no relation whatsoever to the message signal $m(t)$. In any case, it may be removed simply by means of a blocking capacitor. Thus, if we neglect the term A_c in Eq. 9.23, we find that the remainder has, except for scaling factors, the same form as the output of a DSBSC receiver using coherent detection. Accordingly, the output signal-to-noise ratio of an AM receiver using an envelope detector is approximately

$$(SNR)_{O,AM} \simeq \frac{A_c^2 k_a^2 P}{2WN_0} \tag{9.24}$$

This expression is, however, valid only if:

1. The noise, at the receiver input, is small compared to the signal.
2. The amplitude sensitivity k_a is adjusted for a percentage modulation less than or equal to 100%.

Using Eqs. 9.20 and 9.24, we obtain the figure of merit

$$\left.\frac{(SNR)_O}{(SNR)_C}\right|_{AM} \simeq \frac{k_a^2 P}{1 + k_a^2 P} \tag{9.25}$$

Thus, whereas the figure of merit of a DSBSC or SSB receiver using coherent detection is always unity, the corresponding figure of merit of an

AM receiver using envelope detection is always less than unity. In other words, *the noise performance of an AM receiver is always inferior to that of a DSBSC or SSB receiver.* This is owing to the wasteage of transmitted power that results from transmitting the carrier as a component of the AM wave.

EXAMPLE 1 SINGLE-TONE MODULATION

Consider the special case of a sinusoidal wave of frequency f_m and amplitude A_m as the modulating wave, as shown by

$$m(t) = A_m \cos(2\pi f_m t)$$

The corresponding AM wave is

$$s(t) = A_c[1 + \mu \cos(2\pi f_m t)] \cos(2\pi f_c t)$$

where $\mu = k_a A_m$ is the modulation factor. The average power of the modulating wave $m(t)$ is

$$P = \tfrac{1}{2}A_m^2$$

Therefore, using Eq. 9.25, we get

$$\left.\frac{(SNR)_O}{(SNR)_C}\right|_{AM} \simeq \frac{\tfrac{1}{2}k_a^2 A_m^2}{1 + \tfrac{1}{2}k_a^2 A_m^2} = \frac{\mu^2}{2 + \mu^2} \tag{9.26}$$

When $\mu = 1$, which corresponds to 100% modulation, we get a figure of merit equal to 1/3. This means that, other factors being equal, this AM system must transmit three times as much average power as a suppressed-carrier system in order to achieve the same quality of noise performance.

EXERCISE 4 The *carrier-to-noise ratio* of a communication receiver is defined by

$$\rho = \frac{\text{Average carrier power}}{\left(\begin{array}{c}\text{Average noise power in bandwidth of the}\\ \text{modulated wave at the receiver input}\end{array}\right)} \tag{9.27}$$

Show that for a standard AM receiver,

$$\rho = \frac{A_c^2}{4WN_0} \tag{9.28}$$

Express the output signal-to-noise ratio of Eq. 9.24 in terms of the carrier-to-noise ratio ρ.

THRESHOLD EFFECT

When the carrier-to-noise ratio at the receiver input of a standard AM system is small compared with unity, the noise term dominates and the performance of the envelope detector changes completely from that just described. In this case it is more convenient to represent the narrow-band noise $n(t)$ in terms of its envelope $r(t)$ and phase $\psi(t)$, as shown by

$$n(t) = r(t)\cos[2\pi f_c t + \psi(t)] \tag{9.29}$$

The phasor diagram for the detector input $x(t) = s(t) + n(t)$ is shown in Fig. 9.6 where we have used the noise as reference, because it is now the dominant term. To the noise phasor $r(t)$ we have added a phasor representing the signal term $A_c[1 + k_a m(t)]$, with the angle between them equal to $\psi(t)$, the phase of the noise $n(t)$. In Fig. 9.6 it is assumed that the carrier-to-noise ratio is so low that the carrier amplitude A_c is small compared with the noise envelope $r(t)$, most of the time. Then we may neglect the quadrature component of the signal with respect to the noise, and thus find directly from Fig. 9.6 that the envelope detector output is approximately

$$y(t) \simeq r(t) + A_c\cos[\psi(t)] + A_c k_a m(t)\cos[\psi(t)] \tag{9.30}$$

This relation reveals that when the carrier-to-noise ratio is low, the detector output has no component strictly proportional to the message signal $m(t)$. The last term of the expression defining $y(t)$ contains the message signal $m(t)$ multiplied by noise in the form of $\cos[\psi(t)]$. The phase $\psi(t)$ of a narrow-band noise $n(t)$ is uniformly distributed over 2π radians; that is, it can assume a value anywhere between 0 and 2π with equal probability. It follows therefore that we have a complete loss of information in that the

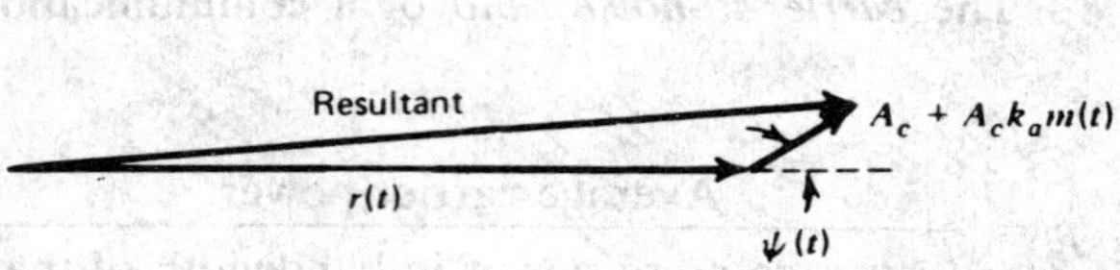

Figure 9.6
Phasor diagram for AM wave plus narrow-band noise for the case of low carrier-to-noise ratio.

detector output does not contain the message signal $m(t)$ at all. The loss of a message in an envelope detector that operates at a low carrier-to-noise ratio is referred to as the *threshold effect. By threshold we mean a value of the carrier-to-noise ratio below which the noise performance of a detector deteriorates much more rapidly than that predicted by Eq. 9.24 assuming a high carrier-to-noise ratio.* It is important to recognize that every nonlinear detector (e.g., envelope detector) exhibits a threshold effect. On the other hand, such an effect does not occur in a coherent detector.

A detailed analysis of the threshold effect in envelope detectors is complicated.[2] We may develop some insight into the threshold effect, however, by using the following qualitative approach.[3] Let R denote the random variable obtained by observing the envelope process, with sample function $r(t)$, at some fixed time. Intuitively, an envelope detector is expected to be operating well into the threshold region if the probability that the random variable R exceeds the carrier amplitude A_c is, say, 0.5. On the other hand, if this same probability is only 0.01, the envelope detector is expected to be relatively free of loss of message and threshold effects. The evaluation of the carrier-to-noise ratios, corresponding to these probabilities, is best illustrated by way of an example.

EXAMPLE 2

From Section 8.14 we recall that the envelope $r(t)$ of a narrow-band Gaussian noise $n(t)$ is Rayleigh-distributed. Specifically, the probability density function of the random variable R obtained by observing the envelope $r(t)$ at some fixed time, is given by

$$f_R(r) = \frac{r}{\sigma_N^2}\exp\left(-\frac{r^2}{2\sigma_N^2}\right) \tag{9.31}$$

where σ_N^2 is the variance of the noise $n(t)$. For an AM system, we have $\sigma_N^2 = 2WN_0$. Therefore the probability of the event $R \geq A_c$ is defined by

$$\begin{aligned} P(R \geq A_c) &= \int_{A_c}^{\infty} f_R(r)\,dr \\ &= \int_{A_c}^{\infty} \frac{r}{2WN_0}\exp\left(-\frac{r^2}{4WN_0}\right) dr \\ &= \exp\left(-\frac{A_c^2}{4WN_0}\right) \end{aligned} \tag{9.32}$$

[2]See Middleton (1960), pp. 563–574.
[3]See Downing (1964), p. 71.

Using Eq. 9.28 for the carrier-to-noise ratio of an AM receiver, we may rewrite Eq. 9.32 in the compact form

$$P(R \geq A_c) = \exp(-\rho) \tag{9.33}$$

Solving for $P(R \geq A_c) = 0.5$, we get

$$\rho = \ln 2 = 0.69 = -1.6 \text{ dB}$$

Similarly, for $P(R \geq A_c) = 0.01$, we get

$$\rho = \ln 100 = 4.6 = 6.6 \text{ dB}$$

Thus with a carrier-to-noise ratio of -1.6 dB the envelope detector is expected to be well into the threshold region, whereas with a carrier-to-noise ratio of 6.6 dB the detector is expected to be operating satisfactorily. We ordinarily need a signal-to-noise ratio considerably greater than 6.6 dB for satisfactory fidelity, which means therefore that threshold effects are seldom of great importance in AM receivers using envelope detection.

EXERCISE 5 Given a carrier-to-noise ratio of 6.6 dB for which the envelope detector of an AM receiver operates satisfactorily, what is the corresponding value of channel signal-to-noise ratio for the case of sinusoidal modulation with 100% modulation?

9.5 FM RECEIVER MODEL

We turn next to study the effects of noise on the performance of FM receivers. Here again we require a receiver model to carry out the analysis. Figure 9.7*a* shows the details of an idealized FM receiver model that satisfies our requirement. As before, the noise $w(t)$ is modeled as white noise of zero mean and power spectral density $N_0/2$. The received FM signal $s(t)$, translated in frequency and amplitude, has a carrier frequency f_c and transmission bandwidth B, so that only a negligible amount of power lies outside the frequency band $f_c - B/2 \leq |f| \leq f_c + B/2$. The FM transmission bandwidth B is in excess of twice the message bandwidth W by an amount that depends on the deviation ratio of the incoming frequency modulated wave; see Section 7.11.

As in the AM case, the IF filter in the model of Fig. 9.7a represents the combined filtering effects of the RF and IF sections of an FM receiver of the superheterodyne type. This filter has a midband frequency f_c and bandwidth B, and therefore passes the FM signal essentially without distortion. We assume that the IF filter in Fig. 9.7a has an ideal bandpass characteristic, with the bandwidth B small compared with the midband

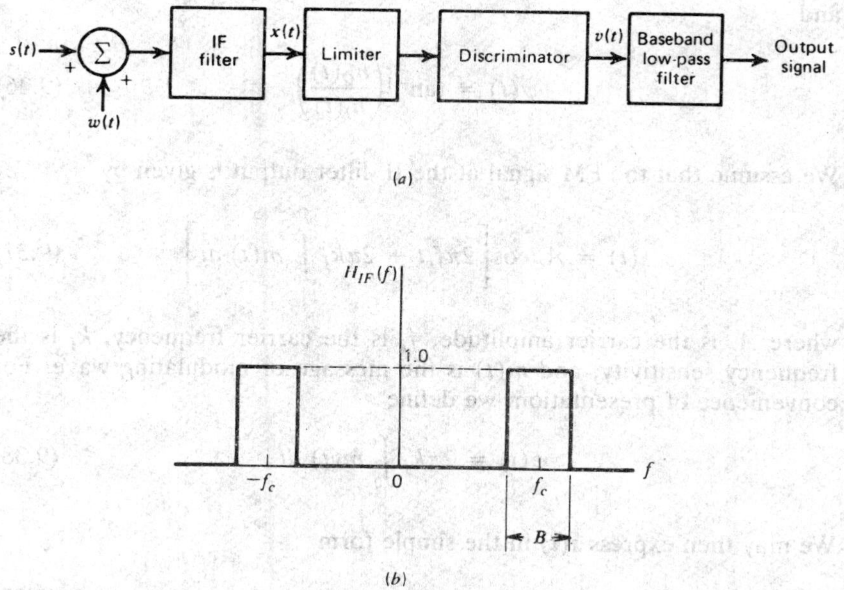

Figure 9.7
Modeling of an FM receiver. (a) Model. (b) Idealized IF filter characteristic.

frequency f_c, as in Fig. 9.7*b*. We may thus use the usual narrow-band representation for the filtered noise $n(t)$ in terms of its in-phase and quadrature components.

The limiter is included in Fig. 9.7a to remove any amplitude variations at the IF output. The discriminator is assumed to be ideal in the sense that its output is proportional to the deviation in the instantaneous frequency of the carrier away from f_c. Also, the postdetection filter is assumed to be an ideal low-pass filter with a bandwidth equal to the message bandwidth W.

9.6 NOISE IN FM RECEPTION

For the noise analysis of FM receivers, we find it convenient to express the narrow-band noise $n(t)$ at the IF filter output in terms of its envelope and phase as in Eq. 9.29. This relation is reproduced here for convenience:

$$n(t) = r(t)\cos[2\pi f_c t + \psi(t)] \tag{9.34}$$

The envelope $r(t)$ and phase $\psi(t)$ are themselves defined in terms of the in-phase component $n_I(t)$ and quadrature component $n_Q(t)$ as follows:

$$r(t) = [n_I^2(t) + n_Q^2(t)]^{1/2} \tag{9.35}$$

and

$$\psi(t) = \tan^{-1}\left(\frac{n_Q(t)}{n_I(t)}\right) \tag{9.36}$$

We assume that the FM signal at the IF filter output is given by

$$s(t) = A_c \cos\left[2\pi f_c t + 2\pi k_f \int_0^t m(t)\, dt\right] \tag{9.37}$$

where A_c is the carrier amplitude, f_c is the carrier frequency, k_f is the frequency sensitivity, and $m(t)$ is the message or modulating wave. For convenience of presentation, we define

$$\phi(t) = 2\pi k_f \int_0^t m(t)\, dt \tag{9.38}$$

We may then express $s(t)$ in the simple form

$$s(t) = A_c \cos[2\pi f_c t + \phi(t)] \tag{9.39}$$

The total signal (i.e., signal plus noise) at the IF section output is therefore

$$\begin{aligned} x(t) &= s(t) + n(t) \\ &= A_c \cos[2\pi f_c t + \phi(t)] + r(t)\cos[2\pi f_c t + \psi(t)] \end{aligned} \tag{9.40}$$

It is informative to represent $x(t)$ by means of a phasor diagram, as in Fig. 9.8. In this diagram we have used the signal term as reference. The relative phase $\theta(t)$ of the resultant phasor representing $x(t)$ is obtained directly from Fig. 9.8 as

$$\theta(t) = \phi(t) + \tan^{-1}\left\{\frac{r(t)\sin[\psi(t) - \phi(t)]}{A_c + r(t)\cos[\psi(t) - \phi(t)]}\right\} \tag{9.41}$$

The envelope of $x(t)$ is of no interest to us, because any envelope variations at the IF section output are removed by the limiter.

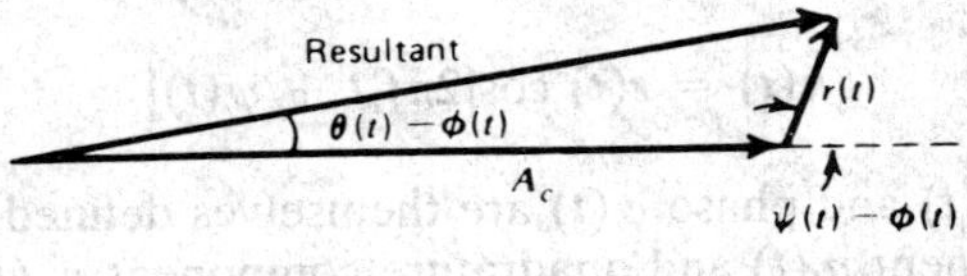

Figure 9.8
Phasor diagram for FM wave plus narrow-band noise for the case of high carrier-to-noise ratio.

Our motivation is to determine the error in the instantaneous frequency of the carrier wave caused by the presence of the narrow-band noise $n(t)$. With the discriminator assumed ideal, its output is proportional to $\dot{\theta}(t)$, where $\dot{\theta}(t)$ is the derivative of $\theta(t)$ with respect to time. In view of the complexity of the expression defining $\theta(t)$, however, we need to make certain simplifying approximations so that our analysis may yield useful results.

We assume that the carrier-to-noise ratio measured at the discriminator input is large compared with unity. Then, most of the time, the expression for the relative phase $\theta(t)$ simplifies as

$$\theta(t) \simeq \underbrace{\phi(t)}_{\text{signal term}} + \underbrace{\frac{r(t)}{A_c}\sin[\psi(t) - \phi(t)]}_{\text{noise term}} \tag{9.42}$$

The signal term $\phi(t)$ is proportional to the integral of the message signal $m(t)$, as in Eq. 9.38. Hence, using Eqs. 9.38 and 9.42, we find that the discriminator output is

$$\begin{aligned} v(t) &= \frac{1}{2\pi}\frac{d\theta(t)}{dt} \\ &\simeq k_f m(t) + n_d(t) \end{aligned} \tag{9.43}$$

where the noise term $n_d(t)$ is defined by

$$n_d(t) = \frac{1}{2\pi A_c}\frac{d}{dt}\{r(t)\sin[\psi(t) - \phi(t)]\} \tag{9.44}$$

We thus see that provided the carrier-to-noise ratio is high, the discriminator output $v(t)$ consists of a scaled version of the original message signal $m(t)$, plus an additive noise component $n_d(t)$. Accordingly, we may use the output signal-to-noise ratio as previously defined to assess the quality of performance of the FM receiver.

The output signal-to-noise ratio is defined as the ratio of the average output signal power to the average output noise power. From Eq. 9.43, the signal component at the discriminator output, and therefore the post-detection filter output, is $k_f m(t)$. Hence, the average output signal power is $k_f^2 P$, where P is the average power of the message signal $m(t)$.

Unfortunately, the calculation of the average output noise power is complicated by the presence of the factor $\sin[\psi(t) - \phi(t)]$ in Eq. 9.44. Since the phase $\psi(t)$ is uniformly distributed over 2π radians, the mean-square value of the noise $n_d(t)$ in Eq. 9.44 will be biased by the message-dependent phase $\phi(t)$. The presence of $\phi(t)$ produces components in the power spectrum of the noise $n_d(t)$ at frequencies that lie outside the message band. However, such frequency components do not appear at the receiver output as they are rejected by the post-detection filter.[4] Hence, insofar as the calculation of inband noise power at the receiver output due to $n_d(t)$ is concerned, we may simplify our task by setting the message-dependent phase $\phi(t)$ equal to zero. Under this condition, Eq. 9.44 simplifies as

$$n_d(t) = \frac{1}{2\pi A_c} \frac{d}{dt} \{r(t) \sin[\psi(t)]\} \tag{9.45}$$

From the definitions of the noise envelope $r(t)$ and phase $\psi(t)$ givèn by Eqs. 9.35 and 9.36, we note that the quadrature component of the narrow-band noise $n(t)$ is

$$n_Q(t) = r(t) \sin[\psi(t)] \tag{9.46}$$

Correspondingly, Eq. 9.45 may be rewritten as

$$n_d(t) = \frac{1}{2\pi A_c} \frac{dn_Q(t)}{dt} \tag{9.47}$$

We may thus state that, *under the condition of high carrier-to-noise ratio, the calculation of the average output noise power in an FM receiver depends only on the carrier amplitude A_c and the quadrature noise component $n_Q(t)$.* Stated in another way, we may use an *unmodulated* carrier to calculate the output signal-to-noise ratio of an FM receiver, provided that the carrier-to-noise ratio is high.

From Section 2.3, we recall that differentiation of a function with respect to time corresponds to multiplication of its Fourier transform by $j2\pi f$. It follows therefore that we may obtain the noise process $n_d(t)$ by passing $n_Q(t)$ through a linear filter with a transfer function equal to

$$\frac{j2\pi f}{2\pi A_c} = \frac{jf}{A_c} \tag{9.48}$$

This means that the power special density $S_{N_d}(f)$ of the noise $n_d(t)$ is related to the power spectral density $S_{N_Q}(f)$ of the quadrature noise component $n_Q(t)$ as follows:

$$S_{N_d}(f) = \frac{f^2}{A_c^2} S_{N_Q}(f) \tag{9.49}$$

With the IF filter in Fig. 9.7*a* assumed to have an ideal band-pass characteristic of bandwidth B and midband frequency f_c, it follows that the

[4] See Downing (1964), pp. 96-98.

narrow-band noise $n(t)$ will have a power spectral density characteristic that is similarly shaped. This means that the quadrature component $n_Q(t)$ of the narrow-band noise $n(t)$ will have the ideal low-pass characteristic shown in Fig. 9.9*a*. The corresponding power spectral density of the noise $n_d(t)$ is shown in Fig. 9.9*b*. That is,

$$S_{N_d}(f) = \begin{cases} \dfrac{N_0 f^2}{A_c^2}, & |f| \leqslant \dfrac{B}{2} \\ 0, & \text{otherwise} \end{cases} \tag{9.50}$$

The discriminator output is followed by a low-pass filter with a bandwidth equal to the message bandwidth W. For wideband FM, by definition,

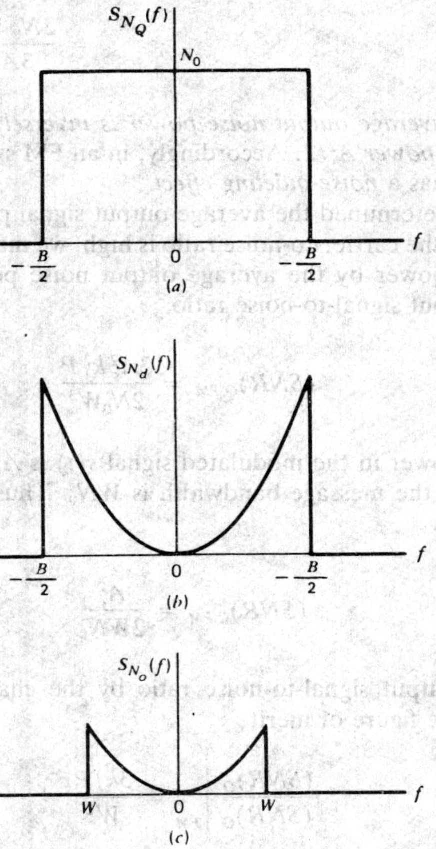

Figure 9.9
Noise analysis of FM receiver. (a) Power spectral density of quadrature component $n_Q(t)$ of narrow-band noise n(t). (b) Power spectral density of noise $n_d(t)$ at discriminator output. (c) Power spectral density of noise $n_o(t)$ at receiver output.

W is much smaller than $B/2$ where B is the transmission bandwidth of the FM signal. This means that the out-of-band components of noise $n_d(t)$ will be rejected. Therefore, the power spectral density $S_{N_o}(f)$ of the noise $n_o(t)$ appearing at the receiver output is defined by

$$S_{N_O}(f) = \begin{cases} \dfrac{N_0 f^2}{A_c^2}, & |f| \leqslant W \\ 0, & \text{otherwise} \end{cases} \tag{9.51}$$

as shown in Fig. 9.9*c*. The average output noise power is determined by integrating the power spectral density $S_{N_O}(f)$ from $-W$ to W. We thus get

$$\begin{aligned} \text{Average power of output noise} &= \frac{N_0}{A_c^2} \int_{-W}^{W} f^2 \, df \\ &= \frac{2N_0 W^3}{3A_c^2} \end{aligned} \tag{9.52}$$

Note that *the average output noise power is inversely proportional to the average carrier power* $A_c^2/2$. Accordingly, in an FM system, increasing the carrier power has a *noise-quieting effect.*

Earlier we determined the average output signal power as $k_f^2 P$. Therefore, provided the carrier-to-noise ratio is high, we may divide this average output signal power by the average output noise power of Eq. 9.52 to obtain the output signal-to-noise ratio

$$(SNR)_{O,FM} = \frac{3A_c^2 k_f^2 P}{2N_0 W^3} \tag{9.53}$$

The average power in the modulated signal $s(t)$ is $A_c^2/2$, and the average noise power in the message bandwidth is WN_0. Thus the channel signal-to-noise ratio is

$$(SNR)_{C,FM} = \frac{A_c^2}{2WN_0} \tag{9.54}$$

Dividing the output signal-to-noise ratio by the channel signal-to-noise ratio, we get the figure of merit

$$\left.\frac{(SNR)_O}{(SNR)_C}\right|_{FM} = \frac{3k_f^2 P}{W^2} \tag{9.55}$$

The frequency deviation Δf is proportional to the frequency sensitivity k_f of the modulator. Also, by definition, the deviation ratio D is equal to the frequency deviation Δf divided by the message bandwidth W. Therefore, it follows from Eq. 9.55 that the figure of merit of a wideband FM system is a quadratic function of the deviation ratio. Now, in wideband FM, the transmission bandwidth B is approximately proportional to the deviation ratio D. Accordingly, we may state *that when the carrier-to-noise ratio is high, an increase in the transmission bandwidth B provides a corresponding quadratic increase in the output signal-to-noise ratio or figure of merit of the FM system.*

EXAMPLE 3 SINGLE-TONE MODULATION

Consider the case of a sinusoidal wave of frequency f_m as the modulating wave, and assume a frequency deviation Δf. The modulated wave is thus defined by

$$s(t) = A_c \cos\left[2\pi f_c t + \frac{\Delta f}{f_m}\sin(2\pi f_m t)\right]$$

where we have made the substitution:

$$2\pi k_f \int_0^t m(t)\,dt = \frac{\Delta f}{f_m}\sin(2\pi f_m t)$$

Differentiating both sides with respect to time:

$$m(t) = \frac{\Delta f}{k_f}\cos(2\pi f_m t)$$

Hence, the average power of the message signal $m(t)$ is

$$P = \frac{(\Delta f)^2}{2k_f^2}$$

Substituting this result into the formula for the output signal-to-noise ratio given by Eq. 9.53, we get·

$$(SNR)_{O,FM} = \frac{3A_c^2(\Delta f)^2}{4N_0W^3}$$
$$= \frac{3A_c^2\beta^2}{4N_0W} \tag{9.56}$$

where $\beta = \Delta f/W$ is the modulation index. Using Eq. 9.55 to evaluate the corresponding figure of merit, we get

$$\left.\frac{(SNR)_O}{(SNR)_C}\right|_{FM} = \frac{3}{2}\left(\frac{\Delta f}{W}\right)^2 = \frac{3}{2}\beta^2 \tag{9.57}$$

It is important to note that the modulation index $\beta = \Delta f/W$ is determined by the bandwidth W of the postdetection low-pass filter and is not related to the sinusoidal message frequency f_m, except insofar as this filter is chosen so as to pass the spectrum of the desired message. For a specified bandwidth W the sinusoidal message frequency f_m may lie anywhere between 0 and W and would yield the same output signal-to-noise ratio.

It is of particular interest to compare the performance of AM and FM systems. One way of making this comparison is to consider the figures of merit of the two systems based on a sinusoidal modulating signal. For an AM system operating with a sinusoidal modulating signal and 100% modulation, we have (from Example 1):

$$\left.\frac{(SNR)_O}{(SNR)_C}\right|_{AM} = \frac{1}{3} \tag{9.58}$$

Comparing this figure of merit with the corresponding result obtained for an FM system, we see that the use of frequency modulation offers the possibility of improved signal-to-noise ratio over amplitude modulation when

$$\tfrac{3}{2}\beta^2 > \tfrac{1}{3}$$

that is,

$$\beta > 0.5$$

We may therefore consider $\beta = 0.5$ *as defining roughly the transition from narrow-band FM to wideband FM*. This statement, based on noise considerations, further confirms a similar observation that was made in Chapter 7 when considering the bandwidth of FM waves.

EXERCISE 6 Consider an FM receiver with an IF filter of bandwidth B. The incoming FM wave is produced by a sinusoidal modulation that produces a frequency deviation Δf equal to $B/2$, so that the carrier swings back and forth across the entire passband of the IF filter. Using the definition of the *carrier-to-noise ratio*

$$\rho = \frac{A_c^2}{2BN_0} \tag{9.59}$$

show that for the situation described herein the output signal-to-noise ratio of the FM receiver is

$$(SNR)_o = 3p\left(\frac{B}{W}\right)^3 \tag{9.60}$$

where W is the message bandwidth and B is the IF filter bandwidth.

COMPARISON OF FM WITH PCM

In this subsection, we compare the capabilities of wideband FM and PCM for exchanging an increase in transmission bandwidth for an improvement in noise performance. With wideband FM, the improvement in signal-to-noise ratio produced by increased transmission bandwidth effectively follows a square law (see Eq. 9.55). That is, by doubling the bandwidth in an FM system that operates above threshold, the signal-to-noise ratio is improved by 6 dB. With binary PCM limited by quantizing noise, on the other hand, doubling the transmission bandwidth permits twice the number of binary digits n in a code word, and therefore increases the signal-to-noise ratio by $6n$ dB (see Eq. 5.24). It follows therefore that FM is less efficient than PCM in exchanging increased bandwidth for improved signal-to-noise ratio.

CAPTURE EFFECT

The inherent ability of an FM system to minimize the effects of unwanted signals (e.g., noise, as discussed earlier) also applies to *interference* produced by another frequency-modulated signal with a frequency content close to the carrier frequency of the desired FM wave. However, interference suppression in an FM receiver works well only when the interference is weaker than the desired FM input. When the interference is the stronger one of the two, the receiver locks on to the stronger signal and thereby suppresses the desired FM input. When they are of nearly equal strength, the receiver fluctuates back and forth between them. This phenomenon is known as the *capture effect.*

9.7 FM THRESHOLD EFFECT

The formula of Eq. 9.53, defining the output signal-to-noise ratio of an FM receiver, is valid only if the carrier-to-noise ratio, measured at the discriminator input, is high compared with unity. It is found experimentally that as the input noise is increased so that the carrier-to-noise ratio is decreased, the FM receiver *breaks*. At first, individual clicks are heard in the receiver output, and as the carrier-to-noise ratio decreases still further, the clicks rapidly merge into a *crackling* or *sputtering sound.* Near the breaking point, Eq. 9.53 begins to fail by predicting values of output signal-to-noise ratio larger than the actual ones. This phenomenon is known as the *threshold effect. The threshold is defined as the minimum carrier-to-noise ratio yielding an FM improvement that is not significantly deteriorated from the value predicted by the signal-to-noise formula of Eq. 9.53 assuming a small noise power.*

For a qualitative discussion of the FM threshold effect, consider first the case when there is no signal present, so that the carrier wave is unmodulated. Then the composite signal at the frequency discriminator input is

$$x(t) = [A_c + n_I(t)] \cos(2\pi f_c t) - n_Q(t) \sin(2\pi f_c t) \qquad (9.61)$$

where $n_I(t)$ and $n_Q(t)$ are the in-phase and quadrature components of the narrow-band noise $n(t)$ with respect to the carrier wave $\cos(2\pi f_c t)$. The phasor diagram of Fig. 9.10 shows the phase relations between the various components of $x(t)$ in Eq. 9.61. As the amplitudes and phases of $n_I(t)$ and $n_Q(t)$ change with time in a random manner, the point P wanders around the point Q. When the carrier-to-noise ratio is large, $n_I(t)$ and $n_Q(t)$ are usually much smaller than the carrier amplitude A_c, so the wandering point P in Fig. 9.10 spends most of its time near point Q. Thus the angle $\theta(t)$ is approximately $n_Q(t)/A_c$ to within a multiple of 2π. The wandering point P occasionally sweeps around the origin and $\theta(t)$ increases or decreases by 2π radians. Figure 9.11 illustrates how, in a rough way, these excursions in $\theta(t)$ produce impulse-like components in $\dot{\theta}(t) = d\theta/dt$. The discriminator output $v(t)$ is equal to $\dot{\theta}(t)/2\pi$. These impulse-like components have different heights depending on how close the wandering point P comes to the origin O, but all have areas nearly equal to $\pm 2\pi$ radians. When the signal shown in Fig. 9.11*b* is passed through the postdetection low-pass filter, corresponding but wider impulse-like components are excited in the receiver output and are heard as clicks. The clicks are produced only when $\theta(t)$ changes by $\pm 2\pi$.

From the phasor diagram of Fig. 9.10, we may deduce the conditions required for clicks to occur. A positive-going click occurs when the en-

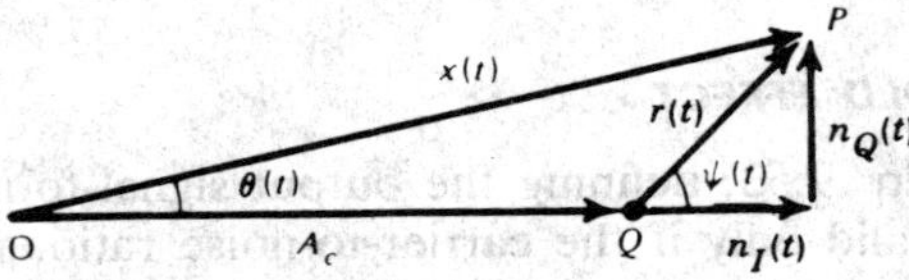

Figure 9.10
A phasor diagram interpretation of Eq. 9.61.

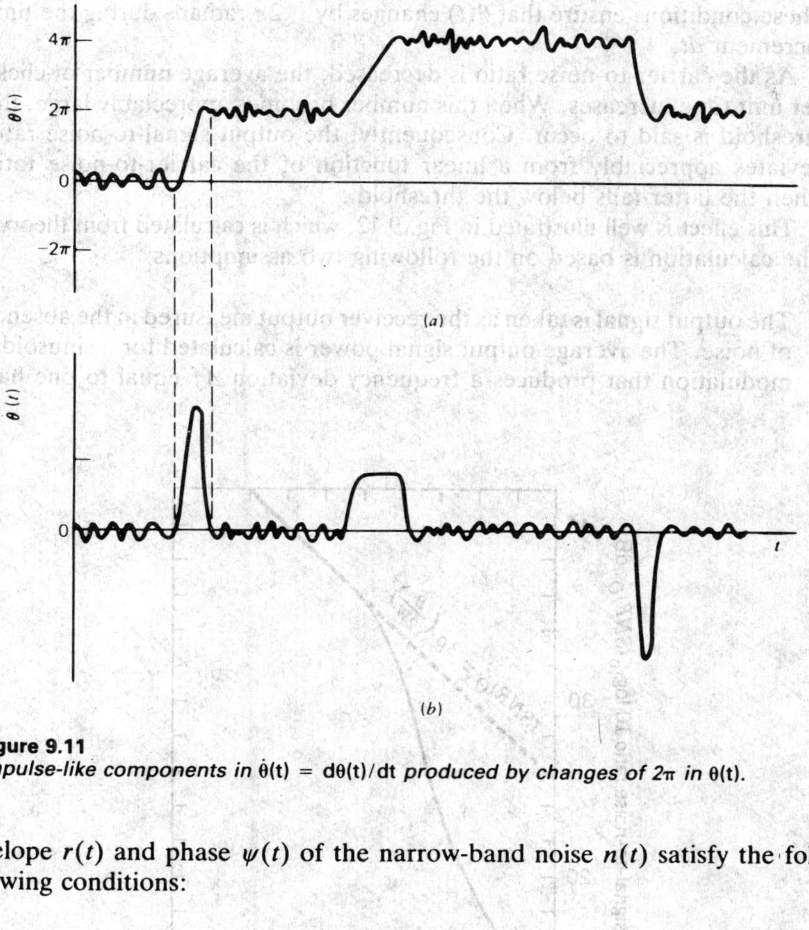

Figure 9.11
Impulse-like components in $\dot{\theta}(t) = d\theta(t)/dt$ *produced by changes of* 2π *in* $\theta(t)$.

velope $r(t)$ and phase $\psi(t)$ of the narrow-band noise $n(t)$ satisfy the following conditions:

$$r(t) > A_c$$
$$\psi(t) < \pi < \psi(t) + d\psi(t)$$
$$\frac{d\psi(t)}{dt} > 0$$

These conditions ensure that the phase $\theta(t)$ of the resultant phasor $x(t)$ changes by 2π radians in the time increment dt, during which the phase of the narrow-band noise increases by the incremental amount $d\psi(t)$. Similarly, the conditions for a negative-going click to occur are

$$r(t) > A_c$$
$$\psi(t) > -\pi > \psi(t) + d\psi(t)$$
$$\frac{d\psi(t)}{dt} < 0$$

These conditions ensure that $\theta(t)$ changes by -2π radians during the time increment dt.

As the carrier-to-noise ratio is decreased, the average number of clicks per unit time increases. When this number becomes appreciably large, the threshold is said to occur. Consequently, the output signal-to-noise ratio deviates appreciably from a linear function of the carrier-to-noise ratio when the latter falls below the threshold.

This effect is well illustrated in Fig. 9.12, which is calculated from theory.[5] The calculation is based on the following two assumptions:

1. The output signal is taken as the receiver output measured in the absence of noise. The average output signal power is calculated for a sinusoidal modulation that produces a frequency deviation Δf equal to one half

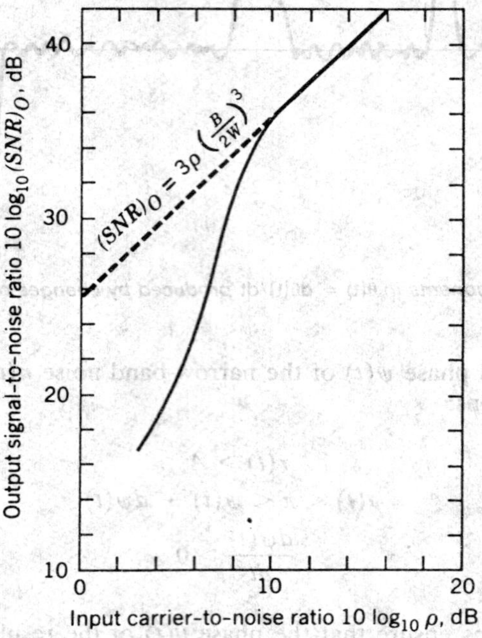

Figure 9.12
Variation of output signal-to-noise ratio with input carrier-to-noise ratio, demonstrating the FM threshold effect.

[5]For a detailed theoretical account of noise in FM receivers, see the classic papers by Rice (1948) and Stumpers (1948). Figure 9.12 is adapted from another paper by Rice (1963).

of the IF filter bandwidth B; the carrier is thus enabled to swing back and forth across the entire IF band.

2. The average output noise power is calculated when there is no signal present; that is, the carrier is unmodulated, with no restriction placed on the value of the carrier-to-noise ratio.

The curve plotted in Fig. 9.12 is for the ratio $(B/2W) = 5$. The linear part of the curve corresponds to the limiting value of $3\rho(B/2W)^3$; see Exercise 6. Figure 9.12 shows that, owing to the threshold phenomenon, the output signal-to-noise ratio deviates appreciably from a linear function of the carrier-to-noise ratio ρ when ρ becomes less than a threshold of 10 dB.

The *threshold carrier-to-noise ratio*, ρ_{th}, depends on the ratio of IF filter bandwidth-to-message bandwidth, B/W. Also, the value of ρ_{th} is influenced by the presence of modulation. Nevertheless, these variations are usually small enough to justify taking ρ_{th} as about 10 dB for most practical cases of interest. We may thus state that the loss of message at an FM receiver output is negligible if the carrier-to-noise ratio satisfies the condition

$$\frac{A_c^2}{2BN_0} \geq 10 \tag{9.62}$$

Since the channel signal-to-noise ratio $(\text{SNR})_C = A_c^2/2WN_0$, we may reformulate this condition as

$$(SNR)_C \geq \frac{10B}{W} \tag{9.63}$$

The IF filter bandwidth B is ordinarily designed to equal the FM transmission bandwidth. Hence, we may use Carson's rule to relate B to the message bandwidth W as follows (see Section 7.11)

$$B = 2W(1 + D)$$

where D is the deviation ratio; for sinusoidal modulation, the modulation index β is used in place of D. Accordingly, we may restate the condition for ensuring no significant loss of message at an FM receiver output as

$$(SNR)_C \geq 20(1 + D) \tag{9.64}$$

or, in terms of decibels,

$$10\log_{10}(SNR)_C \geq 13 + 10\log_{10}(1 + D), \text{ dB} \tag{9.65}$$

EXERCISE 7 Calculate the condition on the channel signal-to-noise ratio to avoid the FM threshold effect for the following values of deviation ratio:

(a) $D = 2$
(b) $D = 5$

Suppose that the FM receiver operates with the following parameters:

$$W = 15 \text{ kHz}$$

$$\frac{N_0}{2} = 0.5 \times 10^{-8} \text{ W/Hz}$$

Find the corresponding condition on the average transmitted power for case (a) and case (b).

FM THRESHOLD REDUCTION

In certain applications such as space communications, there is a particular interest in reducing the noise threshold in an FM receiver so as to satisfactorily operate the receiver with the minimum signal power possible. *Threshold reduction* in FM receivers may be achieved by using an FM demodulator with negative feedback (commonly referred to as an FMFB demodulator), or by using a phase-locked loop demodulator.

Figure 9.13 is a block diagram of an FMFB demodulator. We see that the local oscillator of the conventional FM receiver has been replaced by a voltage-controlled oscillator (VCO) with an instantaneous output frequency that is controlled by the demodulated signal. To understand the operation of this receiver, suppose for the moment that the VCO is removed from the circuit and the feedback loop is left open.[6] Assume that a wideband FM wave is applied to the receiver input, and a second FM wave, from the same source but with a modulation index a fraction smaller, is applied to the VCO terminal of the *product modulator*. The output of the product modulator consists of two components: a sum-frequency component and a difference-frequency component. The IF filter (following the product modulator) is designed to pass only the difference-frequency component. (The combination of the product modulator and the IF filter in Fig. 9.13 constitutes a *mixer*.) The frequency deviation of the IF filter (mixer) output would be small, although the frequency deviation of both input FM waves is large, since the difference between their instantaneous deviations is small. Hence, the modulation indices would subtract, and the resulting FM wave at the IF filter (mixer) output would have a smaller

[6]Our treatment of the FMFB demodulator is based on Enole (1962). See also Roberts (1977), pp. 166–181.

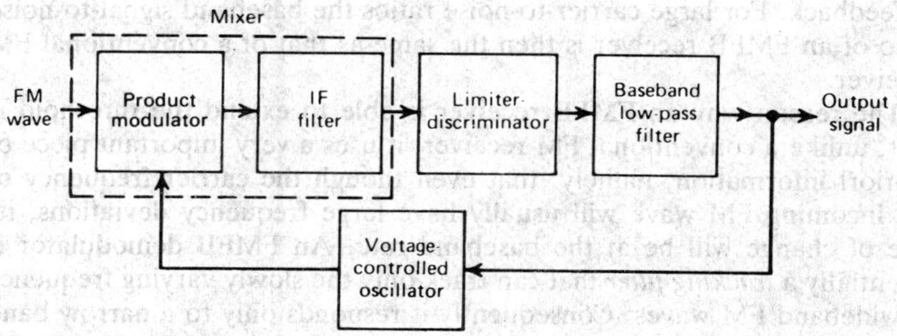

Figure 9.13
FMFB demodulator.

modulation index than the input FM waves. This means that the IF filter bandwidth in Fig. 9.13 need only be a fraction of that required for either wideband FM wave. The FM wave with reduced modulation index passed by the IF filter is then frequency-demodulated by the combination of limiter/discriminator and finally processed by the baseband filter. It is now apparent that the second wideband FM wave applied to the product modulator may be obtained by feeding the output of the baseband low-pass filter back to the VCO, as in Fig. 9.13.

It will now be shown that the signal-to-noise ratio of an FMFB receiver is the same as that of a conventional FM receiver with the same input signal and noise power if the carrier-to-noise ratio is sufficiently large. Assume for the moment that there is no feedback around the demodulator. In the combined presence of an unmodulated carrier $A_c \cos(2\pi f_c t)$ and a narrow-band noise

$$n(t) = n_I(t) \cos(2\pi f_c t) - n_Q(t) \sin(2\pi f_c t),$$

the phase of the composite signal $x(t)$ at the limiter–discriminator input is approximately equal to $n_Q(t)/A_c$. This assumes that the carrier-to-noise ratio is high. The envelope of $x(t)$ is of no interest to us, because the limiter removes all variations in the envelope. Thus the composite signal at the frequency discriminator input consists of a small index phase-modulated wave with the modulation derived from the component $n_Q(t)$ of noise that is in phase quadrature with the carrier. When feedback is applied, the VCO generates a wave that reduces the phase-modulation index of the wave at the IF filter output, that is, the quadrature component $n_Q(t)$ of noise. Thus we see that as long as the carrier-to-noise ratio is sufficiently large, the FMFB receiver does not respond to the in-phase noise component $n_I(t)$, but that it would demodulate the quadrature noise component $n_Q(t)$ in exactly the same fashion as it would demodulate the signal. Signal and quadrature noise are reduced in the same proportion by the applied feedback, with the result that the baseband signal-to-noise ratio is independent of feedback. For large carrier-to-noise ratios the baseband signal-to-noise ratio of an FMFB receiver is then the same as that of a conventional FM receiver.

The reason why an FMFB receiver is able to extend the threshold is that, unlike a conventional FM receiver, it uses a very important piece of a priori information, namely, that even though the carrier frequency of the incoming FM wave will usually have large frequency deviations, its rate of change will be at the baseband rate. An FMFB demodulator is essentially a *tracking filter* that can track only the slowly varying frequency of wideband FM waves. Consequently it responds only to a narrow band of noise centered about the instantaneous carrier frequency. The bandwidth of noise to which the FMFB receiver responds is precisely the band of noise that the VCO tracks. The net result is that an FMFB receiver is capable of realizing a threshold reduction on the order of 5–7 dB, which represents a significant improvement in the design of minimum-power FM systems.

The phase-locked loop demodulator, which was described in Section 7.12, exhibits threshold reduction properties that are similar to those of the FMFB demodulator. Thus, like the FMFB demodulator, a phase-locked loop is a tracking filter and, as such, the bandwidth of noise to which it responds is precisely the band of noise that the VCO tracks. However, although the thresholds of the phase-locked loop and FMFB demodulators occur because of the same basic mechanism, the details by which they occur are, of course, different.[7] Practical experience with the phase-locked loop, however, confirms the conclusion that very comparable performance with the FMFB demodulator is obtained in many situations, so that the choice between these two types of threshold-extension devices is often made in favor of the phase-locked loop because of its simpler construction.

9.8 PRE-EMPHASIS AND DE-EMPHASIS IN FM

In Section 9.6 we showed that the power spectral density of the noise at the receiver output has a square-law dependence on the operating frequency; this is illustrated in Fig. 9.14*a*. In part *b* of this figure we have included the power spectral density of a typical message source; audio and video signals typically have spectra of this form. We see that the power spectral density of the message usually falls off appreciably at higher frequencies. On the other hand, the power spectral density of the output noise increases rapidly with frequency. Thus, at $f = \pm W$, the relative spectral density of the message is quite low, whereas that of the output noise is high in comparison. Clearly, the message is not using the frequency band allowed to it in an efficient manner. It may appear that one way of improving the noise performance of the system is to slightly reduce the bandwidth of

[7] See Roberts (1977), pp. 200–202.

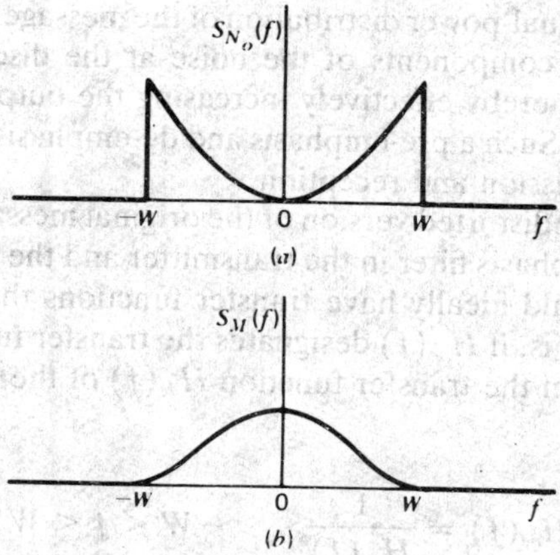

Figure 9.14
(a) Power spectral density of noise at FM receiver output. (b) Power spectral density of a typical message source.

the postdetection low-pass filter so as to reject a large amount of noise power while losing only a small amount of message power. Such an approach, however, is usually not satisfactory because the distortion of the message caused by the reduced filter bandwidth, even though slight, may not be tolerable. For example, in the case of music we find that although the high-frequency notes contribute only a very small fraction of the total power, nonetheless, they contribute a great deal from an aesthetic viewpoint.

A more satisfactory approach to the efficient use of the allowed frequency band is based on the use of *pre-emphasis* in the transmitter and *de-emphasis* in the receiver, as illustrated in Fig. 9.15. In this method, we artificially emphasize the high-frequency components of the message signal prior to modulation in the transmitter, and therefore before the noise is introduced in the receiver. In effect, the low-frequency and high-frequency portions of the power spectral density of the message are equalized in such a way that the message fully occupies the frequency band allotted to it. Then, at the discriminator output in the receiver, we perform the inverse operation by de-emphasizing the high-frequency components, so as to re-

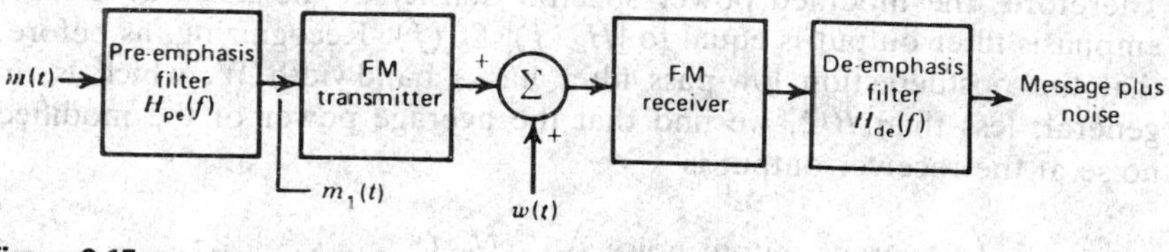

Figure 9.15
Use of pre-emphasis and de-emphasis in an FM system.

store the original signal-power distribution of the message. In such a process the high-frequency components of the noise at the discriminator output are also reduced, thereby effectively increasing the output signal-to-noise ratio of the system. Such a pre-emphasis and de-emphasis process is widely used in FM transmission and reception.

To produce an undistorted version of the original message at the receiver output, the pre-emphasis filter in the transmitter and the de-emphasis filter in the receiver would ideally have transfer functions that are the *inverse* of each other. That is, if $H_{pe}(f)$ designates the transfer function of the pre-emphasis filter, then the transfer function $H_{de}(f)$ of the de-emphasis filter would ideally be

$$H_{de}(f) = \frac{1}{H_{pe}(f)}, \qquad -W < f < W \tag{9.66}$$

This choice of transfer functions makes the average message power at the receiver output independent of the pre-emphasis and de-emphasis procedure.

The pre-emphasis filter is selected so that the average power of the emphasized message signal $m_1(t)$ in Fig. 9.15 has the same average power as the original message $m(t)$. Thus, given the power spectral density $S_M(f)$ of the message signal $m(t)$, we may write

$$\int_{-\infty}^{\infty} |H_{pe}(f)|^2 S_M(f)\, df = \int_{-\infty}^{\infty} S_M(f)\, df \tag{9.67}$$

This *constraint* on the transfer function $H_{pe}(f)$ of the pre-emphasis filter ensures that the bandwidth of the transmitted FM signal remains the same, with or without pre-emphasis.

From our previous noise analysis in FM systems, assuming a high carrier-to-noise ratio, the power spectral density of the noise $n_d(t)$ at the discriminator output is

$$S_{N_d}(f) = \begin{cases} \dfrac{N_0 f^2}{A_c^2}, & |f| \leq \dfrac{B}{2} \\ 0, & \text{otherwise} \end{cases} \tag{9.68}$$

Therefore, the modified power spectral density of the noise at the de-emphasis filter output is equal to $|H_{de}(f)|^2 S_{N_d}(f)$. Recognizing, as before, that the postdetection low-pass filter has a bandwidth W, which is, in general, less than $B/2$, we find that the average power of the modified noise at the receiver output is

$$\begin{pmatrix} \text{Average output noise} \\ \text{power with de-emphasis} \end{pmatrix} = \frac{N_0}{A_c^2} \int_{-W}^{W} f^2 |H_{de}(f)|^2\, df \tag{9.69}$$

Because the average message power at the receiver output is ideally unaffected by the pre-emphasis and de-emphasis procedure, it follows that the improvement in output signal-to-noise ratio produced by the use of pre-emphasis in the transmitter and de-emphasis in the receiver is defined by

$$I = \frac{\text{average output noise power without pre-emphasis and de-emphasis}}{\text{average output noise power with pre-emphasis and de-emphasis}}$$

Earlier we showed that the average output noise power without pre-emphasis and de-emphasis is equal to $2N_0W^3/3A_c^2$; see Eq. 9.52. Therefore, after cancellation of common terms, we may write

$$I = \frac{2W^3}{3\int_{-W}^{W} f^2|H_{de}(f)|^2\, df} \tag{9.70}$$

Note that this improvement factor assumes a high carrier-to-noise ratio at the discriminator input.

EXAMPLE 4

A simple pre-emphasis filter that emphasizes high frequencies and that is commonly used in practice is defined by the transfer function

$$H_{pe}(f) = k\left(1 + \frac{jf}{f_0}\right) \tag{9.71}$$

This transfer function is closely realized by the *RC*-amplifier network shown in Fig. 9.16*a*, provided that $R \ll r$ and $2\pi fCR \ll 1$ inside the frequency band of interest. The amplifier in Fig. 9.16*a* is intended to make up for the attenuation introduced by the *RC* network at low frequencies. The frequency parameter f_0 is $1/(2\pi Cr)$. The corresponding de-emphasis filter in the receiver is defined by the transfer function

$$H_{de}(f) = \frac{1/k}{1 + jf/f_0} \tag{9.72}$$

which can be realized using the RC-amplifier network of Fig. 9.16*b*.

The constant k in Eqs. 9.71 and 9.72 is chosen to satisfy the *constraint* of Eq. 9.67, which requires that *the average power of the pre-emphasized message signal be the same as the average power of the original message signal.* Assume that the power spectral density of the original message

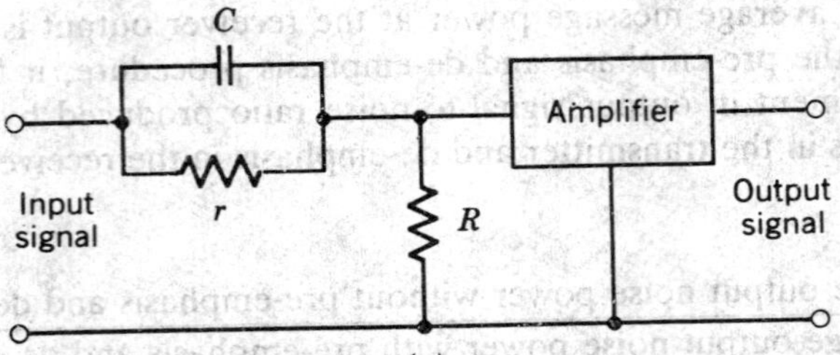

(a)

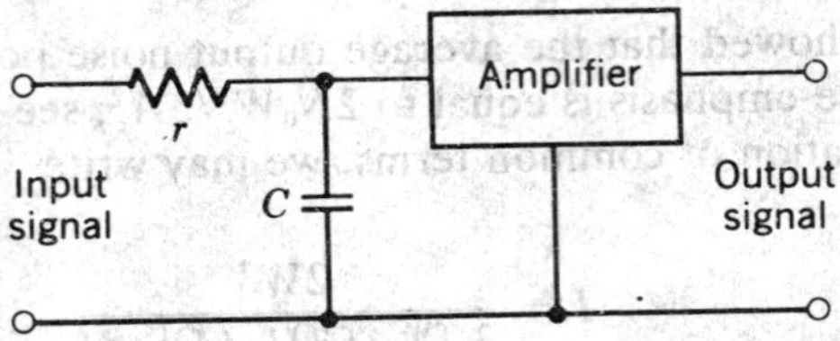

(b)

Figure 9.16
(a) Pre-emphasis filter. (b) De-emphasis filter.

signal $m(t)$ is

$$S_M(f) = \begin{cases} \dfrac{1}{1 + (f/f_0)^2} & |f| < W \\ 0 & \text{elsewhere} \end{cases} \tag{9.73}$$

Then, the use of Eqs. 9.71 and 9.73 in 9.67 yields

$$\int_{-W}^{W} \frac{df}{1 + (f/f_0)^2} = \int_{-W}^{W} k^2 \, df$$

or

$$k^2 = \frac{f_0}{W} \tan^{-1}\left(\frac{W}{f_0}\right) \tag{9.74}$$

Equation 9.70 defines the improvement in output signal-to-noise ratio of the FM receiver, resulting from the combined use of pre-emphasis and de-emphasis. For the pre-emphasis and de-emphasis filters of Fig. 9.16, the use of this equation yields the improvement

$$I = \frac{2W^3}{3\displaystyle\int_{-W}^{W} \frac{k^2 f^2}{1 + (f/f_0)^2} \, df}$$
$$= \frac{(W/f_0)^2 \tan^{-1}(W/f_0)}{3[(W/f_0) - \tan^{-1}(W/f_0)]} \tag{9.75}$$

Typical values for commercial FM broadcasting are

$$f_0 = 2.1 \text{ kHz}$$
$$W = 15 \text{ kHz}$$

The use of this set of values in Eq. 9.75 yields the result

$$I = 4.7$$

Expressing the improvement in decibels, we have

$$I = 6.7 \text{ dB}$$

The output signal-to-noise ratio of an FM receiver without pre-emphasis and de-emphasis is typically 40–50 dB. We thus see that by using the simple pre-emphasis and de-emphasis filters shown in Fig. 9.16, we can obtain a significant improvement in the noise performance of the receiver.

EXERCISE 8 Sketch the power spectral density of the de-emphasized noise, assuming that the shape of the power spectral density of the noise at the de-emphasis filter input is as shown in Fig. 9.14*a* and the de-emphasis filter is as shown in Fig. 9.16*b*.

NONLINEAR TECHNIQUES

The use of the simple *linear* pre-emphasis and de-emphasis filters described herein is an example of how the performance of an FM system may be improved by using the differences between characteristics of signals and noise in the system. These simple filters also find application in audio tape-recording. In recent years *nonlinear* pre-emphasis and de-emphasis techniques have been applied successfully to tape-recording. These techniques (known as *Dolby-A, Dolby-B, Dolby-C,* and *DBX* systems) use a combination of filtering and dynamic range compression to reduce the effects of noise, particularly when the signal level is low.[8]

9.9 DISCUSSION

We conclude the noise analysis of analog modulation systems by presenting a comparison of the relative merits of the different modulation techniques. For the purpose of this comparison, we assume that the modulation is

[8]For a detailed description of Dolby systems, see Stremler (1982), pp. 671–673.

produced by a single sine wave. For the comparison to be meaningful, we also assume that all the different modulation systems operate with exactly the same channel signal-to-noise ratio. In making the comparison, it is informative to keep in mind the transmission bandwidth requirement of the modulation system in question. In this regard, we use a *normalized transmission bandwidth* defined by

$$B_n = \frac{B}{W} \tag{9.76}$$

where B is the transmission bandwidth of the modulated wave and W is the message bandwidth. We may thus make the following observations:

1. In a standard AM system using envelope detection, the output signal-to-noise ratio, assuming sinusoidal modulation, is given by (see Eq. 9.26)

$$(SNR)_O = \frac{\mu^2}{2 + \mu^2}(SNR)_C$$

   This relation is plotted as curve I in Fig. 9.17, assuming $\mu = 1$. In this curve we have also included the AM threshold effect, based on the result of Exercise 4. Since in a standard AM system both sidebands are transmitted, the normalized transmission bandwidth B_n equals 2.

2. In the case of a DSBSC or SSB modulation system using coherent detection, the output signal-to-noise ratio is given by (see Eqs. 9.12 and 9.18):

$$(SNR)_O = (SNR)_C$$

   This relation is plotted as curve II in Fig. 9.17. We see, therefore, that the noise performance of a DSBSC or SSB system, using coherent detection, is superior to that of a standard AM system using envelope detection by 4.8 dB. It should also be noted that neither the DSBSC nor the SSB system exhibits a threshold effect. With regard to transmission bandwidth requirement, we have $B_n = 2$ for the DSBSC system and $B_n = 1$ for the SSB system. Thus, among the family of AM systems, SSB modulation is optimum with regard to noise performance as well as bandwidth conservation.

3. In an FM system using a conventional discriminator, the output signal-to-noise ratio, assuming sinusoidal modulation, is given by (see Eq. 9.57)

$$(SNR)_O = \tfrac{3}{2}\beta^2(SNR)_C$$

   where β is the modulation index. This relation is shown as curves III and IV in Fig. 9.17, corresponding to $\beta = 2$ and $\beta = 5$, respectively. In each case, we have included a 6.7-dB improvement that is typically obtained by using pre-emphasis in the transmitter and de-emphasis in the receiver. To determine the transmission bandwidth requirement,

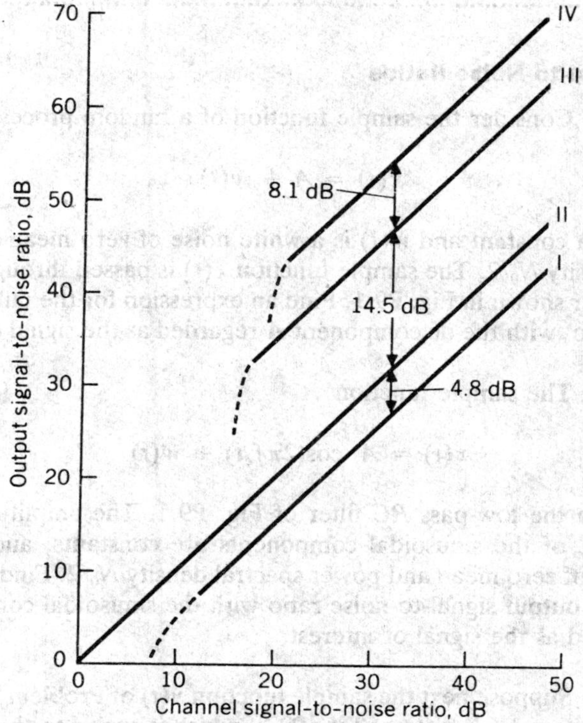

Figure 9.17
Comparison of the noise performance of various analog modulation systems. Curve 1: Full AM, μ = 1. Curve II: DSBSC, SSB. Curve III: FM. β = 2. Curve IV: FM. β = 5. (Curves III and IV include 13-dB pre-emphasis, de-emphasis improvement.)

we use Carson's rule and thus write

$$B_n = 6 \quad \text{for } \beta = 2$$
$$B_n = 12 \quad \text{for } \beta = 5$$

We therefore see that, compared with the SSB system, which is the optimum form of linear modulation, by using wideband FM we obtain an improvement in output signal-to-noise ratio equal to 14.5 dB for a normalized bandwidth $B_n = 6$, and an improvement of 22.6 dB for $B_n = 12$. This clearly illustrates the improvement in noise performance that is achievable by using wideband FM. However, the price that we have to pay for this improvement is increased transmission bandwidth. It is, of course, assumed that the FM system operates above threshold for the noise improvement to be realizable as described herein. The curves III and IV of Fig. 9.17 include the FM threshold effect, based on the results of Exercise 7. Note that the threshold effect in FM manifests itself at a channel signal-to-noise ratio much greater than that in standard AM.

PROBLEMS

P9.1 Signal-to-Noise Ratios

Problem 1 Consider the sample function of a random process

$$x(t) = A + w(t)$$

where A is a constant and $w(t)$ is a white noise of zero mean and power spectral density $N_0/2$. The sample function $x(t)$ is passed through the low-pass RC filter shown in Fig. P9.1. Find an expression for the output signal-to-noise ratio, with the dc component A regarded as the signal of interest.

Problem 2 The sample function

$$x(t) = A_c \cos(2\pi f_c t) + w(t)$$

is applied to the low-pass RC filter of Fig. P9.1. The amplitude A_c and frequency f_c of the sinusoidal components are constants, and $w(t)$ is a white noise of zero mean and power spectral density $N_0/2$. Find an expression for the output signal-to-noise ratio with the sinusoidal component of $x(t)$ regarded as the signal of interest.

Problem 3 Suppose next the sample function $x(t)$ of Problem 2 is applied to the band-pass LCR filter of Fig. P9.2, which is tuned to the frequency f_c of the sinusoidal component. Assume that the Q factor of the filter is high compared with unity. Find an expression for the output signal-to-noise ratio, by treating the sinusoidal component of $x(t)$ as the signal of interest.

Problem 4 The input to the low-pass RC filter of Fig. P9.1 consists of a white noise of zero mean and power spectral density $N_0/2$, plus a signal that is a sequence of constant-amplitude rectangular pulses. The pulse amplitude is A, the pulse duration is T, and the period of the sequence is T_0, where $T \ll T_0$. Derive an expression for the output signal-to-noise ratio of the filter, defined as the ratio of the square of the maximum amplitude of the output signal with no noise at the input to the average power of the output noise.

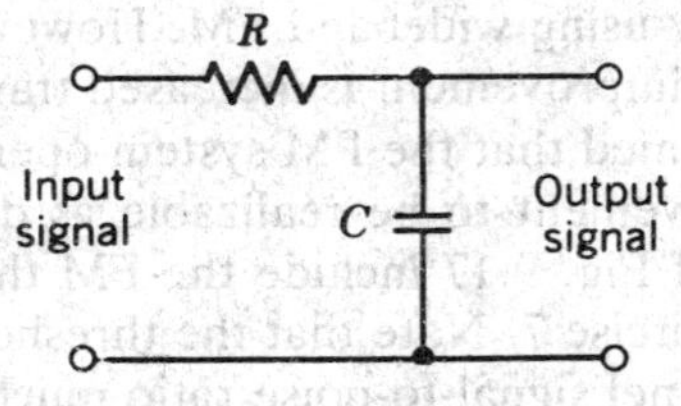

Figure P9.1

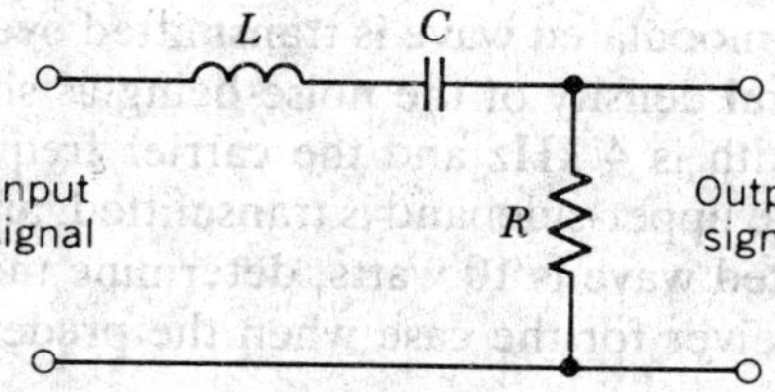

Figure P9.2

P9.3 Signal-to-Noise Ratios for Coherent Reception

Problem 5 Calculate the output signal-to-noise ratio of the coherent receiver of Fig. 9.3, assuming that the modulated signal $s(t)$ is produced by the sinusoidal modulating wave

$$m(t) = A_m \cos(2\pi f_m t)$$

Perform your calculation for the following two receiver types:

(a) Coherent DSBSC receiver
(b) Coherent SSB receiver.

Problem 6 Let a message signal $m(t)$ be transmitted using SSB modulation. The power spectral density of $m(t)$ is

$$S_M(f) = \begin{cases} a\dfrac{|f|}{W}, & |f| \leq W \\ 0, & \text{otherwise} \end{cases}$$

where a and W are constants. White noise of zero mean and power spectral density $N_0/2$ is added to the SSB-modulated wave at the receiver input. Find an expression for the output signal-to-noise ratio of the receiver.

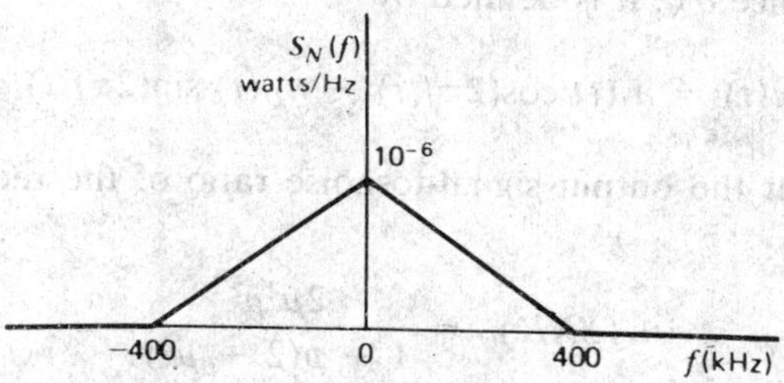

Figure P9.3

Problem 7 An SSB-modulated wave is transmitted over a noisy channel, with the power spectral density of the noise being as shown in Fig. P9.3. The message bandwidth is 4 kHz and the carrier frequency is 200 kHz. Assuming that only the upper-sideband is transmitted, and that the average power of the modulated wave is 10 watts, determine the output signal-to-noise ratio of the receiver for the case when the predetection filter characteristic is ideal.

P9.4 Noise in AM Receivers Using Envelope Detection

Problem 8 The average noise power per unit bandwidth measured at the front end of an AM receiver is 10^{-3} watts per hertz. The modulating wave is sinusoidal, with a carrier power of 80 kilowatts and a sideband power of 10 kilowatts per sideband. The message bandwidth is 4 kHz. Assuming the use of an envelope detector in the receiver, determine the output signal-to-noise ratio of the system. By how many decibels is this system inferior to a DSBSC modulation system?

Problem 9 An unmodulated carrier of amplitude A_c and frequency f_c and band-limited white noise are summed and then passed through an ideal envelope detector. Assume the noise spectral density to be of height $N_0/2$ and bandwidth $2W$, centered about the carrier frequency f_c. Determine the output signal-to-noise ratio for the case when the carrier-to-noise ratio is high.

Problem 10 An AM receiver, operating with a sinusoidal modulating wave and 80% modulation, has an output signal-to-noise ratio of 30 dB. What is the corresponding carrier-to-noise ratio?

Problem 11 Consider an AM receiver using a square-law detector with output proportional to the square of the input, as indicated in Fig. P9.4. The AM wave is defined by

$$s(t) = A_c[1 + \mu \cos(2\pi f_m t)] \cos(2\pi f_c t)$$

Assume that the additive noise at the detector input is Gaussian with zero mean and variance σ_N^2; it is defined by

$$n(t) = n_I(t) \cos(2\pi f_c t) - n_Q(t) \sin(2\pi f_c t)$$

(a) Show that the output signal-to-noise ratio of the receiver is given by

$$(SNR)_O = \frac{2\mu^2\rho^2}{1 + \rho(2 + \mu^2)}$$

where ρ is the carrier-to-noise ratio.

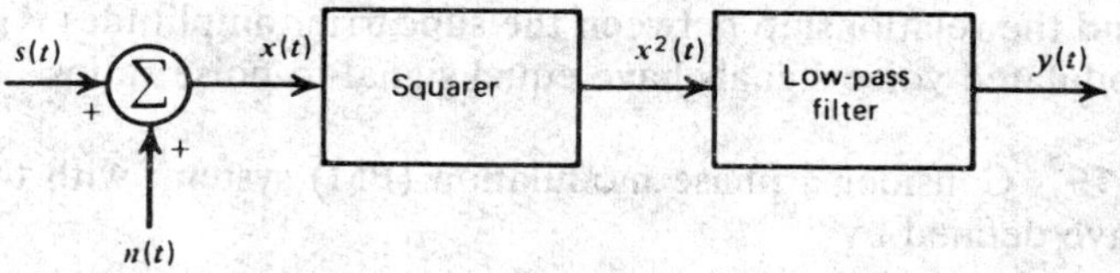

Figure P9.4

(b) Evaluate the asymptotic behavior of $(SNR)_O$ with respect to ρ.
(c) Plot the dependence of $(SNR)_O$ on ρ for the case of 100% modulation.

P9.5 FM Receiver Model

Problem 12 Assume that the FM receiver model of Fig. 9.7*a* and the AM receiver model of Fig. 9.2*a* have the same additive white noise $w(t)$ of zero mean and power spectral density $N_0/2$. Compare the average noise power at the output of the IF filter in Fig. 9.7*a* with that in Fig. 9.2*a*.

P9.6 Noise in FM Reception

Problem 13 Suppose that the spectrum of a modulating signal occupies the frequency band $f_1 \leq |f| \leq f_2$. To accommodate this signal, the receiver of an FM system (without pre-emphasis) uses an ideal band-pass filter connected to the output of the frequency discriminator; the filter passes frequencies in the interval $f_1 \leq |f| \leq f_2$. Determine the output signal-to-noise ratio and figure of merit of the system in the presence of additive white noise at the receiver input.

Problem 14 An FDM system uses single-sideband modulation to combine 12 independent voice signals and then uses frequency modulation to transmit the composite baseband signal. Each voice signal has a power P and occupies the frequency band 0.3–3.4 kHz; the system allocates it a bandwidth of 4 kHz. For each voice signal, only the lower sideband is transmitted. The subcarrier waves used for the first stage of modulation are defined by

$$c_k(t) = A_k \cos(2\pi k f_0 t), \qquad 0 \leq k \leq 11$$

The received signal consists of the transmitted FM signal plus white noise of zero mean and power spectral density $N_0/2$.

(a) Sketch the power spectral density of the signal produced at the frequency discriminator output, showing both the signal and noise components.

(b) Find the relationship between the subcarrier amplitudes A_k so that the modulated voice signals have equal signal-to-noise ratios.

Problem 15 Consider a phase modulation (PM) system, with the modulated wave defined by

$$s(t) = A_c \cos[2\pi f_c t + k_p m(t)]$$

where k_p is a constant and $m(t)$ is the message signal. The additive noise $n(t)$ at the phase detector input is

$$n(t) = n_I(t) \cos(2\pi f_c t) - n_Q(t) \sin(2\pi f_c t)$$

Assuming that the carrier-to-noise ratio at the detector input is high compared with unity, determine: (a) the output signal-to-noise ratio, and (b) the figure of merit of the system. Compare your results with the FM system for the case of sinusoidal modulation.

P9.7 FM Threshold Effect

Problem 16 The results reported in Section 9.7 indicate that the threshold point is defined by the carrier-to-noise ratio.

$$\rho_{th} = 10$$

(a) Show that the output signal-to-noise ratio at the threshold point is given by

$$(SNR)_{O,th} = 30\, \beta^2(\beta + 1)$$

where β is the modulation index (assuming sinusoidal modulation).
(b) Find the modulation index β that produces an output signal-to-noise ratio equal to 34.6 dB at the threshold point. Hence, find the corresponding value of the channel signal-to-noise ratio.

P9.8 Pre-emphasis and De-emphasis in FM

Problem 17 By using the pre-emphasis filter shown in Fig. 9.16*a* and with a voice signal as the modulating wave, an FM transmitter produces a signal that is essentially frequency-modulated by the lower audio frequencies and phase-modulated by the higher audio frequencies. Explain the reasons for this phenomenon.

Problem 18 A phase modulation (PM) system uses a pair of pre-emphasis and de-emphasis filters defined by the transfer functions

$$H_{pe}(f) = k\left(1 + \frac{jf}{f_0}\right)$$

and

$$H_{de}(f) = \frac{1/k}{1 + (jf/f_0)}$$

The constant k is chosen to make the average power of the pre-emphasized message equal to that of the original message signal.

(a) Determine the improvement in output signal-to-noise ratio produced by the use of this pair of filters.
(b) Compare this improvement with that produced in the corresponding FM system.
(c) Given that the message bandwidth $W = 15$ kHz and the cutoff frequency $f_0 = 2.1$ kHz, how do the improvements in SNR for the PM and FM systems compare with each other?

CHAPTER 10

OPTIMUM RECEIVERS FOR DATA COMMUNICATION

A basic issue in the design of receivers is that of detecting a weak signal embedded in a background of *additive noise*. Broadly speaking, *the purpose of detection is to establish the presence or absence of a signal in noise*. In order to enhance the strength of the signal relative to that of the noise, and thereby facilitate the detection process, a detection system usually consists of a *predetection filter* followed by a *decision device*. When the additive noise is white, that is, the power spectral density of the noise is constant, it turns out that the optimum solution to the predetection filter is a *matched filter*, which is so-called because its characterization is matched to that of the signal component in the received signal. A matched filter is optimum in the sense that it maximizes the output signal-to-noise ratio defined in a special way. It is thus apparent that a matched filter is useful in the design of digital communication systems where the concern is to enhance the received pulses so as to maximize the signal-to-noise ratio. In these applications we are primarily interested in improving our ability to recognize a pulse signal in the presence of additive noise and not in preserving the fidelity of the pulse shape.

In this chapter we study the theory and applications of matched filters. We begin the study by formulating the *optimum receiver problem.*

10.1 FORMULATION OF THE OPTIMUM RECEIVER PROBLEM

Consider the situation depicted in Fig. 10.1. Suppose that we have received a signal $x(t)$ that consists of either *white Gaussian noise* $w(t)$ or the noise $w(t)$ *plus a signal* $s(t)$ *of known form.* The implication of the noise being "white" is that its power spectral density has a constant value $N_0/2$, say. The implication of the noise being "Gaussian" is that a sample drawn from such a process has a Gaussian probability distribution for its amplitude. We further assume that the noise has zero mean. We wish to estimate which of the two hypotheses, noise alone or noise plus signal, is true. We do this by operating on the received signal $x(t)$ with a *linear time-invariant receiver* in such a way that if the signal $s(t)$ is present, the receiver output at some arbitrary time $t = T$ will be considerably greater than if $s(t)$ is absent.

For example, in a pulse-code modulation system using on–off signaling, a pulse $s(t)$ may represent symbol 1, whereas its absence may represent symbol 0. We thus have the problem of specifying the input–output relation of the receiver according to some criterion, so as to enhance the detection process as much as possible.

We present two approaches to the solution of this basic optimization problem. One approach is based on *maximization of the signal-to-noise ratio at the receiver output.* The other approach is based on a *probabilistic criterion* directly related to performance ratings of digital communication systems in which we are interested. We will show that: (1) maximization of the output signal-to-noise ratio yields the so-called *matched filter receiver,* which involves a filter matched to the signal component of the received

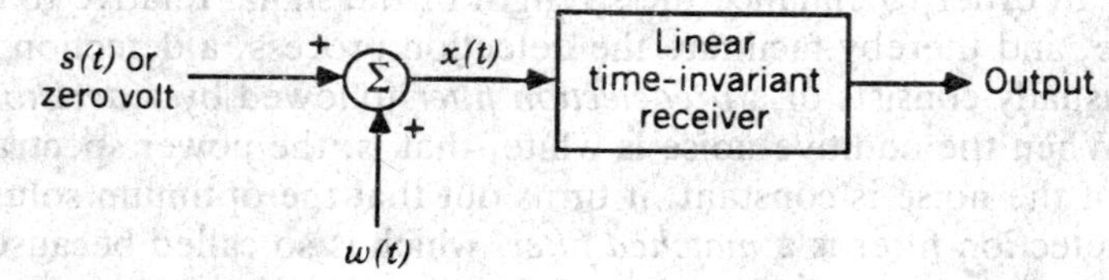

Figure 10.1
Processing of noisy signal.

signal, and (2) the probabilistic approach yields the so-called *correlation receiver,* which involves a correlation of the received signal with a stored replica of the transmitted signal. Furthermore, we will show that these two receiver structures are indeed equivalent for the case of additive white Gaussian noise.

10.2 MAXIMIZATION OF OUTPUT SIGNAL-TO-NOISE RATIO

Consider a linear time-invariant filter of impuse response $h(t)$ or, equivalently, transfer function $H(f)$, with $x(t)$ as input and $y(t)$ as output. Let $s_o(t)$ and $n_o(t)$ denote the signal and noise components of the filter output $y(t)$ produced by the signal component $s(t)$ and white noise component $w(t)$ of the input, respectively. Since the filter is linear, and the signal $s(t)$ and noise $w(t)$ appear additively at the filter input, we may invoke the principle of superposition and thus evaluate their effects at the filter output by considering them separately.

Let $S(f)$ denote the Fourier transform of the input signal component $s(t)$. Then, the Fourier transform of the corresponding output signal $s_o(t)$ is equal to $H(f)S(f)$, and $s_o(t)$ is itself given by the inverse Fourier transform:

$$s_o(t) = \int_{-\infty}^{\infty} H(f)S(f) \exp(j2\pi ft)\, df \tag{10.1}$$

Consider next the effect of the noise $w(t)$ alone on the filter output. The power spectral density $S_{N_o}(f)$ of the output noise $n_o(t)$ is equal to the power spectral density of the input noise $w(t)$ times the squared magnitude of the transfer function $H(f)$ (see Section 8.9). Since $w(t)$ is white with constant power spectral density $N_0/2$, it follows that

$$S_{N_o}(f) = \frac{N_0}{2} |H(f)|^2 \tag{10.2}$$

The average power $\mathcal{N}$ of the output noise $n_o(t)$ equals the total area under the curve of $S_{N_o}(f)$. We may therefore write

$$\begin{aligned}\mathcal{N} &= \int_{-\infty}^{\infty} S_{N_o}(f)\, df \\ &= \frac{N_0}{2} \int_{-\infty}^{\infty} |H(f)|^2\, df \end{aligned} \tag{10.3}$$

A simple way of describing the requirement that the filter output be considerably greater when the input signal $s(t)$ is present than when $s(t)$ is absent, is to ask that, at time $t = T$, the filter make the instantaneous power in the output signal $s_o(t)$ as large as possible compared with the average power in output noise $n_o(t)$. This is equivalent to maximizing the *output signal-to-noise ratio,* defined as

$$(SNR)_O = \frac{|s_o(T)|^2}{\mathcal{N}} \tag{10.4}$$

Using Eqs. 10.2 and 10.3 in 10.4, we get

$$(SNR)_O = \frac{\left| \int_{-\infty}^{\infty} H(f)S(f) \exp(j2\pi fT)\, df \right|^2}{\frac{N_0}{2} \int_{-\infty}^{\infty} |H(f)|^2\, df} \tag{10.5}$$

Our problem is to find, while holding the Fourier transform $S(f)$ of the input signal fixed, the form of the transfer function $H(f)$ of the filter that makes $(SNR)_O$ a maximum. To find the solution to this constrained optimization problem, we may apply a mathematical result known as *Schwarz's inequality* to the numerator of Eq. 10.5.

We will digress from our task briefly to introduce this important inequality, using a notation consistent with that used herein.

SCHWARZ'S INEQUALITY

Consider the complex-valued frequency function $H(f)S(f) \exp(j2\pi fT)$. This function may be viewed as the product of two functions, namely, $H(f)$ and $S(f)\exp(j2\pi fT)$. *Schwarz's inequality* for integrals of complex functions states that the squared magnitude of the total area under the product of two such functions is less than or equal to the product of the total area under the squared magnitude of each of the two functions. In mathematical terms, Schwarz's inequality states that[1]

$$\left| \int_{-\infty}^{\infty} H(f)S(f) \exp(j2\pi fT)\, df \right|^2 \leq \int_{-\infty}^{\infty} |H(f)|^2\, df \int_{-\infty}^{\infty} |S(f)|^2\, df \tag{10.6}$$

[1]Schwarz's inequality, stated in Eq. 10.6 is just an extension of an inequality for real functions described by

$$\left[\int_{-\infty}^{\infty} a(t)b(t)\, dt \right]^2 \leq \int_{-\infty}^{\infty} a^2(t)\, dt \int_{-\infty}^{\infty} b^2(t)\, dt$$

where $a(t)$ and $b(t)$ denote a pair of real-time functions of finite energy. As such, it may be viewed as a generalization of the well-known "distance" relation among vectors, which states that the magnitude of the sum of two vectors is less than or equal to the sum of the magnitudes of the two vectors. For a formal proof of Schwarz's inequality, see Haykin (1988), pp. 574–76.

Here we have used the fact that the exponential term $\exp(j2\pi fT)$ has a magnitude of unity; therefore

$$|S(f)\exp(j2\pi fT)| = |S(f)|$$

Schwarz's inequality also states that Eq. 10.6 is satisfied with equality if, and only if, the first function $H(f)$ is the complex conjugate of the second function $S(f)\exp(j2\pi fT)$. This statement is valid to within a scaling factor. Let $H_{opt}(f)$ denote the special value of $H(f)$ that satisfies this condition. We may then write

$$H_{opt}(f) = S^*(f)\exp(-j2\pi fT) \tag{10.7}$$

where $S^*(f)$ is the complex conjugate of $S(f)$.

Having equipped ourselves with this new mathematical tool, we are ready to resume our task of finding a solution to the optimum receiver problem.

MATCHED FILTER

Using Schwarz's inequality of Eq. 10.6 in the formula for the output signal-to-noise ratio given in Eq. 10.5, we get

$$(SNR)_O \leq \frac{2}{N_0}\int_{-\infty}^{\infty} |S(f)|^2\, df \tag{10.8}$$

The right side of this relation does not depend on the transfer function $H(f)$ of the filter but only on the signal energy and the noise spectral density. Consequently, the output signal-to-noise ratio will be a maximum when $H(f)$ is chosen so that the equality holds, that is,

$$(SNR)_{O,opt} = \frac{2}{N_0}\int_{-\infty}^{\infty} |S(f)|^2\, df \tag{10.9}$$

This condition is fulfilled when the transfer function $H(f)$ assumes its *optimum value* $H_{opt}(f)$, defined by Eq. 10.7.

According to Eq. 10.7, except for the exponential factor $\exp(-j2\pi fT)$ representing a constant time delay T, *the transfer function of the optimum filter is the same as the complex conjugate of the spectrum of the input signal.* Such a filter is called a *matched filter.*

Equation 10.7 specifies the matched filter in the frequency domain. To characterize it in the time domain, we take the inverse Fourier transform of $H_{opt}(f)$ in Eq. 10.7 to obtain the impulse response of the matched filter as

$$h_{opt}(t) = \int_{-\infty}^{\infty} S^*(f)\exp[-j2\pi f(T-t)]\, df$$

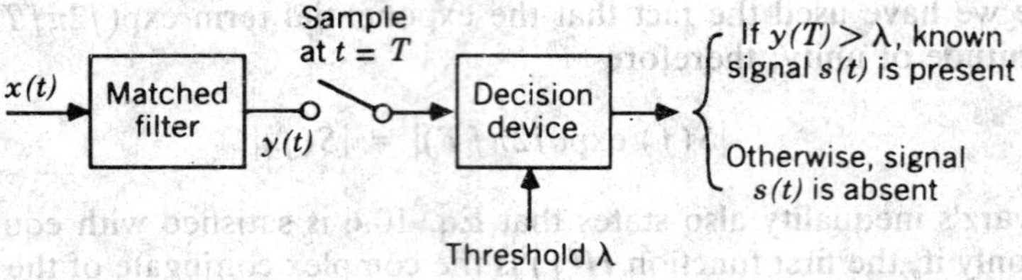

Figure 10.2
Matched filter receiver.

Since for a real-valued signal $s(t)$, we have $S^*(f) = S(-f)$, it follows that

$$\begin{aligned} h_{opt}(t) &= \int_{-\infty}^{\infty} S(-f) \exp[-j2\pi f(T - t)]\, df \\ &= s(T - t) \end{aligned} \tag{10.10}$$

Equation 10.10 shows that the *impulse response of the matched filter*[2] *is a time-reversed and delayed version of the input signal* $s(t)$. Note that in deriving this result the only assumption we have made about the statistics of the input noise $w(t)$ is that it is white with zero mean and a power spectral density $N_0/2$.

The optimum receiver for detecting the presence of the signal $s(t)$ in the received waveform is thus as shown in Fig. 10.2. It consists of a filter matched to $s(t)$, *a sampler,* and a *decision-device.* At time $t = T$, the matched filter output is sampled and the amplitude of this sample is compared with a preset *threshold* λ. If the threshold is exceeded, the receiver decides that the known signal $s(t)$ is present; otherwise, it will decide that it is absent. The receiver of Fig. 10.2 is called a *matched-filter receiver.*

Thus far we have ignored the problem of the physical realizability of a matched filter. For a matched filter operating in *real time* to be physically realizable, it must be causal. That is, its impulse response must be zero for negative time, as shown by

$$h_{opt}(t) = 0, \qquad t < 0$$

In terms of Eq. 10-10, the causality condition becomes

$$h_{opt}(t) = \begin{cases} 0, & t < 0 \\ s(T - t), & t \geq 0 \end{cases} \tag{10.11}$$

[2]The characterization of a matched filter in terms of its transfer function was first derived by North in a classified report (RCA Laboratories Report PTR-6C, June 1943), which was published 20 years later (North, 1963). For a review of the matched filter and its properties, see Turin (1960).

If all the input signal $s(t)$ is to contribute to the output signal component $s_o(t)$, it is apparent from Eq. 10.11 that we must have

$$s(t) = 0, \quad t > T \tag{10.12}$$

This relation simply states that *all the input signal $s(t)$ must have entered the filter by the time $t = T$ at which it is desired to obtain a sample with the maximum output signal-to-noise ratio.*

For Eq. 10.11 to be dimensionally correct, the term $s(T - t)$ should be multiplied by a scaling factor k that makes the impulse response $h_{opt}(t)$ of the matched filter assume a dimension that is the inverse of time. This has the effect of making the transfer function $H_{opt}(f)$ of the matched filter in Eq. 10.7 dimensionless. We have chosen to ignore the use of such a scaling factor merely for convenience of mathematical presentation.

EXERCISE 1 Show that multiplication of the optimum transfer function $H_{opt}(f)$ of Eq. 10.7 by a scaling factor k leaves the maximum signal-to-noise ratio unchanged.

10.3 PROPERTIES OF MATCHED FILTERS

From the results of the preceding section, we may state that a filter, which is matched to an input signal $s(t)$, is characterized in the time domain by the impulse response

$$h_{opt}(t) = s(T - t)$$

which is a time-reversed and delayed version of the input $s(t)$, as illustrated in Fig. 10.3. In the frequency domain, it is characterized by the transfer function

$$H_{opt}(f) = S^*(f) \exp(-j2\pi f T)$$

which is, except for a delay factor, the complex conjugate of the spectrum of the input $s(t)$. Based on this fundamental pair of relations, we may derive some important properties of matched filters, which should help you develop an intuitive grasp of how a matched filter operates.

PROPERTY 1

The spectrum of the output signal of a matched filter with the matched signal as input is, except for a time delay factor, proportional to the energy spectral density of the input signal.

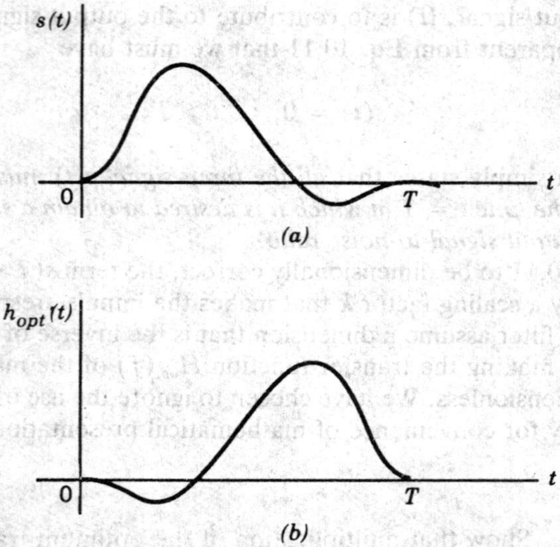

Figure 10.3
(a) Input signal. (b) Impulse response of matched filter.

Let $S_o(f)$ denote the Fourier transform of the filter output $s_o(t)$. Then,

$$\begin{aligned} S_o(f) &= H_{opt}(f)S(f) \\ &= S^*(f)S(f) \exp(-j2\pi f T) \\ &= |S(f)|^2 \exp(-j2\pi f T) \end{aligned} \tag{10.13}$$

This is the desired result, since $|S(f)|^2$ is the energy spectral density of the input signal $s(t)$.

EXAMPLE 1 MATCHED FILTER FOR A RECTANGULAR PULSE

Consider a rectangular pulse $s(t)$ of duration T and amplitude A, as in Fig. 10.4*a*:

$$s(t) = \begin{cases} A, & 0 \le t \le T \\ 0, & \text{otherwise} \end{cases} \tag{10.14}$$

For convenience of presentation, we assume that the pulse $s(t)$ has unit area; that is $AT = 1$. Then, the Fourier transform of $s(t)$ is

$$S(f) = \text{sinc}(fT) \exp(-j\pi f T)$$

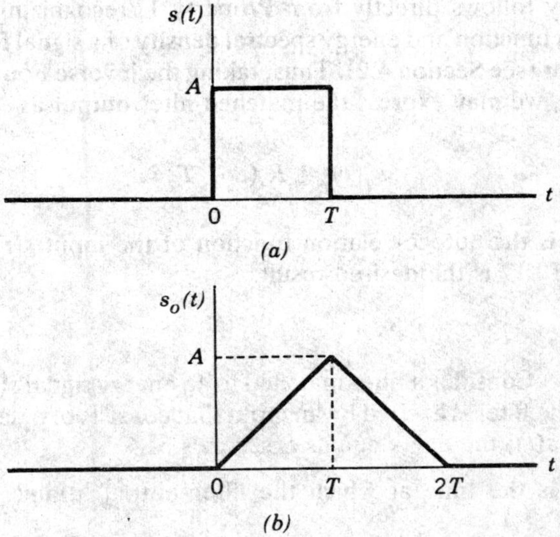

Figure 10.4
(a) Rectangular pulse input. (b) Matched filter output, assuming AT = 1.

The impulse response of a filter matched to the rectangular pulse $s(t)$ is also a rectangular pulse, as shown by

$$h_{opt}(t) = \begin{cases} A, & 0 \leq t \leq T \\ 0, & \text{otherwise} \end{cases} \tag{10.15}$$

The transfer function of this matched filter is (assuming $AT = 1$)

$$H_{opt}(f) = \text{sinc}(fT)\exp(-j\pi fT) \tag{10.16}$$

which, in this example, is the same as $S(f)$. The Fourier transform of the matched filter output is therefore

$$\begin{aligned} S_o(f) &= H_{opt}(f)S(f) \\ &= \text{sinc}^2(fT)\exp(-j2\pi fT) \end{aligned}$$

The factor $\text{sinc}^2(fT)$ is recognized as the energy spectral density of the rectangular pulse $s(t)$, assumed to be of unit area. Thus, $S_o(f)$ is in accord with Property 1.

PROPERTY 2

The output signal of a matched filter is proportional to a shifted version of the autocorrelation function of the input signal to which the filter is matched.

This property follows directly from Property 1, recognizing that the autocorrelation function and energy spectral density of a signal form a Fourier transform pair (see Section 4.2). Thus, taking the inverse Fourier transform of Eq. 10.13, we may express the matched-filter output as

$$s_o(t) = R_s(t - T) \tag{10.17}$$

where $R_s(\tau)$ is the autocorrelation function of the input $s(t)$ for time lag τ. Equation 10.17 is the desired result.

EXERCISE 2 Consider a filter matched to an energy signal $s(t)$ of duration T seconds. The filter is excited by an input that consists of a delayed version of the signal $s(t)$; the delay equals t_0 seconds.

(a) What is the time at which the filter output attains its maximum value?

(b) What is the maximum value of the filter output?

EXAMPLE 2 MATCHED FILTER FOR A RECTANGULAR PULSE (CONTINUED)

Consider again the matched filter for the rectangular pulse $s(t)$ of amplitude A and duration T, as shown in Fig. 10.4*a*. The rectangular pulse $s(t)$ is defined in Eq. 10.14, and the impulse response $h_{opt}(t)$ of the corresponding matched filter is defined in Eq. 10.15. Convolving $s(t)$ with $h_{opt}(t)$, we find that the matched filter output $s_o(t)$ has a triangular waveform. Specifically, for $AT = 1$ we have

$$s_o(t) = \begin{cases} \dfrac{At}{T}, & 0 < t \leq T \\ A\left(2 - \dfrac{t}{T}\right), & T \leq t < 2T \\ 0, & \text{otherwise} \end{cases}$$

This waveform is plotted in Fig. 10.4*b*, which is recognized as the autocorrelation function of the rectangular pulse $s(t)$, shifted by T seconds. Note that the matched filter output $s_o(t)$ attains its maximum value at time $t = T$, and that its duration is twice that of the input signal.

EXERCISE 3 Consider an RF pulse $s(t)$ of amplitude A, duration T, and frequency f_c, as shown in Fig. 10.5*a*. The frequency f_c is an integer multiple

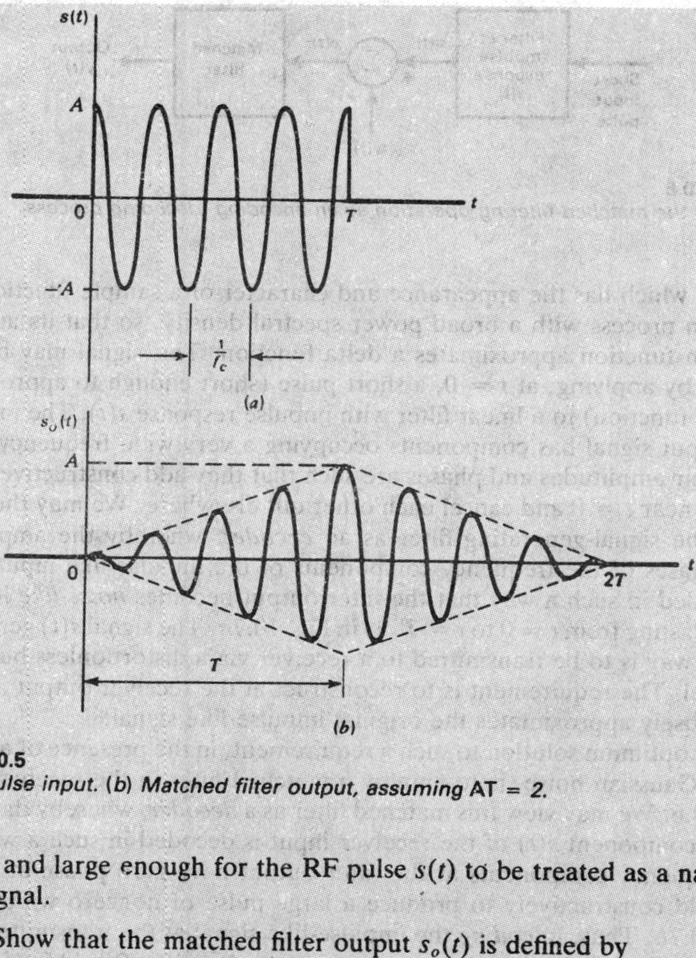

Figure 10.5
(a) RF pulse input. (b) Matched filter output, assuming AT = 2.

of $1/T$, and large enough for the RF pulse $s(t)$ to be treated as a narrow-band signal.

(a) Show that the matched filter output $s_o(t)$ is defined by

$$s_o(t) = \begin{cases} \dfrac{At}{T}\cos(2\pi f_c t), & 0 < t \leq T \\ A\left(2 - \dfrac{t}{T}\right)\cos(2\pi f_c t), & T \leq t < 2T \\ 0, & \text{otherwise} \end{cases}$$

where, for convenience, it is assumed that $AT = 2$.

(b) Verify that the matched filter output $s_o(t)$ has the waveform shown in Fig. 10.5*b*.

EXAMPLE 3 MATCHED FILTER PAIR

A possible exploitation of property 2 of a matched filter is illustrated in Fig. 10.6. Let us suppose that we have a signal $s(t)$, lasting from $t = 0$ to

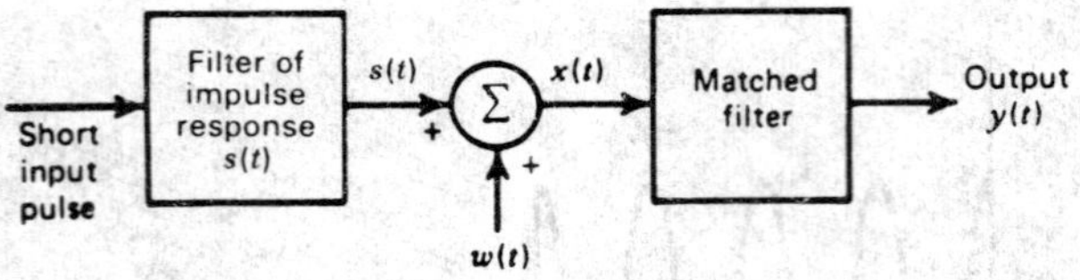

Figure 10.6
Viewing the matched filtering operation as an encoding–decoding process.

$t = T$, which has the appearance and character of a sample function of a random process with a broad power spectral density, so that its autocorrelation function approximates a delta function. This signal may be generated by applying, at $t = 0$, a short pulse (short enough to approximate a delta function) to a linear filter with impulse response $s(t)$. The impulse-like input signal has components occupying a very wide frequency band, but their amplitudes and phases are such that they add constructively only at and near $t = 0$ and cancel each other out elsewhere. We may therefore view the signal-generating filter as an *encoder,* whereby the amplitudes and phases of the frequency components of the impulse-like input signal are coded in such a way that the filter output becomes *noise-like* in character, lasting from $t = 0$ to $t = T$, as in Fig. 10.7*a*. The signal $s(t)$ generated in this way is to be transmitted to a receiver via a distortionless but noisy channel. The requirement is to reconstruct at the receiver output a signal that closely approximates the original impulse-like signal.

The optimum solution to such a requirement, in the presence of additive white Gaussian noise, is to employ a matched filter in the receiver, as in Fig. 10.6. We may view this matched filter as a *decoder,* whereby the useful signal component $s(t)$ of the receiver input is decoded in such a way that all frequency components at the filter output have zero phase at $t = T$, and add constructively to produce a large pulse of nonzero width, as in Fig. 10.7*b*. Thus, in *coding* the impulse-like signal at the transmitter input we have spread the signal energy out over a duration T, and in *decoding* the noise-like signal at the receiver input we are able to concentrate this energy into a relatively narrow pulse. The extent to which the receiver output $s_o(t)$ approximates the original impulse-like signal is simply a reflection of the extent to which the autocorrelation function of the transmitted signal $s(t)$ approximates a delta function. The signal generating and reconstruction filters in Fig. 10.6 are said to constitute a *matched-filter pair.*

The idea of a matched filter pair is basic to a secure communication technique known as *spread spectrum modulation.*[3] In this method of modulation, the *noise-like character of the transmitted signal is produced by having an information-bearing binary sequence modulate a bandwidth-*

[3]For an introductory discussion of *spread spectrum modulation,* see Haykin (1988), pp. 445–73.

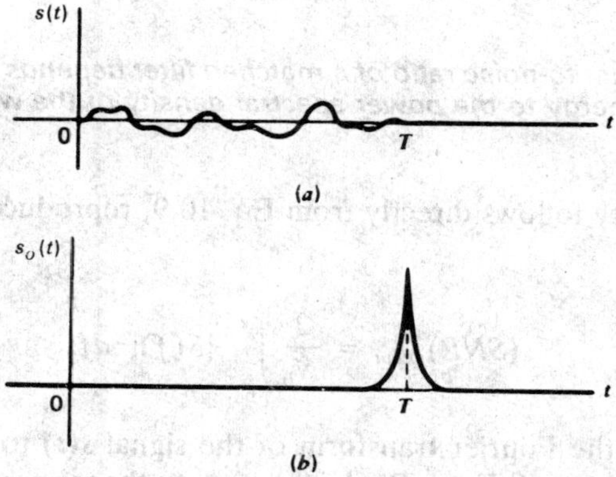

Figure 10.7
(a) Noise-like input signal. (b) Matched filter output.

spreading sequence that acts as a carrier, and the information-bearing sequence is recovered at the receiver by means of a filter matched to the spreading sequence employed in the transmitter. In a popular type of spread spectrum modulation, a *pseudonoise* (PN) sequence is used as the spreading sequence, and each block of pulses constituting a period of the PN sequence is multiplied in the transmitter by $+1$ or -1, depending on whether the particular binary symbol of the information-bearing sequence is a 1 or a 0. The receiver uses a filter matched to the PN sequence employed in the transmitter. From Chapter 8 we recall that the autocorrelation function of a PN sequence (also known as a maximal length sequence) consists of a periodic train of short triangular pulses that have the appearance of an impulse; see Fig. 8.21*a*. Hence, the matched filter output due to the information-bearing sequence consists of a periodic train of short triangular pulses, with the polarity of each pulse being determined by the identity of the corresponding binary symbol of the information-bearing sequence. On the other hand, an *interfering (jamming) signal,* unmatched to the PN sequence, is rejected by the matched filter receiver. The level of this rejection is determined by the ratio T_b/T_c, where T_b is the bit duration of the information-bearing sequence, and T_c is the duration of a basic pulse of the PN sequence; the ratio T_b/T_c, expressed in decibels, is called the *processing gain* of the system. Hence, by assigning a large value (on the order of 1000) to this ratio, a secure communication link is established between the transmitter and the receiver. Moreover, the 1's and 0's of the original information-bearing sequence are detected by sampling the matched filter output every T_b seconds: If the polarity of a sample under test is positive, a decision is made in favor of symbol 1; otherwise, a decision is made in favor of symbol 0.

PROPERTY 3

The output signal-to-noise ratio of a matched filter depends only on the ratio of the signal energy to the power spectral density of the white noise at the filter input.

This property follows directly from Eq. 10.9, reproduced here for convenience

$$(SNR)_{O,opt} = \frac{2}{N_0} \int_{-\infty}^{\infty} |S(f)|^2 \, df$$

where $S(f)$ is the Fourier transform of the signal $s(t)$ to which the filter of interest is matched. From Rayleigh's energy theorem, the signal energy E is given by

$$E = \int_{-\infty}^{\infty} s^2(t) \, dt = \int_{-\infty}^{\infty} |S(f)|^2 \, df$$

Accordingly, we may rewrite the expression for the output signal-to-noise ratio of the matched filter as

$$(SNR)_{O,opt} = \frac{2E}{N_0} \tag{10.18}$$

which is the desired result.

Equation 10.18 is perhaps the most important result in the evaluation of the performance of signal processing systems using matched filters. From Eq. 10.18 we see that dependence on the waveform of the input $s(t)$ has been completely removed by the matched filter. Accordingly, *in evaluating the ability of a matched-filter receiver to combat white Gaussian noise, we find that all signals that have the same energy are equally effective.* Note that the signal energy E is in joules and the noise spectral density $N_0/2$ is in watts per hertz, so that the ratio $2E/N_0$ is dimensionless; however, the two quantities have different physical meaning.

EXERCISE 4 Consider a rectangular pulse of amplitude A and duration T. Show that the output signal-to-noise ratio of a filter matched to this pulse is

$$(SNR)_O = \frac{2A^2T}{N_0} \tag{10.19}$$

PROPERTY 4

The matched-filtering operation may be separated into two matching conditions; namely, spectral phase matching that produces the desired output peak at time T, and spectral amplitude matching that maximizes the output signal-to-noise ratio at time t = T.

In polar form, the spectrum of the signal $s(t)$ being matched may be expressed as

$$S(f) = |S(f)| \exp[j\theta(f)]$$

where $|S(f)|$ is the amplitude spectrum and $\theta(f)$ is the phase spectrum of the signal. The filter is said to be *spectral phase matched* to the signal $s(t)$ if the transfer function of the filter is defined by[4]

$$H(f) = |H(f)| \exp[-j\theta(f) - j2\pi f T]$$

where $|H(f)|$ is real and nonnegative. The output of such a filter is

$$\begin{aligned} s_o'(t) &= \int_{-\infty}^{\infty} H(f)S(f) \exp(j2\pi ft)\, df \\ &= \int_{-\infty}^{\infty} |H(f)|\,|S(f)| \exp[j2\pi f(t - T)]\, df \end{aligned}$$

where the product $|H(f)|\,|S(f)|$ is real and nonnegative. The spectral phase matching ensures that all the spectral components of the output $s_o'(t)$ add constructively at time $t = T$, thereby causing the output to attain its maximum value, as shown by

$$s_o'(t) \leq s_o'(T) = \int_{-\infty}^{\infty} |S(f)|\,|H(f)|\, df$$

For *spectral amplitude matching,* we choose the amplitude response $|H(f)|$ of the filter to maximize the output signal-to-noise ratio at $t = T$ by using

$$|H(f)| = |S(f)|$$

and the standard matched filter is the result.

10.4 APPROXIMATIONS IN MATCHED FILTER DESIGN

In considering the design of a matched filter, we have to take account of two aspects of the problem—*physical realizability* and *practical feasibility.*

[4] Birdsall (1976).

For a matched filter operating in real time to be physically realizable, its impulse response must be zero for negative time. In Section 10.2 we showed that if the signal $s(t)$ to which the filter is to be matched lasts from $t = 0$ to $t = T$, then the physical realizability requirement is satisfied by introducing a finite delay, equal to T, in the impulse response of the filter. Then, of course, we must wait until time $t = T$ for the output signal component $s_o(t)$ of the matched filter to reach its peak value $s_o(T)$. In other words, we cannot expect the output signal component $s_o(t)$ to contain the full information about the input signal $s(t)$ until the signal has been fully received by the filter. Suppose, however, that the signal duration T is too large and we cannot afford to wait until time $t = T$ before extracting information about the signal $s(t)$. Then, in order to maximize the output signal-to-noise ratio at some instant $t = T'$, where $T' < T$, we should use the part of the optimum impulse response $h_{opt}(t)$ that extends from $t = 0$, to $t = T'$, and delete the remainder. The resulting output signal-to-noise ratio, measured at time $t = T'$, is still of the form of Eq. 10.18 except that now E must be interpreted not as the total signal energy, but rather as that part of the signal energy having been received by the filter at time $t = T'$. Obviously, in such a case, we are no longer dealing with a true matched filter, but rather an approximation to it, with the nature of the approximation determined by what fraction of the signal energy is received by time $t = T'$.

Another problem encountered in the construction of a matched filter is that it is often difficult to realize a filter with a transfer function exactly equal to the complex conjugate of the spectrum of the input signal $s(t)$. We may, then, have to apply some form of approximation to the optimum transfer function $H_{opt}(f)$ in order to arrive at a practical realization. Such an approximation results in some loss in performance compared with a true matched filter. This procedure is best illustrated by examples.

EXAMPLE 4 APPROXIMATIONS FOR A MATCHED FILTER FOR A RECTANGULAR PULSE

Consider again the rectangular pulse $s(t)$ of Fig. 10.4*a*. The pulse has amplitude A and duration T; let $AT = 1$ for convenience of presentation. In this example, we examine two different low-pass structures for approximating the matched filter for this rectangular pulse. The two structures are an ideal low-pass filter and an RC low-pass filter, which are considered in turn.

1. *Ideal low-pass filter with variable bandwidth:* The transfer function $H_{opt}(f)$ of the matched filter of interest is given in Eq. 10.16, which is reproduced here for convenience:

$$H_{opt}(f) = \text{sinc}(fT)\exp(-j\pi fT)$$

The amplitude response $|H_{opt}(f)|$ of the matched filter is plotted in Fig. 10.8*a*. We wish to approximate this amplitude response with an *ideal low-pass filter* of bandwidth B. The amplitude response of this approximating filter is shown in Fig. 10.8*b*. The requirement is to determine the particular value of bandwidth B that will provide the best approximation to the matched filter.

From Example 4 of Chapter 3, we recall that the maximum value of the output signal, produced by an ideal low-pass filter in response to the rectangular pulse of Fig. 10.4*a*, occurs at $t = T/2$ for $BT \leq 1$. This maximum value, expressed in terms of the sine integral, is equal to $(2A/\pi)\text{Si}(\pi BT)$. The average noise power at the output of the ideal low-pass filter is equal to BN_0. The maximum output signal-to-noise ratio of the ideal low-pass filter is therefore

$$(SNR)'_O = \frac{(2A/\pi)^2\,\text{Si}^2(\pi BT)}{BN_0} \tag{10.20}$$

Thus, using Eqs. 10.19 and 10.20, and assuming that $AT = 1$, we get

$$\frac{(SNR)'_O}{(SNR)_O} = \frac{2}{\pi^2 BT}\,\text{Si}^2(\pi BT)$$

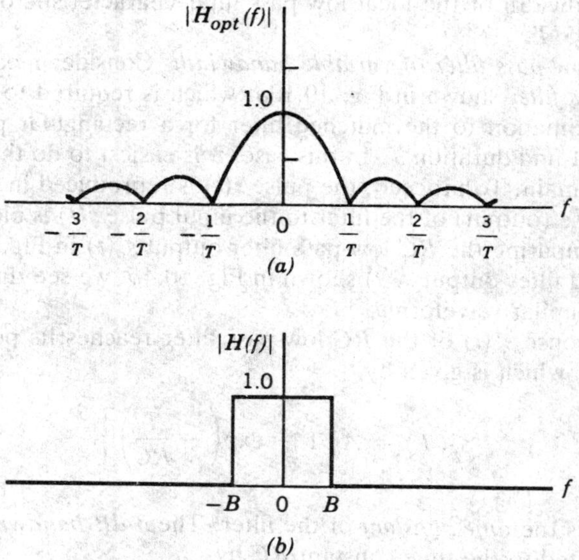

Figure 10.8
(a) Amplitude response of a filter matched to a rectangular pulse. (b) Amplitude response of an ideal low-pass filter approximating the matched filter.

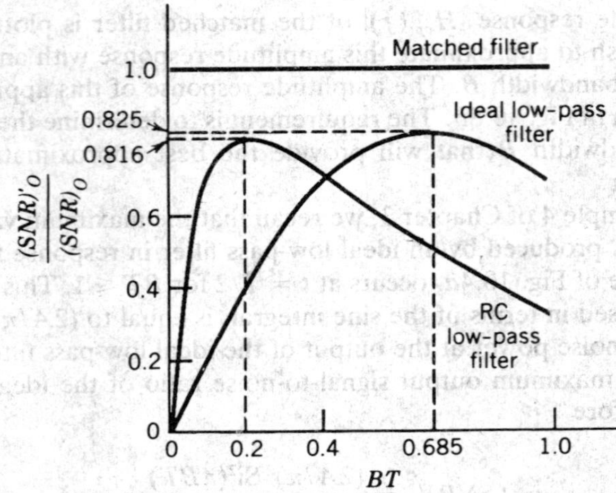

Figure 10.9
The effect of varying the time-bandwidth product BT *on the output signal-to-noise ratio of an ideal low-pass filter and that of RC low-pass filter.*

This ratio is plotted in Fig. 10.9 as a function of the time-bandwidth product BT. The peak value on this curve occurs for $BT = 0.685$, for which we find that the maximum signal-to-noise ratio of the ideal low-pass filter is 0.84 dB below that of the true matched filter. Therefore, the "best" value for the bandwidth of the ideal low-pass filter characteristic of Fig. 10.8*b* is $B = 0.685/T$.

2. *RC Low-pass filter of variable bandwidth:* Consider next the simple *RC low-pass filter* shown in Fig. 10.10*a*, which is required to provide the best approximation to the matched filter for a rectangular pulse $s(t)$ of amplitude A and duration T. In this case, it is easiest to do the analysis in the time domain. To proceed, the pulse $s(t)$ is reproduced in Fig. 10.10*b*. The response (output) of the filter to the input pulse $s(t)$ is plotted in Fig. 10.10*c*. Comparing the RC low-pass filter output $s_o'(t)$ in Fig. 10.10*c* with the matched filter output $s_o(t)$ shown in Fig. 10.4*b*, we see that they have somewhat similar waveforms.

The response $s_o'(t)$ of the RC low-pass filter reaches its peak value at time $t = T$, which is given by

$$s_o'(T) = A\left[1 - \exp\left(-\frac{T}{RC}\right)\right] \tag{10.21}$$

where RC is the *time constant* of the filter. The *3-dB bandwidth B* of the filter is related to the time constant RC by

$$B = \frac{1}{2\pi RC}$$

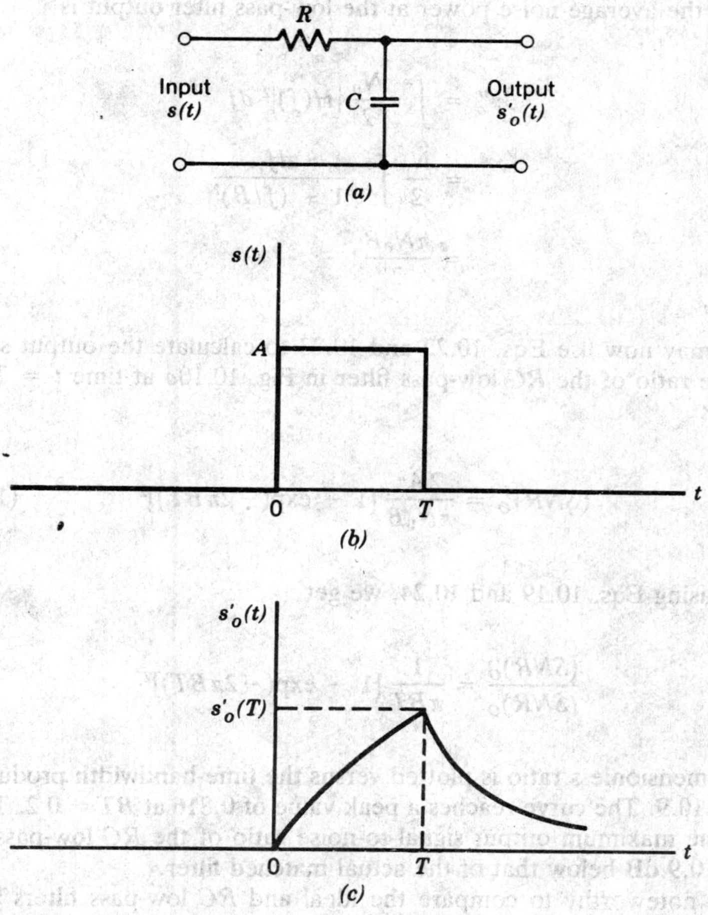

Figure 10.10
(a) RC low-pass filter. (b) Rectangular pulse input. (c) Response of the filter.

We may therefore rewrite Eq. 10.21 in terms of the bandwidth B as

$$s_o'(T) = A[1 - \exp(-2\pi BT)] \tag{10.22}$$

Our next task is to calculate the average power at the RC low-pass filter output produced in response to a white noise input of zero mean and power spectral density $N_0/2$. The transfer function of the filter is

$$H(f) = \frac{1}{1 + j2\pi fRC}$$

$$= \frac{1}{1 + (jf/B)}$$

Hence, the average noise power at the low-pass filter output is

$$\begin{aligned} \mathcal{N}'_o &= \int_{-\infty}^{\infty} \frac{N_0}{2} |H(f)|^2 \, df \\ &= \frac{N_0}{2} \int_{-\infty}^{\infty} \frac{df}{1 + (f/B)^2} \\ &= \frac{\pi N_0 B}{2} \end{aligned} \tag{10.23}$$

We may now use Eqs. 10.22 and 10.23 to calculate the output signal-to-noise ratio of the *RC* low-pass filter in Fig. 10.10*a* at time $t = T$; the result is

$$(SNR)'_O = \frac{2A^2}{\pi N_0 B} [1 - \exp(-2\pi BT)]^2 \tag{10.24}$$

Thus, using Eqs. 10.19 and 10.24, we get

$$\frac{(SNR)'_O}{(SNR)_O} = \frac{1}{\pi BT} [1 - \exp(-2\pi BT)]^2$$

This dimensionless ratio is plotted versus the time-bandwidth product BT in Fig. 10.9. The curve reaches a peak value of 0.816 at $BT \simeq 0.2$. Therefore, the maximum output signal-to-noise ratio of the *RC* low-pass filter is only 0.9 dB below that of the actual matched filter.

It is noteworthy to compare the ideal and *RC* low-pass filters as approximate realizations of the matched filter for a rectangular pulse. Despite its simplicity, the *RC* low-pass filter is worse than the ideal low-pass filter by only 0.06 dB; this degradation in performance is small enough to be ignored in practice. Accordingly, the *RC* low-pass filter is the preferred solution.

10.5 PROBABILISTIC APPROACH

The filter optimization criterion based on maximization of the output signal-to-noise ratio, described in Section 10.2, has the advantage of requiring knowledge of only the power spectral density of the noise $w(t)$ at the receiver input. Although such a criterion has a strong intuitive justification, nevertheless, we should prefer to use criteria directly related to probabilistic performance ratings of the system under study. For example, in a pulse-code modulation system with on–off signaling, symbol 1 is represented by the presence of a pulse $s(t)$, whereas symbol 0 is represented by the absence of the pulse. The presence of noise at the front end of the receiver causes two kinds of error to arise:

1. An error that occurs when symbol 0 is transmitted and the receiver decides in favor of symbol 1.
2. An error that occurs when symbol 1 is transmitted and the receiver decides in favor of symbol 0.

For a choice of criterion to optimize the performance of this system, we may wish to minimize the *average probability of error* involving both kinds of error. This brings us into the realm of classic *statistical hypothesis-testing procedures.*

LIKELIHOOD RATIO

In the simplest hypothesis-testing problem, the observed signal $x(t)$ is either due solely to white Gaussian noise $w(t)$ of zero mean and power spectral density $N_0/2$, which constitutes the *null hypothesis,* or due to both an exactly known signal $s(t)$ and noise $w(t)$, which constitutes the *alternative hypothesis.* Denoting the null hypothesis as H_0 and the alternative hypothesis as H_1, we may write:

$$\begin{aligned} H_0&: x(t) = w(t) \\ H_1&: x(t) = s(t) + w(t) \end{aligned} \tag{10.25}$$

The problem is to observe the received signal $x(t)$ over an interval from zero to T seconds and then decide whether H_0 or H_1 is true, according to some criterion.

To get a probabilistic description of the continuous received signal $x(t)$, we first assume that m amplitude samples of $x(t)$ are available, and then take the limit as m approaches infinity. At time t_k, we thus have

$$\begin{aligned} H_0&: x_k = w_k \\ H_1&: x_k = s_k + w_k \end{aligned} \tag{10.26}$$

where x_k, s_k, and w_k refer to sample values of $x(t)$, $s(t)$, and $w(t)$ at time t_k, respectively; the time index $k = 1, 2, \ldots, m$. We may then define an m-by-1 *observation vector* $\mathbf{x}$ that consists of the sample values $x_1, x_2, \ldots, x_m$, as shown by

$$\mathbf{x} = \begin{bmatrix} x_1 \\ x_2 \\ \vdots \\ x_m \end{bmatrix}$$

The vector **x** represents a single realization of the signal observed (measured) at the receiver input. Let the *random vector* **X** denote the ensemble of all such realizations; naturally, the randomness arises because of the additive white Gaussian noise at the receiver input.

Let $f_0(\mathbf{x})$ denote the conditional probability density function of the random vector **X** given that H_0 is true, and let $f_1(\mathbf{x})$ denote the conditional probability density function of **X** given that H_1 is true.[5] These two conditional probability density functions are basic to the probabilistic approach to receiver design.

In the *binary hypothesis-testing problem,* we know that either H_0 or H_1 is true. Thus, assuming that a choice has to be made each time the experiment is conducted, one of four things can happen:

1. H_0 is true: choose H_0.
2. H_0 is true: choose H_1.
3. H_1 is true: choose H_1.
4. H_1 is true: choose H_0.

It is apparent that alternatives (1) and (3) correspond to correct choices, whereas alternatives (2) and (4) correspond to errors. The purpose of a *decision rule* is to attach some relative importance to the four possible courses of action. To implement the decision rule, we divide the total observation space Z into two parts, Z_0 and Z_1. In particular, when an observation falls in Z_0 we choose hypothesis H_0, and when an observation falls in Z_1 we choose hypotheses H_1. Accordingly, we may identify two important probabilities:

1. The *conditional probability of correct reception,* defined as the m-fold integral

$$\int_{Z_i} f_i(\mathbf{x})\, d\mathbf{x}, \qquad i = 0, 1.$$

where the m-dimensional decision region Z_i corresponds to hypothesis H_i.

2. The *conditional probability of error,* defined as the m-fold integral

$$\int_{\bar{Z}_i} f_i(\mathbf{x})\, d\mathbf{x}, \qquad i = 0, 1.$$

[5]According to the notation described in Chapter 8, the conditional probability density function of the random vector **X**, given that hypothesis H_0 is true, is written as $f_{\mathbf{X}}(\mathbf{x}|H_0)$. In the material presented herein, the notation is simplified by denoting this conditional probability density function as $f_0(\mathbf{x})$. Similar remarks hold for $f_1(\mathbf{x})$.

where $\overline{Z}_i$ denotes "the not Z_i" decision region; that is,

$$\overline{Z}_i = \begin{cases} Z_1, & i = 0 \\ Z_0, & i = 1 \end{cases}$$

In a digital communication system, we are specifically interested in minimizing the *average probability of error*. Let p and q denote the *a priori probabilities of hypotheses* H_0 and H_1, respectively. These probabilities represent the observer's information about the source that generates the observation vector $\mathbf{x}$ before the experiment is conducted. Then, we may express the average probability of error as

$$P_e = p \int_{Z_1} f_0(\mathbf{x})\, d\mathbf{x} + q \int_{Z_0} f_1(\mathbf{x})\, d\mathbf{x} \tag{10.27}$$

On the right side of Eq. 10.27, the first integral represents the conditional probability of an *error of the first kind,* and the second integral represents the conditional probability of an *error of the second kind.* Since the total observation space $Z = Z_0 + Z_1$, we may rewrite Eq. 10.27 as

$$P_e = p \int_{Z - Z_0} f_0(\mathbf{x})\, d\mathbf{x} + q \int_{Z_0} f_1(\mathbf{z})\, d\mathbf{x} \tag{10.28}$$

We note, however, that the probability of an observation falling in the total observation space Z is equal to 1, because it is a certain event; that is,

$$\int_Z f_0(\mathbf{x})\, d\mathbf{x} = 1$$

Hence, we may simplify Eq. 10.28 as

$$P_e = p + \int_{Z_0} [q f_1(\mathbf{x}) - p f_0(\mathbf{x})]\, d\mathbf{x} \tag{10.29}$$

On the right side of Eq. 10.29 the first term is fixed whereas the integral represents the error probability controlled by those points $\mathbf{x}$ that we assign to Z_0. Therefore, all values of $\mathbf{x}$ for which $p f_0(\mathbf{x})$ is greater than $q f_1(\mathbf{x})$ should be assigned to Z_0 because they contribute a negative amount to the integral. Similarly, all values of $\mathbf{x}$ for which the reverse is true should be assigned to Z_1 (i.e., excluded from Z_0) because they would contribute a positive amount to the integral. Values of $\mathbf{x}$ where the two terms are equal have no effect on the average probability of error P_e and may be assigned arbitrarily. We will assume that such points are assigned to Z_0. We may thus define the decision regions as

If $qf_1(\mathbf{x})$ is greater than $pf_0(\mathbf{x})$,
assign $\mathbf{x}$ to Z_1 and accordingly choose hypothesis H_1.
Otherwise, assign $\mathbf{x}$ to Z_0 and choose hypothesis H_0. (10.30)

Equivalently, we may write

$$\frac{f_1(\mathbf{x})}{f_0(\mathbf{x})} \underset{H_0}{\overset{H_1}{\gtrless}} \frac{p}{q} \tag{10.31}$$

The quantity on the left side of Eq. 10.31 is called the *likelihood ratio.* Denoting this ratio by $\Lambda(\mathbf{x})$, we have

$$\Lambda(\mathbf{x}) = \frac{f_1(\mathbf{x})}{f_0(\mathbf{x})} \tag{10.32}$$

Note that since the likelihood ratio $\Lambda(\mathbf{x})$ is a ratio of two functions of a random variable, it is itself a random variable. However, regardless of the dimensionality of $\mathbf{x}$, the likelihood ratio $\Lambda(\mathbf{x})$ is a one-dimensional random variable. In terms of $\Lambda(\mathbf{x})$, we may thus rewrite Eq. 10.31 simply as

$$\Lambda(\mathbf{x}) \underset{H_0}{\overset{H_1}{\gtrless}} \frac{p}{q} \tag{10.33}$$

This test is called the *minimum probability of error criterion.*[6]

Since the natural logarithm is a monotonic function, and both sides of Eq. 10.33 are positive, it follows that an equivalent test is

$$\ln\Lambda(\mathbf{x}) \underset{H_0}{\overset{H_1}{\gtrless}} \ln\left(\frac{p}{q}\right) \tag{10.34}$$

[6]Equation 10.33 is a special case of the *Bayes' test:*

$$\Lambda(\mathbf{x}) \underset{H_0}{\overset{H_1}{\gtrless}} \eta$$

where η is called the *threshold* of the test. According to the Bayes' test, the threshold η is determined by two sets of factors: (a) the a priori probabilities p and q, and (b) the individual costs assigned to the four possible outcomes of the binary hypothesis testing problem. For a detailed treatment of Bayes' test and related issues, see the following references: van Trees (1968), Helstrom (1968), and Whalen (1971).

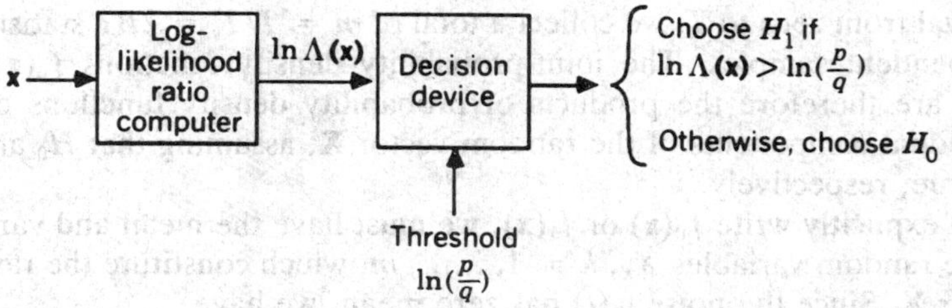

Figure 10.11
Likelihood ratio receiver.

We refer to $\ln \Lambda(\mathbf{x})$ as the *log-likelihood ratio*. When the two hypotheses H_0 and H_1 are equally likely (i.e., $p = q$), the decision level against which the log-likelihood ratio is compared is zero. This assumption is usually true in digital communications.

The optimum receiver based on Eq. 10.34 is known as the *likelihood ratio receiver*, shown in Fig. 10.11. We see that all the data processing required for the test is involved in computing the log-likelihood ratio $\ln \Lambda(\mathbf{x})$, based on the observation vector $\mathbf{x}$, and it is not affected by the a priori probabilities p and q. This invariance property of the likelihood ratio test is of considerable practical importance. The values of the a priori probabilities affect only the decision level. This means that we can construct a processor based only on the log-likelihood ratio and accommodate any subsequent changes in our estimates of the a priori probabilities, if ever required, by simply varying the decision level.

EXERCISE 5 Justify the statement that the likelihood ratio $\Lambda(\mathbf{x})$ and the ratio p/q are both positive.

CORRELATION RECEIVER

Let us momentarily assume that the noise is band-limited white Gaussian noise with power spectral density:

$$S_W(f) = \begin{cases} \dfrac{N_0}{2}, & |f| < B \\ 0, & |f| > B \end{cases} \tag{10.35}$$

If the signal $x(t)$ containing such a noise (as an additive component) is sampled with sampling interval $T_s = 1/2B$, the samples are uncorrelated, and being Gaussian, they are statistically independent. In the observation interval from zero to T, we collect a total of $m = T/T_s = 2BT$ statistically independent samples. The joint-probability density functions $f_0(\mathbf{x})$ and $f_1(\mathbf{x})$ are therefore the products of probability density functions of the individual components of the random vector $\mathbf{X}$, assuming that H_0 and H_1 are true, respectively.

To explicitly write $f_0(\mathbf{x})$ or $f_1(\mathbf{x})$, we must have the mean and variance of the random variables X_k, $k = 1, \ldots, m$, which constitute the random vector $\mathbf{X}$. Since the noise $w(t)$ has zero mean, we have

$$H_0\text{: mean of } X_k = 0$$

$$H_1\text{: mean of } X_k = s_k \tag{10.36}$$

where s_k is the value of the signal $s(t)$ at time t_k. The variance of X_k is the same under both H_0 and H_1, as shown by

$$\sigma^2(X_k) = \sigma^2(W_k) = \sigma^2 \tag{10.37}$$

where W_k is the random variable obtained by observing the band-limited white noise process $w(t)$ at time t_k. The variance σ^2 is simply that of the noise component:

$$\sigma^2 = N_0 B = \frac{N_0}{2T_s} \tag{10.38}$$

The equality $\sigma^2 = N_0 B$ follows from the fact that the average noise power (represented by σ^2) equals the total area under its power spectral density curve. We may therefore express the conditional probability density function $f_0(\mathbf{x})$ as

$$f_0(\mathbf{x}) = \prod_{k=1}^{m} f_0(x_k)$$

where

$$f_0(x_k) = \frac{1}{\sqrt{2\pi}\sigma} \exp\left(-\frac{x_k^2}{2\sigma^2}\right)$$

Hence, we have

$$f_0(\mathbf{x}) = \frac{1}{(2\pi\sigma^2)^{m/2}} \exp\left(-\sum_{k=1}^{m} \frac{x_k^2}{2\sigma^2}\right) \tag{10.39}$$

Similarly, we may write

$$f_1(\mathbf{x}) = \frac{1}{(2\pi\sigma^2)^{m/2}} \exp\left[-\sum_{k=1}^{m} \frac{(x_k - s_k)^2}{2\sigma^2}\right] \tag{10.40}$$

Substituting Eqs. 10.39 and 10.40 in 10.34, and simplifying, we get

$$\sum_{k=1}^{m} \frac{1}{2\sigma^2}(2s_k x_k - s_k^2) \underset{H_0}{\overset{H_1}{\gtrless}} \ln\left(\frac{p}{q}\right)$$

or, equivalently,

$$\sum_{k=1}^{m} \frac{s_k x_k}{\sigma^2} \underset{H_0}{\overset{H_1}{\gtrless}} \ln\left(\frac{p}{q}\right) + \frac{1}{2}\sum_{k=1}^{m} \frac{s_k^2}{\sigma^2} \tag{10.41}$$

From Eq. 10.38, the variance σ^2 of the noise is equal to $N_0/2T_s$. Therefore, substituting this value for σ^2 in Eq. 10.41, we get

$$\frac{2}{N_0}\sum_{k=1}^{m} s_k x_k T_s \underset{H_0}{\overset{H_1}{\gtrless}} \ln\left(\frac{p}{q}\right) + \frac{1}{N_0}\sum_{k=1}^{m} s_k^2 T_s \tag{10.42}$$

The decision rule in Eq. 10.42 is expressed in terms of m uniformly spaced samples of the received signal $x(t)$ and of the known signal $s(t)$. To obtain the corresponding decision rule in terms of the continuous functions $x(t)$ and $s(t)$, we allow the sampling interval T_s to approach zero and m (and therefore B) to approach infinity in such a way that the observation interval mT_s remains a constant, T. In the limit, the summations in Eq. 10.42 become integrals, yielding

$$\frac{2}{N_0}\int_0^T s(t)x(t)\,dt \underset{H_0}{\overset{H_1}{\gtrless}} \ln\left(\frac{p}{q}\right) + \frac{1}{N_0}\int_0^T s^2(t)\,dt \tag{10.43}$$

Equivalently, we may write

$$\int_0^T s(t)x(t)\,dt \underset{H_0}{\overset{H_1}{\gtrless}} \lambda \tag{10.44}$$

where λ is a new threshold defined by

$$\lambda = \frac{1}{2}N_0 \ln\left(\frac{p}{q}\right) + \frac{1}{2}\int_0^T s^2(t)\,dt \tag{10.45}$$

We also note that

$$E = \int_0^T s^2(t)\,dt$$

is the energy of the known signal $s(t)$. That is, the threshold λ in the test described by Eq. 10.44 depends on the a priori probabilities p and q, the noise spectral density N_0, and the signal energy E; see Eq. 10.45.

EXERCISE 6 Discuss the ways in which the threshold λ is affected by the following values of a priori probabilities:

(a) Symbols 0 and 1 occur with equal probability.
(b) Symbol 0 occurs twice as frequently as symbol 1.
(c) Symbol 1 occurs twice as frequently as symbol 0.

IMPLEMENTATION CONSIDERATIONS

The decision rule described by Eq. 10.44 may be implemented as shown in Fig. 10.12. This receiver is called a *correlation receiver.*[7] *It correlates the received signal $x(t)$ with a stored replica of the known signal $s(t)$. If the correlator output is larger than the predetermined threshold λ, we choose H_1; otherwise, we choose H_0.*

Consider the integral term on the left side of Eq. 10.43. Substituting the time difference $T - \tau$ for t in this integral, we have

$$\int_0^T s(t)x(t)\,dt = -\int_T^0 s(T-\tau)x(T-\tau)\,d\tau$$

$$= \int_0^T s(T-\tau)x(T-\tau)\,d\tau$$

However, from Eq. 10.10, we note that $s(T - t)$ is simply the impulse response $h_{opt}(t)$ of a linear time-invariant filter matched to the known signal $s(t)$. Therefore,

$$\int_0^T s(t)x(t)\,dt = \int_0^T h_{opt}(\tau)x(T-\tau)\,d\tau \qquad (10.46)$$

The term on the right side of Eq. 10.46 is the output of a matched filter of impulse response $h_{opt}(t)$ due to an input $x(t)$, evaluated at time T. *This means that the matched-filter receiver of Fig. 10.2 and the correlation receiver of Fig. 10.12 are equivalent.* That a criterion based on maximization of the output signal-to-noise ratio and a probabilistic criterion should lead to exactly the same result in the case of additive white Gaussian noise is no coincidence; indeed, it is testimony to the intimate connection between

[7]The derivation of the correlation receiver using a probabilistic criterion is historically credited to Woodward (1964).

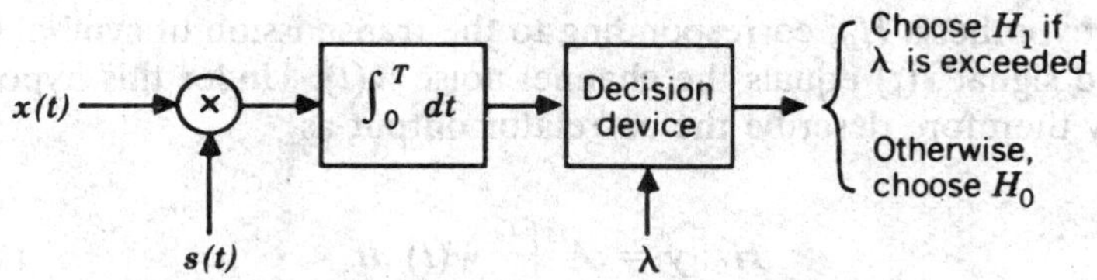

Figure 10.12
Correlation receiver.

the two types of criteria in this special case. Note, however, that the correlator output in Fig. 10.12 and the matched-filter output in Fig. 10.2 are equivalent only at time T.

10.6 PROBABILITY OF ERROR FOR BINARY PCM

As an application of the binary hypothesis-testing procedure described by Eq. 10.44, we consider the performance of a binary PCM system in the presence of channel noise; the receiver is depicted in Fig. 10.12. We do so by evaluating the average probability of error for such a system under the following assumptions:

1. The PCM system uses an on–off format, in which symbol 1 is represented by A volts and symbol 0 by zero volt.
2. The symbols 1 and 0 occur with equal probability.
3. The channel noise $w(t)$ is white and Gaussian with zero mean and power spectral density $N_0/2$.

To determine the average probability of error, we consider the two possible kinds of error separately. We begin by considering the first kind of error that occurs when symbol 0 is sent and the receiver chooses symbol 1. In this case, the probability of error is just the probability that the correlator output in Fig. 10.12 will exceed the threshold λ owing to the presence of noise, so the transmitted symbol 0 is mistaken for symbol 1. Since the a priori probabilities of symbols 1 and 0 are equal, we have $p = q$. Correspondingly, the expression for the threshold λ given in Eq. 10.45 simplifies as follows

$$\lambda = \frac{A^2 T_b}{2} \tag{10.47}$$

where T_b is the bit duration, and $A^2 T_b$ is the signal energy consumed in the transmission of symbol 1. Let y denote the correlator output:

$$y = \int_0^{T_b} s(t)x(t)\, dt \tag{10.48}$$

Under hypothesis H_0, corresponding to the transmission of symbol 0, the received signal $x(t)$ equals the channel noise $w(t)$. Under this hypothesis we may therefore describe the correlator output as

$$H_0: y = A \int_0^{T_b} w(t)\, dt \tag{10.49}$$

Since the white noise $w(t)$ has zero mean, the correlator output under hypothesis H_0 also has zero mean. In such a situation, we speak of a *conditional mean,* which (for the situation at hand) we describe by writing

$$m_0 = E[Y|H_0] = E\left[\int_0^{T_b} W(t)\, dt\right] = 0 \tag{10.50}$$

where the random variable Y represents the correlator output with y as its sample value and $W(t)$ is a white-noise process with $w(t)$ as its sample function. The subscript 0 in the conditional mean m_0 refers to the condition that hypothesis H_0 is true. Correspondingly, let σ_0^2 denote the *conditional variance* of the correlator output, given that hypothesis H_0 is true. We may therefore write

$$\begin{aligned}\sigma_0^2 &= E[Y^2|H_0] \\ &= E\left[\int_0^{T_b}\int_0^{T_b} W(t_1)W(t_2)\, dt_1\, dt_2\right]\end{aligned} \tag{10.51}$$

The double integration in Eq. 10.51 accounts for the squaring of the correlator output. Interchanging the order of integration and expectation in Eq. 10.51, we may write

$$\begin{aligned}\sigma_0^2 &= \int_0^{T_b}\int_0^{T_b} E[W(t_1)W(t_2)]\, dt_1\, dt_2 \\ &= \int_0^{T_b}\int_0^{T_b} R_W(t_1 - t_2)\, dt_1\, dt_2\end{aligned} \tag{10.52}$$

The parameter $R_W(t_1 - t_2)$ is the *ensemble-averaged autocorrelation function* of the white-noise process $W(t)$. From random process theory, it is recognized that the autocorrelation function and power spectral density of a random process form a Fourier transform pair. Since the white-noise process $W(t)$ is assumed to have a constant power spectral density of $N_0/2$, it follows that the autocorrelation function of such a process consists of a delta function weighted by $N_0/2$. Specifically, we may write

$$R_W(t_1 - t_2) = \frac{N_0}{2}\,\delta(\tau - t_1 + t_2) \tag{10.53}$$

Substituting Eq. 10.53 in 10.52, and using the property that the total area under the Dirac delta function $\delta(\tau - t_1 + t_2)$ is unity, we get

$$\sigma_0^2 = \frac{N_0 T_b A^2}{2} \tag{10.54}$$

The statistical characterization of the correlator output is completed by noting that it is Gaussian distributed, since the white noise at the correlator input is itself Gaussian (by assumption). In summary, we may state that under hypothesis H_0 the correlator output is a Gaussian random variable with zero mean and variance $N_0 T_b A^2/2$, as shown by

$$f_0(y) = \frac{1}{\sqrt{\pi N_0 T_b A}} \exp\left(-\frac{y^2}{N_0 T_b A^2}\right) \tag{10.55}$$

where the subscript in $f_0(y)$ signifies the condition that symbol 0 was sent.

Figure 10.13*a* shows the bell-shaped curve for the probability density function of the correlator output, given that symbol 0 was transmitted. The probability of the receiver deciding in favor of symbol 1 is given by the area shown shaded in Fig. 10.13*a*. The part of the *y*-axis covered by this

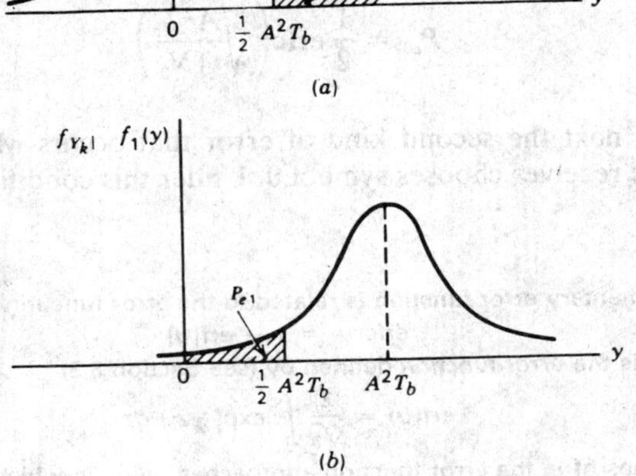

Figure 10.13
Conditional probability of error calculations. (a) Conditional probability of error, given that symbol was sent. (b) Conditional probability of error, given that symbol 1 was sent.

area corresponds to the condition that the correlator output y is in excess of the threshold λ defined by Eq. 10.47. Pet P_{e0} denote the *first conditional probability of error, given that symbol 0 was sent.* Hence, we may write

$$P_{e0} = \int_{\lambda}^{\infty} f_0(y)\, dy = \frac{1}{\sqrt{\pi N_0 T_b A}} \int_{A^2T_b/2}^{\infty} \exp\left(-\frac{y^2}{N_0 T_b A^2}\right) dy \tag{10.56}$$

Define

$$z = \frac{y}{\sqrt{N_0 T_b A}} \tag{10.57}$$

We may then rewrite Eq. 10.56 in terms of the new variable z as

$$P_{e0} = \frac{1}{\sqrt{\pi}} \int_{\sqrt{A^2T_b/4N_0}}^{\infty} \exp(-z^2)\, dz \tag{10.58}$$

Equation 10.58 can only be solved using numerical methods. A similar integral has been tabulated and is known as the *complementary error function*[8]

$$\text{erfc}(u) = \frac{2}{\sqrt{\pi}} \int_{u}^{\infty} \exp(-z^2)\, dz \tag{10.59}$$

Accordingly, we may redefine the conditional probability of error P_{e0} in terms of the complementary error function as

$$P_{e0} = \frac{1}{2}\, \text{erfc}\left(\sqrt{\frac{A^2 T_b}{4N_0}}\right) \tag{10.60}$$

Consider next the second kind of error that occurs when symbol 1 is sent and the receiver chooses symbol 0. Under this condition, correspond-

[8]The complementary error function is related to the error function as

$$\text{erfc}(u) = 1 - \text{erf}(u)$$

where erf(u) is the *error function* defined by (see Section 8.3)

$$\text{erf}(u) = \frac{2}{\sqrt{\pi}} \int_{0}^{u} \exp(-z^2)\, dz$$

For large values of u, the error function approaches unity, in which case it is numerically more convenient to work with the complementary error function. A short table of values of the error function erf(u) for u in the range 0 to 3.3 is given in Table 6 of Appendix D.

ing to hypothesis H_1, the correlator input consists of a rectangular pulse of amplitude A and duration T_b plus the channel noise $w(t)$. We may thus apply Eq. 10.48 to write

$$H_1: y = A \int_0^{T_b} [A + w(t)]\, dt \tag{10.61}$$

The fixed quantity A in the integrand of Eq. 10.61 serves to shift the correlator output from a mean value of zero volt under hypothesis H_0 to a mean value of A^2T_b under hypothesis H_1. However, the conditional variance of the correlator output under hypothesis H_1 has the same value as that under hypothesis H_0. Moreover, the correlator output is Gaussian distributed as before. In summary, the correlator output under hypothesis H_1 is a Gaussian random variable with mean A^2T_b and variance $N_0T_b^2/2$, as depicted in Fig. 10.13*b*. Let P_{e1} denote the *second conditional probability of error, given that symbol 1 was sent.* This probability equals the area shown shaded in Fig. 10.13*b*, which corresponds to those values of the correlator output less than the threshold λ set at $A^2T_b/2$. From the symmetric nature of the Gaussian density function, it is clear that

$$P_{e1} = P_{e0} \tag{10.62}$$

Note that this statement is only true when the a priori probabilities p and q are equal; this assumption was made in calculating the threshold λ.

AVERAGE PROBABILITY OF ERROR

To determine the average probability of error of the PCM receiver, we note that the two possible kinds of error just considered are mutually exclusive events. Thus, with the a priori probability of transmitting a 0 equal to p, and the a priori probability of transmitting a 1 equal to q, we find that the *average probability of error,* P_e, is given by

$$P_e = pP_{e0} + qP_{e1} \tag{10.63}$$

Since $P_{e1} = P_{e0}$, and $p + q = 1$, Eq. 10.63 simplifies as

$$P_e = P_{e0} = P_{e1}$$

or

$$P_e = \frac{1}{2}\,\text{erfc}\left(\frac{1}{2}\sqrt{\frac{A^2T_b}{N_0}}\right) \tag{10.64}$$

The term A^2T_b equals the *signal energy* when symbol 1 is sent. Let the dimensionless parameter η_1 denote the *signal energy-to-noise power spectral density ratio* under this condition, as defined by

$$\eta_1 = \frac{A^2 T_b}{N_0} \tag{10.65}$$

We may interpret the parameter η_1 in another way. We observe that A^2 denotes the *peak signal power*. The ratio N_0/T_b denotes the *average noise power measured in a bandwidth equal to the bit rate $1/T_b$*. The parameter η_1 is therefore the ratio of the peak signal power to the average noise power so defined.

We may thus express the average probability of error of the optimum binary PCM system using on–off signaling in terms of η_1 as

$$P_e = \frac{1}{2}\,\mathrm{erfc}\left(\frac{1}{2}\sqrt{\eta_1}\right) \tag{10.66}$$

This formula shows that the average probability of error P_e depends solely on the signal energy-to-noise power spectral density ratio η_1. Figure 10.14 shows P_e plotted versus η_1 in decibels. We see that the average probability of error P_e decreases very rapidly as the signal energy-to-noise power spectral density ratio η_1 is increased, so that eventually a very small increase in the signal energy will make the reception of binary data over a white Gaussian noise channel almost error free. Clearly, there is an *error thresh-*

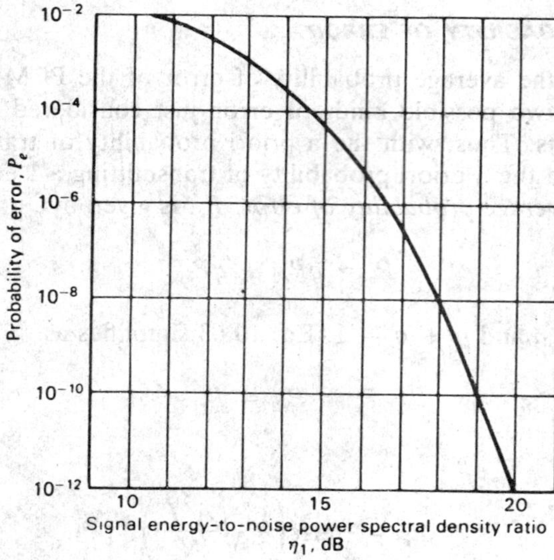

Figure 10.14
Probability of error in a PCM receiver.

old (at about 17 dB) below which the receiver performance may involve significant numbers of errors and above which the effect of channel noise is practically negligible. In other words, provided that the signal energy-to-noise power spectral density ratio η_1 exceeds the error threshold, channel noise has virtually no effect on the receiver performance, which is precisely the goal of PCM. When, however, η_1 drops below the error threshold, there is a sharp increase in the rate at which errors occur in the receiver. Because decision errors result in the construction of incorrect code words, we find that when the errors are frequent, the reconstructed message at the receiver output bears little resemblance to the original message. In such a situation, we say the message has become mutilated by *decoding noise.*

EXERCISE 7 Using a procedure similar to that described for deriving P_{e0}, show that the conditional probability of error of the second kind, P_{e1}, has the same value as that given by Eq. 10.60.

EXERCISE 8 Consider the *suboptimum* binary PCM receiver shown in Fig. 10.15. It involves the use of an ideal low-pass filter, followed by a sampler. The output of the sampler is compared to a threshold, and then a decision is made in favor of binary symbol 1 or 0. The transmitted PCM signal uses an on–off format with symbol 1 represented by A volts, and symbol 0 represented by zero volt. The symbols 1 and 0 occur with equal probability, thereby justifying the use of a threshold of $A/2$ volts.

(a) Show that the average probability of error of this receiver is given by

$$P_e = \frac{1}{2}\,\mathrm{erfc}\left(\frac{A}{2\sqrt{2}\sigma}\right)$$

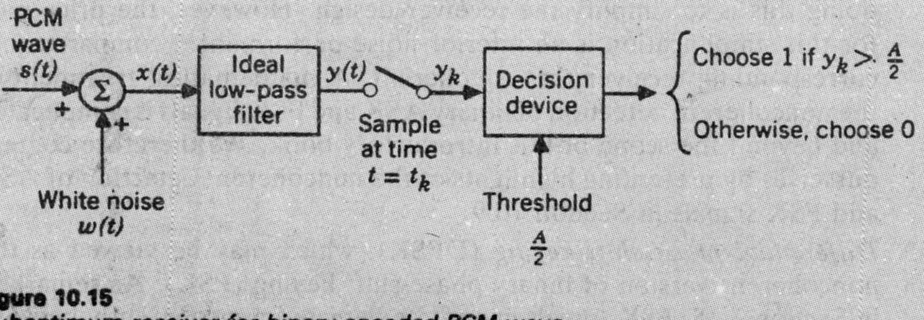

Figure 10.15
Suboptimum receiver for binary-encoded PCM wave.

where $\sigma^2 = N_0B$, with B equal to the low-pass filter bandwidth and $N_0/2$ equal to the power spectral density of the channel noise (assumed to be white and Gaussian with zero mean).
(b) Set the bandwidth B equal to the bit rate $1/T_b$. By how many decibels is the receiver of Fig. 10.15 inferior to the matched filter receiver of Fig. 10.2?
(c) Repeat your calculation in part b for the bandwidth B set at the optimum value $0.685/T_b$, where T_b is the bit duration (see Fig. 10.9).

10.7 *NOISE IN DIGITAL MODULATION SCHEMES*

The binary hypothesis-testing procedure described by Eq. 10.44 may also be applied to evaluate the noise performance of the various digital modulation schemes described in Section 7.15. In this context, we may identify the following detection scenarios in the presence of additive white Gaussian noise at the receiver input:

1. *Coherent detection of binary amplitude-shift keying* (ASK), *phase-shift keying* (PSK), and *frequency-shift keying* (FSK) *signals,* assuming that the receiver has perfect knowledge of the phase of the received signal: In other words, there is *phase synchronization* between the receiver and transmitter. The phase may be estimated from the received signal. For example, in the case of binary PSK we may achieve phase synchronization by using a *Costas loop* (see Section 7.2). Alternatively, provision for phase synchronization may be made by sending a *pilot carrier* at the cost of some wastage in transmitted power. The coherent detection of binary ASK, PSK, and FSK signals is considered in Section 10.8. The treatment of binary ASK signals follows directly from Eq. 10.44. For the treatment of binary PSK and FSK signals, we consider them as special cases of a generalized binary hypothesis-testing procedure that involves a pair of arbitrary signals with equal energy, which represent binary symbols 1 and 0.
2. *Noncoherent detection of binary* ASK *and* FSK *signals,* ignoring the phase information contained in the received signal: The motivation for doing this is to simplify the receiver design. However, the price paid for this simplification is an inferior noise performance, compared to a corresponding receiver that is coherent. A mathematical treatment of the noncoherent detection of binary ASK and FSK signals is complicated and beyond the scope of this introductory book. We therefore content ourselves by presenting highlights of the noncoherent detection of ASK and FSK signals in Section 10.9.
3. *Differential phase-shift keying* (DPSK), which may be viewed as the noncoherent version of binary phase-shift keying (PSK). As remarked in Section 7.15, PSK signals cannot be detected noncoherently because they use a single carrier frequency and have a constant envelope. Thus,

DPSK may be used as an alternative to PSK. This form of digital modulation is considered in Section 10.9.

4. *Coherent detection of quadriphase-shift keying* (QPSK) *and minimum shift keying* (MSK) *signals.* This issue is considered in Section 10.10.

In all these schemes, it is assumed that the receiver is properly *bit-timed,* so that the receiver may perform its decisions on received symbols in synchronism with the interbit transition points in the original binary data stream. Bit-timing information may be extracted from the received signal by the use of appropriate circuitry.[9]

The generation and coherent detection of M-ary ASK and M-ary PSK signals (for all M) involve only the use of linear operations. These digital modulation schemes are therefore said to be linear. Also, it is of interest to note that M-ary ASK and M-ary PSK signals are sometimes combined to produce hybrid *amplitude-phase keying* (APK); this is done in order to provide a more efficient use of channel bandwidth. A popular form of APK is M-ary *quadrature-amplitude modulation* (QAM), which consists of the quadrature multiplexing of two M-ary ASK signals. However, unlike M-ary PSK and M-ary FSK signals, we find that APK signals do not have a constant envelope and therefore require the use of a linear channel for their transmission.[10]

10.8 COHERENT DETECTION OF BINARY MODULATED WAVES

Let $s_0(t)$ and $s_1(t)$ denote the signals used to represent binary symbols 0 and 1, respectively. We may then distinguish between binary ASK, binary FSK, and binary PSK signals, as follows:

1. *Binary ASK signals*

$$\begin{aligned} s_1(t) &= A_c \cos(2\pi f_c t), && \text{symbol 1} \\ s_0(t) &= 0, && \text{symbol 0} \end{aligned} \tag{10.67}$$

2. *Binary PSK signals*

$$\begin{aligned} s_1(t) &= A_c \cos(2\pi f_c t), && \text{symbol 1} \\ s_0(t) &= A_c \cos(2\pi f_c t + \pi) && \text{symbol 0} \end{aligned} \tag{10.68}$$

3. *Binary FSK signals*

$$\begin{aligned} s_1(t) &= A_c \cos(2\pi f_1 t), && \text{symbol 1} \\ s_0(t) &= A_c \cos(2\pi f_2 t), && \text{symbol 0} \end{aligned} \tag{10.69}$$

[9] For a discussion of the synchronization problem in digital communications, see Lindsey and Simon (1973, Chapters 2 and 9).

[10] For a discussion of M-ary PSK and M-ary FSK, and M-ary QAM schemes, see Haykin (1988, pp. 313–38).

In both parts of Eqs. 10.67 through 10.69, we have $0 \leq t \leq T_b$, where T_b is the *bit duration.* We usually find that in the case of FSK signals, the frequencies f_1 and f_2 are both large compared with the bit rate $1/T_b$, whereas in the case of ASK and PSK signals, f_c is large compared with $1/T_b$. Moreover, in both PSK and FSK signals, the same *signal energy per bit* is transmitted, as shown by

$$\begin{aligned} E_b &= \int_0^{T_b} s_0^2(t)\, dt = \int_0^{T_b} s_1^2(t)\, dt \\ &= \frac{A_c^2 T_b}{2} \end{aligned} \tag{10.70}$$

In ASK signals, on the other hand, the transmitted signal energy alternates between 0 (when symbol 0 is sent) and the value $A_c^2 T_b/2$ (when symbol 1 is sent). In this case, we define the *average signal energy per bit* as

$$E_{av} = \frac{A_c^2 T_b}{4} \tag{10.71}$$

Throughout the discussion, we assume that symbols 0 and 1 are sent with equal probability; that is,

$$p = q = \frac{1}{2} \tag{10.72}$$

COHERENT DETECTION OF BINARY ASK SIGNALS

The receiver for coherent detection of binary ASK signals is shown in Fig. 10.16, which follows directly from the single-path correlation receiver of Fig. 10.12. Assuming that symbols 1 and 0 are equally likely, the threshold λ is calculated from Eq. 10.45 as

$$\begin{aligned} \lambda &= \frac{1}{2}\left(\frac{1}{2} A_c^2 T_b\right) \\ &= \frac{1}{4} A_c^2 T_b \end{aligned} \tag{10.73}$$

where we have made use of Eq. 10.70. To calculate the average probability of error for the coherent binary ASK receiver of Fig. 10.16, we may follow a procedure similar to that used for the binary (unipolar) PCM receiver in Section 10.6. A more expedient approach, however, is to recognize that the operation of a matched filter (correlation) receiver in additive white Gaussian noise depends only on the ratio of signal energy-to-noise power

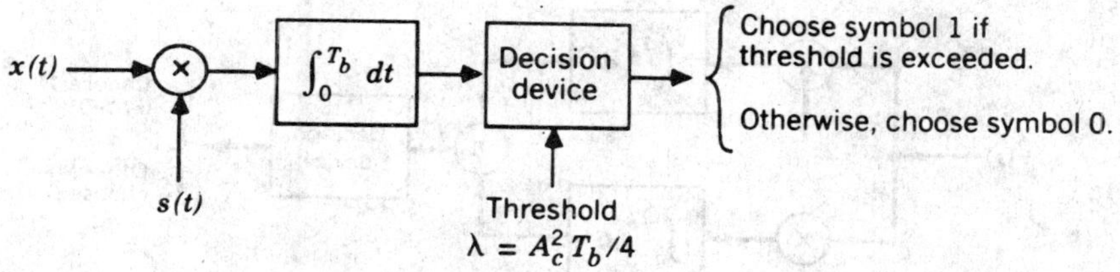

Figure 10.16
Coherent receiver for the detection of binary ASK signals.

spectral density and not on the signal waveform. (See Property 3 of matched filters in Section 10.3.) Accordingly, we may calculate the average probability of error for coherent binary ASK by substituting $A_c^2T_b/2$ (signal energy for symbol 1 in binary ASK) for A^2T_b (signal energy for symbol 1 in binary PCM) in Eq. 10.64. We may thus express the average probability of error in coherent binary ASK as

$$P_e = \frac{1}{2}\,\mathrm{erfc}\left(\frac{1}{2}\sqrt{\frac{A_c^2T_b}{2N_0}}\right) \tag{10.74}$$

Using the definition of average signal energy per bit given in Eq. 10.71, we may rewrite the formula of Eq. 10.74 as

$$P_e = \frac{1}{2}\,\mathrm{erfc}\left(\sqrt{\frac{E_{av}}{2N_0}}\right) \tag{10.75}$$

GENERALIZED COHERENT RECEIVER FOR BINARY DECISION-MAKING

The optimum receiver for the coherent detection of binary FSK and PSK signals may be viewed as special cases of the two-path *coherent correlation receiver* shown in Fig. 10.17. This receiver represents a generalization of the single-path correlation receiver of Fig. 10.12. We assume that the receiver of Fig. 10.17 is synchronized to the transmitter, which is equivalent to saying that (1) the receiver is equipped with replicas of the transmitted signals $s_0(t)$ and $s_1(t)$, and (2) the timing of the decision-making process performed by the receiver is coincident with the bit timing of the transmitted signal.

The receiver output l in Fig. 10.17 is given by

$$l = \int_0^{T_b} x(t)[s_1(t) - s_0(t)]\,dt \tag{10.76}$$

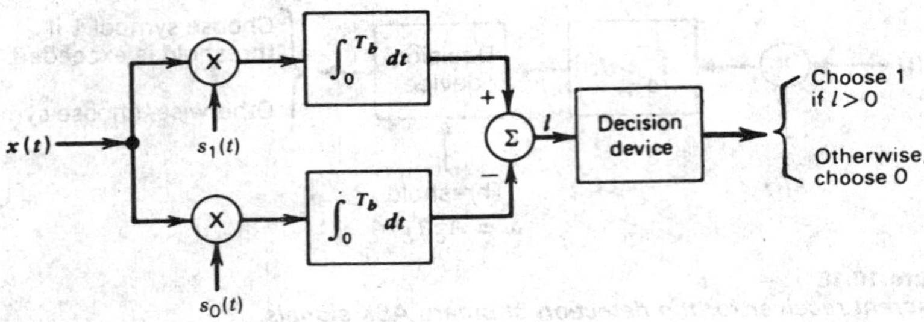

Figure 10.17
Generalized two-path correlation receiver.

where $x(t)$ is the noise-contaminated received signal. The output l is compared with a threshold of zero volt; this threshold is chosen assuming that symbols 1 and 0 occur with equal probability. If l is greater than zero, the receiver chooses symbol 1; otherwise, it chooses symbol 0. Since the channel noise $w(t)$ is Gaussian, it follows that the receiver output is likewise Gaussian distributed. The mean value of the receiver output is conditional on whether symbol 0 or 1 was actually sent. However, the variance of the correlator output is the same, regardless of whether symbol 0 or 1 was sent.

Following a procedure similar to that described in Section 10.6, we may show that the conditional mean of the correlator output, given that symbol 0 was sent, is defined by

$$\begin{aligned} m_0 &= \int_0^{T_b} s_0(t)[s_1(t) - s_0(t)]\, dt \\ &= -E_b(1-\rho) \end{aligned} \tag{10.77}$$

where E_b is the signal energy per bit. The parameter ρ is the *correlation coefficient* of the signals $s_0(t)$ and $s_1(t)$, defined by

$$\begin{aligned} \rho &= \frac{\int_0^{T_b} s_0(t)s_1(t)\, dt}{\left[\int_0^{T_b} s_0^2(t)\, dt\right]^{1/2}\left[\int_0^{T_b} s_1^2(t)\, dt\right]^{1/2}} \\ &= \frac{1}{E_b}\int_0^{T_b} s_0(t)s_1(t)\, dt \end{aligned} \tag{10.78}$$

The correlation coefficient ρ has an absolute value less than or equal to unity. The conditional mean of the correlator output, given that symbol 1 was sent, is defined by

$$m_1 = E_b(1 - \rho) \tag{10.79}$$

We may also show that the conditional variance σ_0^2 of the correlator output given that symbol 0 was sent, and its conditional variance σ_1^2 given that symbol 1 was sent, have a common value defined by

$$\begin{aligned} \sigma_0^2 &= \sigma_1^2 \\ &= \frac{N_0}{2} \int_0^{T_b} [s_1(t) - s_0(t)]^2 \, dt \\ &= N_0 E_b(1 - \rho) \end{aligned} \tag{10.80}$$

An error of the first kind occurs when we send symbol 0, but the correlator output l is greater than zero volt, and the receiver therefore chooses symbol 1. An error of the second kind occurs when we send symbol 1, but the correlator output l is less than zero volt, and the receiver therefore chooses symbol 0. From the symmetry of the receiver of Fig. 10.17, it is apparent that the (conditional) probabilities of both kinds of error are equal. Thus, recognizing that the correlator output is Gaussian distributed with conditional means $\pm E_b(1 - \rho)$ and variance $N_0E_b(1 - \rho)$, and assuming that symbols 0 and 1 occur with equal probability (which justifies the use of a threshold equal to zero volt), we find that the average probability of error in the receiver of Fig. 10.17 is given by

$$P_e = \frac{1}{2} \operatorname{erfc}\left(\sqrt{\frac{E_b(1 - \rho)}{2N_0}}\right) \tag{10.81}$$

We may now consider the following two special cases:

1. *Coherent detection of binary PSK signals.* In the case of binary PSK signals, the coherent receiver reduces to a single path as in Fig. 10.18*a*. This follows from the fact that $s_0(t)$ is the negative of $s_1(t)$. Moreover, the correlation coefficient $\rho = -1$. A pair of equienergy signals for which the correlation coefficient equals -1 are called *antipodal signals.* Thus, putting $\rho = -1$ in Eq. 10.81 gives the average probability of symbol error in a coherent binary PSK receiver as

$$P_e = \frac{1}{2} \operatorname{erfc}\left(\sqrt{\frac{E_b}{N_0}}\right) \tag{10.82}$$

2. *Coherent detection of binary FSK signals.* The coherent receiver for binary FSK signals is shown in Fig. 10.18*b*. The frequencies f_2 and f_1 of the FSK signal are usually spaced far enough apart to justify treating the signals $s_0(t)$ and $s_1(t)$ as *orthogonal* with each other. This condition

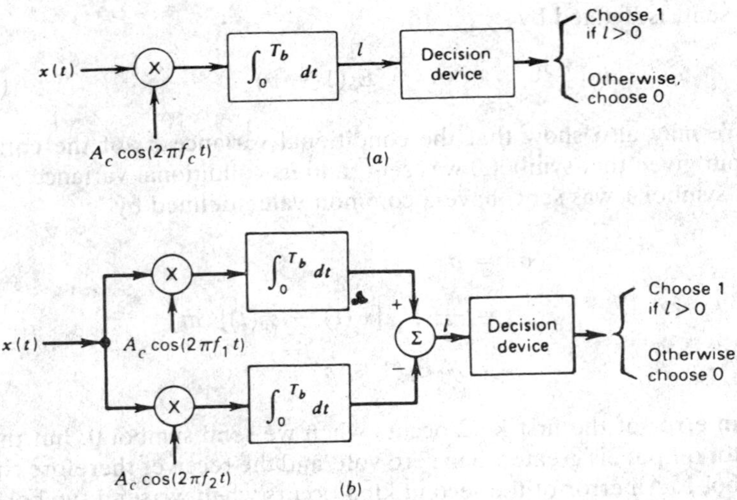

Figure 10.18
(a) Coherent receiver for PSK signals. (b) Coherent receiver for FSK signals.

corresponds to having the correlation parameter $\rho = 0$. Therefore, putting $\rho = 0$ in Eq. 10.81, we find that the average probability of error in a coherent binary FSK receiver is given by

$$P_e = \frac{1}{2} \operatorname{erfc}\left(\sqrt{\frac{E_b}{2N_0}}\right) \tag{10.83}$$

Comparing Eqs. 10.75 and 10.83, we see that the coherent receivers for binary ASK and binary FSK signals exhibit the same average probability of error when the average signal energy per bit, E_{av}, in binary ASK is the same as the signal energy per bit, E_b, in binary FSK.

EXERCISE 9 For the generalized correlation binary receiver shown in Fig. 10.17, show that the conditional means m_0 and m_1, and the common value of the conditional variances σ_0^2 and σ_1^2 are given by Eqs. 10.77, 10.79, and 10.80, respectively.

EXERCISE 10 Using the formula for the Gaussian probability density function, derive Eq. 10.81 for the average probability of error in the generalized receiver of Fig. 10.17.

10.9 NONCOHERENT DETECTION OF BINARY MODULATED WAVES

Up to this point in our discussion, we have assumed that the information-bearing signal is completely known at the receiver. In practice, however, it is often found that in addition to the uncertainty due to the channel noise, there is an additional uncertainty due to the randomness of signal parameters. The usual cause of this uncertainty is distortion in the transmission medium. Perhaps the most common random signal parameter is the phase, which is especially true for narrow-band signals. For example, transmission over a multiplicity of paths of different and variable lengths, or rapidly varying delays in the propagating medium from transmitter to receiver, may cause the phase of the received signal to change in a way that the receiver cannot follow. Synchronization with the phase of the transmitted carrier may then be too costly, and the designer may simply choose to disregard the phase information in the received signal at the expense of some degradation in the noise performance of the system.

In this section, we first consider the noncoherent detection of binary FSK signals and then address differential phase-shift keying (DPSK).

NONCOHERENT DETECTION OF BINARY FSK SIGNALS

In this case, the receiver is composed of a pair of matched filters followed by envelope detectors, as in Fig. 10.19. One of the two filters is matched to the signal $s_0(t) = A_c \cos(2\pi f_2 t)$ corresponding to the transmission of symbol 0, and the other filter is matched to the signal $s_1(t) = A_c \cos(2\pi f_1 t)$ corresponding to the transmission of symbol 1. The envelope detectors serve the purpose of destroying the dependence of the matched filter outputs on the unknown phase of the received signal. The resulting envelopes are sampled once every T_b seconds. Let l_0 and l_1 denote the envelope samples of the lower and upper paths of the receiver, respectively. Then, if $l_1 > l_0$, the receiver chooses symbol 1; otherwise, it chooses symbol 0.

The receiver commits an error of the first kind when symbol 0 is transmitted but the presence of channel noise makes l_1 greater than l_0 and the receiver therefore chooses symbol 1. It commits an error of the second kind when symbol 1 is transmitted but owing to channel noise l_0 is greater than l_1 and the receiver therefore chooses symbol 0. The inclusion of envelope detectors in the receiver of Fig. 10.19 makes the evaluation of these two conditional probabilities of error rather complicated. The reason for the complication is that envelope detection is a nonlinear operation, with the result that the random variables obtained by sampling the envelope detector outputs are no longer Gaussian distributed. For the present discussion, we simply state the formula for the average probability of error P_e in the noncoherent FSK receiver of Fig. 10.19, assuming that symbols

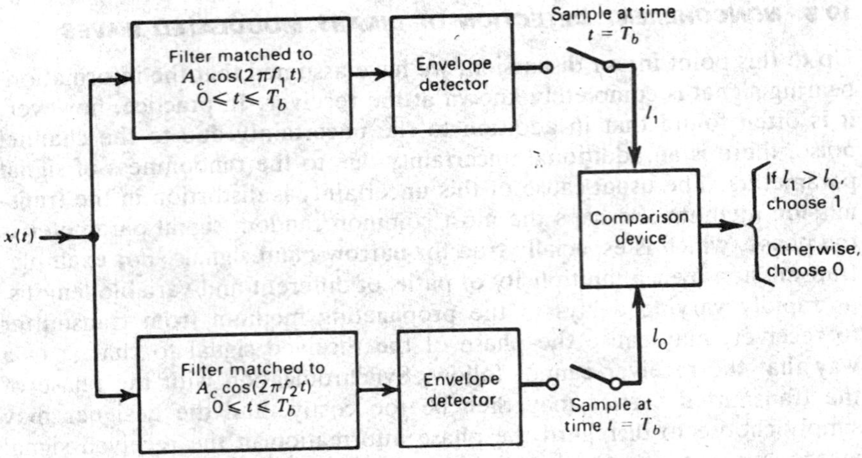

Figure 10.19
Noncoherent receiver for the detection of FSK signals.

1 and 0 occur with equal probability. The formula for P_e is[11]

$$P_e = \frac{1}{2}\exp\left(-\frac{E_b}{2N_0}\right) \tag{10.84}$$

where E_b is the transmitted FSK signal energy per bit and $N_0/2$ is the power spectral density of the channel noise (assumed to be white and zero-mean Gaussian).

We also note that when the signal-to-noise ratio is high, the average probability of error for the noncoherent detection of binary ASK signals is the same as that for the noncoherent detection of binary FSK signals, provided that the average signal energy per bit in binary ASK is the same as the signal energy per bit in binary FSK.

DIFFERENTIAL PHASE-SHIFT KEYING

The method of differential phase-shift keying (DPSK) may be viewed as the "noncoherent" version of phase-shift keying. It operates on the assumption that the unknown phase θ of the received signal remains essentially constant over two-bit intervals. In the context of noise performance, the major difference between a DPSK system and a coherent binary PSK system is not in the differential encoding, which can be used in any case, but rather it lies in the way in which the reference signal is derived for the phase detection of the received signal. Specifically, in a DPSK receiver the

[11] For a derivation of Eq. 10.84, see Haykin (1988, pp. 300–307).

reference is contaminated by additive noise to the same extent as the information pulse; that is, they have the same signal-to-noise ratio. This makes the determination of the overall probability of error in DPSK receivers somewhat complicated. Therefore, it will not be given here. However, the result is[12]

$$P_e = \frac{1}{2}\exp\left(-\frac{E_b}{N_0}\right) \tag{10.85}$$

It is of interest to note that, since in a DPSK receiver decisions are made on the basis of the signal received in two successive bit intervals, there is a tendency for bit errors to occur in pairs.

10.10 COHERENT DETECTION OF QUATERNARY MODULATED WAVES

A limitation of binary modulated waves is that they do not make the most efficient use of *channel bandwidth*, which represents a precious communication resource. One way of improving bandwidth utilization is to use *quadrature multiplexing*. Two important examples of such an approach are *quadriphase-shift keying* (QPSK) *and minimum shift keying*, (MSK), which were considered in Section 7.15. In QPSK, a special form of phase modulation, the carrier assumes one of four equispaced phase shifts (e.g., $\pm\pi/4$, $\pm 3\pi/4$) in response to one of the four possible (Gray encoded) dibits 00, 10, 11, and 01. In MSK, a special form of frequency modulation, phase continuity is maintained at the interbit transition points of the incoming binary data stream, and the change in carrier frequency from symbol 1 to symbol 0 is chosen to be equal to one half the bit rate of the binary data.

Although QPSK and MSK have different waveforms and employ different methods for their generation and detection, they exhibit the same performance when they are coherently detected in the presence of additive white Gaussian noise at the receiver input. In the sequel, we present a derivation for the average probability of symbol error for QPSK, which represents an extension of the result obtained previously for the coherent detection of binary PSK signals.

In Fig. 10.20 we show a coherent receiver for the detection of QPSK. We assume that the receiver (channel) noise is zero-mean white Gaussian with power spectral density $N_0/2$. We also assume that all four Gray-encoded dibits 00, 10, 11, and 01 occur with equal probability. We note that this receiver may be viewed as a quadrature-multiplexed version of two coherent binary PSK receivers; one receiver operates with the carrier $\cos(2\pi f_c t)$, and the other receiver operates with the carrier $\sin(2\pi f_c t)$. We may therefore equate the probability of error P_{eI} for the in-phase channel

[12]For a derivation of Eq. 10.85, see Haykin (1988), pp. 307–309.

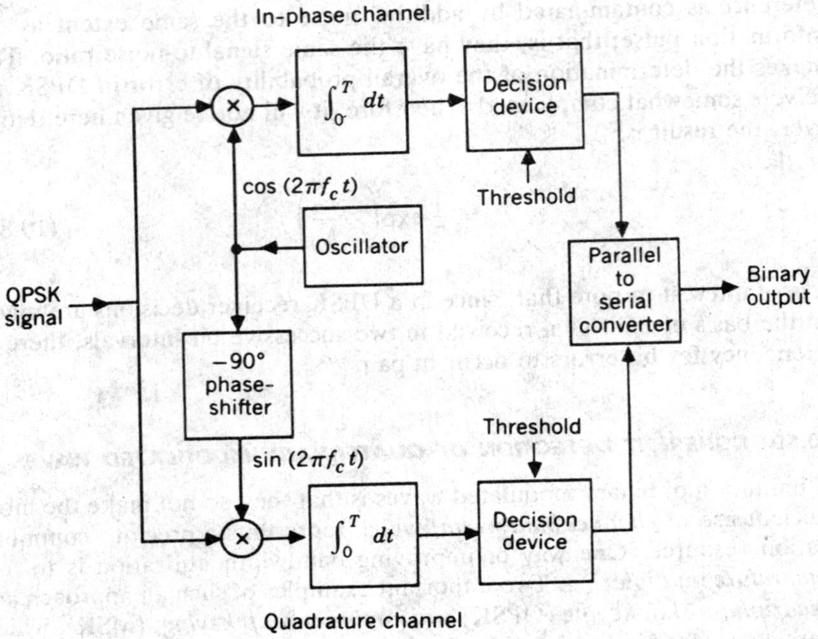

Figure 10.20
QPSK receiver.

and the probability of error P_{eQ} for the quadrature channel in Fig. 10.20 to that for the coherent detection of binary PSK signals. In particular, we may write

$$P_{eI} = P_{eQ} = \frac{1}{2} \operatorname{erfc}\left(\sqrt{\frac{E}{2N_0}}\right) \tag{10.86}$$

where E is the transmitted *signal energy per symbol.* The probability that the QPSK receiver will *correctly* identify the transmitted data sequence is equal to the probabilities that both correlators in the in-phase and quadrature paths of the receiver yield correct results. Let P_c denote the *average probability of correct reception.* We may then write

$$P_c = (1 - P_{eI})(1 - P_{eQ}) \tag{10.87}$$

We may simplify Eq. 10.87 by noting that $P_{eI} = P_{eQ}$ and that they both usually have a small value compared to unity. Accordingly, we may approximate Eq. 10.87 as

$$P_c \simeq 1 - 2P_{eI} \tag{10.88}$$

The *average probability of symbol error* in the QPSK receiver of Fig. 10.20 is therefore given by

$$\begin{aligned} P_e &= 1 - P_c \\ &\simeq 2P_{el} \\ &= \operatorname{erfc}\left(\sqrt{\frac{E}{2N_0}}\right) \end{aligned} \tag{10.89}$$

In a QPSK system, there are two bits per symbol, so that the signal energy per symbol is twice the signal energy per bit; that is,

$$E = 2E_b \tag{10.90}$$

Thus, expressing the average probability of symbol error in terms of the ratio E_b/N_0, we may write

$$P_e \simeq \operatorname{erfc}\left(\sqrt{\frac{E_b}{N_0}}\right) \tag{10.91}$$

As mentioned previously, MSK has the same noise performance as QPSK when they are both detected coherently in the presence of additive white Gaussian noise. Accordingly, we may also use Eq. 10.91 to calculate the average probability of symbol error in a coherent MSK receiver.

EXERCISE 11 Explain the reason for the use of $2N_0$ (for the effect of noise) in the formula for the average probability of error in the in-phase or quadrature channel given in Eq. 10.86.

10.11 DISCUSSION

Throughout this chapter we have used the overall probability of committing a symbol error as the figure of merit for evaluating the noise performance of a digital communication system. It should be realized, however, that even if two systems yield the same symbol error probability, their performances, from the users' viewpoint, may be quite different. In particular, the greater the number of bits per symbol, the more the bit errors will cluster together. For example, if the symbol error probability is 10^{-3}, the expected number of symbols occurring between any two erroneous symbols is 1000. If each symbol represents 1 bit of information (as in a binary PSK or binary FSK system), the expected number of bits separating two erroneous bits is 1000. If, on the other hand, there are 2 bits per symbol (as in a QPSK system), the expected separation is 2000 bits. Of course, a symbol error generally creates more bit errors in the second case, so that the percentage of bit errors tends to be the same. Nevertheless, this clustering effect may make one system more attractive than another, even at the same symbol error rate. In the final analysis, which system is preferable will depend on the particular situation.

Two systems having an unequal number of symbols may be compared in a meaningful way only if they use the same amount of energy to transmit each bit of information. It is the total amount of energy needed to transmit the complete message that represents the cost of the transmission, not the amount of energy needed to transmit a particular symbol satisfactorily. Accordingly, in comparing the different data transmission systems considered in this chapter, we will use, as the basis of our comparison, the probability of symbol error expressed as a function of the *signal energy per bit-to-average noise power per unit bandwidth ratio;* that is E_b/N_0.

In Table 10.1, we have summarized the expressions for the symbol error probability P_e for coherent binary PSK, coherent binary FSK, noncoherent binary FSK, DPSK, and coherent QPSK and MSK. In Fig. 10.21 we have used these expressions to plot P_e as a function of E_b/N_0. In practice, we generally design a digital communication system for an average probability of symbol error P_e equal to 10^{-4} or less. On the basis of the curves in Fig. 10.21, we may state the following:

1. The error rates for all the systems decrease monotonically with increasing values of E_b/N_0.

TABLE 10.1 Summary of Formulas for the Symbol Error Probability P_e for Different Digital Modulation Techniques

	P_e
Coherent binary PSK	$\frac{1}{2}\,\mathrm{erfc}\left(\sqrt{\frac{E_b}{N_0}}\right)$
Coherent binary FSK	$\frac{1}{2}\,\mathrm{erfc}\left(\sqrt{\frac{E_b}{2N_0}}\right)$
Coherent QPSK Coherent MSK	$\mathrm{erfc}\left(\sqrt{\frac{E_b}{N_0}}\right)$
Noncoherent binary FSK	$\frac{1}{2}\exp\left(-\frac{E_b}{2N_0}\right)$
DPSK	$\frac{1}{2}\exp\left(-\frac{E_b}{N_0}\right)$

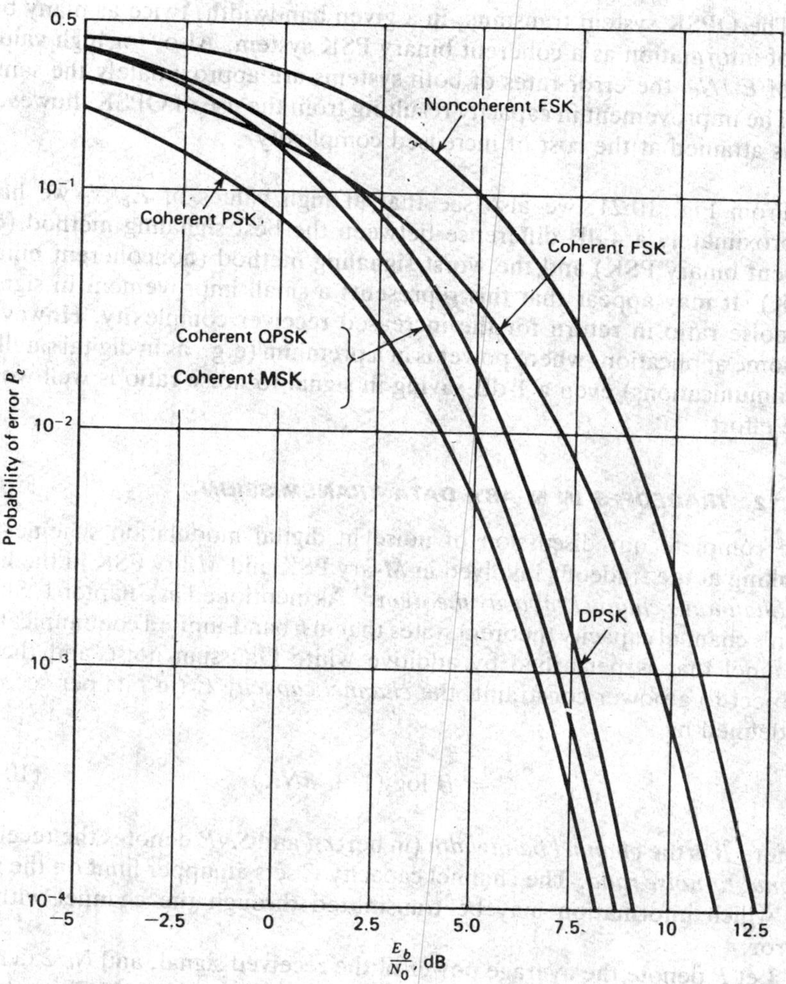

Figure 10.21
Comparison of the noise performances of different PSK and FSK systems.

2. For any value of E_b/N_0, coherent PSK produces a smaller error rate than any of the other systems.
3. The phase modulation systems, coherent binary PSK and DPSK, require an E_b/N_0 that is 3 dB less than their frequency modulation system counterparts—coherent binary FSK and noncoherent binary FSK, respectively—to realize the same error rate.
4. At high values of E_b/N_0, the noncoherent receivers, DPSK and noncoherent binary FSK, perform almost as well (to within about 1 dB) as their coherent counterparts, PSK and coherent binary FSK, respectively, for the same bit rate and signal energy per bit.

5. The QPSK system transmits, in a given bandwidth, twice as many bits of information as a coherent binary PSK system. Also, for high values of E_b/N_0, the error rates of both systems are approximately the same. The improvement in capacity resulting from the use of QPSK, however, is attained at the cost of increased complexity.

From Fig. 10.21, we also see that at high values of E_b/N_0 we have approximately a 4-dB difference between the best signaling method (coherent binary PSK) and the worst signaling method (noncoherent binary FSK). It may appear that this represents a small improvement in signal-to-noise ratio in return for the increased receiver complexity. However, in some applications where power is at a premium (e.g., as in digital satellite communications) even a 1-dB saving in signal-to-noise ratio is well worth the effort.

10.12 TRADEOFFS IN M-ARY DATA TRANSMISSION

We complete our discussion of noise in digital modulation schemes by looking at the tradeoffs involved in M-ary PSK and M-ary FSK in the light of *Shannon's channel capacity theorem.*[13] As mentioned in Chapter 1, Shannon's channel capacity theorem states that in a band-limited communication channel that is perturbed by additive white Gaussian noise and that is subject to a power constraint, the *channel capacity* C (in bits per second) is defined by

$$C = B \log_2(1 + SNR) \tag{10.92}$$

where B is the *channel bandwidth* (in hertz), and SNR denotes the received *signal-to-noise ratio*. The channel capacity C sets an upper limit on the rate at which information may be transmitted through the channel without error.

Let P denote the average power of the received signal, and $N_0/2$ denote the power spectral density of the channel noise. We may express the average signal power P in terms of the ***signal energy per bit*** as follows

$$\begin{aligned} P &= \frac{E_b}{T_b} \\ &= E_b R_b \end{aligned} \tag{10.93}$$

where T_b is the *bit duration* in seconds and R_b (defined as $1/T_b$) is the *bit rate* in bits per second. In a bandwidth B, the average noise power (measured over both negative and positive frequencies) equals $N_0 B$. Hence, we

[13] The discussion presented herein is summarized from Haykin (1988), pp. 334–336.

may express the received signal-to-noise ratio as

$$SNR = \frac{E_b R_b}{N_0 B} = \frac{E_b/N_0}{B/R_b} \tag{10.94}$$

The ratio E_b/N_0 is the *signal energy per bit-to-average noise power per unit bandwidth ratio,* and R_b/B is the *bandwidth efficiency.* The bandwidth efficiency, in units of bits per second per hertz, provides a measure of the extent to which channel bandwidth is being used.

Since the channel capacity C sets an upper limit on the bit rate R_b, we have

$$R_b \leqslant C \tag{10.95}$$

Thus, we may combine Eqs. 10.92, 10.94, and 10.95 to recast Shannon's channel capacity theorem in the form:

$$\frac{R_b}{B} \leqslant \log_2\left(1 + \frac{E_b/N_0}{B/R_b}\right) \tag{10.96}$$

Equivalently, we may write

$$\frac{E_b}{N_0} \leqslant \frac{2^{R_b/B} - 1}{R_b/B} \tag{10.97}$$

This relation states that for a specified bandwidth efficiency R_b/B, the received signal energy per bit-to-noise power spectral density ratio E_b/N_0 must satisfy Eq. 10.97 if transmission over the channel is to be error-free.

In the limiting case when the channel bandwidth B is infinitely large (i.e., R_b/B approaches zero), we find from Eq. 10.97 that the corresponding limit on E_b/N_0 is $\log_e 2 = 0.693$ (i.e., -1.6 dB). This special value is referred to as the *Shannon limit.*

In Fig. 10.22*a*, we show a plot of the bandwidth efficiency R_b/B versus the *signal energy per bit-to-average noise power per unit bandwidth ratio* E_b/N_0 for coherent M-ary PSK signaling for different numbers of phase levels defined by $M = 2^K$, where $K = 1, 2, 3, 4, 5, 6$. Each point corresponds to an average probability of symbol error $P_e = 10^{-5}$.

In Fig. 10.22*b* we show a plot of R_b/B versus E_b/N_0 for coherent M-ary FSK signaling for different numbers of frequency levels $M = 2^K$, where $K = 1, 2, 3, 4, 5, 6$. Each point corresponds to an average probability of symbol error $P_e = 10^{-5}$. It is assumed that the frequency separation of the M transmitted sinusoids is the minimum so that they are *orthogonal* to each other over a signaling interval.

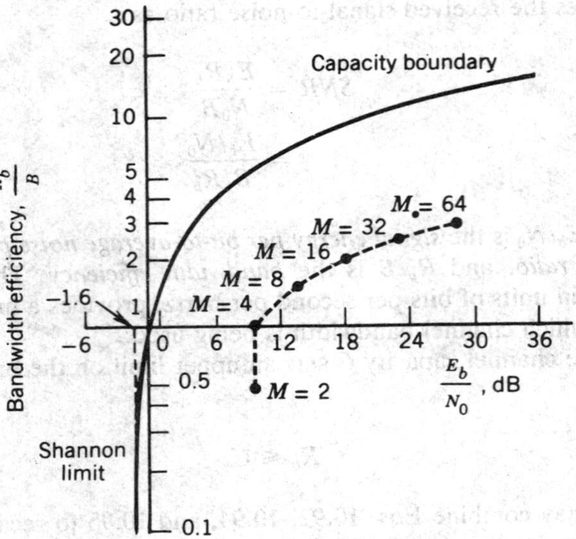

(a)

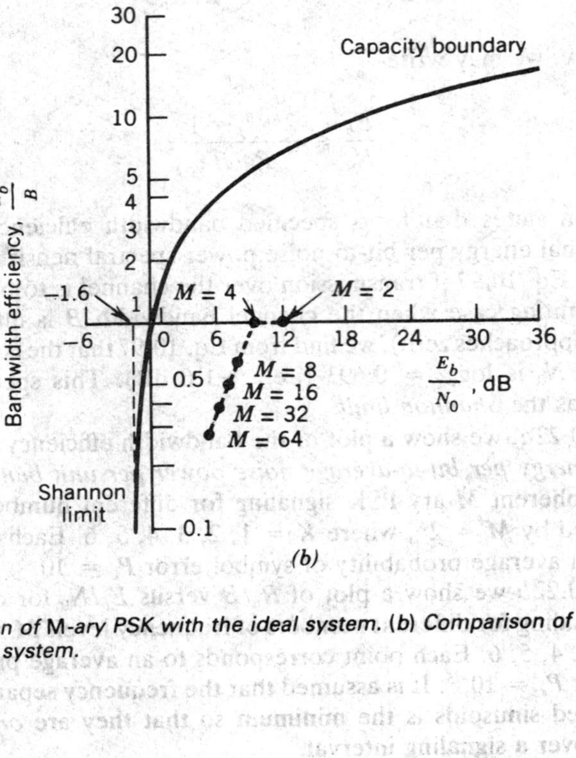

(b)

Figure 10.22
(a) Comparison of M-ary PSK with the ideal system. (b) Comparison of M-ary FSK with the ideal system.

In both parts of Fig. 10.22, we also show the *capacity curve* obtained by plotting Eq. 10.97 when it is satisfied with equality.

Figure 10.22 clearly depicts the tradeoffs involved in the use of M-ary signaling. In particular, we may make the following observations:

1. In the case of M-ary PSK, as the number of phase levels M is increased, the bandwidth efficiency is improved but at the expense of an increase in the required signal energy per bit (for $M > 4$).
2. In the case of M-ary FSK, as the number of frequency levels M is increased, the required signal energy per bit is decreased but at the expense of reduced bandwidth efficiency (for $M > 4$).

PROBLEMS

P10.2 Maximization of Output Signal-to-Noise Ratio

Problem 1 Consider the signal $s(t)$ shown in Fig. P10.1.

(a) Determine the impulse response of a filter matched to this signal and sketch it as a function of time.

(b) Plot the matched filter output as a function of time.

(c) What is the peak value of the output?

Problem 2 The amplitude of the pulse in Fig. P10.1 is doubled. What is the factor by which the pulse duration has to be reduced, so that a filter matched to this new pulse has the same performance as the matched filter in Problem 1, when both filters operate in the same additive white Gaussian noise?

P10.3 Properties of Matched Filters

Problem 3 Consider a filter that is matched to an energy signal $s_1(t)$ of duration T seconds. The filter is excited by another energy signal $s_2(t)$,

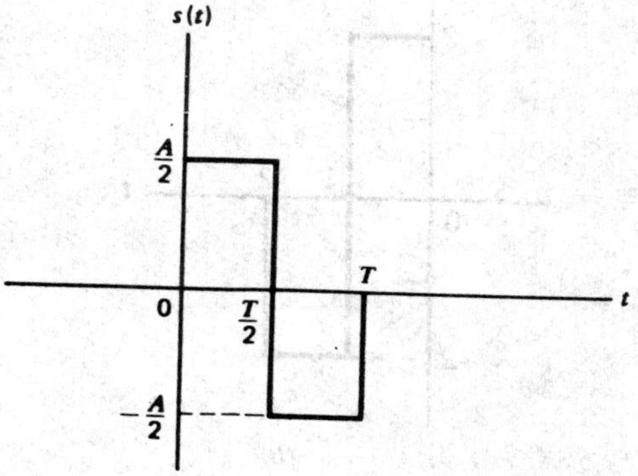

Figure P10.1

also of duration T seconds. Both signals are of equal energy E. Find the filter output sampled at time $t = T$, given the following alternatives:

(a) The signals $s_1(t)$ and $s_2(t)$ are orthogonal to each other over the interval $0 \leq t \leq T$.

(b) The two signals are correlated with each other, and their correlation coefficient is equal to 0.5. (For the definition of correlation coefficient, use may be made of the formula given by the first line of Eq. 10.78).

Problem 4

(a) Let the signal $s_1(t)$ have the waveform shown in Fig. P10.2a. Plot the following waveforms:

(i) The impulse response of the corresponding matched filter.
(ii) The filter ouput in response to $s_1(t)$ as input.

(b) Plot the waveform of the output of a filter matched to $s_1(t)$ but excited with an input having the waveform of $s_2(t)$ shown in Fig. P10.2b. What is the value of the output at time $t = 2$?

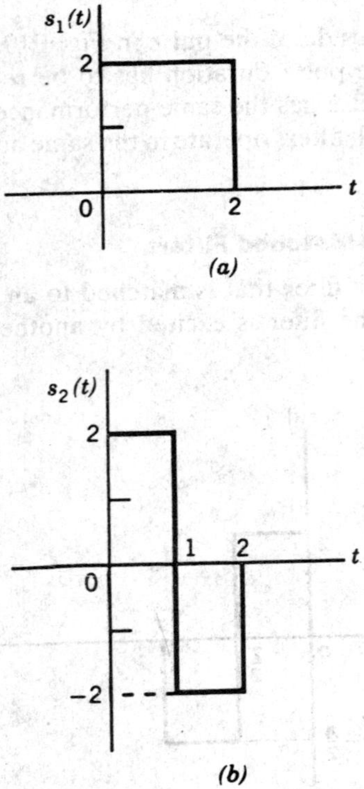

Figure P10.2

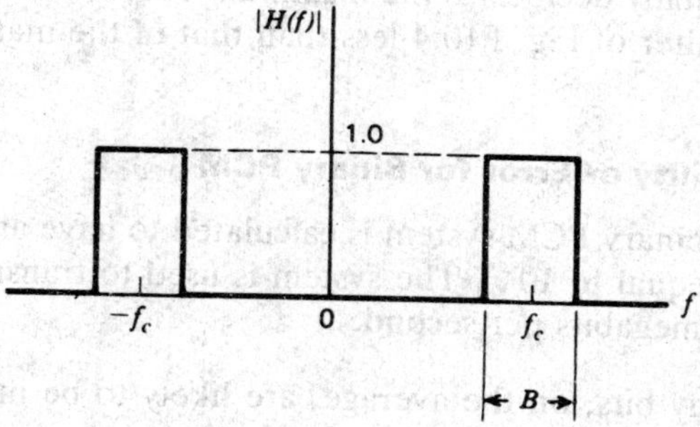

Figure P10.3

P10.4 Approximations in Matched Filter Design

Problem 5 Consider a matched filter for the RF pulse:

$$s(t) = \begin{cases} A\cos(2\pi f_c t), & 0 \leq t \leq T \\ 0, & \text{otherwise} \end{cases}$$

where A is the amplitude, T is the duration, and f_c is the frequency. The frequency $f_c \gg 1/T$, so that the pulse $s(t)$ may be regarded as a narrow-band signal. The requirement is to approximate the matched filter with an ideal band-pass filter of bandwidth B; the amplitude response of the filter is shown in Fig. P10.3. The bandwidth B is chosen to maximize the output signal-to-noise ratio of the filter.

(a) Find the bandwidth B of the filter.
(b) By how many decibels is the maximum output signal-to-noise ratio of the approximating band-pass filter less than that of the matched filter?

Problem 6 Repeat Problem 5 using the LCR filter shown in Fig. P10.4, which is tuned to the frequency f_c of the RF pulse $s(t)$. The requirement is to choose the 3-dB bandwidth of the filter so as to maximize the output signal-to-noise ratio of the filter.

(a) Find the 3-dB bandwidth of the filter. What is the Q-factor of the filter?

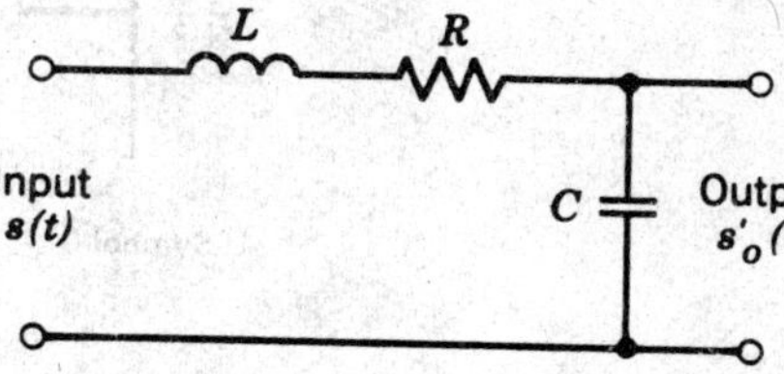

Figure P10.4

(b) By how many decibels is the maximum output signal-to-noise ratio of the *LCR* filter of Fig. P10.4 less than that of the matched filter?

P10.6 Probability of Error for Binary PCM

Problem 7 A binary PCM system is calculated to have an average probability of error equal to 10^{-5}. The system is used to transmit binary data at the rate of 5 megabits per second.

(a) How many bits, on the average, are likely to be in error during a transmission period that lasts 2 sec?
(b) Repeat the calculation for a transmission period of 1 min.

Problem 8 A binary PCM wave uses the Manchester code to describe symbols 1 and 0, as illustrated in Fig. P10.5. The additive noise at the receiver input is white and Gaussian with zero mean and power spectral density $N_0/2$. Assuming that symbols 1 and 0 occur with equal probability, find an expression for the average probability of error at the receiver output, using matched filters.

P10.8 Coherent Detection of Binary Modulated Waves

Problem 9 Consider a phase-locked loop consisting of a multiplier, loop filter, and voltage-controlled oscillator (VCO), as in Fig. P10.6. Let the signal applied to the multiplier input be a binary PSK signal defined by

$$s(t) = A_c \cos[2\pi f_c t + k_p m(t)]$$

Figure P10.5

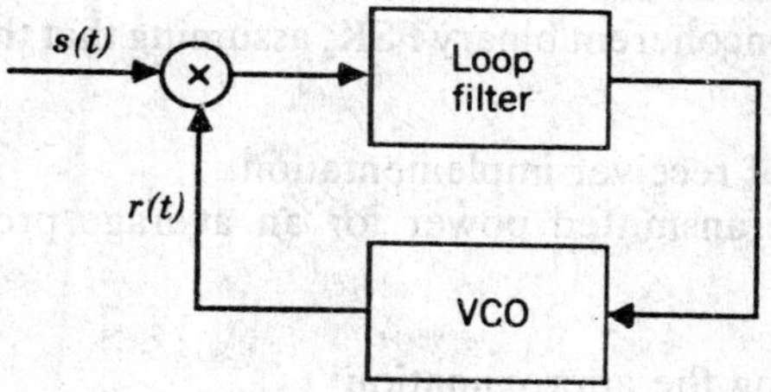

Figure P10.6

where k_p is the phase sensitivity, and the data signal $m(t)$ takes on the value $+1$ V for binary symbol 1 and -1 V for binary symbol 0. The VCO output is

$$r(t) = A_v \sin[2\pi f_c t + \theta(t)]$$

(a) Evaluate the loop filter output, assuming that this filter removes the modulated components with carrier frequency $2f_c$.
(b) Show that this output is proportional to the data signal $m(t)$ when the loop is phase-locked, that is, $\theta(t) = 0$.

Problem 10 A binary PSK signal is applied to a correlation receiver that lacks perfect phase synchronization with the transmitter. Specifically, it is supplied with a local carrier whose phase differs from that of the carrier used in the transmitter by ϕ radians.

(a) Determine the effect of the phase error ϕ on the average probability of error of this receiver.
(b) As a check on the formula derived in part (a), show that when the phase error is zero the formula reduces to the same form as in Eq. 10.82.

Problem 11 A binary FSK system transmits binary data at the rate of 2.5×10^6 bits per second. During the course of transmission, white Gaussian noise of zero mean and power spectral density 10^{-20} watts per hertz is added to the signal. In the absence of noise, the amplitude of the received signal is 1 μV. Determine the average probability of error assuming coherent detection of the binary FSK signal. For this calculation, you may use Table 6 of Appendix D for the error function.

P10.9 Noncoherent Detection of Binary Modulated Waves

Problem 12 An MSK signal is applied to a noncoherent FSK receiver. What is the average probability of error for this system?

Problem 13 Rank DPSK with the coherent versions of binary PSK and binary FSK and noncoherent binary FSK, assuming that the issue of interest is the following:

(a) Simplicity of receiver implementation.
(b) Minimum transmitted power for an average probability of error equal to 10^{-4}.

Problem 14 Using the approximation

$$\text{erfc}(x) \simeq \frac{\exp(-x^2)}{\sqrt{\pi}x},$$

calculate the ratio $(P_e)_{\text{PSK}}/(P_e)_{\text{DPSK}}$, where $(P_e)_{\text{PSK}}$ and $(P_e)_{\text{DPSK}}$ refer to the average probability of error for PSK and DPSK, respectively. What is the value of this ratio for a *signal energy per bit-to-average noise power per unit bandwidth ratio* $E_b/N_0 = 10$ dB?

Problem 15 The binary sequence 101011000 is transmitted over a noisy channel using DPSK, assuming an initial bit of 1. Owing to noise in the channel, an error is made in the fourth bit of the reconstructed binary sequence at the receiver output. Show that the next bit of this sequence will also be in error; that is, *errors in a DPSK receiver tend to occur in pairs.*

P10.10 Coherent Detection of Quaternary Modulated Waves

Problem 16 A QPSK signal is applied to a receiver that is improperly phase-synchronized with respect to the receiver. In particular, the local carrier applied to the correlator in the upper path of the receiver in Fig. 10.20 is $\cos(2\pi f_c t + \phi)$ and that applied to the correlator in the lower path is $\sin(2\pi f_c t + \phi)$, where ϕ is the phase error.

(a) Calculate the average probability of symbol error for this receiver.
(b) As a check on the formula derived in part (a), show that when ϕ is zero it reduces to the same form as in Eq. 10.89.

APPENDIX A

POWER RATIOS AND DECIBELS

In system calculations and measurements involving the use of power ratios, it is customary practice to use a unit called the *decibel*. The decibel, commonly abbreviated as dB, is one tenth of a larger unit, the *bel*.[1] In practice, however we find that for most applications the bel is too large a unit; hence, the wide use of dB as the unit for expressing power ratios. The dB is particularly appropriate for sound measurements because the ear responds to sound in an approximately logarithmic fashion. Thus, equal dB increments are perceived by the ear as equal increments in sound.

Let P denote the power at some point of interest in a system. Let P_0 denote the *reference* power level with which the power P is to be compared. The number of decibels in the power ratio P/P_0 is defined as $10 \log_{10}(P/P_0)$. For example, a power ratio of 2 corresponds to 3 dB, and a power ratio of 10 corresponds to 10 dB.

We may also express the signal power P itself in dB if we divide P by one watt or one milliwatt. In the first case, we express the signal power P in dBW as $10 \log_{10}(P/1\text{W})$, where W is the abbreviation for watt. In the second case, we express the signal power P in dBm as $10 \log_{10}(P/1\text{mW})$, where mW is the abbreviation for milliwatt.

[1]The unit, bel, is named in honor of Alexander Graham Bell. In addition to inventing the telephone, Bell was the first to use logarithmic power measurements in sound and hearing research.

APPENDIX B

BESSEL FUNCTION

A Bessel function of the first kind of order n and argument x, commonly denoted by $J_n(x)$, is defined by

$$J_n(x) = \frac{1}{2\pi}\int_{-\pi}^{\pi} \exp(jx\sin\theta - jn\theta)\, d\theta \tag{B.1}$$

or, equivalently,

$$J_n(x) = \frac{1}{\pi}\int_{0}^{\pi} \cos(x\sin\theta - n\theta)\, d\theta \tag{B.2}$$

Just as the trigonometric functions can be expanded in power series, so can the Bessel function $J_n(x)$ be expanded in a power series:

$$J_n(x) = \sum_{m=0}^{\infty} \frac{(-1)^m(\frac{1}{2}x)^{n+2m}}{m!(n+m)!} \tag{B.3}$$

Evaluating Eq. B.3 for $n = 0, 1, 2,$ for example, we thus have

$$J_0(x) = 1 - \frac{x^2}{2^2} + \frac{x^4}{2^2\cdot 4^2} - \frac{x^6}{2^2\cdot 4^2\cdot 6^2} + \cdots \tag{B.4}$$

$$J_1(x) = \frac{x}{2} - \frac{x^3}{2^2\cdot 4} + \frac{x^5}{2^2\cdot 4^2\cdot 6} - \cdots \tag{B.5}$$

$$J_2(x) = \frac{x^2}{2\cdot 4} - \frac{x^4}{2^2\cdot 4\cdot 6} + \frac{x^6}{2^2\cdot 4^2\cdot 6\cdot 8} - \cdots \tag{B.6}$$

B.1 PROPERTIES OF BESSEL FUNCTION

The Bessel function $J_n(x)$ has the following properties[1]:

1. $$J_n(x) = (-1)^n J_{-n}(x) \tag{B.7}$$

[1]For a mathematical treatment of Bessel functions, see Whittaker and Watson (1927), pp. 355–385.

To prove this relation, we replace θ by $(\pi - \theta)$ in Eq. B.2. Then, noting that $\sin(\pi - \theta) = \sin\theta$, we get

$$J_n(x) = \frac{1}{\pi}\int_0^{\pi} \cos(x \sin\theta + n\theta - n\pi)\, d\theta$$
$$= \frac{1}{\pi}\int_0^{\pi} [\cos(n\pi)\cos(x\sin\theta + n\theta) + \sin(n\pi)\sin(x\sin\theta + n\theta)]\, d\theta$$

For integer values of n, we have

$$\cos(n\pi) = (-1)^n$$
$$\sin(n\pi) = 0$$

Therefore,

$$J_n(x) = \frac{(-1)^n}{\pi}\int_0^{\pi} \cos(x \sin\theta + n\theta)\, d\theta \tag{B.8}$$

From Eq. B.2 we also find that by replacing n with $-n$:

$$J_{-n}(x) = \frac{1}{\pi}\int_0^{\pi} \cos(x\sin\theta - n\theta)\, d\theta \tag{B.9}$$

The desired result follows immediately from Eqs. B.8 and B.9.

2.
$$J_n(x) = (-1)^n J_n(-x) \tag{B.10}$$

This relation is obtained by replacing x with $-x$ in Eq. B.2, and then using Eq. B.8.

3.
$$J_{n-1}(x) + J_{n+1}(x) = \frac{2n}{x} J_n(x) \tag{B.11}$$

This *recurrence formula* is useful in constructing tables of Bessel coefficients. For example, the use of Eq. B.11 for $n = 1$ yields a value for $J_1(x)$ that is in exact agreement with that of Eq. B.5.

4. For small values of x, we have

$$J_n(x) \simeq \frac{x^n}{2^n n!} \tag{B.12}$$

This relation is obtained simply by retaining the first term in the power series of Eq. B.3 and ignoring the higher-order terms. Thus, when x is small, we have

$$J_0(x) \simeq 1$$
$$J_1(x) \simeq \frac{x}{2} \tag{B.13}$$

5. For large values of x, we have

$$J_n(x) \simeq \sqrt{\frac{2}{\pi x}}\cos\left(x - \frac{\pi}{4} - \frac{n\pi}{2}\right) \tag{B.14}$$

This shows that for large values of x, the Bessel function $J_n(x)$ behaves like a sine wave with progressively decreasing amplitude.

6. With x real and fixed, $J_n(x)$ approaches zero as the order n goes to infinity.

7.
$$\sum_{n=-\infty}^{\infty} J_n(x)\exp(jn\phi) = \exp(jx\sin\phi) \tag{B.15}$$

To prove this property, consider the sum $\Sigma_{n=-\infty}^{\infty} J_n(x)\exp(jn\phi)$ and use the formula of Eq. B.1 for $J_n(x)$ to obtain

$$\sum_{n=-\infty}^{\infty} J_n(x)\exp(jn\phi) = \frac{1}{2\pi}\sum_{n=-\infty}^{\infty}\exp(jn\phi)\int_{-\pi}^{\pi}\exp(jx\sin\theta - jn\theta)\,d\theta$$

Interchanging the order of integration and summation:

$$\sum_{n=-\infty}^{\infty} J_n(x)\exp(jn\phi) = \frac{1}{2\pi}\int_{-\pi}^{\pi} d\theta\,\exp(jx\sin\theta)\sum_{n=-\infty}^{\infty}\exp[jn(\phi-\theta)] \tag{B.16}$$

From Example 14 of Chapter 2, we deduce that

$$\delta(\phi-\theta) = \frac{1}{2\pi}\sum_{n=-\infty}^{\infty}\exp[jn(\phi-\theta)], \qquad -\pi \le \phi - \theta \le \pi \tag{B.17}$$

Therefore, substituting Eq. B.17 in B.16, and using the sifting property of a delta function, we get

$$\begin{aligned}\sum_{n=-\infty}^{\infty} J_n(x)\exp(jn\phi) &= \int_{-\pi}^{\pi}\exp(jx\sin\theta)\delta(\phi-\theta)\,d\theta \\ &= \exp(jx\sin\phi)\end{aligned}$$

which is the desired result.

8.
$$\sum_{n=-\infty}^{\infty} J_n^2(x) = 1, \qquad \text{for all } x \tag{B.18}$$

To prove this property, we may proceed as follows. We observe that $J_n(x)$ is real. Hence, multiplying Eq. B.1 by its complex conjugate, and summing over all possible values of n, we get

$$\sum_{n=-\infty}^{\infty} J_n^2(x) = \frac{1}{(2\pi)^2}\sum_{n=-\infty}^{\infty}\int_{-\pi}^{\pi}\int_{-\pi}^{\pi}\exp(jx\sin\theta - jn\theta - jx\sin\phi + jn\phi)\,d\theta\,d\phi$$

Interchanging the order of double integration and summation:

$$\sum_{n=-\infty}^{\infty} J_n^2(x) = \frac{1}{(2\pi)^2}\int_{-\pi}^{\pi}\int_{-\pi}^{\pi} d\theta\,d\phi\,\exp[jx(\sin\theta - \sin\phi)] \times \sum_{n=-\infty}^{\infty}\exp[jn(\phi-\theta)] \tag{B.19}$$

Substituting Eq. B.17 in B.19, and using the sifting property of a delta function, we finally get

$$\sum_{n=-\infty}^{\infty} J_n^2(x) = \frac{1}{2\pi}\int_{-\pi}^{\pi} d\theta = 1$$

which is the desired result.

B.2 TABLE OF VALUES

Table B.1 gives values of the Bessel function $J_n(x)$ for values of the order n from 0 up to 14, and for values of the argument x in the interval (0.5, 12).

This table may be used to illustrate properties 3 through 8 listed in Section B.1. The reader is invited to do this as an exercise.

TABLE B.1 Bessel Functions

n \ x	$J_n(x)$ 0.5	1	2	3	4	6	8	10	12
0	0.9385	0.7652	0.2239	−0.2601	−0.3971	0.1506	0.1717	−0.2459	0.0477
1	0.2423	0.4401	0.5767	0.3391	−0.0660	−0.2767	0.2346	0.0435	−0.2234
2	0.0306	0.1149	0.3528	0.4861	0.3641	−0.2429	−0.1130	0.2546	−0.0849
3	0.0026	0.0196	0.1289	0.3091	0.4302	0.1148	−0.2911	0.0584	0.1951
4	0.0002	0.0025	0.0340	0.1320	0.2811	0.3576	−0.1054	−0.2196	0.1825
5	—	0.0002	0.0070	0.0430	0.1321	0.3621	0.1858	−0.2341	−0.0735
6		—	0.0012	0.0114	0.0491	0.2458	0.3376	−0.0145	−0.2437
7			0.0002	0.0025	0.0152	0.1296	0.3206	0.2167	−0.1703
8			—	0.0005	0.0040	0.0565	0.2235	0.3179	0.0451
9				0.0001	0.0009	0.0212	0.1263	0.2919	0.2304
10				—	0.0002	0.0070	0.0608	0.2075	0.3005
11					—	0.0020	0.0256	0.1231	0.2704
12						0.0005	0.0096	0.0634	0.1953
13						0.0001	0.0033	0.0290	0.1201
14						—	0.0010	0.0120	0.0650

APPENDIX C

SYSTEM NOISE AND CALCULATIONS

The inevitable presence of *noise* in a communication system causes the reliable transmission of electrical signals through the system to be undermined. It is therefore important to know how noise arises in the system. There are many potential sources of noise. The sources of noise may be *external* to the system (e.g., atmospheric noise, galactic noise, man-made noise) or *internal* to the system. The second category includes an important type of noise that arises owing to spontaneous fluctuations of current or voltage in electrical circuits. This type of noise, in one way or another, is present in every communication system and represents a fundamental limitation on the transmission or detection of signals.

In this appendix we briefly discuss the physical sources of noise in electrical circuits and develop quantitative measures for assessing the presence of noise in a system. We finish the discussion by presenting link calculations for line-of-sight propagation through free space.

C.1 ELECTRICAL NOISE

In an electrical circuit, noise is generated owing to various physical phenomena.[1] We have *thermal noise* produced by the random motion of electrons in conducting media, and *shot noise* produced by random fluctuations of current flow in electronic devices. These two are the most common examples of spontaneous fluctuation noise encountered in electrical circuits.

Besides thermal noise and shot noise, transistors exhibit a low-frequency noise phenomenon known as *flicker noise.* The mean-square value of flicker noise is inversely proportional to frequency; hence, it is also referred to as "one-over-f" noise. Yet another type of noise encountered in semiconductor devices is *burst noise,* whose mean-square value falls off as $1/f^2$.

Flicker noise and burst noise are both *nonwhite,* with their degrading effects being observed at low frequencies. Ordinarily, they can be ignored above a few kilohertz. On the other hand, thermal noise and shot noise are both *white* for all practical purposes; hence, their degrading influence on the operation of a communication system extends right across the com-

[1]For a detailed treatment of electrical noise, see Bell (1985), Van Der Ziel (1970) and Robinson (1974).

plete frequency band of interest. A brief discussion of thermal noise and shot noise is therefore in order.

THERMAL NOISE

Thermal noise[2] is an ubiquitous source of noise that arises from thermally induced motion of electrons in conducting media. In a conductor there is a large number of "free" electrons and an equally large number of ions bound by strong molecular forces. The ions vibrate randomly about their normal positions. The free electrons move along randomly oriented paths, owing to continuous collisions with the vibrating ions. The net effect of this random motion is an electric current that is likewise random. However, the mean value of the current is zero since, on the average, there are as many electrons moving in one direction as there are in another.

A thorough analysis of thermal noise requires the use of thermodynamic and quantum mechanical considerations that are beyond the scope of this book. For the purpose of the discussion presented here, it suffices to say that the power spectral density of thermal noise produced by a resistor is given by[3]

$$S_{TN}(f) = \frac{2h|f|}{\exp(h|f|/kT) - 1} \tag{C.1}$$

where T is the *absolute temperature in degrees Kelvin*, k is *Boltzmann's constant,* and h is *Planck's constant.* Note that the power spectral density $S_{TN}(f)$ is measured in watts per hertz. For "low" frequencies defined by

$$f << \frac{kT}{h}$$

we may use the approximation

$$\exp\left(\frac{h|f|}{kT}\right) \simeq 1 + \frac{h|f|}{kT}$$

Correspondingly, we may approximate the formula of Eq. C.1 as follows

$$S_{TN}(f) \simeq 2kT \tag{C.2}$$

[2]Thermal noise was first studied experimentally by Johnson in 1928, and for this reason it is sometimes referred to as "Johnson noise"; see Johnson (1928). Johnson's experiments were confirmed theoretically by Nyquist; see Nyquist (1928).

[3]For a discussion of the physical issues involved in the formulation of Eq. C.1, and for a historical account of the pertinent literature, see Bell (1985).

To develop a feeling for the frequencies for which the use of this approximate formula is justified, we assume operation at a room temperature of 17°C, for which we have $T = 290°K$. Then, using the values of Boltzmann's constant and Planck's constant:

$$k = 1.38 \times 10^{-23} \text{ joules/°K}$$

and

$$h = 6.63 \times 10^{-34} \text{ joule} \cdot \text{second}$$

we find that

$$\frac{kT}{h} = 6 \times 10^{12} \text{ Hz}$$

This upper frequency limit lies in the infrared region that is well above the spectrum of frequencies encountered in conventional electrical communication systems. Therefore, for all practical purposes the use of the approximate formula of Eq. C.2 is perfectly justified.

Thus, given a resistor of R ohms, we find from Eq. C.2 that the mean-square value of the *thermal noise voltage* measured across the terminals of the resistor equals

$$\begin{aligned} E[V_{TN}^2] &= 2RB_N S_{TN}(f) \\ &= 4kTRB_N \text{ volts}^2 \end{aligned} \qquad \text{(C.3)}$$

where B_N is the bandwidth (in hertz) over which the noise voltage is measured. We may thus model a noisy resistor by the *Thévenin equivalent circuit* consisting of a noise voltage generator with a mean-square value of $E[V_{TN}^2]$ in series with a noiseless resistor, as in Fig. C.1*a*. Alternatively,

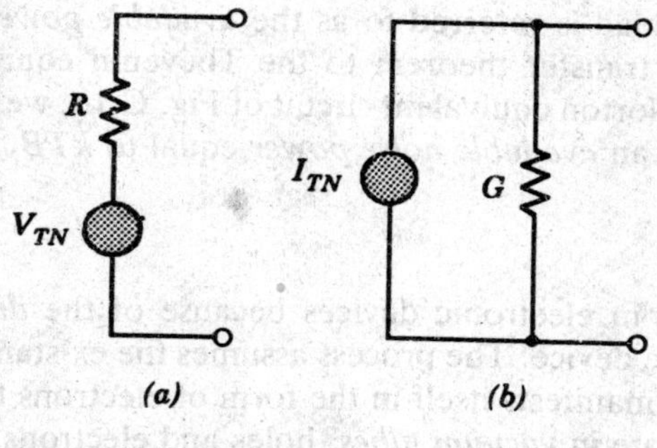

Figure C.1
Models of a noisy resistor. (a) Thévenin equivalent circuit. (b) Norton equivalent circuit.

we may use the *Norton equivalent circuit* consisting of a noise current generator in parallel with a noiseless conductance, as in Fig. C.1*b*. The mean-square value of the noise current generator is

$$E[I_{TN}^2] = \frac{1}{R^2} E[V_{TN}^2]$$
$$= 4kTGB_N \text{ amps}^2 \qquad \text{(C.4)}$$

where $G = 1/R$ is the conductance.

It is also of interest to note that because the number of electrons in a resistor is very large and their random motions inside the resistor are statistically independent of each other, the central limit theorem indicates that thermal noise is Gaussian-distributed with zero mean. Accordingly, for the band of frequencies encountered in electrical communication systems, we may model thermal noise as *white Gaussian noise* of zero mean.

EXAMPLE 1

Consider a system for which the noise bandwidth $B_N = 10$ kHz, the absolute temperature $T = 290°$K (i.e., room temperature), and the resistance $R = 5$ KΩ. Then, the use of Eq. C.3 yields the mean-square value of the thermal noise voltage to be 0.8×10^{-12} volts squared. That is, the thermal noise voltage produced by the resistor has a root mean-square (rms) value equal to 0.89 μV.

AVAILABLE NOISE POWER

Noise calculations involve the transfer of power, so we find that the use of the *maximum-power transfer theorem* is applicable. This theorem states that *the maximum possible power is transferred from a source of internal resistance R to a load of resistance R_l when $R_l = R$*. Under this matched condition, the power produced by the source is divided equally between the internal resistance of the source and the load resistance, and the power delivered to the load is referred to as the *available power*. Applying the maximum-power transfer theorem to the Thévenin equivalent circuit of Fig. C.1*a* or the Norton equivalent circuit of Fig. C.1*b*, we find that a noisy resistor produces an *available noise power* equal to kTB_N watts.

SHOT NOISE

Shot noise arises in electronic devices because of the *discrete* nature of current flow in the device. The process assumes the existance of an *average* current flow that manifests itself in the form of electrons flowing from the cathode to the plate in *vacuum tubes,* holes and electrons flowing in *semiconductor devices,* and photons emitted in *photodiodes*. Although the average number of particles moving across the device per unit time is assumed to be constant, the process of current flow through the device exhibits *fluctuations* about the average value. The manner in which these fluctuations arise varies from one device to another. In a vacuum-tube device, the fluctuations are produced by the random emission of electrons from the cathode. In a semiconductor device, the cause is the random diffusion of electrons or the random recombination of electrons with holes. In a photodiode, it is the random emission of photons. In all these devices, the physical mechanism that controls current flow through the device has built-in statistical fluctuations about some average value. The shot noise produced by these fluctuations is thus *dependent on the average value of the current.*

Consider for example, a *temperature-limited vacuum diode,* shown in Fig. C.2. It consists of two electrodes enclosed in a vacuum: a *cathode,* which is heated so that it emits electrons; and an *anode* or *plate,* which is maintained at a positive potential with respect to the cathode so that it gathers the electrons. We assume that the cathode-plate potential difference is large enough to cause the electrons emitted thermionically by the heated cathode to be pulled to the plate with such high velocities that space-charge effects are negligible. The plate current is then determined effectively by the rate at which electrons are emitted from the cathode. By considering the plate current as the sum of a succession of current pulses, with each pulse caused by the transit of one electron through the cathode-plate space, we find that the mean-square value of the randomly fluctuating component of the current is given by

$$E[I_{SN}^2] = 2qIB_N \text{ amps}^2 \tag{C.5}$$

where q is the *electron charge* equal to 1.60×10^{-19} coulombs, I is the mean value of the current in amperes, and B_N is the bandwidth of the measuring instrument in hertz. Equation C.5 is called the *Schottky formula.* The typical transit time of an electron from cathode to plate is on the order of 10^{-9} sec. The Schottky formula holds provided that the operating frequency is small compared with the reciprocal of the transit time, so that we may neglect transit time effects.

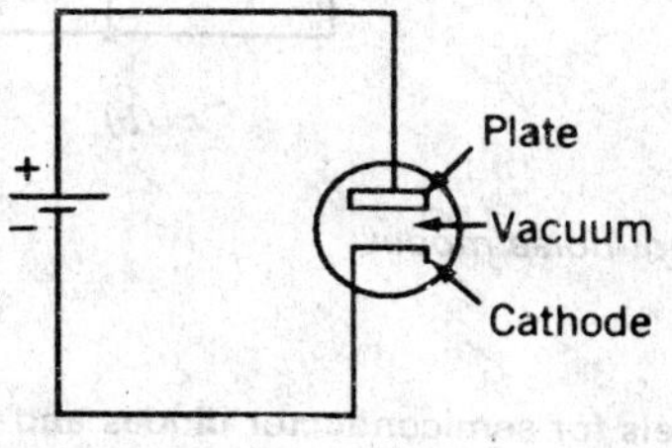

Figure C.2
Temperature-limited vacuum diode.

Another important characteristic of shot noise is that it is Gaussian-distributed with zero mean. This follows from the fact that the number of electrons contributing to the shot noise current is very large, and their random emissions from the cathode are, for practical purposes, statistically independent of each other. Hence, the central limit theorem predicts a Gaussian distribution for shot noise.

The Schottky formula (C.5) also holds for a *semiconductor junction* diode. In this case the mean value I of the current is given by the *diode equation:*

$$I = I_s \exp\left(\frac{qV}{kT}\right) - I_s \tag{C.6}$$

where V is the voltage applied across the diode and I_s is the saturation current; the other constants are as defined previously. Thus, the current I consists of two components that produce statistically independent shot-noise contributions of their own, as shown by

$$\begin{aligned} E[I_{SN}^2] &= 2qI_s \exp\left(\frac{qV}{kT}\right) B_N + 2qI_s B_N \\ &= 2q(I + 2I_s)B_N \end{aligned} \tag{C.7}$$

Figure C.3 shows the *noise model* of a junction diode.[4] The model includes the *dynamic resistance* of the diode, defined by

$$\begin{aligned} r &= \frac{\delta V}{\delta I} \\ &= \frac{kT}{q(I + I_s)} \end{aligned} \tag{C.8}$$

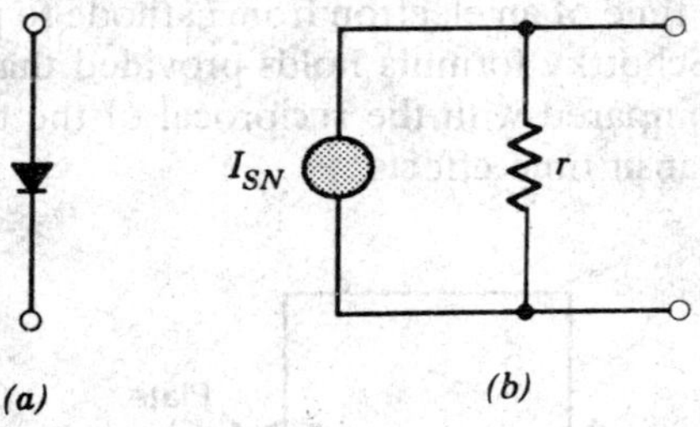

Figure C.3
(a) Junction diode. (b) Shot-noise model.

[4] For details of noise models for semiconductor diodes and transistors, see Robinson (1974), pp. 93–116.

Note, however, that the dynamic resistance r is noiseless since it does not involve power dissipation.

In a *bipolar junction transistor,* shot noise is generated at both the emitter and collector junctions. On the other hand, in a *junction field-effect transistor* the use of an insulated gate structure avoids junction shot noise; nevertheless, shot noise is produced by the flow of gate current. Of course, in both devices thermal noise arises from internal ohmic resistance: *base resistance* in a bipolar transistor and *channel resistance* in a field effect transistor.

C.2 *NOISE FIGURE*

A convenient measure of the noise performance of a linear two-port device is furnished by the *noise figure,*[5] which lends itself to both circuit analysis and measurement. Consider a linear two-port device connected to a signal source of internal impedance $Z(f) = R(f) + jX(f)$ at the input, as in Fig. C.4. The noise voltage $v(t)$ represents the thermal noise associated with the internal resistance $R(f)$ of the source. The output noise of the device is made up of two contributions, one due to the source and the other due to the device itself. We define *the available output noise power in a band of width B_N centered at frequency f as the maximum average noise power in this band obtainable at the output of the device.* The maximum noise power that the two-port device can deliver to an external load is obtained when the load impedance is the complex conjugate of the output impedance of the device, that is, when the resistance is matched and the reactance is tuned out. We define *the noise figure of the two-port device as the ratio of the total available output noise power* (*due to the device and the source*) *per unit bandwidth to the portion thereof due solely to the source.*

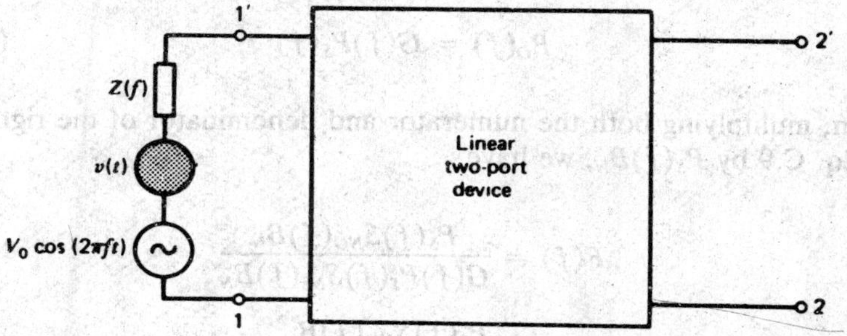

Figure C.4
Linear two-port device.

[5]For a classic paper on noise figure, see Friis (1944).

Let the spectral density of the total available noise power at the device output be $S_{NO}(f)$, and the spectral density of the available noise power due to the source at the device input be $S_{NS}(f)$. Also let $G(f)$ denote the *availabe power gain of the two-port device, defined as the ratio of the available signal power at the output of the device to the available signal power of the source when the signal is a sinusoidal wave of frequency f*. Then we may express the noise figure $F(f)$ of the device as

$$F(f) = \frac{S_{NO}(f)}{G(f)S_{NS}(f)} \tag{C.9}$$

If the device were noise free, $S_{NO}(f) = G(f)S_{NS}(f)$, and the noise figure would then be unity. In a physical device, however, $S_{NO}(f)$ is larger than $G(f)S_{NS}(f)$, so that the noise figure is always larger than unity. The noise figure is commonly expressed in decibels, that is, as $10 \log_{10} F(f)$.

The noise figure may also be expressed in an alternative form. Let $P_S(f)$ denote the available signal power from the source, which is the maximum average signal power that can be obtained. For the case of a source providing a single-frequency signal component with open-circuit voltage $V_0 \cos(2\pi f t)$, the available signal power is obtained when the load connected to the source is $Z^*(f) = R(f) - jX(f)$, yielding the value

$$P_S(f) = \left[\frac{V_0}{2R(f)}\right]^2 R(f)$$

$$= \frac{V_0^2}{4R(f)} \tag{C.10}$$

The available signal power at the output of the device is therefore,

$$P_O(f) = G(f)P_S(f) \tag{C.11}$$

Then, mulitplying both the numerator and denominator of the right side of Eq. C.9 by $P_S(f)B_N$, we have

$$F(f) = \frac{P_S(f)S_{NO}(f)B_N}{G(f)P_S(f)S_{NS}(f)B_N}$$

$$= \frac{P_S(f)S_{NO}(f)B_N}{P_O(f)S_{NS}(f)B_N}$$

$$= \frac{\rho_S(f)}{\rho_O(f)} \tag{C.12}$$

where

$$\rho_S(f) = \frac{P_S(f)}{S_{NS}(f)B_N} \tag{C.13}$$

$$\rho_O(f) = \frac{P_O(f)}{S_{NO}(f)B_N} \tag{C.14}$$

We refer to $\rho_S(f)$ as the *available signal-to-noise ratio of the source* and to $\rho_O(f)$ as the *available signal-to-noise ratio at the device output,* both measured in a narrow band of width B_N centered at f. Since the noise figure is always greater than unity, it follows from Eq. C.12 that the signal-to-noise ratio always decreases with amplification, which is a significant result.

The noise figure $F(f)$, as defined herein, is a function of the operating frequency and hence is referred to as the *spot noise figure.* In contrast, we define an *average noise figure* F of a two-port device as the ratio of the total noise power at the device output to the output noise power due solely to the source. That is,

$$F = \frac{\int_{-\infty}^{\infty} S_{NO}(f)\, df}{\int_{-\infty}^{\infty} G(f)S_{NS}(f)\, df} \tag{C.15}$$

It is apparent that in the case of thermal noise in the input circuit with $R(f)$ constant, and constant gain throughout a fixed band with zero gain at other frequencies, the spot noise figure $F(f)$ and the average noise figure F are identical.

C.3 EQUIVALENT NOISE TEMPERATURE

A disadvantage of the noise figure F is that when it is used to compare low-noise devices, the values obtained are all close to unity. This makes the comparison rather difficult. In such cases, it is preferable to use the *equivalent noise temperature.* Consider a linear two-port device with its input resistance matched to the internal resistance of the source as shown in Fig. C.5. In this diagram, we have also included the noise voltage generator associated with the internal resistance R_s of the source. The mean-square value of this noise voltage is $4kTR_sB_N$. Hence, the available noise power at the device input is

$$\mathcal{N}_s = kTB_N \tag{C.16}$$

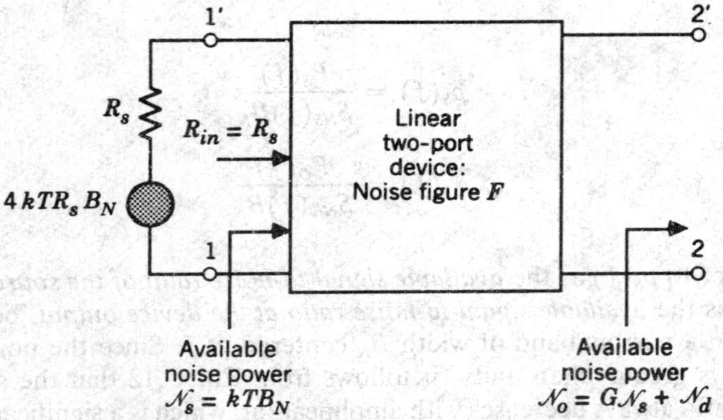

Figure C.5
Linear two-port device matched to the internal resistance of a source connected to the input.

Let $\mathcal{N}_d$ denote the noise power contributed by the two-port device to the total available output noise power $\mathcal{N}_o$. We define $\mathcal{N}_d$ as

$$\mathcal{N}_d = GkT_eB_N \tag{C.17}$$

where G is the available power gain of the device and T_e is its equivalent noise temperature. Then it follows that the total output noise power is

$$\begin{aligned}\mathcal{N}_o &= G\mathcal{N}_s + \mathcal{N}_d \\ &= Gk(T + T_e)\,B_N\end{aligned} \tag{C.18}$$

The noise figure of the device is therefore

$$\begin{aligned}F &= \frac{\mathcal{N}_o}{G\mathcal{N}_s} \\ &= \frac{T + T_e}{T}\end{aligned}$$

Solving for the equivalent noise temperature:

$$T_e = T(F - 1) \tag{C.19}$$

where F is the noise figure of the device measured under matched input conditions, and with the noise source at temperature T. Equation C.19 is the desired relation between the equivalent noise temperature and noise figure of a two-port network.

NOISE SPECTRAL DENSITY

A composite two-port network with equivalent noise temperature T_e (referred to the input) produces the available noise power

$$\mathcal{N}_{av} = kT_e B_N \tag{C.20}$$

Hence, recognizing that $\mathcal{N}_{av} = N_0 B_N$, we find that the noise may be modeled as white Gaussian noise with zero mean and power spectral density $N_0/2$, where

$$N_0 = kT_e \tag{C.21}$$

Note that the power spectral density of the noise so modeled depends only on Boltzmann's constant and the equivalent noise temperature T_e. It is the simplicity of this model that makes the equivalent noise temperature of a composite network such a useful concept.

C.4 CASCADE CONNECTION OF NOISY NETWORKS

Consider next a pair of noisy two-port networks with available power gains G_1 and G_2 that are connected in cascade, as depicted in Fig. C.6. It is assumed that the equivalent noise temperatures of the individual networks are T_1 and T_2, respectively. The total noise power $\mathcal{N}_o$ at the system output is made up of three contributions:

1. *The source noise power $\mathcal{N}_s$ generated at the input of network 1 and amplified by both networks*: The contribution of $\mathcal{N}_s$ to the total noise power $\mathcal{N}_o$ is

$$G_1 G_2 \mathcal{N}_s = G_1 G_2 (kTB_N)$$

where we have made use of the expression for $\mathcal{N}_s$ given in Eq. C.16. This contribution is represented by the top paths in Fig. C.6.

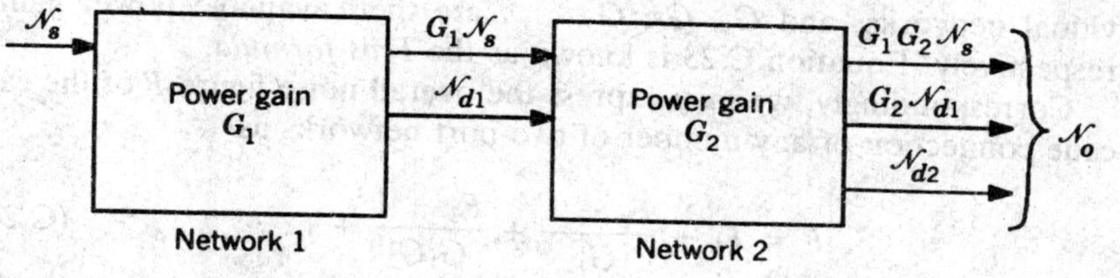

Figure C.6
A cascade of two noisy networks.

2. *The noise power $\mathcal{N}_{d1}$ introduced in network 1 and amplified by network 2*: The contribution to $\mathcal{N}_o$ produced by $\mathcal{N}_{d1}$ is

$$G_2\mathcal{N}_{d1} = G_2(G_1kT_1B_N)$$

where we have made use of Eq. C.17, adapted to the situation at hand. This contribution is represented by the middle paths in Fig. C.6.

3. *The noise power $\mathcal{N}_{d2}$ introduced in network 2*: This final contribution to the total noise power $\mathcal{N}_o$ is

$$\mathcal{N}_{d2} = G_2kT_2B_N$$

where again we have made use of Eq. C.17. This contribution is represented by the bottom path in Fig. C.6.

Adding these three contributions, we therefore get

$$\begin{aligned}\mathcal{N}_o &= G_1G_2\mathcal{N}_s + G_2\mathcal{N}_{d1} + \mathcal{N}_{d2}\\ &= G_1G_2(kTB_N) + G_2(G_1kT_1B_N) + G_2kT_2B_N\\ &= G_1G_2k\left(T + T_1 + \frac{T_2}{G_1}\right)B_N\end{aligned}$$

Here we recognize the product G_1G_2 as the overall value of the available power gain of the cascaded pair of networks shown in Fig. C.6. Accordingly, by analogy with Eq. C.18, we may define an equivalent noise temperature T_e for this network as

$$T_e = T_1 + \frac{T_2}{G_1} \tag{C.22}$$

The result of Eq. C.22 may be readily generalized to the cascade connection of any number of noisy two-port networks, as shown by

$$T_e = T_1 + \frac{T_2}{G_1} + \frac{T_3}{G_1G_2} + \cdots \tag{C.23}$$

where T_1, T_2, T_3, . . . are the equivalent noise temperatures of the individual networks, and G_1, G_2, G_3, . . . are their available power gains, respectively. Equation C.23 is known as the *Friis formula*.

Correspondingly, we may express the overall noise figure F of the cascade connection of any number of two-port networks as

$$F = F_1 + \frac{F_2 - 1}{G_1} + \frac{F_3 - 1}{G_1G_2} + \cdots \tag{C.24}$$

where F_1, F_2, F_3, . . . are the noise figures of the individual networks.

From Eq. C.23 we see that if the first stage of a cascade connection of noisy two-port networks has a high available power gain, then the overall value of the equivalent noise temperture T_e is practically the same as that of the first stage; Eq. C.24 reveals a similar result formulated in terms of the noise figure. It is for this reason that we find in a low-noise receiver, extra care is taken in the design of the pre-amplifier at the front end of the receiver.

C.5 TELECOMMUNICATION LINK CALCULATIONS

In this section we present signal and noise power calculations for telecommunication links that rely on *line-of-sight propagation* through space. Such calculations are encountered in a *satellite communication system,*[6] for example. In this system, a message signal is transmitted from a ground station via the *uplink* to a synchronous satellite, amplified in a *transponder* therein, and then retransmitted from the satellite via the *downlink* to another ground station. The satellite is positioned in a *geostationary orbit* (around the earth) so that it is always visible to different ground stations located inside the satellite antenna's coverage zones on the earth's surface. In effect, the satellite acts as a powerful *repeater* in the sky. Another important application is that of a deep-space telecommunication system[7] used for the transmission of information between a spacecraft and a ground station. In this application the system is provided with a *tracking* capability such that the spacecraft is always visible to the ground station. For the analysis presented here, we will consider a telecommunication link that is illustrative of space applications. Nevertheless, the results of the analysis are equally applicable to a *line-of-sight radio microwave link* between a transmitting source and a receiver located a known distance apart.

CALCULATION OF RECEIVED SIGNAL POWER

Figure C.7 illustrates a link between a spacecraft and a ground station. The link includes a *transmitting source* (on the spacecraft) with its output radiated through the spacecraft's antenna. At the ground station, a *receiving antenna* is used to collect signal power from the incoming electromagnetic wave and feed it to the *low-noise receiver* through a piece of *waveguide*.

Let the transmitting source radiate a total power P_T. If this power were radiated *isotropically* (i.e., uniformly in all directions), then the *power flux density* at a distance r from the source is $P_T/4\pi r^2$, where $4\pi r^2$ is the surface area of a sphere of radius r. In practice, we use a highly directional *antenna*

[6]For a detailed treatment of link calculations in satellite communications, see Pratt and Bostian (1986), Chapter 4.

[7]For a detailed treatment of link calculations in deep-space communications, see Yuen (1983), Section 1.2.

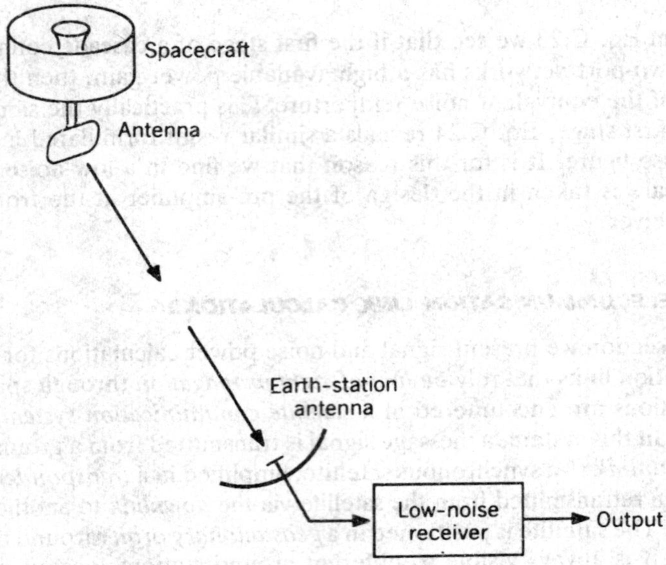

Figure C.7
Space communication link.

so that the transmitted power is radiated primarily along a particular direction of interest. The antenna has a *gain* that is defined as the *ratio of power radiated per unit solid angle in a given direction to the average power radiated per unit solid angle.* Let G_T denote the gain of the transmitting antenna in the direction in which maximum power is radiated; this direction is called the *boresight* of the antenna. The gain G_T is a measure of the increase in power radiated by the antenna over that radiated from an isotropic source. Thus, for a transmitter of total power P_T driving a *lossless* antenna with gain G_T, the power flux density at distance r in the direction of the antenna boresight is given by

$$\mathfrak{F} = \frac{P_T G_T}{4\pi r^2} \tag{C.25}$$

Let A_{eff} denote the *effective aperture area* of the receiving antenna. This area is related to the *physical aperture area* A of the antenna by

$$A_{eff} = \eta A \tag{C.26}$$

where η is the *aperture efficiency*. Typically, η is in the range of 40 to 90%, depending on the type of antenna used. The gain of the receiving antenna G_R is defined in terms of the effective aperture area A_{eff} by

$$G_R = \frac{4\pi A_{eff}}{\lambda^2} \tag{C.27}$$

where λ is the *wavelength* of the transmitted electromagnetic wave; λ is equal to c/f where c is the *speed of propagation* (which is the same as the *speed of light*), and f is the transmitting frequency. Equivalently, we have

$$A_{eff} = \frac{\lambda^2 G_R}{4\pi} \tag{C.28}$$

Hence, given the power flux density $\mathfrak{F}$ at the receiving antenna with effective aperture area A_{eff}, the *received power* is

$$P_R = A_{eff}\mathfrak{F} \tag{C.29}$$

Thus, substituting Eqs. C.25 and C.28 in C.29, we get the desired result:

$$P_R = P_T G_T G_R \left(\frac{\lambda}{4\pi r}\right)^2 \tag{C.30}$$

Equation C.30 is known as the *Friis transmission formula*.

The product $P_T G_T$ in this equation is called the *effective isotropic radiated power* (EIRP). It describes the combination of the transmitting source and antenna in terms of an effective isotropic source with power $P_T G_T$ radiated uniformly in all directions. The term $(4\pi r/\lambda)^2$ is called the *path loss* or *space loss*; it may be viewed as the ratio of received power to transmitted power between two antennas that are separated by a distance r.

From Eq. C.30 we see that for given values of wavelength λ and distance r, the received signal power P_R can be increased by three methods:

1. *The spacecraft-transmitted power P_T is increased.* Typically, P_T is 20 W or less. Even though this transmitted power may appear low, the input power required for its generation represents a substantial fraction of the total power available on the spacecraft. Hence, there is a physical limit on how large a value we can assign to the transmitted power P_T.
2. *The gain G_T of the transmitting antenna is increased.* This will help concentrate the transmitted power more intensely in the direction of the receiving antenna. However, a large value of G_T requires the use of a large antenna. The choice of G_T is therefore limited by size and weight constraints permissible on the spacecraft.
3. *The gain G_R of the receiving antenna is increased.* This will enable the receiver to collect as much of the radiated signal power as possible. Here again, size and weight constraints place a physical limit on the size of the ground-station antenna, although these constraints are far less demanding than those on the spacecraft antenna; we typically have $G_R >> G_T$.

Let the receiving antenna gain and the space loss be expressed in *decibels* (dB). Likewise, let the effective radiated power and the received power be expressed in *decibels relative to 1 watt* (dBW). Then, we may restate the Friis transmission formula in the form:

$$P_R = \text{EIRP} + G_R - L_S \tag{C.31}$$

where

$$\text{EIRP} = 10 \log_{10}(P_T G_T)$$

$$G_R = 10 \log_{10}\left(\frac{4\pi A_{eff}}{\lambda^2}\right)$$

$$L_S = 20 \log_{10}\left(\frac{4\pi r}{\lambda}\right)$$

In Eq. C.30, G_R appears as a power ratio, whereas in Eq. C.31 it is expressed in decibels.

Equation C.31 is idealized in that it does not account for *losses* in the link. To correct for this, it is customary to include a term that represents the combined effect of losses in the atmosphere due to rain attenuation, losses in the transmitting and receiving antennas, and possible loss of gain due to mispointing of the antennas. Let L_0 denote the overall value of this loss expressed in decibels; this term is sometimes called the *system margin*. Then we may modify the expression for the received signal power as

$$P_R = \text{EIRP} + G_R - L_S - L_0 \tag{C.32}$$

Equation C.32 represents a *link power budget* in that it allows the system designer of a telecommunication link to adjust controllable parameters such as the EIRP or the receiving antenna gain G_R and make quick calculations of the received power.

The received power P_R is commonly called the *carrier power*. This is because in a space communication link the method of modulation commonly used for transmitting message signals maintains the envelope of the sinusoidal carrier wave constant; hence, the carrier power is typically equal to the received power.

EXAMPLE 2

A spacecraft is located at a distance of 40,000 km from a ground station on the earth's surface. A transmitting source of frequency 4 GHz radiates a power of 10 W through an antenna with a gain of 20 dB. Assume that the effective aperture area of the receiving antenna is 10 m². We want to calculate the received signal power, ignoring losses in the links.

The effective radiated power equals

$$\begin{aligned} \text{EIRP} &= 10 \log_{10}(P_T G_T) \\ &= 10 \log_{10} P_T + 10 \log_{10} G_T \\ &= 10 \log_{10} 10 + 20 \\ &= 30 \text{ dBW} \end{aligned}$$

The speed of propagation equals the speed of light: $c = 3 \times 10^8$ m/s. The wavelength of the transmitted electromagnetic wave is therefore

$$\begin{aligned} \lambda &= \frac{c}{f} \\ &= \frac{3 \times 10^8}{4 \times 10^9} \\ &= 0.075 \text{ m} \end{aligned}$$

Hence, the space loss equals

$$\begin{aligned} L_S &= 20 \log_{10}\left(\frac{4\pi r}{\lambda}\right) \\ &= 20 \log_{10}\left(\frac{4\pi \times 4 \times 10^7}{0.075}\right) \\ &= 196.5 \text{ dB} \end{aligned}$$

The gain of the receiving antenna equals

$$\begin{aligned} G_R &= 10 \log_{10}\left(\frac{4\pi A_{eff}}{\lambda^2}\right) \\ &= 10 \log_{10}\left(\frac{4\pi \times 10}{0.075^2}\right) \\ &= 43.5 \text{ dB} \end{aligned}$$

Thus, using Eq. C.32 and ignoring the system margin L_0, we find that the received signal power equals

$$\begin{aligned} P_R &= 30 + 43.5 - 196.5 \\ &= -123 \text{ dBW} \end{aligned}$$

Equivalently, we have

$$P_R = 5 \times 10^{-13} \text{ watts}$$

The low value of power P_R indicates that the received signal at the ground station is extremely weak. The inclusion of system margin L_0 (to account for losses in the link) will make the received signal even weaker.

CALCULATION OF SYSTEM NOISE TEMPERATURE

In a space communication link the receiver has to cope with extremely weak signals, as illustrated in Example 2; the weakness results from transmission over long distances. It is therefore imperative that we keep the available noise power low so as to permit a high quality of communication. From Eq. C.20 we see that the available noise power at the receiver input is kT_eB_N, where k is Boltzmann's constant, T_e is the equivalent noise temperature of the receiver, and B_N is the noise bandwidth. Accordingly, the effect of noise is minimized by adopting a combination of two independent strategies:

1. *The equivalent noise temperature of the receiver is maintained as low as possible.* An effective way of accomplishing this requirement is to employ *cryogenically cooled* parametric amplifiers. An ordinary amplifier converts power from a dc source (e.g., power supply or battery) into power at some signal frequency of interest. A *parametric amplifier*, on the other hand, converts power at one frequency (from a source generally known as the *pump*) into power at another frequency, the signal frequency.[8] With liquid helium cooling at 4°K above absolute zero, parametric amplifiers are able to achieve a noise temperature of 20° to 40°K at 4 GHz. Another way of providing low-noise amplification at microwave frequencies is to use cryogenically cooled maser amplifiers. A *maser* (*m*icrowave *a*mplification by the *s*timulated *e*mission of *r*adiation) performs amplification by a quantum-mechanical process, adding almost no noise to the signal it amplifies. Cryogenically cooled parametric and maser amplifiers, however, are expensive to install and maintain; their use can therefore be justified only in large ground stations. Without physical cooling, a noise temperature from 70 to 200°K can be achieved by using gallium arsenide field effect transistors (GaAsFET) or uncooled parametric amplifiers.
2. *The noise bandwidth is minimized, while still permitting the information-carrying (modulated) signals to pass through unaltered.* The value used for the noise bandwidth B_N should be the band-pass version of the equivalent noise bandwidth. For a rough calculation, we may use instead the overall 3-dB bandwidth of the amplifier in the receiver; the error introduced by so doing is usually small.

[8]For a discussion of parametric amplifiers, see Angelakos and Everhart (1968), pp. 82–93. This reference also presents a description of masers (pp. 93–98).

To calculate the *system noise temperature,* we have to include, in addition to the noise generated in the receiver, two other sources of noise:

1. *Antenna noise due to random radiation picked up by the ground-station antenna.* This random radiation includes that from the atmosphere, hot bodies in the field of view of the antenna, the omnipresent 2.7°K thermal background of the universe, and that portion of the ground seen by the sidelobes of the antenna. Let T_a denote the *antenna noise temperature.* Let L_w denote the *loss factor* in the antenna feed and waveguide, where $L_w \geqslant 1$. Accordingly, the input noise is attenuated by the factor L_w. That is, the antenna noise temperature referred to the amplifier input equals T_a/L_w.
2. *Noise generated due to resistance in the antenna feed and waveguide:* Let T_{amb} denote the *ambient temperature.* Then the equivalent noise generator that represents resistance in the antenna feed and waveguide has the noise temperature

$$T_l = \left(1 - \frac{1}{L_w}\right)T_{amb} \tag{C.33}$$

Note that when there is no loss, L_w is unity and T_l is zero.

Figure C.8 shows the components responsible for noise in the receiving system. Let T_s denote the *system noise temperature* referred to the receiver input. From Fig. C.8, we see that T_s is given by

$$T_s = \frac{T_a}{L_w} + \left(1 - \frac{1}{L_w}\right)T_{amb} + T_e \tag{C.34}$$

where, as defined previously,

T_a = antenna noise temperature

L_w = loss factor in the antenna feed and waveguide

T_{amb} = ambient temperature of the antenna feed and waveguide

T_e = equivalent noise temperature of the receiver

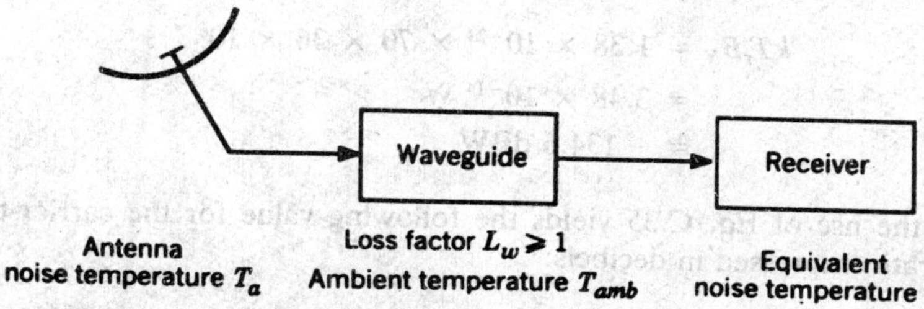

Figure C.8
Components responsible for receiving system noise in a space communication link.

CARRIER-TO-NOISE RATIO

The *carrier-to-noise ratio* (CNR) is defined as the ratio of the carrier power to the available noise power, with both measured at the receiver input. As mentioned previously, the carrier power is the same as the received signal power P_R. The formula for P_R is described by the Friis transmission equation (C.30). To calculate the available noise power at the receiver input, we use the expression kT_sB_N, where k is Boltzmann's constant, T_s is the system noise temperature, and B_N is the noise bandwidth. We therefore have

$$\text{CNR} = \frac{P_R}{kT_sB_N} \tag{C.35}$$

Here again, the simplicity of this formula stems from the use of noise temperature as the measure of how noisy the system is.

EXAMPLE 3

Consider a receiver with a cryogenically cooled amplifier. The equivalent noise temperature of the receiver is 20°K. The ground station uses a large antenna operating at a frequency of 4 GHz and an elevation of 5°; the antenna noise temperature is estimated to be 50°K. Calculate the system noise temperature, assuming no loss in the antenna feed and waveguide. Hence, calculate the carrier-to-noise ratio, assuming a carrier power of −123 dBW (as in Example 2) and a noise bandwidth of 36 MHz.

From Eq. C.34 we see that with $L_w = 1$ the system noise temperature equals

$$\begin{aligned} T_s &= T_a + T_e \\ &= 50 + 20 \\ &= 70°\text{K} \end{aligned}$$

Hence, the available noise power equals

$$\begin{aligned} kT_sB_N &= 1.38 \times 10^{-23} \times 70 \times 36 \times 10^6 \\ &= 3.48 \times 10^{-14}\ \text{W} \\ &= -134.6\ \text{dBW} \end{aligned}$$

Thus, the use of Eq. C.35 yields the following value for the carrier-to-noise ratio expressed in decibels:

$$\begin{aligned} 10 \log_{10}(\text{CNR}) &= -123 + 134.6 \\ &= 11.6\ \text{dB} \end{aligned}$$

APPENDIX D

TABLES

In Tables 1 through 8 of Appendix D, we present the following material:

1. Summary of properties of the Fourier transform.
2. Short table of Fourier transform pairs.
3. Table of trigonometric identities.
4. Short table of series expansions and summations.
5. Short table of integrals.
6. Values of the error function erf(u) for u in the range 0 to 3.30.
7. List of useful constants.
8. List of recommended unit prefixes.

TABLE 1 Properties of the Fourier Transform

Property	Mathematical Description
1. Linearity	$ag_1(t) + bg_2(t) \rightleftharpoons aG_1(f) + bG_2(f)$ where a and b are constants
2. Time scaling	$g(at) \rightleftharpoons \frac{1}{\lvert a \rvert} G\left(\frac{f}{a}\right)$ where a is a constant
3. Duality	If $g(t) \rightleftharpoons G(f)$, then $G(t) \rightleftharpoons g(-f)$
4. Time shifting	$g(t - t_0) \rightleftharpoons G(f) \exp(-j2\pi f t_0)$
5. Frequency shifting	$\exp(j2\pi f_c t) g(t) \rightleftharpoons G(f - f_c)$
6. Area under $g(t)$	$\int_{-\infty}^{\infty} g(t)\, dt = G(0)$
7. Area under $G(f)$	$g(0) = \int_{-\infty}^{\infty} G(f)\, df$
8. Differentiation in the time domain	$\frac{d}{dt} g(t) \rightleftharpoons j2\pi f G(f)$
9. Integration in the time domain	$\int_{-\infty}^{t} g(\tau)\, d\tau \rightleftharpoons \frac{1}{j2\pi f} G(f) + \frac{G(0)}{2} \delta(f)$
10. Conjugate functions	If $g(t) \rightleftharpoons G(f)$, then $g^*(t) \rightleftharpoons G^*(-f)$
11. Multiplication in the time domain	$g_1(t) g_2(t) \rightleftharpoons \int_{-\infty}^{\infty} G_1(\lambda) G_2(f - \lambda)\, d\lambda$
12. Convolution in the time domain	$\int_{-\infty}^{\infty} g_1(\tau) g_2(t - \tau)\, d\tau \rightleftharpoons G_1(f) G_2(f)$

TABLE 2 Fourier Transform Pairs

Time Function	Fourier Transform
$\text{rect}\left(\frac{t}{T}\right)$	$T\,\text{sinc}(fT)$
$\text{sinc}(2Wt)$	$\frac{1}{2W}\,\text{rect}\left(\frac{f}{2W}\right)$
$\exp(-at)u(t), \quad a > 0$	$\frac{1}{a + j2\pi f}$
$\exp(-a\lvert t\rvert), \quad a > 0$	$\frac{2a}{a^2 + (2\pi f)^2}$
$\exp(-\pi t^2)$	$\exp(-\pi f^2)$
$\begin{cases} 1 - \frac{\lvert t\rvert}{T}, & \lvert t\rvert < T \\ 0, & \lvert t\rvert \geq T \end{cases}$	$T\,\text{sinc}^2(fT)$
$\delta(t)$	1
1	$\delta(f)$
$\delta(t - t_0)$	$\exp(-j2\pi f t_0)$
$\exp(j2\pi f_c t)$	$\delta(f - f_c)$
$\cos(2\pi f_c t)$	$\frac{1}{2}[\delta(f - f_c) + \delta(f + f_c)]$
$\sin(2\pi f_c t)$	$\frac{1}{2j}[\delta(f - f_c) - \delta(f + f_c)]$
$\text{sgn}(t)$	$\frac{1}{j\pi f}$
$\frac{1}{\pi t}$	$-j\,\text{sgn}(f)$
$u(t)$	$\frac{1}{2}\delta(f) + \frac{1}{j2\pi f}$
$\sum_{i=-\infty}^{\infty} \delta(t - iT_0)$	$\frac{1}{T_0}\sum_{n=-\infty}^{\infty} \delta\left(f - \frac{n}{T_0}\right)$

TABLE 3 Trigonometric Identities

$$\exp(\pm j\theta) = \cos\theta \pm j\sin\theta$$

$$\cos\theta = \tfrac{1}{2}[\exp(j\theta) + \exp(-j\theta)]$$

$$\sin\theta = \frac{1}{2j}[\exp(j\theta) - \exp(-j\theta)]$$

$$\sin^2\theta + \cos^2\theta = 1$$

$$\cos^2\theta - \sin^2\theta = \cos(2\theta)$$

$$\cos^2\theta = \tfrac{1}{2}[1 + \cos(2\theta)]$$

$$\sin^2\theta = \tfrac{1}{2}[1 - \cos(2\theta)]$$

$$2\sin\theta\cos\theta = \sin(2\theta)$$

$$\sin(\alpha \pm \beta) = \sin\alpha\cos\beta \pm \cos\alpha\sin\beta$$

$$\cos(\alpha \pm \beta) = \cos\alpha\cos\beta \mp \sin\alpha\sin\beta$$

$$\tan(\alpha \pm \beta) = \frac{\tan\alpha \pm \tan\beta}{1 \mp \tan\alpha\tan\beta}$$

$$\sin\alpha\sin\beta = \tfrac{1}{2}[\cos(\alpha - \beta) - \cos(\alpha + \beta)]$$

$$\cos\alpha\cos\beta = \tfrac{1}{2}[\cos(\alpha - \beta) + \cos(\alpha + \beta)]$$

$$\sin\alpha\cos\beta = \tfrac{1}{2}[\sin(\alpha - \beta) + \sin(\alpha + \beta)]$$

TABLE 4 Series Expansions and Summations

1. *Expansions*

Taylor series

$$f(x) = f(a) + \frac{f'(a)}{1!}(x - a) + \frac{f''(a)}{2!}(x - a)^2 + \cdots + \frac{f^{(n)}(a)}{n!}(x - a)^n + \cdots$$

where

$$f^{(n)}(a) = \left.\frac{d^n f(x)}{dx^n}\right|_{x=a}$$

MacLaurin series

$$f(x) = f(0) + \frac{f'(0)}{1!}x + \frac{f''(0)}{2!}x^2 + \cdots + \frac{f^{(n)}(0)}{n!}x^n + \cdots$$

where

$$f^{(n)}(0) = \left.\frac{d^n f(x)}{dx^n}\right|_{x=0}$$

Binomial series

$$(1 + x)^n = 1 + nx + \frac{n(n-1)}{2!}x^2 + \frac{n(n-1)(n-2)}{3!}x^3 + \cdots, \qquad |nx| < 1$$

Exponential series

$$\exp x = 1 + x + \frac{1}{2!}x^2 + \frac{1}{3!}x^3 + \cdots$$

Logarithmic series

$$\ln(1 + x) = x - \tfrac{1}{2}x^2 + \tfrac{1}{3}x^3 - \cdots$$

Trigonometric series

$$\sin x = x - \frac{1}{3!}x^3 + \frac{1}{5!}x^5 - \cdots$$

$$\cos x = 1 - \frac{1}{2!}x^2 + \frac{1}{4!}x^4 - \cdots$$

$$\tan x = x + \tfrac{1}{3}x^3 + \tfrac{2}{15}x^5 + \cdots$$

$$\sin^{-1} x = x + \tfrac{1}{6}x^3 + \tfrac{3}{40}x^5 + \cdots$$

$$\tan^{-1} x = x - \tfrac{1}{3}x^3 + \tfrac{1}{5}x^5 - \cdots, \qquad |x| < 1$$

$$\text{sinc}\, x = 1 - \frac{1}{3!}(\pi x)^2 + \frac{1}{5!}(\pi x)^4 - \cdots$$

2. *Summations*

Arithmetic series

$$\sum_{n=1}^{N} n = \frac{N(N+1)}{2}$$

Geometric series

$$\sum_{n=0}^{N} r^n = \frac{1 - r^{N+1}}{1 - r}$$

TABLE 5 Integrals

Indefinite integrals

$$\int x \sin(ax)\, dx = \frac{1}{a^2}[\sin(ax) - ax\cos(ax)]$$

$$\int x \cos(ax)\, dx = \frac{1}{a^2}[\cos(ax) + ax\sin(ax)]$$

$$\int x \exp(ax)\, dx = \frac{1}{a^2}\exp(ax)(ax - 1)$$

$$\int x \exp(ax^2)\, dx = \frac{1}{2a}\exp(ax^2)$$

$$\int \exp(ax)\sin(bx)\, dx = \frac{1}{a^2 + b^2}\exp(ax)[a\sin(bx) - b\cos(bx)]$$

$$\int \exp(ax)\cos(bx)\, dx = \frac{1}{a^2 + b^2}\exp(ax)[a\cos(bx) + b\sin(bx)]$$

$$\int \frac{dx}{a^2 + b^2x^2} = \frac{1}{ab}\tan^{-1}\left(\frac{bx}{a}\right)$$

$$\int \frac{x^2\, dx}{a^2 + b^2x^2} = \frac{x}{b^2} - \frac{a}{b^3}\tan^{-1}\left(\frac{bx}{a}\right)$$

Definite integrals

$$\int_0^\infty \frac{x\sin(ax)}{b^2 + x^2}\, dx = \frac{\pi}{2}\exp(-ab), \qquad a > 0,\ b > 0$$

$$\int_0^\infty \frac{\cos(ax)}{b^2 + x^2}\, dx = \frac{\pi}{2b}\exp(-ab), \qquad a > 0,\ b > 0$$

$$\int_0^\infty \frac{\cos(ax)}{(b^2 - x^2)^2}\, dx = \frac{\pi}{4b^3}[\sin(ab) - ab\cos(ab)], \qquad a > 0,\ b > 0$$

$$\int_0^\infty \operatorname{sinc} x\, dx = \int_0^\infty \operatorname{sinc}^2 x\, dx = \tfrac{1}{2}$$

$$\int_0^\infty \exp(-ax^2)\, dx = \frac{1}{2}\sqrt{\frac{\pi}{a}}, \qquad a > 0$$

$$\int_0^\infty x^2\exp(-ax^2)\, dx = \frac{1}{4a}\sqrt{\frac{\pi}{a}}, \qquad a > 0$$

Integration by parts

$$\int f(t)g'(t)\, dt = f(t)g(t) - \int f'(t)g(t)\, dt$$

where

$$f'(t) = \frac{df(t)}{dt}$$

and

$$g'(t) = \frac{dg(t)}{dt}$$

TABLE 6 Error Function

u	erf(u)	u	erf(u)
0.00	0.00000	1.10	0.88021
0.05	0.05637	1.15	0.89612
0.10	0.11246	1.20	0.91031
0.15	0.16800	1.25	0.92290
0.20	0.22270	1.30	0.93401
0.25	0.27633	1.35	0.94376
0.30	0.32863	1.40	0.95229
0.35	0.37938	1.45	0.95970
0.40	0.42839	1.50	0.96611
0.45	0.47548	1.55	0.97162
0.50	0.52050	1.60	0.97635
0.55	0.56332	1.65	0.98038
0.60	0.60386	1.70	0.98379
0.65	0.64203	1.75	0.98667
0.70	0.67780	1.80	0.98909
0.75	0.71116	1.85	0.99111
0.80	0.74210	1.90	0.99279
0.85	0.77067	1.95	0.99418
0.90	0.79691	2.00	0.99532
0.95	0.82089	2.50	0.99959
1.00	0.84270	3.00	0.99998
1.05	0.86244	3.30	0.999998

The *error function* erf(u) is defined by

$$\text{erf}(u) = \frac{2}{\sqrt{\pi}} \int_0^u \exp(-z^2)\, dz$$

Note that

$$\text{erf}(0) = 0$$
$$\text{erf}(\infty) = 1$$
$$\text{erf}(-u) = -\text{erf}(u)$$

The *complementary error function* erfc(u) is defined by

$$\text{erfc}(u) = \frac{2}{\sqrt{\pi}} \int_u^\infty \exp(-z^2)\, dz$$

These two functions are related by

$$\text{erfc}(u) = 1 - \text{erf}(u)$$

TABLE 7 Useful Constants

Physical Constants	
Boltzmann's constant	$k = 1.38 \times 10^{-23}$ joule/degree Kelvin
Planck's constant	$h = 6.626 \times 10^{-34}$ joule·second
Electron (fundamental) charge	$q = 1.602 \times 10^{-19}$ coulomb
Speed of light in vacuum	$c = 2.998 \times 10^{8}$ meters/second
Standard (absolute) temperature	$T_0 = 273$ degree Kelvin
Thermal voltage	$V_T = 0.026$ volt at room temperature
Thermal energy kT at standard temperature	$kT_0 = 3.77 \times 10^{-21}$ joule
One hertz (Hz) = 1 cycle/second	
One watt (W) = 1 joule/second	
Mathematical Constants	
Base of natural logarithm	$e = 2.7182818$
Logarithm of e to base 2	$\log_2 e = 1.442695$
Logarithm of 2 to base e	$\ln 2 = 0.693147$
Logarithm of 2 to base 10	$\log_{10} 2 = 0.30103$
Pi	$\pi = 3.1415927$

TABLE 8 Recommended Unit Prefixes

Multiples and Submultiples	Prefixes	Symbols
10^{12}	tera	T
10^{9}	giga	G
10^{6}	mega	M
10^{3}	kilo	k
10^{-3}	milli	m
10^{-6}	micro	μ
10^{-9}	nano	n
10^{-12}	pico	p

GLOSSARY

CONVENTIONS AND NOTATIONS

1. The symbol $|\ |$ means the magnitude of the complex quantity contained within.
2. The symbol arg( ) means the phase angle of the complex quantity contained within.
3. The symbol Re[ ] means the "real part of," and Im[ ] means the "imaginary part of."
4. The symbol ln($\cdot$) denotes the natural logarithm of the quantity contained within, whereas the logarithm to the base a is denoted by $\log_a(\ )$. The symbol exp( ) denotes the exponential function; for example, $\exp(x)$ denotes e^x, where e is the base of the natural logarithm.
5. The use of an asterisk as superscript denotes complex conjugate; for example, x^* is the complex conjugate of x.
6. The symbol $\rightleftharpoons$ indicates a Fourier transform pair. For example, let $g(t)$ denote a time function and $G(f)$ denote its Fourier transform; we then write $g(t) \rightleftharpoons G(f)$.
7. The symbol F[ ] indicates the Fourier transform operation; for example $F[g(t)] = G(f)$. The symbol $F^{-1}[\ ]$ indicates the inverse Fourier transform operation; for example $F^{-1}[G(f)] = g(t)$
8. The symbol ☆ denotes convolution; for example

$$x(t) \star h(t) = \int_{-\infty}^{\infty} x(\tau)h(t-\tau)\,d\tau$$

9. The symbol $\oplus$ denotes modulo-two addition.
10. The use of subscript p indicates that the pertinent function is periodic; for example, the function $g_p(t)$ is a periodic function of time t.
11. The use of a caret (hat) over a function indicates one of two things:
    (a) The Hilbert transform of a function. For example, the function $\hat{g}(t)$ is the Hilbert transform of $g(t)$.
    (b) The estimate of an unknown parameter. For example, the quantity $\hat{\alpha}(\mathbf{x})$ is an estimate of the unknown parameter α, based on the observation vector $\mathbf{x}$.
12. The use of a tilde over a function indicates the complex envelope of a narrow-band signal. For example, the function $\tilde{g}(t)$ is the complex envelope of the narrow-band signal $g(t)$.
13. The use of subscript + indicates the pre-envelope of a signal. For example, the function $g_+(t)$ is the pre-envelope of the signal $g(t)$. We may thus write $g_+(t) = g(t) + j\hat{g}(t)$, where $\hat{g}(t)$ is the Hilbert transform of $g(t)$.

14. The use of subscripts I and Q denotes the in-phase and quadrature components of a narrow-band signal or a narrow-band random process with respect to the carrier $\cos(2\pi f_c t)$.
15. The term "baseband" refers to the band of frequencies representing the original signal delivered by a source of information.
16. In the context of a probability system, $\mathcal{S}$ denotes a sample space of elementary events, and $\mathcal{E}$ denotes a class of events that is a subset of the sample space $\mathcal{S}$.
17. Random variables are uppercase (e.g., X or $\mathbf{X}$), whereas sample values of random variables are lowercase (e.g., x or $\mathbf{x}$).
18. The symbol $E[\ ]$ means the expected value of the random variable enclosed within.
19. The symbol Var[ ] means the variance of the random variable enclosed within.
20. The symbol Cov [ ] means the covariance of the two random variables enclosed within.
21. The average probability of symbol error is denoted by P_e. In the case of binary signaling techniques, P_{e0} denotes the conditional probability of error given that symbol 0 was transmitted, and P_{e1} denotes the conditional probability of error given that symbol 1 was transmitted. The a priori probabilities of symbols 0 and 1 are denoted by p and q, respectively.
22. The symbol $\langle\ \rangle$ denotes the time average of the sample function enclosed within.
23. A boldface lowercase letter denotes a vector, and a boldface uppercase letter denotes a matrix.

FUNCTIONS

1. Rectangular function $\quad \text{rect}(t) = \begin{cases} 1, & -\frac{1}{2} < t < \frac{1}{2} \\ 0, & |t| > \frac{1}{2} \end{cases}$

2. Unit step function $\quad u(t) = \begin{cases} 1, & t > 0 \\ 0, & t < 0 \end{cases}$

3. Signum function $\quad \text{sgn}(t) = \begin{cases} 1, & t > 0 \\ -1, & t < 0 \end{cases}$

4. Dirac delta function $\quad \delta(t) = 0, \quad t \neq 0$

$$\int_{-\infty}^{\infty} \delta(t)\, dt = 1$$

or equivalently

$$\int_{-\infty}^{\infty} g(t)\, \delta(t - t_0)\, dt = g(t_0)$$

5. Sinc function $\mathrm{sinc}(x) = \frac{\sin(\pi x)}{\pi x}$

6. Sine integral $\mathrm{Si}(u) = \int_0^u \frac{\sin x}{x}\, dx$

7. Error function $\mathrm{erf}(u) = \frac{2}{\sqrt{\pi}} \int_0^u \exp(-z^2)\, dz$

   Complementary error function $\mathrm{erfc}(u) = 1 - \mathrm{erf}(u)$

8. Bessel function of the first kind of order n $J_n(x) = \frac{1}{2\pi} \int_{-\pi}^{\pi} \exp(jx \sin\theta - jn\theta)\, d\theta$

ABBREVIATIONS

ac: alternating current
ADM: adaptive delta modulation
APK: amplitude-phase keying
AWGN: additive white Gaussian noise
AM: amplitude modulation
ASK: amplitude-shift keying
BPF: band-pass filter
CDMA: code-division multiple access
CNR: carrier-to-noise ratio
CPFSK: continuous-phase frequency-shift keying
CW: continuous wave
dB: decibel
dBm: decibel milliwatt
dBW: decibel watt
dc: direct current
DM: delta modulation
DPCM: differential pulse-code modulation
DPSK: differential phase-shift keying
DSBSC: double-sideband suppressed-carrier
EIRP: effective isotropic radiated power
FDM: frequency-division multiplexing
FDMA: frequency-division multiple access
FET: field-effect transistor
FFT: fast Fourier transform
FIR: finite-duration impulse response
FM: frequency modulation
FMFB: frequency modulator with feedback

FSK: frequency-shift keying
HDTV: high-definition television
Hz: hertz
IF: intermediate frequency
iid: independent and identically distributed
ISI: intersymbol interference
LMS: least mean-square
LPF: low-pass filter
modem: modulator-demodulator
MSK: minimum shift keying
NRZ: nonreturn-to-zero
NTSC: National Television System Committee
PAM: pulse-amplitude modulation
PCM: pulse-code modulation
PDM: pulse-duration modulation
PLL: phase-locked loop
PM: phase modulation
PN: pseudonoise
PPM: pulse-position modulation
PSK: phase-shift keying
QAM: quadrature-amplitude modulation
QPSK: quadriphase-shift keying
rad: radian
RF: radio frequency
rms: root mean-square
RZ: return-to-zero
s: second
SLSC: split-luminance and split-chrominance
SNR: signal-to-noise ratio
$(SNR)_C$: channel signal-to-noise ratio
$(SNR)_O$: output signal-to-noise ratio
SSB: single sideband
TDM: time-division multiplexing
TDMA: time-division multiple access
TV: television
V: volt
VCO: voltage-controlled oscillator
VLSI: very large-scale integration
VSB: vestigial sideband
WSS: wide-sense stationary
W: watt

REFERENCES AND BIBLIOGRAPHY

M. Abramowitz and I. A. Stegun, *Handbook of Mathematical Functions,* Dover, New York, 1965.

D. J. Angelakos and T. E. Everhart, *Microwave Communications,* McGraw-Hill New York, 1968.

E. H. Armstrong, "A Method of Reducing Disturbances in Radio Signaling by a System of Frequency Modulation," *Proc. IRE,* vol. 24, pp. 689–740, May 1936.

Bell Telephone Laboratories, *Transmission Systems for Communications,* Western Electric, Winston-Salem, North Carolina, 1970.

D. A. Bell, *Noise and the Solid State,* Pentech Press, Plymouth, England, 1985.

W. R. Bennett, "*Spectra of Quantized Signals,*" Bell Syst. Tech. J., vol. 27, pp. 446–472, July 1948.

W. R. Bennett, *Introduction to Signal Transmission,* McGraw-Hill, New York, 1970.

L. V. Berkner, "Communications" in C. F. J. Overhage (ed.), *The Age of Electronics,* McGraw-Hill, New York, 1962, pp. 35–50.

T. G. Birdsall, "On Understanding the Matched Filter in the Frequency Domain," *IEEE Trans. Educ.,* vol. E.19, pp. 168–169, November 1976.

H. S. Black, *Modulation Theory,* Van Nostrand, Princeton, N.J., 1953.

R. B. Blackman and J. W. Tukey, *The Measurement of Power Spectra,* Dover, New York, 1958.

I. F. Blake, *An Introduction to Applied Probability,* Wiley, New York, 1979.

R. Bracewell, *The Fourier Transform and its Applications,* 2nd ed., McGraw-Hill, New York, 1978.

J. Brown and E. V. D. Glazer, *Telecommunications,* 2nd ed., Wiley, New York, 1974.

G. A. Campbell and R. M. Foster, *Fourier Integrals for Practical Applications,* Van Nostrand, Princeton, N.J., 1948.

A. B. Carlson, *Communication Systems,* 3rd ed., McGraw-Hill, New York, 1986.

C. R. Carson and T. C. Fry, "Variable Frequency Electric Circuit Theory with Application to the Theory of Frequency Modulation," *Bell Syst. Tech. J.,* vol. 16, pp. 513–540, October 1937.

R. V. Churchill, *Fourier Series and Boundary Value Problems,* 2nd ed., McGraw-Hill, New York, 1963.

J. P. Costas, "Synchronous Communications," *Proc. IRE,* vol. 44, pp. 1713–1718, December 1956.

W. B. Davenport and W. L. Root, *Random Signals and Noise,* McGraw-Hill, New York, 1958.

P. A. M. Dirac, *The Principles of Quantum Mechanics,* 3rd ed., Oxford University Press, London, 1947.

J. J. Downing, *Modulation Systems and Noise,* Prentice-Hall, Englewood Cliffs, N.J., 1964.

A. Einstein, "Method for the Determination of the Statistical Values of Observations Concerning Quantities Subject to Irregular Fluctuations," *Archives des Sciences Physiques et Naturalles,* vol. 37, pp. 254–256, 1914.

I. H. Enole, "Decreasing the Threshold in FM by Frequency Feedback," *Proc. IRE,* vol. 50, pp. 18–30, January 1962.

W. Feller, *An Introduction to Probability Theory and Its Applications,* 3rd ed., Wiley, New York, 1968.

T. L. Fines, *Theories of Probability,* Academic Press, New York, 1973.

J. B. J. Fourier, *The Analytical Theory of Heat* (trans. A. Freeman), Cambridge University Press, London, 1878.

H. T. Friis, "Noise Figures in Radio Receivers," *Proc. IRE,* vol. 32, pp. 419–422, July 1944.

T. C. Fry, *Probability and Its Engineering Uses,* Van Nostrand-Reinhold, Princeton, N.J., 1965.

D. Gabor, "Theory of Communication," *J. IEE* (London), vol. 93, Part 3, pp. 429–441, November 1946.

R. M. Gagliardi, *Introduction to Communications Engineering,* Wiley, New York, 1978.

F. M. Gardner, *Phaselock Techniques,* 2nd ed., Wiley, New York, 1979.

W. A. Gardner, *Statistical Spectral Analysis: A Non-Probabilistic Theory,* Prentice-Hall, Englewood, N.J., 1987.

S. Goldman, *Frequency Analysis, Modulation, and Noise,* McGraw-Hill, New York, 1948.

G. W. Golomb (ed.), *Digital Communications with Space Applications,* Prentice-Hall, Englewood Cliffs, N.J., 1964.

S. Haykin, *Communication Systems,* 2nd ed., Wiley, New York, 1983.

S. Haykin, *Adaptive Filter Theory,* Prentice-Hall, Englewood Cliffs, N.J., 1986.

S. Haykin, *Digital Communications,* Wiley, New York, 1988.

C. W. Helstrom, *Statistical Theory of Signal Detection,* Pergamon, Elmsford, N.Y., 1968.

F. S. Hill, Jr., "On Time Domain Representations for Vestigial Sideband Signals," *Proc. IEEE,* vol. 62, pp. 1032–1033, July 1974.

N. S. Jayant and P. Noll, *Digital Coding of Waveforms,* Prentice-Hall, Englewood Cliffs, N.J., 1984.

A. J. Jerri, "The Shannon Sampling Theorem—Its Various Extensions and Applications: A Tutorial Review," *Proc. IEEE,* vol. 65, pp. 1565–1596, 1977.

J. B. Johnson, "Thermal Agitation of Electricity in Conductors," *Physical Review,* vol. 32, pp. 97–109, July 1928.

A. I. Khintchine, "Correlation Theory of Stochastic Processes," *Math. Ann.,* vol. 109, pp. 604–615, 1934.

B. P. Lathi, *Modern Digital and Analog Communication Systems,* Holt, Rinehart & Winston, New York, 1983.

A. Lender, "The Duobinary Technique for High Speed Data Transmission," *IEEE Trans. Commun. Elec.*, vol. 82, pp. 214–218, 1963.

M. J. Lighthill, *An Introduction to Fourier Analysis and Generalized Functions,* Cambridge University Press, London, 1959.

W. C. Lindsey and M. K. Simon, *Telecommunication Systems Engineering,* Prentice-Hall, Englewood Cliffs, N.J., 1973.

J. M. Manley, "The Concept of Frequency in Linear System Analysis," *IEEE Communications Magazine,* vol. 20, pp. 26–35, January 1982.

S. J. Mason and H. J. Żimmerman, *Electronic Circuits, Signals, and Systems,* Wiley, New York, 1960.

D. Mennie, "AM Stereo: Five Competing Options," *IEEE Spectrum,* vol. 15, pp. 56–58, 1978.

D. Middleton, *An Introduction to Statistical Communication Theory,* McGraw-Hill, New York, 1960.

D. O. North, "An Analysis of the Factors which Determine Signal/Noise Discrimination in Pulsed-carrier Systems," Proc. IEEE, vol. 51, pp. 1016–1027, July 1963.

H. Nyquist, "Certain Topics in Telegraph Transmission Theory," *Trans. AIEE,* vol. 47, pp. 617–644, 1928.

H. Nyquist, "Thermal Agitation of Electric Charge in Conductors," *Physical Review,* vol. 32, pp. 110–113, July 1928.

B. M. Oliver, J. R. Pierce, and C. E. Shannon, "The Philosophy of PCM," *Proc. IRE,* vol. 36, pp. 1324–1332, 1948.

A. V. Oppenheim and R. W. Schafer, *Digital Signal Processing,* Prentice-Hall, Englewood Cliffs, N.J., 1975.

P. F. Panter, *Modulation, Noise, and Spectral Analysis,* McGraw-Hill, New York, 1965.

A. Papoulis, *The Fourier Integral and Its Applications,* McGraw-Hill, New York, 1962.

A. Papoulis, *Probability, Random Variables, and Stochastic Processes,* 2nd ed., McGraw-Hill, New York, 1984.

S. Pasupathy, "Correlative Coding: A Bandwidth-efficient Signaling Scheme," *IEEE Communications Magazine,* vol. 15, pp. 4–11, July 1977.

J. R. Pierce, *Symbols, Signals and Noise,* Harper, New York, 1961.

J. R. Pierce and E. C. Posner, *Introduction to Communication Science and Systems,* Plenum, New York, 1980.

T. Pratt and C. W. Bostian, *Satellite Communications,* Wiley, New York, 1986.

A. H. Reeves, "The Past, Present, and Future of PCM," *IEEE Spectrum,* vol. 12, pp. 58–63, May 1975.

S. O. Rice, "Mathematical Analysis of Random Noise," *Bell Syst. Tech. J.*, vol. 23, pp. 282–333, July 1944; vol. 24, pp. 96–157, January 1945.

S. O. Rice, "Statistical Properties of a Sine-Wave Plus Random Noise," *Bell Syst. Tech. J.*, vol. 27, pp. 109–157, 1948.

S. O. Rice, "Noise in FM Receivers," in M. Rosenblatt (ed.), *Symposium Proceedings of Time Series Analysis*, (Wiley, New York, 1963), pp. 395–422.

S. O. Rice, "Envelopes of Narrow-band Signals," *Proc. IEEE*, vol. 70, pp. 692–699, July 1982.

J. H. Roberts, *Angle Modulation: The Theory of System Assessment*, IEE Communication Series 5, Institution of Electrical Engineers, London, 1977.

R. A. Roberts and C. P. Mullis, *Digital Signal Processing*, Addison-Wesley, Reading, Mass., 1987.

F. N. Robinson, *Noise and Fluctuations in Electronic Devices and Circuits*, Clarendon Press, Oxford, 1974.

H. E. Rowe, *Signals and Noise in Communication Systems*, Van Nostrand, Princeton, N.J., 1965.

T. S. Rzeszewski, "A compatible high-definition television system," *Bell Syst. Tech. J.*, vol. 62, pp. 2091–2111, Sept. 1983.

T. S. Rzeszewski, (ed.), *Television Technology Today*, IEEE Press, New York, 1985.

D. J. Sakrison, *Communication Theory: Transmission of Waveforms and Digital Information*, Wiley, New York, 1968.

D. V. Sarwate and M. B. Pursley, "Crosscorrelation Properties of Pseudo Random and Related Sequences," *Proc. IEEE*, vol. 68, pp. 593–619, May 1980.

M. Schwartz, W. R. Bennett, and S. Stein, *Communication Systems and Techniques*, McGraw-Hill, New York, 1966.

M. Schwartz, *Information Transmission, Modulation, and Noise*, 3rd ed., McGraw-Hill, New York, 1980.

K. S. Shanmugam, *Digital and Analog Communication Systems*, Wiley, New York, 1979.

C. E. Shannon, "A Mathematical Theory of Communication," *Bell Syst. Tech. J.*, vol. 27, (Part I), pp. 379–423, (Part II), pp. 623–656, 1948.

C. E. Shannon, "Communication in the Presence of Noise," *Proc. IRE*, vol. 37, pp. 10–21, January 1949.

D. Slepian, "On Bandwidth," *Proc. IEEE*, vol. 64, pp. 292–300, March 1976.

H. Stark and F. B. Tuteur, *Modern Electrical Communications: Theory and Systems*, Prentice-Hall, Englewood Cliffs, N.J., 1979.

F. G. Stremler, *Introduction to Communication Systems*, 2nd ed., Addison–Wesley, Reading, Mass., 1982.

F. L. Stumpers, "Theory of Frequency-Modulation Noise," *Proc. I.R.E.*, vol. 36, pp. 1081–1902, September 1948.

H. Taub and D. L. Schilling, *Principles of Communication Systems*, 2nd ed., McGraw-Hill, New York, 1986.

F. E. Terman, *Electronic and Radio Engineering*, 4th ed., McGraw-Hill, New York, 1955.

J. B. Thomas, *Introduction to Probability,* Springer-Verlag, New York, 1986.

G. L. Turin, "An Introduction to Matched Filters," *IRE Trans. Information Theory,* vol. IT-6, pp. 311–329, June 1960.

A. Van Der Ziel, *Noise: Sources, Characterization, Measurement,* Prentice-Hall, Englewood Cliffs, N.J., 1970.

H. L. van Trees, *Detection, Estimation, and Modulation Theory, Part I,* Wiley, New York, 1968.

A. D. Whalen, *Detection of Signals in Noise,* Academic Press, New York, 1971.

E. T. Whittaker and G. N. Watson, *A Course of Modern Analysis,* 4th ed., Cambridge University Press, London, 1927.

B. Widrow and S. D. Stearns, *Adaptive Signal Processing,* Prentice-Hall, Englewood Cliffs, N.J., 1985.

N. Wiener, "Generalized Harmonic Analysis," *Ann. Math.*, vol. 55, pp. 117–258, 1930.

R. A. Williams, *Communication Systems Analysis and Design,* Prentice-Hall, Englewood Cliffs, N.J., 1987.

P. M. Woodward, *Probability and Information Theory, with Applications to Radar,* 2nd ed., Pergamon, Elmsford, N.Y., 1964.

J. M. Wozencroft and I. M. Jacobs, *Principles of Communication Engineering,* Wiley, New York, 1965.

J. H. Yuen (ed.), *Deep Space Telecommunications Systems Engineering,* Plenum, New York, 1983.

ANSWERS TO EXERCISES

Many of the exercises have their answers integrated into them. The answers to the remaining exercises are presented below on a chapter-by-chapter basis.

CHAPTER 2

Exercise 1: The amplitude spectra have a form similar to the amplitude of a sinc function. In case (a), the main lobe will be one-quarter the height and 2.5 times the width of that in case (b).
Exercise 4: $g(t) = 1/(1 - j2\pi t) = 1/[1 + (2\pi t)^2] + j2\pi t/[1 + (2\pi t)^2]$
Exercise 5: $G(f) = (1/2j)\,[1/(1 + j2\pi(f - f_c)) - 1/(1 + j2\pi(f + f_c))]$
Exercise 9: On the left side of the relation, $g_2(t)$ is replaced by $g_2^*(t)$.
Exercise 10: On the right side of the relation, $G_2(f)$ is replaced by $G_2^*(f)$. The result of changes in Exercise 10 will be the same as those in Exercise 9, as shown by $\int_{-\infty}^{\infty} g_1(t)g_2^*(t)\,dt = \int_{-\infty}^{\infty} G_1(f)\,G_2^*(f)\,df$
Exercise 11: Bandwidth $= 1/T$.
Exercise 12: Bandwidth $= a/2\pi$.
Exercise 15: $\cos(2\pi f_c t)u(t) \rightleftharpoons f/[j2\pi(f^2 - f_c^2)] + \frac{1}{4}[\delta(f - f_c) + \delta(f + f_c)]$

CHAPTER 3

Exercise 1: The system is noncausal because $h(t)$ is nonzero for negative time. The system, however, is stable because $\int_{-\infty}^{\infty} |h(t)|\,dt = \int_{-\infty}^{0} \exp(at)\,dt = 1/a$.
Exercise 4: The modification may arise through the use of a phase-inverting amplifier or transformer.
Exercise 5: $[1/H_c(f)] = \sum_{m=-M}^{M} C_m \exp(-j2\pi fm\,\Delta\tau)$ where M is some finite integer.
Exercise 6: In theory, the delay τ_0 would have to be infinitely large for the ideal low-pass filter to be causal.
Exercise 8: In-phase component is $m(t)$, and quadrature component is zero.
Exercise 9: $y(0) = \int_{-\infty}^{\infty} H(f)X(f)\,df = \int_{-\infty}^{\infty} \text{Re}[H(f)X(f)]df$; $\tilde{y}(0) = \int_{-\infty}^{\infty} \tilde{H}(f)\tilde{X}(f)\,df$; $y(0) = \text{Re}[\tilde{y}(0)]$. Hence, $\text{Re}[H(f)X(f)] = \text{Re}[\tilde{H}(f)\tilde{X}(f)]$. The product $H(f)X(f)$ occupies the band $f_c - W \leqslant f \leqslant f_c + W$ for positive frequencies, and its image for negative frequencies. On the other hand, the product $\tilde{H}(f)\tilde{X}(f)$ occupies the band $-W \leqslant f \leqslant W$. With $\tilde{X}(0) = 2X(f_c)$, it follows that we must have $\tilde{H}(0) = H(f_c)$.
Exercise 10: Phase delay $\tau_p = \beta(f)/f$, where $\beta(f)$ is the phase. If $\beta(f)$ is constant for all f, the phase delay becomes inversely proportional to frequency, and phase distortion results.

CHAPTER 4

Exercise 4: $R_g(0) = \int_{-\infty}^{\infty} \Psi_g(f)\,df$; $\Psi_g(0) = \int_{-\infty}^{\infty} R_g(\tau)\,d\tau$
Exercise 10: *Direct procedure:* Use the formula for the autocorrelation function of the energy signal $g(t)$. *Indirect procedure:* 1. Compute the Fourier transform $G(f)$

of the signal $g(t)$. 2. Compute the energy spectral density $\Psi_g(f) = |G(f)|^2$. 3. Compute the inverse Fourier transform of $\Psi_g(f)$.

Exercise 11: *Direct procedure:* Use the formula for the autocorrelation function of the power signal $g(t)$. *Indirect procedure:* 1. Compute the Fourier transform $G_T(f)$ of the truncated signal $g_T(t)$ for large T. 2. Compute $|G_T(f)|^2$. 3. Compute the inverse Fourier transform of $|G_T(f)|^2$.

Exercise 15:

$$R_{g_p}(\tau) = \begin{cases} \dfrac{A^2}{2}\left(1 + \dfrac{2\tau}{T_0}\right) & -\dfrac{T_0}{4} \leq \tau \leq 0 \\ \dfrac{A^2}{2}\left(1 - \dfrac{2\tau}{T_0}\right) & 0 \leq \tau \leq \dfrac{T_0}{4} \\ 0, & \text{for the remainder of the period} \end{cases}$$

$$S_{g_p}(f) = \frac{A^2}{16} \sum_{n=-\infty}^{\infty} \text{sinc}^2\left(\frac{n}{4}\right) \delta\left(f - \frac{n}{T_0}\right)$$

Exercise 16: $R_N(0) = N_0/4RC$

Exercise 17: $R_N(0) = (N_0/2) \int_{-\infty}^{\infty} df/[1 + (2\pi fRC)^2] = N_0/4RC$

Exercise 18: $B_N = 1/(4RC) = \pi B/2$

CHAPTER 5

Exercise 1: Constant 1

Exercise 2: Number of representation levels = 512

Exercise 3: Code word length = 9; Bit rate = 75.6 Mb/s; Bandwidth = 75.6 MHz

Exercise 4: 768 kb/s.

Exercise 5: 1. Four-level Gray coding. 2. On-off signaling, polar signaling, and bipolar signaling. 3. Return-to-zero signaling, and Manchester coding.

Exercise 6: $\delta^2/3$.

Exercise 7:

System Level	Number of Voice Channels	Number of Picturephone Channels	Number of Television Channels
T1	24		
T2	96	1	
T3	672	7	1
T4	4032	42	6

CHAPTER 6

Exercise 2: 6.312 MHz.

Exercise 3:

b_k		0	0	1	0	1	1	0
a_k	0	0	0	1	1	0	1	1

Polar representation of a_k, volts	−1	−1	−1	+1	+1	−1	+1	+1
c_k, volts		−2	−2	0	+2	0	0	+2
$\lvert c_k \rvert$		2	2	0	2	0	0	2
$\hat{b}_k$		0	0	1	0	1	1	0

Exercise 4: The duobinary technique has correlated digits, while the other two methods have independent digits.

Exercise 5:

b_k			0	0	1	0	1	1	0
a_k	1	1	1	1	0	1	1	0	1
Polar representation of a_k, volts	+1	+1	+1	+1	−1	+1	+1	−1	+1
c_k, volts			0	0	−2	0	+2	−2	0
$\lvert c_k \rvert$, volts			0	0	2	0	2	2	0
$\hat{b}_k$			0	0	1	0	1	1	0

Exercise 6:

b_k			0	0	1	0	1	1	0
a_k	0	0	0	0	1	0	0	1	0
Polar representation of a_k, volts	−1	−1	−1	−1	+1	−1	−1	+1	−1
c_k, volts			0	0	+2	0	−2	+2	0
$\lvert c_k \rvert$, volts			0	0	2	0	2	2	0
$\hat{b}_k$			0	0	1	0	1	1	0

CHAPTER 7

Exercise 11: The low- and high-frequency ends of the amplitude response of the sideband shaping filter are the opposite of those in Fig. 7.24.
Exercise 13: The wide band 90° phase shifter (Hilbert transformer) of Fig. 7.20 is replaced by a filter whose transfer function is defined by Eq. 7.64.
Exercise 14: $f_l = f_o - f_c$
Exercise 15: $f_i(t) = 5 + 2\cos(4\pi t)$, hertz
Exercise 16: The input of the phase modulator has a triangular waveform. The output is the same as the FM wave shown in Fig. 7.37*b*.
Exercise 17: $\Delta f = 0.2$ kHz; $\beta = 0.2$. The carrier frequency f_c plays no role in these two calculations.
Exercise 18: The lower side-frequency component of a narrow-band FM wave is shifted in phase by 180° compared to that of a standard amplitude-modulated wave.
Exercise 19: A standard AM contains a carrier of fixed amplitude and two side-frequencies. A wideband FM contains a carrier of varying amplitude and an infinite number of side-frequencies.
Exercise 20: For $\beta = 0.2$, the amplitude spectrum of the FM wave consists essentially of a carrier at f_c and a pair of side-frequencies at $f_c \pm f_m$.
Exercise 21: $B = 130$ kHz
Exercise 22: Mid-band frequency of the filter is $2f_1$; Bandwidth of the filter is $2B_1$; where f_1 is the carrier frequency of $s_1(t)$ and B_1 is its bandwidth.
Exercise 23: $\Delta f = 60$ kHz; $\beta = 12$. Separation of adjacent side-frequencies is 5 kHz.

CHAPTER 8

Exercise 1: The probability of getting one head and one tail is 1/2 if no distinction is made as to which coin turns up head.
Exercise 2: The probability that the two dice add up to 6 is 5/36.
Exercise 3: $P(A_0|B_1) = pp_0/[pp_0 + (1-p)p_1]$;
$P(A_1|B_0) = pp_1/[(1-p)p_0 + pp_1]$.
Exercise 4: (a) $P(B_0) = \frac{1}{2}$; (b) $P(B_1) = \frac{1}{2}$; (c) $P(A_0|B_0) = 1 - 10^{-4}$;
(d) $P(A_1|B_1) = 1 - 10^{-4}$
Exercise 5: $P(|X - m_X| < \varepsilon) = 1 - (\sigma_X^2/\varepsilon^2)$
Exercise 7: $P(-3 < X \leqslant 13) = \text{erf}(1/\sqrt{2})$
Exercise 8: Mean-square value $= A^2$
Exercise 9: First, $\sigma(f)$ is nonnegative for all f. Second, the total area under the curve of $\sigma(f)$ is one.
Exercise 11: As N becomes large, the power spectral density of the *PN* sequence assumes the same form as that of the corresponding random binary wave.
Exercise 12: $S_Y(f) = 4\sin^2(\pi f T)S_X(f) \simeq 4\pi^2 f^2 T^2 S_X(f)$. That is, for low-frequency inputs, the comb filter acts as a differentiator.
Exercise 14: $f_X(x) = 1/(2\sqrt{\pi N_0 W})\exp(-x^2/4N_0W)$
Exercise 15: (a) For the narrow-band random process $X(t)$, $R_X(\tau) = 2N_0W\,\text{sinc}(2W\tau)\cos(2\pi f_c\tau)$. (b) For the in-phase component $X_I(t)$ and quadrature component $X_Q(t)$: $R_{X_I}(\tau) = R_{X_Q}(\tau) = 2N_0W\,\text{sinc}(2W\tau)$.

CHAPTER 9

Exercise 1: (d) When using DSBSC modulation and coherent detection, the translated signal sidebands add coherently, whereas the translated noise sidebands add incoherently.
Exercise 4: $(SNR)_{O,AM} = 2\rho k_a^2 P$.
Exercise 5: 11.4 dB
Exercise 7: The values of threshold channel signal-to-noise ratio are (a) $(SNR)_C \geqslant$ 17.8 dB; (b) $(SNR)_C \geqslant$ 20.8 dB. The corresponding values of average transmitted power P_c are (a) $P_c \geqslant 4.5$ mW; (b) $P_c \geqslant 9$ mW.
Exercise 8: The power spectral density of de-emphasized noise varies with frequency as $f^2/[1 + (f^2/f_0^2)]$, where f_0 is the cutoff frequency of the de-emphasis filter. Thus, starting at the initial value of zero at zero frequency, the de-emphasized noise power spectral density increases with frequency, reaches a peak, and then decreases with frequency.

CHAPTER 10

Exercise 2: (a) The matched filter output reaches its peak at time $t = t_0 + T$.
(b) The maximum value of the output is proportional to signal energy.
Exercise 6: (a) $\lambda = \frac{1}{2}E$; (b) $\lambda = (N_0/2)\ ln\ (2) + \frac{1}{2}E$;
(c) $\lambda = -(N_0/2)\ ln\ (2) + \frac{1}{2}E$
Exercise 8: (b) 1.5 dB; (c) 0.68 dB.

INDEX

Absolute temperature, 630
Adaptive equalization, 250
  adaptation constant, 253
  decision-directed mode, 254
  least mean-square algorithm, 253
  training mode, 253
Adaptive prediction, 212
Adaptive quantization, 212
Aliasing, 181
Amplitude distortion, 99
Amplitude modulation (AM), 260
  envelope detector, 272
  envelope distortion, 263
  frequency-domain description, 263
  percentage modulation, 261
  sidebands, 264
  single-tone modulation, 265
  square-law detector, 271
  square-law modulator, 268
  switching modulator, 269
  time-domain description, 261
  transmission bandwidth, 264
Amplitude-shift keying:
  binary, 369
  detection, 370, 574, 576
  generation, 370
  noncoherent detection, 372
  probability of error, 577
AM radio, 311
  image signal, 312
  intermediate frequency, 311
  superheterodyne operation, 312
AM receiver, noise in, 503
  channel signal-to-noise ratio, 503
  figure of merit, 504
  model, 496
  output signal-to-noise ratio, 504
  threshold effect, 506
AM stereo, 392
Analog communications, 8
Analog pulse modulation, 182
Analog signals, 4
Angle modulation, 322
  angular argument, 322
  instantaneous frequency, 323
  sinusoidal modulation, 325
  square modulating wave, 326
  types, 323
  zero crossings, 324
Angular argument, *see* Phase
Angular frequency, 23
Antenna gain, 616
Antenna noise temperature, 621
Aperiodic signals, 3
Aperture effect, 186
Arithmetic series, 627
Audio signals:
  studio quality, 196
  telephone quality, 5
Autocorrelation function:
  energy signals, 138
  periodic signals, 153
  power signals, 147
  random processes, 439
Available noise power, 606

Balanced frequency modulator, 347
Balanced modulator, 276
Band-limited signal, strictly, 51, 70
Band-pass transmission, 106
Bandwidth, signal:
  band-pass, 51
  low-pass, 51
  message, 9, 263
  null-to-null bandwidth, 52
  3-dB bandwidth, 52
Bandwidth, system, 3-dB bandwidth, 96
Bandwidth-signal-to-noise ratio tradeoff, 10, 196
Baseband, 111, 228
Baseband data transmission, 228
  channel bandwidth, 235
  ideal solution, 231
  practical considerations, 232
  timing error, 232
Baud, 247
Bayes' rule, 410
Bel, 597
Bessel function of first kind, 598
Binary symmetric channel, 410
  transition probability diagram, 411
Binomial series, 627
Bipolar signaling, 197
Bit, 10, 195
Bit duration, 228
Bit rate, 232
Bit stuffing, 219
Bolzmann's constant, 604, 630

Burst noise, 603
Butterworth low-pass filter, 174

Carrier, 8, 260
Carrier delay, *see* Phase delay
Carrier-to-noise ratio, 505, 622
Carrier power, 505, 618
Carson's rule, 336
Causality:
  signals, 73
  systems, 90
Central limit theorem, 428
Channel capacity theorem, 10, 588
Channels, 7, 84
  band-limited, 9
  broadcast, 7
  point-to-point, 7
  power-limited, 9
Chi-squared distribution, 434
Comb filter, 463
Coding, 194
  binary code, 195
Coherent detection, *see* Detection
Communication, 2
Communication system:
  elements, 6
  limitations, 9
  resources, 9
Companding, 192
  A-law, 193
  μ-law, 192
Complementary error function, 428, 570, 629
Complex envelope, 108
Complex exponential function, 57
Compressor, 192
Conjugate function, 43
Convolution:
  in frequency domain, 46
  notation, 46
  properties, 79
  in time domain, 49
Convolution integral, 46, 84
  derivative, 49
  graphical interpretation, 86
Convolution sum, 89
Correlation, *see* Cross-correlation functions
Correlation coefficient, 579
Correlation receiver, 373, 563
  generalized two-path, 578
Correlation theorem, 141
Correlative coding, 237
  decision feedback, 240
  duobinary signaling, 237
  modified duobinary signaling, 243
  precoding, 241, 245
Costas loop, 281
Cross-correlation functions:
  energy signals, 140
  power signals, 149
  random signals, 463
Cross-spectral densities, 465

dc signal, 56
Decibel, 52, 94, 597
Decoding, 200
Delay distortion, 99
Delta function, 53
  applications, 56
  Fourier transform, 56
  properties, 54, 79
  replication property, 56
  sifting property, 56
Delta modulation, 206
  adaptive, 213
  comparison with PCM, 211
  granular noise, 209
  linear, 210
  slope-overload distortion, 209
Demodulation, 8, 260
  AM waves, 271
  DSBSC waves, 279
  FM waves, 346
  SSB waves, 293
  VSB waves, 303
Detection, 370, 539. *See* also Demodulation
  coherent detection, 370, 574
  noncoherent detection, 372, 574
Deterministic signals, 3
Dibit, 198, 374
Differential encoding, 372
Differential phase-shift keying, 372
  detection, 372, 574
  probability of error, 583
Differential pulse-code modulation, 202
  adaptive, 212
  comparison with PCM, 211
  prediction gain, 205
Differentiation in time domain, 38
Digital communications, 7, 177
Digital formats, 197
Digital modulation techniques, 368
  comparison, 586
Digital pulse modulation, 178
Digital satellite communications, 382
Digital signals, 5
Digital telephony, 215
Dirac delta function, *see* Delta function

Discrete Fourier transform, 71
inverse, 72
Discrete-time Fourier transform, 68
Distortionless transmission, conditions for, 97
Double-exponential distribution, 481
Double-sideband suppressed-carrier modulation, 274
channel signal-to-noise ratio, 498
demodulation, 279
figure of merit, 500
frequency-domain description, 274
generation, 276
output signal-to-noise ratio, 500
quadrature null effect, 280
single-tone modulation, 280
time-domain description, 274
transmission bandwidth, 274
Duality property of Fourier transform, 33
Duty cycle, 20

Einstein–Wiener–Khintchine relations, 148, 453
Electron charge, 607, 630
Energy, 4
Energy density spectrum, *see* Energy spectral density
Energy signals, 4
Energy spectral density, 128
interpretation, 135
measurement, 137
properties, 130
Envelope, 176, 261
Envelope delay, *see* Group delay
Envelope detection, 272
Equalization, 100, 186
adaptive, 250
Equivalent noise temperature, 611
Ergodicity, 446
in autocorrelation function, 448
in mean, 447
Error function, 426, 570, 629
Even part, 77
Expander, 194
Expectation operator, 159, 419
Exponential pulse, 26
damped sinusoid, 38
decaying, 27
double, 30, 80
rising, 27
Exponential series, 627
Eye pattern, 248

Fast Fourier transform algorithm, 73
Feedback shift register, 458
Figure of merit, 495
Filters, 84
finite-duration impulse response, 90
frequency response, 94
ideal band-pass, 110
ideal low-pass, 102, 104, 123, 132
time response, 84
Flat-top samples, 182
Flicker noise, 603
FM radio, 364
FM receiver, noise in, 509
capture effect, 517
channel signal-to-noise ratio, 514
compared to PCM, 517
de-emphasis, 524
Dolby, 529
figure of merit, 514
model, 508
output signal-to-noise ratio, 514
pre-emphasis, 524
threshold carrier-to-noise ratio, 521
threshold effect, 517
threshold extension (reduction), 522
FM stereo, 366
Fold-over, *see* Aliasing
Fourier analysis, 5, 13
Fourier series, 14
basis functions, 14
complex exponential form, 15
Dirichlet's conditions, 15
properties, 75
real form, 14
Fourier transform, 20
Dirichlet's conditions, 22
frequency shifting, 36
inverse, 22
notations, 23
numerical computation, 71
properties, 29, 78
relation to Laplace transform, 73
Frequency, 5
Frequency changing (translation), *see* Mixing
Frequency modulation (FM), 323
Carson's rule, 336
demodulation, 346
direct method of generation, 344
frequency deviation, 327
frequency sensitivity, 323
harmonic distortion, 340
indirect method of generation, 339
limiting, 363
modulation index, 328
narrow-band, 330
ninety percent bandwidth, 336

Frequency modulation (FM) (*Continued*)
relation to phase modulation, 324
single-tone modulation, 327
transmission bandwidth, 335
wideband, 332
Frequency multiplication, 340
Frequency-shifting property of Fourier transform, 36
Frequency-shift keying, binary, 369
coherent detection, 370, 574, 580
generation, 370
noncoherent detection, 574, 580
probability of error, 580
Frequency-shift keying, *M*-ary, 589
Frequency synthesizer, 391
Friis formula for noisy networks, 614
Friis transmission formula, 617

Gain, 94
Gaussian distribution, 425
Gaussian process, 468
properties, 469
Gaussian pulse, 39, 54, 78
Generalized distributions, 53
Geometric series, 627
Gibbs phenomenon, 49
Gray code, 198
Group delay, 114

Hartley modulator, 292
Hartley oscillator, 344
Hertz, 630
Heterodyning, *see* Mixing
Hilbert transform, 95
inverse, 96

Ideal sampling function, 65
Imaginary part, 44
Impulse response, 7
Integration:
by parts, 628
in time domain, 41, 62, 80
Intersymbol interference, 230
correlative coding, 237
ideal solution for, 231
raised cosine spectrum, 233
Inverse relation between time and frequency, 33, 50

Kell effect, 315
Kell factor, 315

Likelihood ratio, 562
Limiting, hard, 363
band-pass, 364
Linearity, 5, 30, 84
Linear maximal sequences, *see* Pseudonoise sequences
Linear modulation, 230, 250, 304, 374
Line codes, *see* Digital formats
Link calculations, 615
Link power budget, 618
Logarithmic series, 627
Log-likelihood ratio, 563

MacLaurin series, 627
Manchester code, 197, 594
M-ary data transmission, baseband, 247
frequency-shift keying, 589
phase-shift keying, 589
Maser, 620
Matched filter, 543
design approximations, 553
pair, 549
properties, 545
spectral amplitude matching, 553
spectral phase matching, 553
Maximum power transfer theorem, 606
Minimum probability of error criterion, 562
Minimum shift keying, 378
comparison with quadriphase-shift keying, 381
probability of error, 585
Mixing, 308
Modulation, 8, 260
amplitude, 260
angle, 322
digital, 368
linear, 230, 250, 304, 374
Modulo-2 addition, 458
M12 multiplexer, 219
Multipath, 123
Multiple access, 382
code-division, 382
frequency-division, 382
time-division, 382
Multiplexing, 8
frequency division, 309, 390
time division, 213
Multiplication in time domain, 45

Narrow-band filters, 109
baseband (low-pass) equivalent, 110
complex low-pass transfer function, 109
Narrow-band random processes, 471
envelope and phase components, 478
in-phase and quadrature components, 471
properties, 472

Narrow-band signals, 106
complex envelope, 108
in-phase and quadrature components, 109
pre-envelope, 106
National Television System Committee Standards for TV, 316
Natural (binary) code, 198
Neper, 94
Noise, 160, 493, 603
Noise-equivalent bandwidth:
band-pass filter, 167
low-pass filter, 166
Noise figure, 609
Noise spectral density, 613
Noncoherent detection, *see* Detection
Nonlinear distortion, 116
harmonic distortion, 117
intermodulation distortion, 119
Nonlinearity, 116
memoryless, 116, 340
strong, 341
weak, 341, 395
Nonperiodic signals, *see* Aperiodic signals
Nyquist bandwidth, 232
Nyquist interval, 71, 180
Nyquist rate, 71, 180

Odd part, 77
Optimum receiver, 540
hypothesis testing, 559
likelihood ratio, 562
receiver, 563
log-likelihood ratio, 563
minimum probability of error criterion, 562
probabilistic approach, 558
signal-to-noise ratio maximization, 541
Orthogonal signals, 14, 150, 580

Parametric amplifier, 620
Parseval's power theorem, 152
Partial response signaling, *see* Correlative coding
Path loss, 617
Periodic pulse train, 18, 155
Periodic signals, 3, 14
ac power, 152
average power, 152
dc power, 152
generating function, 64
Fourier series representation, 14
Fourier transform representation, 63
root mean-square value, 153
Periodogram, 143
Phase, 322, 478
Phase delay, 114
Phase distortion, 99
Phase-locked loop, 353
closed-loop transfer function, 358
first-order, 358
FM threshold reduction, 524
frequency demodulation, 358
hold-in frequency range, 361
linearized model, 355
loop parameter, 355
nonlinear model, 355
open-loop transfer function, 357
phase demodulation, 594
phase-plane plot, 361
practical considerations, 362
second-order, 362
Phase modulation, 323
relation to frequency modulation, 324
Phase-shift keying:
binary, 369
detection, 370, 574, 579
generation, 370
phase error, 595
probability of error, 580
M-ary, 589
Planck's constant, 604, 630
Poisson's sum formula, 65
Power, average, 4
Power ratio, 597
Power signals, 4
Power spectral density, 141, 452
interpretation, 145
measurement, 146
modulated wave (process), 143, 461
power signals, 143
random signals, 453
Power spectrum, *see* Power spectral density
Prediction filter, 202
adaptive, 212
Pre-envelope, 106
Principle of superposition, 5, 84
Probability of correct reception:
average, 585
conditional, 560
Probability of error:
average, 561
conditional, 560
Probability theory, 404
a posteriori probability, 411
a priori probability, 410
axioms of probability, 406
conditional probability, 408
event, 404
impossible, 405

Probability theory, event (*Continued*)
joint, 408
probability, 405
joint probability, 408
relative frequency approach, 404
sample point, 405
sample space, 405
statistically independent events, 410
statistical regularity, 404
sure event, 405
Product modulator, 276
Pseudonoise sequences, 460
autocorrelation function, 459
power spectral density, 460
Pulse-amplitude modulation, 182
Pulse-code modulation, 8, 179
compared:
with delta modulation, 211
with differential pulse code modulation, 211
with frequency modulation, 310
error threshold, 572
M-ary, 247
probability of error, 567
sub-optimum receiver, 573
Pulse-duration modulation, 182
Pulse-position modulation, 182

Quadrature-amplitude modulation, 283, 381, 575
Quadrature-carrier multiplexing, *see* Quadrature-amplitude modulation
Quadrature-modulated process, 464
Quadriphase-shift keying, 374
compared with minimum shift keying, 381
detection, 378, 575, 583
generation, 376
phase error, 596
probability of error, 585
Quantizing (quantizer), 187
dynamic range, 189
midriser type, 222
midtread type, 189, 222
overload distortion, 189
representation levels, 189
step size, 189
Quantizing noise (error), 189
average power, 190, 422

Radar, 36
frequency-modulated, 392
Radio frequency pulse, 36, 548
Raised cosine pulse, 74
Raised cosine spectrum, 233
Random binary wave, 444
autocorrelation function, 445
power spectral density, 456
Random processes, 3, 158, 403, 434
autocorrelation function, 439
autocovariance function, 439
filtering, 450, 462
mean, 439
partial description, 440
power spectral density, 452
strict stationarity, 437
wide-sense stationarity, 440
Random variables, 414
conditional probability density function, 418
cumulative distribution function, 415
joint distribution function, 417
joint probability density function, 418
probability density function, 416
square-law transformation, 432
transformation, one-to-one, 430
Random vectors, 435
joint distribution function, 436
Rayleigh distribution, 480, 481, 507
Rayleigh's energy theorem, 128
Real part, 44
Received power, 617
Receiver, 7
Reconstruction circuit, 186
Rectangular function, 25
Rectangular pulse, 24, 31, 35, 54, 546, 548
doublet, 41
Regenerative repeaters, 201
overall probability of error for chain of, 413

Sample-and-hold circuit, 182
Sampling period, 67
Sampling process, 66, 180
aperture effect, 186
instantaneous, 67
Sampling rate, 67
Sampling theorem, 70, 180
Schottky formula, 607
Schwarz's inequality, 542
Scrambler, 390
Semiconductor junction diode, noise model, 608
Shannon limit, 589
Shot noise, 606
Signal, 2
distortion, 96
amplitude, 99
delay, 99
linear, 96

Signal energy per bit-to-average noise power per unit bandwidth, 586
Signal-to-noise ratios, 494
  channel, 495
  output, 494
Signal-to-quantizing noise ratio, 191, 195
Signum function, 59
Sinc function, 18
Sinc pulse, 33, 34, 80, 129, 231
  truncated, 46
Sine integral, 48, 104
Single sideband modulation, 284
  channel signal-to-noise ratio, 501
  compared with PCM, 310
  demodulation, 293
  figure of merit, 502
  frequency description, 285
  generation, 286, 292
  output signal-to-noise ratio, 502
  phase distortion, 294
  single-tone modulation, 291
  time-domain description, 288
  transmission bandwidth, 286
Sinusoidal functions, 58
Sinusoidal process, 442, 448, 455
Slope circuit, 347
Space loss, *see* Path loss
Spectral characteristics:
  bandwidth, 51
  main lobe, 25
  sidelobes, 26
Spectrum, 13, 17, 23
  amplitude, 17, 24
  phase, 17, 24
Spectrum analyzer, 391, 400
Speed of light, 617, 630
Spread spectrum modulation, 550
Square-law detector, 271
Square-law modulator, 268
Statistical averages, 419
  characteristic function, 421
  correlation coefficient, 424
  covariance, 424
  mean, 419
  mean-square value, 420
  standard deviation, 420
  variance, 420
Stochastic processes, *see* Random processes
Synchronization in digital communications:
  bit timing, 372
  phase synchronization, 372
Synchronous detection, *see* Detection
System, 84
  amplitude distortion, 99
  amplitude response, 93
  bandwidth, 96
  causality, 90
  conjugate symmetry, 93
  delay distortion, 99
  frequency response, 91
  impulse response, 84
  linearity, 5, 84
  noise temperature, 621
  phase response, 93
  signal distortion, 96
  stability, 91
  time-invariance, 5

Tapped-delay-line filter, 89, 100, 205, 251
Taylor series, 627
Television, 312
  aspect ratio, 315
  bandwidth, 314
  black-and-white, 312
  blanking pulses, 313
  chrominance signals, 318
  chrominance subcarrier, 320
  color, 318
  encoding, 196
  high definition, 320
  line-scanning frequency, 313, 320
  luminance signal, 318
  modulation format, 316
  pixel, 315
  raster scan, 312
  resolution, 314
Temperature-limited vacuum diode, 607
Thermal energy, 630
Thermal noise, 604
Thermal voltage, 385, 630
Time-bandwidth product, 52
Time scaling, 31
Time shifting property of Fourier transform, 34
T1 system, 217, 236
Transfer function, 92
Transmission codes, *see* Digital formats
Transmitted power, 9
Transmitter, 6
Transversal filter, *see* Tapped-delay-line filter
Triangular pulse, 41
Trigonometric identities, 626
Trigonometric series, 627

Unbiased estimate, 447
Uniform distribution, 416
  mean, 422
  mean-square value, 422
  variance, 422

Unit impulse, *see* Delta function
Unit step function, 26, 61
Unvoiced sounds, 212

Varactor, 344
Vestigial sideband modulation, 295
  demodulation, 303
  frequency-domain description, 295
  generation, 296
  side-band shaping filter, 296
  time-domain description, 299
  transmission bandwidth, 295
Video signal, 312

Voiced sounds, 212
Voltage-controlled oscillator, 344

Watt, 630
Wavelength, 617
White noise, 160, 476
  ideal band-pass filtered, 476, 497
  ideal low-pass filtered, 162
  RC low-pass filtered, 163
  Sinusoidal wave plus, 164

Zero crossing detector, 353
Zero crossings, *see* Angle modulation